Date Due

Energy for a Sustainable World

Energy for a Sustainable World

Jose Goldemberg
Thomas B. Johansson
Amulya K.N. Reddy
Robert H. Williams

JOHN WILEY & SONS

New York   Chichester   Brisbane   Toronto   Singapore

First Published in 1988
Reprinted 1988, 1990
WILEY EASTERN LIMITED
4835/24 Ansari Road, Daryaganj
New Delhi 110 002, India

Distributors:

Australia and New Zealand
JACARANDA WILEY LTD.
P O Box 1226, Milton Old 4064, Australia

Canada:
JOHN WILEY & SONS CANADA LIMITED
22 Worcester Road, Rexdale, Ontario, Canada

Europe and Africa:
JOHN WILEY & SONS LIMITED
Baffins Lane, Chichester, West Sussex, England

South East Asia:
JOHN WILEY & SONS (PTE) LTD.
05-04, Block B, Union Industrial Building
37 Jalan Pemimpin, Singapore 2057

Africa and South Asia:
WILEY EASTERN LIMITED
4835/24 Ansari Road, Daryaganj
New Delhi 110 002, India

North and South America and rest of the world:
JOHN WILEY & SONS, INC.
605 Third Avenue, New York, NY 10158 USA

Library of Congress Cataloging-in-Publication Data

Energy for a sustainable world.
    1. Power resources.   2. Energy policy.
1. Goldemberg, Jose, 1928–
TJ163.2.E4755   1988   333.79   87-23099 10-3-90

ISBN 0-470-20983-6 John Wiley & Sons, Inc.
ISBN 81-224-0000-0 Wiley Eastern Limited

Printed in India at Ramprintograph, New Delhi.

Preface

What initiated and sustained the collaboration among the four of us is of considerable relevance to the theme of this book.

Each of us started his career as a physical scientist and turned eventually to energy research. However, two of us are from developing countries, and two from industrialized countries. Also, we live and work in different countries—Brazil, Sweden, India and the United States—located in four continents. And, our cultural backgrounds and experiences are diverse. Despite all this, we have forged bonds and functioned as a well-knit team achieving together what no one of us could have achieved individually.

Our interactions began about a decade ago when various international meetings and visits gave us opportunities to exchange views on the energy problem. These brief meetings soon revealed a remarkable measure of shared values and concerns about the interaction of technology and society. They also showed an identity of outlook and similarity of approach on matters concerning energy in society. These early interactions showed that the four of us could work together with mutual respect and equality and avoid the hierarchical modes of functioning which nearly always vitiate international collaborations.

The formal decision to collaborate was greatly facilitated by the encouragement of several scientists at Princeton University's Center for Energy and Environmental Studies—particularly Robert Socolow, Frank von Hippel, Gautam Dutt, Harold Feiveson and Theodore Taylor.

The analyses of energy issues that each of us had carried out prior to our collaboration had convinced us that the multifaceted problems of our countries and our world could not be solved by the conventional approach to the energy problem. As we combined our efforts, we were led from a critique of conventional wisdom to a new approach to energy. When significant progress had been made we felt that we should expound and elaborate the new approach—and that is how this book came to be written.

The new view is best appreciated in the context of a brief description of the conventional approach to the energy problem. In that approach, economic growth is the principal objective. The link to energy is fashioned by assuming that total energy use and Gross Domestic Product (as the measure of growth) are correlated. To the extent that structural changes and opportunities for making more efficient use of energy are considered at all, they are treated as minor perturbations on the economy. Thus, the level of energy use becomes an indicator of human welfare, the higher the better. Consequently, the central task of energy analysis is to make projections of future energy demand and then determine appropriate mixes of energy supplies for satisfying that demand. The expansion of supplies to the levels projected involves formidable engineering challenges; hence, the primary goal of energy planning is to make this energy supply expansion possible, that is, to bring about *a sustainable energy system*.

Independently and jointly, we realized that this approach to energy was becoming increasingly dangerous both for individual countries and for the world at large. The implementation of conventional wisdom was making energy unaffordable for crucial development needs, aggravating societal inequalities, causing serious near-and long-term environmental and security problems, and eroding self-reliance. Such problems in energy planning could only be avoided by fundamentally modifying the conventional approach to the energy problem. Accordingly, we explored a normative approach to energy planning by incorporating from the start, broad societal goals aimed at facilitating the achievement of, not merely a sustainable energy system, but what is more crucial, *a sustainable world*.

This approach involves economic growth as a necessary, but not a sufficient, condition. At the most fundamental level the goals of society should be equity, economic efficiency, environmental harmony, long-term viability, self-reliance, and peace. Energy production and use should be compatible with, and if possible contribute to, these societal goals. These goals are relevant for both developing countries (for which they define the objectives of development) and industrialized countries, as well as for the relationship between these countries and for the global community.

Can such energy plans be formulated? And even if they can be formulated, are they achievable? To address this challenge, we adopted an "end-use methodology" for the energy problem. In this framework, the level of energy use is not assumed to be an indicator of human welfare. Instead, inputs of energy are regarded only as means to the end of providing a wide range of energy services like lighting, cooking, comfortable space conditions, transport services, etc. Understanding the human needs which energy serves and exploring more effective ways of directing energy resources towards the satisfaction of these needs become the important tasks in end-use-oriented energy analysis.

Our new approach has led us to replace the perspective proposed in more conventional energy analyses by a vision of the global future that is dramatically different. A world adopting the energy strategy we describe would be more equitable, economically viable and environmentally sound. It would also be more conducive to achieving self-reliance and peace. And, it would offer hope for the long term future.

This book suggests that, contrary to widely held beliefs, the future for energy is very much more a matter of choice than of destiny. Energy futures compatible with the achievement of a sustainable world are within grasp. The choices we urge require imaginative political leadership, but they represent far less difficult and hazardous options for this leadership than those demanded by the conventional projections of the world's energy future.

Our work is only a preliminary effort to work out the implications of the new approach. Many other efforts are needed. The joy in this endeavour comes from the feeling of being harbingers of hope rather than prophets of doom.

JOSE GOLDEMBERG
THOMAS B. JOHANSSON
AMULYA K.N. REDDY
ROBERT H. WILLIAMS

Acknowledgements

This book presents the finding of the End-use Oriented Global Energy Project (EUOGEP), of which the authors are co-organizers.

The authors thank Roberto Hukai, Eric Larson, Jose Roberto Moreira, and Frank von Hippel for technical assistance in the preparation of some of the material presented here and for their comments on early drafts. The authors also thank Erik Bogren, Gautam Dutt, Harold Feiveson, Roger Fredriksson, Howard Geller, Leif Gustavsson, Sten Karlsson, Ole Leissner, Robert Socolow, Peter Steen, Ingrid Stjernquist, Per Svenningsson, and Theodore Taylor for their encouragement, review comments, and suggestions in the course of this writing. The authors are indebted to Mr. David Sheridan for his diligent editorial efforts in helping to bring the manuscript to its final form.

The authors acknowledge support for the End-use Oriented Global Energy Project and the research leading to this book from the Alida and Mark Dayton Charitable Trust, the Changing Horizons Charitable Trust, the Energy Research Commission of Sweden, the Indian Institute of Science, the International Labor Organization, the Max and Anna Levinson Foundation, the Rockefeller Brothers Fund, the Swedish International Development Authority, the Macauley and Helen Dow Whiting Foundation, and the World Resources Institute.

The authors are also grateful to the World Resources Institute for publishing the reports *Energy for a Sustainable World* and *Energy for Development,* which summarize and disseminate the major findings of this book.

Finally, the authors would like to thank their families for their sustained encouragement and moral support during the course of writing this book.

Contents

Energy for a Sustainable World

1. Introduction

1.1 Energy—A Major Global Concern

1.1.1 *Introduction*

Energy is inseparable from matter—all material phenomena are associated with energy changes. Energy is also essential to life. And human society cannot survive without a continuous supply of energy.

The original source of energy for social activities was human energy; the energy of human muscle provided the mechanical power necessary in the dawn of history. Then came the control and use of fire from the combustion of wood, and with this the ability to exploit chemical transformations brought about by heat energy and thereby to cook food, heat dwellings, and extract metals (bronze and iron). The energy of draught animals also began to play a role in agriculture, transport and even industry. So did water power. Later, in rapid succession, human societies acquired control over coal, steam, and electricity. Thus, from one perspective, history is the story of the control over energy sources for the benefit of society.

So important is energy to human society that the magnitude of energy consumed per capita became one of the indicators of a country's "modernisation". In the process, the appetite for energy often exceeded the capacity of local sources of supply; the energy supplies of some modern countries had to be shipped from halfway round the world. At the same time, the sheer intensity of energy production and use began to result in deleterious impacts on the environment through pollution. These strategic and environmental consequences of the pattern of energy consumption were virtually ignored for a long time. Modern societies were preoccupied with satisfying their "need" for energy.

Over the last decade or so, the whole picture changed, however. The gravity of the environmental risks became clear by the late 1960's. And then came the "oil shocks" of late 1973 and 1979, when price increases caused economic distress at international, national and local levels. The oil shocks awakened people and governments to the problems of continuing energy supply expansion.

Hit hardest by the oil-price increases have been the populations of industrialized countries and the rich in developing countries. The poor in developing countries felt the impact of oil-price increases indirectly through price increases of essential goods which are carried by oil-dependent modes of transport. In developing countries, the poor's energy problem is a fuelwood problem—a long-standing problem—every bit as acute but little noticed until the oil shocks made the non-poor pay closer attention to energy.

The energy crisis has persisted for over a decade. If it appears now to have diminished in intensity somewhat, this lull is deceptive. Most industrialized countries continue to rely heavily on imported oil. They remain vulnerable to the supply disruptions and economy-destabilizing price hikes which characterized the 1970's. At the current consumption rate, the world's remaining supply of recoverable oil will last only about 100 years. And many developing countries continue to devote a substantial portion of their limited export earnings to pay for the oil they import. Meanwhile, the fuelwood crisis worsens. Some 100 million people now suffer acute scarcity of fuelwood and about one billion are experiencing a fuelwood deficit. Equally important, global energy concerns extend beyond the availability and cost of energy. Acid rain, the climate-changing build-up of carbon dioxide in the atmosphere, nuclear weapons proliferation, and other major unresolved global problems are rooted, partially or entirely, in energy.

This book is devoted to the problem of energy. In the rest of this chapter, we explore the main dimensions of the energy problem. The crises with respect to oil, fuelwood, and electricity are introduced. The conventional approaches to the global studies carried out in recent years are described. Then, we examine the impacts of such energy strategies upon global problems that are as serious as or more so than the energy problem.

We do not treat energy supply or consumption as an end in itself. Rather, we focus on the *end-uses* of energy and the services that energy performs. Energy use is, after all, only a means of providing illumination, heat, mechanical power, and the other *energy services* associated with satisfying human needs. Here we are especially interested in understanding how patterns of energy use might be shaped so as to promote the achievement of certain basic societal goals—equity, economic efficiency, environmental soundness, long-term viability, self-reliance, and peace. Ours is, in short, a normative analysis.*

In Chapter 2, we examine energy strategies for industrialized countries in terms of their economic viability, environmental integrity and strategic security. In Chapter 3, we examine energy strategies for developing countries within the context of those countries' need-oriented, environmentally-sound, self-reliant development. The focus in both Chapters is on *end-uses* of energy, the management of energy demand, and the exploitation of synergisms. In Chapter 4, we provide rough estimates of global demand and an illustrative energy supply scenario which, unlike others that have been proposed, *is* compatible with the values of equity, environmental soundness, economic efficiency, long-term viability, self-reliance, and peace. In Chapter 5, we outline the policies necessary to implement end-use-oriented strategies. And in the concluding Chapter 6 we discuss the political feasibility of implementing the kind of energy future we envisage.

1.1.2 *Main Findings*

Our most important finding is that it *is* possible to formulate energy strategies which are not only compatible with, but even contribute to, the solution of the other major global problems—including North-South disparities, the poverty of the majorities in the developing countries and of minorities in the industrialized countries, food scarcities and undernutrition, environmental degradation in both the industrialized and developing countries, the threat of global climatic change, the pressure from population growth, and global insecurity and risks of nuclear weapons proliferation and thus the threat of nuclear war. Thus it appears that the energy problem can be turned into a powerful and positive force for improving the human condition on this globe. Instead of being the destabilizing force that it is today, energy can become an instrument for contributing to the achievement of a sustainable world.

The formulation of such energy strategies is made possible by shifting the focus of energy analysis from the traditional preoccupation with energy supplies to the end-uses of energy. In this end-use approach, much closer attention is paid to present and future human needs served

---

*Those who instead seek to understand the future of energy based on extrapolations of "inherent tendencies" in the energy economy also make the normative judgement that the existing taxes, subsidies, and regulations that now shape the evolution of the energy system represent the way the world ought to work. Thus our analysis is certainly not unique in being normative, although it is not customary to call one's analysis normative explicitly.

by energy, the technical and economic details of how energy is being used, and alternative technological options for providing the energy services that are needed. In addition, the end-use approach reveals opportunities for energy-efficient improvements and synergistic solutions to energy problems which usually go unidentified by supply energy analysts.

For developing countries, our end-use-focused analysis shows that it is feasible with final energy use of approximately 1 kilowatt (kW)* per capita—roughly the same as the present level—to provide enough energy services not only to satisfy the basic human needs of the whole population but also to raise their standard of living to the level of Western Europe. Energy need *not* be a constraint on the satisfaction of basic human needs or the improvement of living standards in developing countries if available energy sources are used more efficiently and if there is a shift to modern energy carriers, especially electricity and biomass-based gaseous and liquid fuels.

For industrialized countries, our end-use focused analysis highlights ongoing structural changes that indicate a far lower level of demand for energy-intensive activities such as steel making than is projected in conventional energy forecasts. Also, activities requiring relatively little energy, such as electronics, computers, biotechnology, and all kinds of services are rapidly growing in importance. In addition, we identify enormous opportunities for energy efficiency improvements in all energy using sectors. By taking into account the ongoing structural changes and these energy efficiency opportunities, it appears technically and economically feasible by the year 2020 to *reduce* the average final energy use per capita in industrialized countries by about 50 per cent— from 4.9 to 2.5 kW, while *continuing economic growth*—increasing per capita Gross Domestic Product (GDP) by 50-100 per cent.

At the global level, energy services in 2020 can be provided for a world population nearly double that of today and having a much higher standard of living, with only 11 terawatts (TW) of primary energy input. This 11 TW is only slightly more than the 10 TW used in 1980 and far less than the 20 to 40 TW envisaged for 2020 in most conventional energy forecasts. In addition, the great disparity in the distribution of energy services between industrialized and developing countries narrows dramatically. Today, the developing countries account for three-fourths of the people, and have two-thirds of the energy demand. In our analysis, by 2020 the developing countries' share increases to two-thirds and the industrialized countries' share decreases to one-third.

The lower demand levels identified through the end-use-oriented approach result in an increase of energy supply options. Because of the overall reduction in energy supply requirements, it is possible to become less dependent both on oil as well as nuclear power and coal, thereby reducing their economic and external costs. And various supply options involving renewable energy sources that would be of limited value in the high-demand futures, due to land-use and other physical constraints, can contribute more significantly to overall energy supply.

An energy system compatible with the solutions of other global problems can be evolved in the decades to come. There is such a wide range of feasible outcomes for global energy that *the future of energy is much more a matter of choice than of prediction.*

1.1.3 *The Influence of Cheap Oil on the Economic Bases of Industrialized Countries*
The predominance of oil in industrialized countries is a comparatively recent phenomenon.

---

* In this book annual average rates of energy use are expressed in Watts. Thus 1 kilowatt (kW) is an abbreviation for 1 kW-year/year. Whenever a power unit refers instead to an installed electric power generating capacity or to a peak demand, it will be explicitly identified as such. See Appendix B for definitions of the energy terms used in this book.

Other fuels or energy carriers dominated the earlier phases of development of these countries. Wood played an overwhelming role as an energy source from the dawn of human history until around 1850, after which it was increasingly replaced with coal (Figure 1-1). The coal age lasted through the first half of the 20th century. It was followed by what may be called the *oil era,*, because modern industrialization has since then been based primarily on oil and its sister fuel, natural gas.

The energy budgets of industrialized countries shifted between 1925 and 1972 (Note 1.1-A) from an overwhelming dependence (over 80 per cent) on coal to a predominant reliance (to the extent of about 70 per cent) on oil and natural gas. During this period, oil consumption in industrialized countries grew from 2 to 45 million barrels per day (Note 1.1.-B), at the spectacular average annual rate of 6.5 per cent per year—about 85% faster than total energy use (Note 1.1-C).

Because oil is ubiquitous, its remarkable properties are taken for granted. Oil has a high energy density, is easy to transport, store, and use, and it is much cleaner than coal. And, for decades, it was incredibly cheap. In 1970, for instance, it required only $1.0 (1984 $)* to purchase from a U.S. wellhead 20 liters (5 U.S. gallons) of oil weighing about 18 kg (40 lb.) and containing the energy equivalent of the work a man could do in one year! Nature completed the "manufacture" of oil several hundred million years ago, a pre-historic achievement that made possible the low "marginal production costs" of oil. With technological advances, the costs of discovery, extraction, processing and distribution came down over time, and consequently, oil prices remained stable or even declined until 1973.

Low petroleum prices were a major factor in the rapid increase in affluence in industrialized countries. Inexpensive oil enabled an extraordinary degree of private mobility via the automobile; it made long-distance freight transport so cheap that households could consume out-of-season fresh vegetables and fruits transported thousands of kilometres from remote farms in completely different climatic zones; it meant that workers, especially in the United States, could live in suburban homes a considerable distance from their workplaces; it made possible widely available, low-cost, central heating; for industry it meant cleaner, more economical and easier-to-use fuels.

While oil was playing an important role in making all this possible, decades went by with very little thought given to the long-term sustainability and limitation of this resource. Until 1973, the future was widely believed to be but a bigger and better version of the past. Business, it was thought, could continue as usual.

1.1.4 *The Influence of Cheap Oil on the Economic Bases of Developing Countries*

Industrialization has been taking place very much later in the developing countries and, in most instances, really began only after World War II. Despite this lag, oil quickly assumed a position of dominance in newly industrializing countries. Even in the few countries such as Brazil and India, which already had made some progress in industrialization before World War II, a marked shift towards oil occurred after the war.

The shift to oil in Brazil (Figure 1.2) was at the expense of the traditional biomass fuels (fuelwood, charcoal and agricultural wastes), which had accounted in 1940 for over 75 per cent of the total primary energy in comparison with petroleum, which represented only 9 per cent of the total consumption. Oil's share began to rise, though, with the large increase in the number of automobiles and gasoline-fueled trucks and the shift to liquefied petroleum gas (LPG) for cooking. By 1975, oil consumption in Brazil had risen to almost 50 per cent of the total energy consumption.

---

* See Appendix B for a Table of US GNP deflators converting US dollar values to other base years.

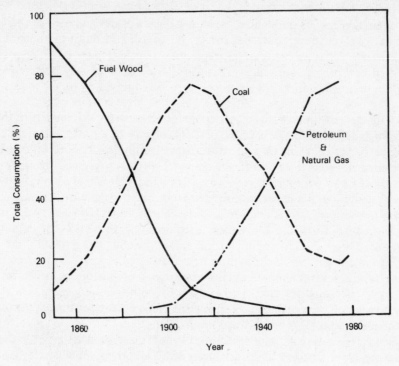

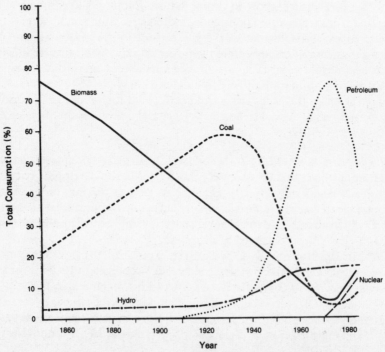

Figure 1-1. The changing sources of energy in the US energy economy (top) and the Swedish energy economy (bottom). *Sources:* Top from Executive Office of the President, 1977; bottom from A. Jarnegren et al., 1980.

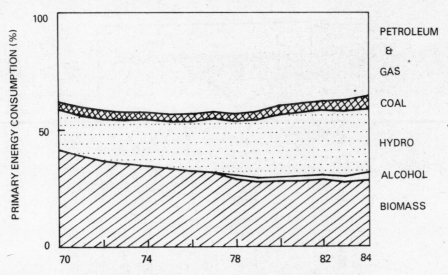

Figure 1.2. Changing sources of energy in the Brazilian economy.
Source: Goldemberg *et al.,* 1984.

In the case of India, too, the increased percentage of oil consumption has been primarily because of the increased share of freight borne by trucks and of passengers carried by automobiles and buses. In addition, kerosene, a petroleum product, replaced vegetable oils in lamps for domestic illumination.

The net effect of such fuel shifts is that, in their so-called "modern" sectors which use commercial energy sources, developing countries have become nearly as dependent on oil as the industrial countries.

1.1.5 *The Impacts of the Oil Shocks*

In late 1973, Arab oil producers, reacting to U.S. steps to re-supply Israel during the October 1973 Arab-Israeli War, embargoed oil shipments to the United States and reduced oil output and shipments to other nations. For the first time, the oil-exporting countries set the price of the oil *obtained from their own territories* without consulting the foreign oil companies which processed and/or distributed this oil. The oil embargo created such a scarcity and panic in oil markets that the Organization of Petroleum Exporting Countries (OPEC) was able to raise the world oil price three-fold (Figure 1.3). In 1972, The Shah of Iran described the situation thus:

> "In 1947, the posted price for a barrel of oil in the Persian Gulf was $2.17. Then it was brought down to $1.79, and that lasted until 1969. So there were 22 years of cheap fuel that made Europe what it is, that made Japan what it is. Then the price of wheat jumped 300 per cent, vegetables the same, and sugar in the past six years increased 16 times. So we charged experts to study what prices we should put on oil. Do you know that from oil you have today 70,000 derivatives? When we empty our wells, then you will be denied what I call this noble product. It will take you $8 to extract your shale or tar sands. So I said let us start with the bottom price of $7; that is the government intake."[1]

To understand why in 1973, for the first time, OPEC was able to set the world oil price, it is necessary to look beyond the Persian Gulf for the underlying causes. Why is it that Persian

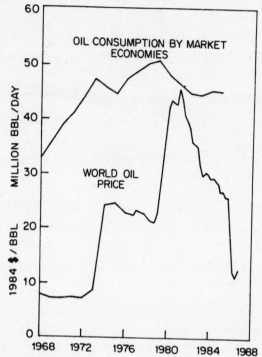

Figure 1.3. The world oil price and oil consumption by countries with market economies.
The world oil price shown is the refiners' acquisition cost for oil imported into the United States. Price data since 1976 are quarterly prices; earlier prices are annual averages.
Sources for prices: Energy Information Administration, *Monthly Energy Review,* various issues, and Energy Information Administration, April 1984.
Source for oil demand data: The British Petroleum Company, 1985.

Gulf producers did not or could not use their oil as a weapon in earlier Arab-Israeli conflicts—for example, the Suez Crisis of 1956 or the June War of 1967? The answer is that during those crises the pattern of production and consumption of oil was very different.

A combination of three factors gave the Persian Gulf OPEC nations the leverage they did not have prior to 1973:

(1) The increased consumption of oil, particularly by the industrialized market countries,*
(2) The levelling off of U.S. oil production, and
(3) The resulting shift in oil dependency onto Persian Gulf producers.

In the 1950's and 1960's, the oil imports of the industrialized countries with market economies were at much lower levels than in 1973, and the U.S. had sufficient spare capacity to supply its allies in the event of supply disruptions. But oil imports by the major importers (the U.S., Japan and Western Europe) grew dramatically in the pre-embargo period, increasing from 6 to 27 million barrels per day between 1960 and 1973 (Figure 1.4).

---

* In discussing the world oil situation, the Organization for Economic Cooperation and Development (OECD) countries (see Appendix B) [essentially the industrialized countries with market economies] are often considered separately from the Council for Mutual Économic Assistance (CMEA) countries (see Appendix B) [mainly the industrialized countries with centrally planned economies], because there is little oil trade between these two parts of the world.

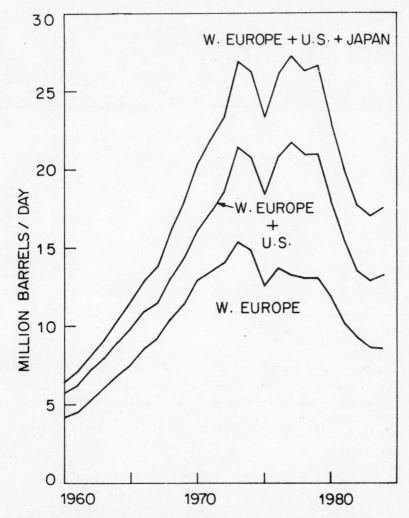

Figure 1.4. Oil imports by the major industrialized countries with market economies.
Source: The British Petroleum Company, 1985.

Also, during the pre-embargo period the United States lost its ability to increase its oil production to make up for supply disruptions. In fact, the United States had to become a major importer of oil to meet its increasing domestic demand. US oil production peaked in 1970, but oil demand continued to grow rapidly, so that US imports rose from 3 to 6 million barrels per day between 1970 and 1973.

At the same time, Persian Gulf oil producers were capturing an ever larger share of the world oil market. Figure 1-5 shows that between 1960 and 1973 Persian Gulf oil production quadrupled, and its share of total oil production in countries with market economies increased from 28 to 44 per cent.

After the first oil shock, the oil markets cooled off for a number of years. Oil consumption by non-Communist countries fell 2 million barrels per day, and during the 1973-75 period, the oil price fell in real terms, as a result of slackening of demand due to conservation efforts, and

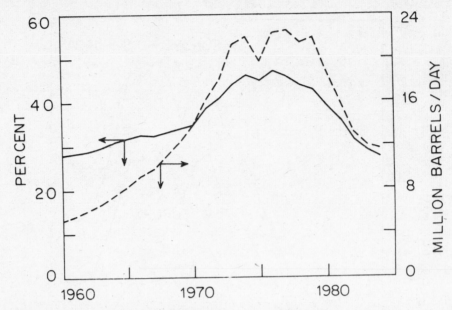

Figure 1.5. Middle East oil production, in million barrels per day (right axis) and as a percent of oil production in countries with market economies (left axis).
Source: The British Petroleum Company, 1985.

the ensuing economic recession induced by the "oil shock". Then, as the world economy pulled out of the recession, oil demand soared once again, increasing 5 million barrels per day between 1975 and 1979. This renewed surge in demand created the tight market conditions that made it possible for OPEC once again to raise the price of oil—this time from $22 a barrel in 1978 to a peak of $45 a barrel in 1981 (1984 $)—in the aftermath of the Iranian revolution (Figure 1.3).

The OECD countries have paid dearly for their voracious oil appetite, developed when oil was cheap. Between 1973 and 1981 these countries paid about 1.5 trillion dollars (1984 $) *more* for their oil imports than they would have paid had the oil price remained at the 1972 level (Note 1.1-D). This expenditure, equivalent to one-sixth of aggregate GDP of OECD countries in 1980, resulted in a loss of purchasing power for other goods and services, and was a major contributor to the 1970's phenomenon of 'stagflation' in which these countries were afflicted simultaneously with inflation and recession.

The economic impacts of the oil shocks on oil-importing *developing* countries have been even more severe than the effects felt in the OECD countries. These countries have had to devote increasing amounts of foreign exchange to pay for their oil import bills, thereby diverting economic resources from investments vitally needed for development. Compounding the direct effects of the higher oil import bills have been the indirect effects of reductions in export earnings due to the recession in the OECD countries and a worsening of the terms of trade as the costs of imported manufactured goods have increased relative to earnings from exports of raw materials. To cover the cost of oil and the reduced growth of exports and also to finance some major projects, many developing countries turned to borrowing and they found banking institutions in North America and Western Europe eager to lend them large sums at easy terms. Between 1974 and 1984 the debt by all developing countries increased from 15 to 34 per cent of GNP and debt servicing requirements increased from 12 to 20 per cent of export earnings.[2]

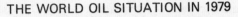

THE WORLD OIL SITUATION IN 1979

Total World Production = 65.7 MBPD

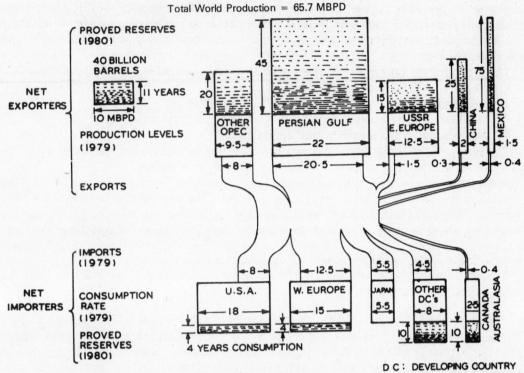

Figure 1.6. The world oil situation in 1979. Production, consumption, imports and exports are expressed in million barrels per day (MBPD). Proved reserves (as of 1980) are expressed as the number of years the reserves would last at the present rate of production (consumption) for oil-importing (oil-exporting) countries.

Source: Frank von Hippel, 1983.

The second oil shock, like the first, is being followed by a decline in oil demand and a softening of the world oil price (Figure 1.3), and by a modest increases in non-OPEC production (Note 1.1-E). In other words, the years 1981 through 1985 have been the years of the oil "glut"—a condition that has led to many news reports suggesting that OPEC has been broken. It may well be that the current "glut" will continue—perhaps even for a few years. But the balance of oil power lies with OPEC and the Persian Gulf. As shown in Figure 1.6, OPEC controls two-thirds of the world's proved oil reserves, and more than half of the proved reserves are controlled by countries around the Persian Gulf.

1.1.6 *The Future of World Oil*

From a fundamental point of view, the oil crises in the 1970's originated from the fact that the world's remaining supplies of oil and natural gas appear to be largely concentrated in the Persian Gulf region. Numerous estimates (Notes 1.1.-Fa, 1.1-Fb, and 1.1-Fc) have been made of how long the world's remaining supplies of oil and gas will last. While these estimates vary from one study to the next, most of them suggest that these supplies will be exhausted in less than one hundred years. If world oil and gas demand were to continue growing at the 1966-73 pre-embargo rate of 7.2 per cent per year, the estimated remaining ultimately recoverable oil and gas resources (Note 1.1-G) would be exhausted in 30 years (Note 1.1-H). And even growth at the more modest post-embargo rate—1.7 percent per year from 1973 to 1981—would lead to exhaustion just 30 years later. Of course, in the real world, exponential growth will not continue until resources are exhausted. Instead, demand tends to reach a plateau, after which there is a continuing decline. But these calculations nevertheless show clearly just how limited oil and gas resources are:

The prognosis for oil supplies in the immediate decades ahead is:

(1)   constant OPEC production capacities,
(2)   declining oil reserves in the industrialized countries,
(3)   at best, modest increases in oil production for the rest of the world, and
(4)   increasing OPEC internal consumption.

The net overall effect is almost sure to be a tightening of oil supplies and higher oil prices in the 1990s and beyond unless the overall demand for oil is curbed.

While there is much uncertainty about just how the future of oil will evolve, one far-reaching conclusion can be drawn: *the oil troubles to date mark the beginning of the transition away from oil.* Hence, the industrialized world must find ways to curb its appetite for oil, and the developing world must try to base its development largely on energy sources other than oil.

1.1.7 *Oil is Not the Only Component of the Global Energy Problem*

Though oil was largely responsible for drawing attention to the energy problem, there are other aspects to this problem.

The Fuelwood Crisis: Developing countries are also in the grip of a fuelwood crisis which is at least as serious as the oil crisis. For the poor, the fuelwood crisis may be even more serious than the increase in oil prices, since biomass (mainly fuelwood) is the main source of energy for at least half world's population. In many rural areas of the developing countries, fuelwood consumption averages about 1 tonne per capita per year—equivalent in energy to the fuel that the average Western European consumes for automobiles. On the average, biomass provides over 40 per cent of the total energy used in developing countries. But some developing countries such as Nepal, Ethiopia, and Tanzania depend on biomass for over 90 per cent of their energy (Note 1.1-I). Biomass, mainly in the form of fuelwood, accounts for 14 percent of world energy consumption—the equivalent of 20 million barrels of oil per day, which is slightly more than the rate of oil use in the United States.

Fuelwood is a major energy source in the traditional sectors of developing countries. Fuelwood use in the domestic sector (cooking, water-heating and space-heating, but mainly cooking) is done with traditional technologies of very low efficiency (of the order of 10 percent) and accounts for about half the trees which are felled in the world. In addition, small-scale and rural industries such as brick-making also depend on fuelwood.

This dependence has put increasing pressure on the biomass resource base that sustains fuelwood demand. The human population which relies on this fuel has been growing exponentially. Between 1961 and 1975, fuelwood consumption (including the wood used to produce charcoal made from this fuelwood) grew at a rate of about 2.1 percent per year. Increasingly in the developing world, fuelwood is being consumed faster than it can be regenerated. The present depleting mode of fuelwood use is not sustainable.

The human consequences of this situation are profound. The fuelwood crisis involves human suffering on a grand scale. It involves backbreaking drudgery and long hours. As the resource base dwindles, fuelwood gatherers must spend more and more time and trek ever increasing distances finding what they need. Women and children in particular must toil daily to keep their families supplied with fuelwoods for cooking and heat. It interferes with the children's schooling and contributes mightily to the misery of poor women. And, amplifying this human suffering is the tremendous ecological damage arising from the deforestation caused by fuelwood demand.

The Electricity Crisis: Electricity is another aspect of the energy problem that is usually given a great deal of attention in energy policy-making circles, both in the industrialized and developing countries. In fact, prior to the oil embargo, the energy problem was often defined as an electricity supply problem. Nevertheless, a discussion of electricity is deferred in this book to the chapters on industrialized and developing countries (Chapters 2 and 3) because the nature of the electricity crisis differs basically from the other energy crises described above. Whereas the oil and fuelwood crises are *fundamental and inherent* in that they stem from natural limits on resource availability, the electricity crisis is more a matter of planning and management. However, the fact that it has a different character does not in any way diminish its intensity and impact. In particular, the extremely high marginal cost of electricity production, especially from nuclear or coal power plants, is a matter of growing concern in many countries.

There is an energy problem. Its chief components are the oil, the fuelwood, and the electricity crises. It is serious, and it is global.

1.2 Conventional Approaches to the Global Energy Problem

Several studies of the global energy problem have been carried out since the late 1970s. The global energy demand forecasts presented in some of these these studies are shown in Figure 1.7 and reviewed elsewhere.[6] Here attention is focused on three of the better known studies carried out since 1977:

- The 1977 Report of the Workshop on Alternative Energy Strategies (WAES study)
- The 1978 and 1983 versions of the World Energy Conference Report on World Energy Demand (WEC Study)[8, 9]
- The 1981 Report of the Energy Systems Program Group of the International Institute for Applied Systems Analysis (IIASA study.[10-12]

The WAES study was a report prepared at the Massachusetts Institute of Technology by participants from 12 industrialized countries (Canada, Denmark, Finland, France, Germany, Italy, Japan, Netherlands, Norway, Sweden, United Kingdom, United States) and three oil-rich developing countries (Iran, Mexico, Venezuela). Most participants were either government officials or representatives of energy industries. The work was carried out over a two-and-a-half year period, during which there were many meetings.

The WAES study made energy projections to the year 2000. The WAES effort nominally addressed the energy problem of the developing countries, but it concentrated mainly on industrialized countries with market economies. The analysis of energy in developing countries was limited largely to work carried out by the World Bank; it was also restricted to commercial energy use—a serious limitation, since nearly half the total energy used in developing countries is non-commercial energy and only a fraction of their populations participate in the commercial energy economy.

The 1978 WEC study was carried out by the Energy Research Group at the Cavendish Laboratory, Cambridge, England, for the Conservation Commission of the World Energy Conference. It is a global analysis that looks to the years 2000 and 2020, and presents its results for seven world regions—North America; Western Europe; Japan, Australia and New Zealand; the USSR and Eastern Europe; China and Centrally Planned Asia; OPEC; and non-OPEC

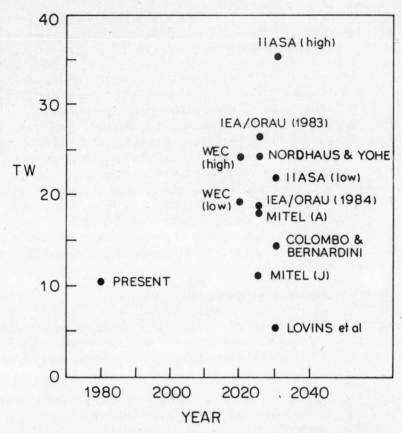

Figure 1.7. Recent projections of global energy use reviewed in J. Goldemberg, T.B. Johansson, A.K.N. Reddy, and R.H. Williams, 1985. The IIASA (high and low) and WEC (high and low) points refer, respectively, to the IIASA and 1983 WEC scenarios discussed in the text. The IEA/ORNL projections were made in studies carried out by researchers at the Institute for Energy Analysis and the Oak Ridge National Laboratory (J. Edmonds and J. Reilly, 1983; and J. Edmonds, J. Reilly, J.R. Trabalka, and D.E. Reichle, 1984). The Nordhaus and Yohe point is the mean of a large number of alternative projections made by the authors for a U.S. National Academy of Sciences study of the problem of atmospheric CO_2 and climatic change (W.D. Nordhaus and G.W. Yohe, 1983). The MITEL points are two of 11 global energy scenarios described in a study by the MIT Energy Laboratory dealing with the problem of atmospheric CO_2 and climatic change—with Scenario J being one of several "relatively CO_2 benign scenarios" emphasizing energy conservation (D.J. Rose, M.M. Miller, and C. Agnew, 1983). The Colombo & Bernardini projection was made in a study for the Panel on Low Energy Growth of the Commission of the European Communities (U. Colombo and O. Bernardini, 1979). The Lovins *et al.* projections was made in a 1981 study exploring the role of energy conservation in coping with the CO_2/climatic change problem (A. Lovins *et al.*,1981), carried out for the German Federal Environmental Agency.

Developing countries. In contrast to the WAES study, an important feature of WEC analysis of developing countries is that it takes non-commercial energy into account.

The 1983 WEC study was the result of a cooperative effort of one central team and 10 regional working teams involving 50 experts (30 from developing countries and 20 from industrialized countries) having varying backgrounds (11 for energy authorities, 9 for oil, 9 for electricity, 14 for international and regional organizations, and 7 for academic and research organizations).

The work was carried out in 17 international meetings and supplemental correspondence. The analysis, carried out for 10 global regions, focused on the supply of and interregional trade in primary energy. Both commercial energy (coal, oil, natural gas, nuclear, hydro, and new sources) and non-commercial energy (fuelwood, and animal and vegetable wastes) were taken into account.

The IIASA study was carried out between 1973 and 1979 by analysts at the International Institute for Applied Systems Analysis in Vienna. In view of the longer time that it took and the larger number of contributors that it involved, the IIASA study represents a much greater effort than the earlier WAES and WEC studies. The IIASA project output includes an 800 + page report, along with an elegantly produced 200 + page summary volume,[10] and articles in the *Scientific American*[11] and *Science*[12] magazines.

The IIASA study makes global projections to the years 2000 and 2030 by aggregating projections for seven regions of the world (Note 1.2-A). The primary emphasis in this study, too, is on commercial energy. Because the IIASA analysts attempted to give a more detailed picture of the global energy future, the IIASA models are much more complex—but this increase in complexity makes it difficult for any but the model's designers to understand the working of the models.[13, 14, 15]

The basic approach in all three studies consists of two steps:

- estimating future energy demand on the basis of (1) assumptions about future demographic and economic trends and (2) historical correlations between such trends and energy demand (see, for example, Box 1.1);
- matching this demand to a mix of energy supplies.

This mix is chosen so that it is compatible with estimates of the energy resources base and, in the case of new energy supply technologies, with judgements about how much energy can be produced by these supply technologies at various future dates.

The energy demand projections of the WAES, WEC, and IIASA studies show many common features (Notes 1.2-B and 1.2-C). All studies envisage that worldwide growth in energy demand will be slower in the future than the long-term historical trend. For the period up to the year 2000, the projected growth rate for global energy demand varies from 2 to 3.5 percent per year. This is considerably less than the pre-1973 growth rate, which was in excess of 4 percent per year, but is comparable to the growth rate between the two oil shocks of the 1970's—2.8 percent per year (Note 1.2-5). Also, all the studies envision that energy demand in developing countries will grow much more rapidly than demand in industrialized countries. And, the studies that look beyond the year 2000 share the perception that energy demand will grow more slowly after the turn of the century in industrialized countries than in the period before 2000.

There are also some important differences between the three studies. The IIASA demand growth rates for industrialized countries tend to be lower than the 1978 WEC growth rates, which in turn are lower than the WAES growth rates. The 1983 version of the WEC study[9] projects global energy demand growing at about the rate projected for the IIASA study—with energy use projected for 2020 less than what was projected in the 1978 WEC study by an amount equal to about 60 percent of global energy use in 1978. Thus, there has been a definite downward trend in energy demand projections over the past several years, probably reflecting the higher energy prices brought on by the second oil shock and the much slower global energy growth rate in recent years.

On the supply side also, there are similarities among these studies. The 1978 WEC and IIASA projections (Note 1.2-E) involve oil supplies increasing annually at only 0 to 1 percent through

---

Box 1.1: The Estimation of Future Energy Demand

The process of estimating future demand involves, for the most part, the use of historical data (time-series) relationships between energy demand and various economic and demographic variables to calculate future energy demand for assumed future values of these economic and demographic variables. The results of these projections are then adjusted in various ways to reflect "professional judgements" about demand saturation, supply constraints, etc.

The simplest and most transparent of these estimations is to be found in the 1978 WEC study. In the WEC model aggregate energy demand E(t) in the year t for each world region is estimated ("to zeroth order") by relating aggregate energy demand to the average energy price P(t) and the gross national product (or gross regional product G(t) via the equation:

$$\ln \left[E(t)/E(1972)\right] = ep \ln \left[P(t)/P(1972)\right] + eg \ln \left[G(t)/G\,(1972)\right]$$

Here *ep* and *eg* are the price and GNP elasticities of energy demand respectively [where the price elasticity *ep* is the percentage increase in energy demand associated with a 1 percent increase in energy price, with GNP held constant; and the GNP elasticity *eg* is the percentage increase in energy demand associated with a 1 percent increase in GNP, with energy price held constant]. The elasticities assumed in the WEC analysis are *ep* = 0.3 and *eg* = 0.95 (1.10) for OECD (other) countries. With these parameters, alternative energy scenarios are derived, assuming alternative price and GNP projections. Specifically, the average energy price in the year 2020 is assumed to range from 1.8 to 2.2 times the 1972 price (in real terms), and economic growth rates to the year 2020 are assumed to be in the ranges:

★ 2.8 to 3.7 percent per year for OECD countries.
★ 3.2 to 4.5 percent per year for centrally planned economies.
★ 3.8 to 5.3 percent per year for developing countries

The resulting projections are then adjusted to reflect (i) some saturation effects that the modellers believe this simple model does not capture, (ii) non-market induced conservation measures (e.g., the mandated automotive fuel economy standards in the U.S.), and (iii) energy supply constraints. The effect of these modifications is minor in the "low growth" scenario (a 7 percent reduction in the demand as projected by the model for the 2020) but substantial with high economic growth. This difference is largely because the modellers think there would be great difficulty in providing the supplies required for high economic growth. Hence, they advance for their "high growth" scenario a demand level for 2020 which is ⅓ less than that predicted with their simple"unconstrained"model.

---

the first quarter of the next century—which is quite low compared to historical rates of 6.5 percent per year 1925-1972 (Note 1.1.-B) and 2 percent per year between 1973-79 (Note 1.2-D). These studies are somewhat more optimistic about natural gas supplies. The 1978 WEC study projects 1 percent annual growth, while the IIASA study expects these supplies to increase between 1.5 and 2.5 percent per year.

The really significant growth projected in these studies is for coal and nuclear power. The 1978 WEC study projects that these sources combined will grow worldwide between 4 and 5 percent per year, while the IIASA study envisions growth of coal and nuclear power 3 to 4 percent per year for the next 40 or 50 years. Coal and nuclear power combined are projected in the 1978 WEC and IIASA studies to account for 70 to 80 percent of the total net increase in energy supplies in this period. Since nuclear power starts from such a small base—121 GW(e) of installed capacity* worldwide in 1978—its expected growth is especially dramatic in these studies, reaching 2600 to 4400 GW(e) by 2030 in the IIASA study (Note 1.2-F).

The more recent (1983) WEC study envisages considerably less growth in coal and nuclear power, but even in this case these sources provide half of the net increase in supplies in this period, and installed nuclear power generating capacity is expected to increase 10- to 15-fold by 2020 (Note 1.2-E and 1.2-F).

These energy studies all satisfy one important long-run objective of energy policy: to shift from energy dependence on oil to more abundant energy resources. The WEC and IIASA energy supply projections (Note 1.2-E) show that 40 to 50 years from now these analysts expect that oil will account for no more than 15 to 25 percent of the total world energy supply, compared to 40 percent in 1978. There certainly are no serious supply constraints on either coal or nuclear fuels during this period. Worldwide coal supplies estimated to be available at prices (in 1984 $) up to $90 per tonne (equivalent on an energy basis to less than half the world oil price in 1980) would last 130 to 260 years at the coal demand levels projected for the period 2020 to 2030 in the IIASA and WEC studies (Notes 1.2-E and 1.2-G). And if nuclear power technology is shifted from present-day light water reactors (LWRs), which can make use of only about one-half of 1 percent of the fission energy stored in natural uranium, to fast breeder reactors (FBRs), which can tap about half of uranium's fission energy potential, then uranium can become virtually an inexhaustible resource. However, the direct economic as well as the external environmental and social costs of this coal-nuclear path way from petroleum dependence are extremely high (Section 1.3).

Despite the fact that these studies all envisage slower energy growth in the future than in the past, they are still basically *supply-oriented* analyses: *the solution to the energy problem is to produce even more energy.* The present level of global energy demand is already so high that meeting the demand levels projected in these studies would require Herculean production efforts.

The scale of the effort can be appreciated by examining some of the important IIASA supply numbers. To achieve IIASA's production projection for 2030 would require the worldwide building up of new production capacity at an average rate, between 1975 and 2030, of:

- one new central-station coal or nuclear power plant of 1 GW (e) electrical generating capacity *every 1½ to 2 days* (Note 1.2-H).
- one new nuclear plant of 1 GW(e) generating capacity *every 4 to 6 days* (Note 1.2-I), plus
- new fossil fuel production capacity equivalent to bringing on line a new Alaska pipeline (2 million barrels of oil equivalent per day) *every 1 to 2 months* (Note 1.2-J).

This is the level of effort that would provide both the net growth and the replacement of retired facilities necessary to supply IIASA's energy future. In light of the expansion problems which have plagued the energy industry throughout the 1970's and the 1980's, it is clear that such production targets as the opening of a new 1 GW(e) nuclear power plant every 4 or 6 days over a 55-year period are unrealistic.

It is generally accepted, in view of the six-fold increase in the world oil price since 1972 and the large increases in other energy prices as well, that improvements in the efficiency with which energy is used must be a key part of any realistic vision of the global energy future. Indeed, all three global energy studies under consideration here claim to emphasize energy efficiency:

---

* 1 GW(e)—1 million kilowatts electrical output capacity. A typical nuclear power plant has a capacity of about 1 GW(e).

WAES: "...our projections show only a relatively small decrease in the ratio of growth in energy use to growth in GWP (Gross World Product), even with the substantial efficiency improvements assumed in the scenarios..."[6]

WEC (1978): "World energy demand in the year 2020 is expected to be between three and four times present consumption if average economic growth is similar to that achieved in the past forty to fifty years and there are vigorous and successful measures to improve the efficiencies with which energy is used",

IIASA: "Strong energy conservation trends were built into the scenario demand projects from the beginning...",

While these studies all envisage total energy demand growing more slowly in the future than in the past, they greatly underestimate the real opportunities for energy efficiency improvements. As will become apparent in the coming chapters, our analysis shows that future energy demand would be far less than projected in these studies if close attention is paid to technically and economically feasible energy efficiency improvements.

1.3 Global Energy Strategies and Other Major Problems of the World

1.3.1 *The Importance of Considering Other Major Global Problems*

Energy is a global problem, but it is certainly not the only one. The world is beset by a host of other major problems, such as the wretched living conditions of the majority of human beings on this planet, undernutrition and food shortages, widespread unemployment and underemployment, massive national indebtedness and scarcity of development capital, environmental degradation, the threat of serious climatic changes associated with the build-up in the atmosphere of carbon-dioxide, global insecurity and nuclear weapons proliferation, both of which increase the danger of nuclear war. These other global issues are irrelevant to discussions of energy only if they have no energy connections. It turns out, though, that many of them are in fact closely linked to energy. It is important, therefore, to understand these global problems and in particular to discern their energy linkages.

1.3.2. *North-South Disparities*[16, 17]*—The Importance of Meeting the Energy Requirements of Developing Countries*

Global Gap: One of the most serious global problems is the tremendous economic and social gap that separates the rich from the poor countries amongst the family of nations, or the North and South as they are loosely called (Figure 1.8). So wide is this gap that it appears to be unbridgeable. The very rich and the very poor nations seem to belong to totally different worlds.

The North-South gap can be described in many ways, but perhaps the most telling is the population-income situation. With only a quarter of the world's population, the industrialized countries account for 80 percent of global income. This means that the three billion people of the South (including China) have an average per capita income only one-tenth of that in the North. And with this enormous disparity in per capita income come enormous disparities in people's lives:

"In the North, the average person can expect to live for more than seventy years; he or she will rarely be hungry, and will be educated at least up to secondary level. In the countries of the South the great majority of people have a life expectancy of closer to fifty years; in the poorest countries one out of every four children dies before the age of five; one-fifth or more of all the people in the South suffer from hunger and malnutrition; fifty percent have no chance to become literate."[16]

This gap between the North and the South is a crucial factor in the worsening crisis in the world economy.

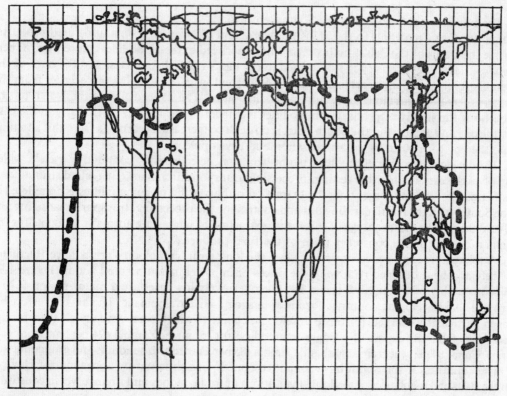

Figure 1.8. This map is based on the Peters Projection, rather than the more familiar Mercator Projection.

The Peters Projection introduces several innovative characteristics: an accurate rendition of the proportion of the land surface area, graphical representations of the entire world surface, including the polar regions; the Equator is placed at the center of the map; the usual grid of 180 Meridians (East and West) and 90 Meridians each (North and South) is replaced by a decimal degree network dividing the earth both East and West and North and South into 100 fields each; angle accuracy in the main North-South, East-West directions.

The surface distortions that do appear are distributed at the Equator and the poles; the more densely settled earth zones, it is claimed, appear in proper proportion to each other. This projection represents an important step away from the prevailing Eurocentric geographical and cultural concept of the world. The dark line separates the "North" from the "South."
Source: Willy Brandt *et al.,* 1980.

An independent international commission—the Brandt Commission—devoted itself wholly to the North-South issue and presented a report on the matter. It is to this report[16] that reference can be made for details; what follows is only a cursory sketch of the subject.

Origins of North-South Disparities: The North-South gap and the international problems associated with it have their historical origins in the era when the world was almost completely divided into imperial powers and their colonies. All this changed during the period following World War II when most of the nations which are now called *developing countries* achieved political independence. But the process of decolonization is not complete even now. In particular, the attainment of economic liberation has not kept pace with political independence. Developing countries are newcomers to the international setting, and they find that the "rules of the global game" were written before their arrival on the scene.

The fundamental handicap of the South stems from its colonial past when the colonies were suppliers of raw materials to the industrial world and receivers of its manufactured products. While the status of the South in this regard has improved in recent decades, more than 90 percent of the world's industry is still located in the industrialized countries, and the bulk of the South's foreign exchange earnings (roughly 60 percent in 1978 excluding oil) accrues from the export of major agricultural and mineral commodities. The dependence of the North on the South for the supply of primary commodities is evident (Note 1.3-A), particularly in the case of Europe and Japan. But the South finds it difficult to increase its earnings from this situation and thereby to finance its industrialization and development.

The developing countries are thwarted by what may be called the "commodity cycle", which involves the extraction or harvesting, processing, transportation, marketing and distribution of commodities. By and large, only the extraction or harvesting of commodities is completed in the South, with the North carrying out and controlling the processing, transport, marketing and distribution. The disadvantage of this is that extraction or harvesting normally accounts for less than a quarter of the consumer price of the final product.

A related problem concerns prices. Over the long term, commodity prices have fluctuated widely and have declined relative to the prices of imported manufactured goods (Figure 1-9). The instability of commodity prices and their relatively low level are a major reason for the trouble that developing countries face in financing their development. The adverse effect has been particularly marked for those developing countries which earn virtually all their foreign exchange from the export of a single commodity: for example, Zambia (94 percent from copper), Cuba (84 percent from sugar) and Gambia (85 percent from groundnuts and groundnut oil).

The countries of the South have sought to remedy this situation by appealing for changes in the rules and regulations and global institutions of trade, money and finance which constitute.

COMPOSITE COMMODITY PRICE INDEX, 1948-82
Index (1977-79 average = 100)

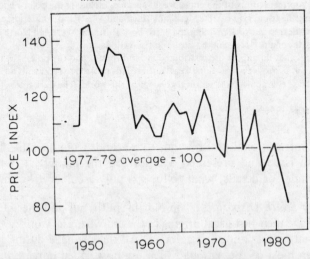

Figure 1.9. This graph shows the trend in aggregate non-oil commodity prices relative to a price index for manufactured goods imported by developing countries.

The commodities are coffee, cocoa, tea, maize, rice, wheat, sorghum, soybeans, groundnuts, palm oil, coconut oil, copra, groundnut oil, soybean meal, sugar, beef, bananas, oranges, cotton, jute, rubber, tobacco, logs, copper, tin, nickel, bauxite, aluminium, iron ore, manganese ore, lead, zinc, and phosphate rock.
Source: The World Bank, 1983.

the international economic system. They are demanding a New International Economic Order involving a restructuring of international trade (reform of tariff structures, stabilization of commodity prices, improved access to markets, etc.); a reorganization of international finance (debt relief, financial cooperation in favour of the developing countries, reform of the international monetary system, etc.); regulation of the activities of multinational corporations; transfer of resources from industrialized to developing countries; better terms for technology transfer; and more voice in international decision-making processes for developing countries. But the South has found the North reluctant to surrender control of this system, which it established in the colonial era.

A basic economic weakness of the South has been its inability to influence world commodity prices. Of course, a group of nations in the South did gain control of the world oil market for a while in the 1970's, and thereby attained a sixfold increase in their income from the sale of that commodity. But efforts to exert similar control, i.e., form cartels in other commodity markets—bauxite, copper, uranium, tin, cotton, cocoa, etc.—have fizzled out. And the prospects for exerting such control in the future do not look all that bright. There are certain preconditions which make possible supplier control of commodity prices to the extent accomplished by OPEC: (1) a product which is relatively scarce and for which there is no immediate substitute: (2) suppliers who are able to cooperate with one another and regulate their output; and (3) buyers who are unable or unwilling to reduce their demand. These preconditions do not appear to exist in any major commodity market, though they could re-appear in the oil market as the worldwide supply of recoverable oil declines and/or if consumer demand increases. This strongly suggests that OPEC's triumph in the 1970's did not in reality signal a fundamental reordering of world economic relations, at least in terms of other commodities, as was hoped at the time by many in the developing world. Income from commodity exports remains, for the foreseeable future, an extremely narrow and shaky base upon which to build development. Countries in the South need alternative income sources—from the export of manufactured goods and from the development of their own internal markets.

The past two decades, in fact, have witnessed a new trend in the South—particularly in the so-called *newly industrializing countries*—the export of manufactured products. Though the export of manufactured goods is crucial for the development of the South, the process of gaining access to Northern markets is hindered by the rising trend of protectionism in the North, which is resulting in a host of tariff and non-tariff barriers.

Transnational Corporations: These corporations have become powers that transcend national boundaries and are virtually independent of the governments of both the industrialized and developing countries. Particularly since World War II, there has been a tremendous increase in the magnitude, complexity and sophistication of their operations. They now have a major influence in the world economy, controlling the flows of capital, technology, manufactured goods, and major commodities such as food.

The preoccupation of these corporations with maximizing short run profits is often in basic conflict with the national interests of the host countries. In addition, the transnational corporations have often used their dominance over the distribution of technology—which is necessary for development—to the disadvantage of developing countries. For example, they have promoted technology that is ill-suited to the socio-economic and environmental conditions in developing countries, sold obsolete technology and overcharged for technology. Such practices jeopardize the development process in the South and the stability of the world economy.

Conventional Global Energy Strategies and North-South Disparities: The disparities between rich and poor countries are likely to increase with conventional energy strategies, which do not address these disparities. The one-quarter of the world's population living in the industrialized world presently consumes energy at an average per capita rate which is six times that in the developing countries, and it accounts for two-thirds of the global energy use. Despite this, the supply-oriented WEC and IIASA studies project that as much as a half to two-thirds of the increment in world energy demand over the next 40 to 50 years will be accounted for by the already-industrialized countries. This projected increase in energy use by the industrialized countries would drive up energy prices, especially oil prices, thereby making the development of the oil-importing developing countries all the more difficult.

Energy Strategies and the Reduction of North-South Disparities: The fundamental solution to this problem is to emphasize and accelerate the development of developing countries, a process requiring significant enhancement of energy *services*. Above all, it is imperative that energy does not become a constraint on development.

Since oil is such a scarce non-renewable resource, developing countries must not continue to base their economic development on oil. Instead, they should move to alternative energy systems. Of course, this transition cannot be accomplished overnight—adequate oil supplies will be required during the transition period. To make oil available for essential purposes at affordable prices during the transition period, it is vital that the North curb its appetite for oil in order to ease the upward pressure on the world oil price.

In gaining access to energy for development, it is imperative that developing countries do not become burdened with energy-wasteful technologies and energy-intensive life-styles—as has often happened. Instead, by pursuing improvements in energy efficiency and alternative fuels, developing countries can achieve a much higher level of energy services per unit of energy input than is currently the norm in either industrialized or developing countries.

1.3.3 *Global Economic Problems—The Importance of Economic Efficiency*

Over the past decade, the global economy has suffered two major recessions, brought on in part by the oil shocks of 1973 and 1979 (Figure 1.10, top). The 1974-75 recession was sharp but relatively brief. The 1980-83 recession was the longest and most severe since the Great Depression of the 1930s'.

The recessions brought sharp increases in unemployment to the OECD countries. Unemployment in the United States rose from about 6 million in 1979 (6 percent unemployment) to about 11 million in 1982 (10 percent unemployment). In Western Europe, the unemployment problem has been worse, with 18 million workers unemployed in 1982. Moreover, unemployment has generally not fallen much in the recovery periods following these recessions (Figure 1.10, bottom).

By far the heaviest impact of the recent recession has been on the developing countries. In 1982, the prices of primary materials fell in relation to those of manufactured goods to a post-1945 low.[18] The poor terms of trade faced by almost all developing countries were due, not only to the weak demand for commodities in industrialized countries, but also to high interest rates which discouraged storage of industrial raw materials. Compounding the historical problem of low commodity prices in recessions is the relatively new phenomenon of protectionism to restrict the import into the North of manufactured goods from the rapidly industrializing countries of the South.

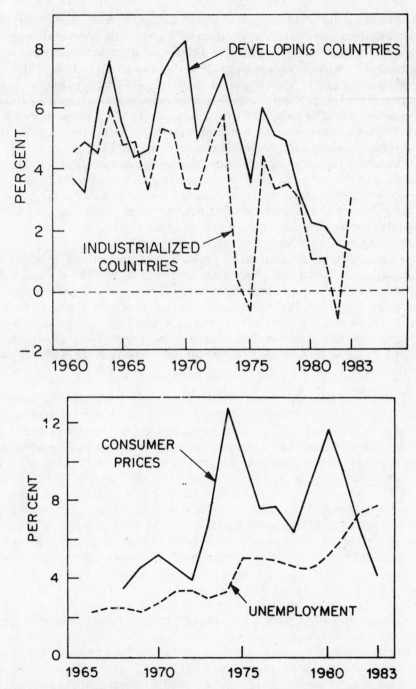

Figure 1.10. Trends in economic indicators: Gross Domestic Product growth rates for industrialized and developing countries with market economies (top); inflation and unemployment rates in seven major industrialized countries with market economies (bottom).

Source: The World Bank, 1984.

The 1974-75 recession did not have nearly so serious an impact on developing countries (Figure 1.10, top), not only because of its shorter duration but also because heavy borrowing permitted continued growth for many of these countries.[8] There was no such cushion in the more recent recession because the availability of foreign capital declined abruptly after 1981.

The 1980-83 recession and the beginning of the recovery in 1984 has been a period of especially high interest rates in the United States. When U.S. interest rates rise, the impact is felt throughout the market-economy world, because of the predominant role of the dollar in global financial transactions. The most serious consequence of high interest rates has been a global debt crisis involving developing country debtors and the international banking system.

The future welfare of the world depends on overcoming the problems that continue to trouble the global economy. The pursuit of development goals to achieve a decent living standard for the majority of the world's population living in the South requires rapid economic growth there for decades to come. Moreover, the economic climate for technical change needs to improve in the North so that these economies will remain viable using much lower inputs of basic resources, which are becoming ever more costly and scarce.

The complex causes of the global economic problems and the courses of action required for economic recovery are beyond the scope of the present analysis. The discussion here will, therefore, concentrate on just one important energy-related aspect of the global economic problem: the global debt crisis. While this is a current economic issue, the dimensions of which are likely to change markedly in the years to come, it is worth describing even in a book with a long-term perspective like the present one, because it illustrates the inter-connectedness of the various parts of the global economic system and the vulnerability of the system to unexpected perturbations.

The Debt Crisis: The debt crisis emerged full-blown in late 1982, when it became apparent that some developing country debtors were going to have great difficulty in meeting their debt-servicing obligations to foreign creditors. In response, rescue operations were launched involving the creditor banks, the International Monetary Fund (IMF) and the creditor governments. In the resulting rescue agreements, much of the debt-servicing was rescheduled in exchange for the debtors commitments to impose austerity measures on their economies. The costs to some have been high. The most hard-pressed debtor countries have had to drastically cut back on imports to improve their trade balances—the result of which have been reduced standards of living (especially for the middle class), depleted inventories of raw materials and spare parts, and an investment standstill. There have been loud protests in debtor countries against the harsh domestic austerity measures imposed on debtors by the IMF as conditions for rescheduling of interest payments. In some cases, it has been argued that a country's government should not be held responsible for debts incurred by previous, discredited regimes which did not always use the loans in ways that served true development goals.

While the debtors have suffered greatly from the debt crisis, its ramifications extend far beyond the national borders of the debtor countries. While the prospect of default on a small loan is a problem for the debtor, the default on a large loan of the kind involved here is a problem not only for the debtor but for the creditor as well. Throughout the debt crisis, a continuing concern has been whether the crisis will cause the global financial system to crumble.

Developing country debtors got into trouble servicing their debt mainly because real interest rates in the US began to surge in the early 1980's. Two-thirds of the debt of developing countries is denominated in dollars and much is at variable interest rates.[18] Debtors have argued that it is the creditors who are responsible for the debt crisis, and therefore, their efforts should be directed at reducing interest rates rather than at trying to squeeze more interest payments out of the debtors.

Several factors have contributed to the surge in U.S. interest rates: the deflationary trend, the erosion of the OPEC petrodollar surpluses in the aftermath of the world oil glut of the early 1980's, the long-term trend of declining savings rates in OECD countries, and the huge U.S. budget deficit.

The 1970's were a period of high but variable inflation rates, with two major inflation surges correlated closely with the oil price shocks of 1973 and 1979 (Figure 1.10,bottom). Normally, the nominal interest rate follows closely changes in the inflation rate. When the inflation rate rose at the time of the second oil shock, the nominal interest rate followed. But subsequently, as tighter monetary policies succeeded in curbing the growth in the inflation rate, the nominal interest rate continued to grow rapidly (Figure 1.11), reflecting the general lack of confidence that monetary controls would really be effective in keeping inflation under control. And although the nominal interest rate then began to fall as inflation subsided, the real interest rate—the difference between the nominal rate and the rate of inflation—remained at a level several times higher than the average real interest rate in the 1970's (Figure 1.11). By the time it seemed clear that inflation in the United States was under control, other factors reducing the overall supply of capital came into play to keep the real interest rate high.

A noteworthy initial cause of the shrinking capital supply was the drying up of the petrodollar surplus. Beginning with the first oil crisis, banks began recycling billions of dollars of savings from oil revenues which OPEC countries could not absorb in the short run. This OPEC surplus contributed to the low real interest rates that persisted throughout the period from the first to the second oil price shock (Figure 1.11). But with the dramatic fall in the world oil price after the onset of the world oil glut in the early 1980's this surplus disappeared.

Another factor contributing to capital scarcity is the declining saving rate in OECD countries. In 1974, the net savings rate in OECD countries peaked at 14 percent of GDP; subsequently, the savings rate declined, falling to below 9 percent by 1981.[18] When the OPEC surplus vanished, the low savings rate became an important factor contributing to the high interest rate problem.

Still another major factor contributing to capital scarcity is the large increase in the U.S. budget deficit which has been ascribed to the tax cut of 1981 and the sharp rise in the defense expenditures (a $100 billion increase between 1981 and 1984). The 1984 U.S. budget deficit of about $200 billion is equivalent to nearly half of the total investment in the United States—up from a deficit level equivalent to only one-sixth of total invesment in 1981. This budget deficit, which must be financed by government borrowing, is expected to stay at an annual rate of some $200 billion for several years—a prospect suggesting that the high interest rates of the early 1980's may persist for years to come.

Despite the sharp rise in the U.S. budget deficit, the real interest rate in the United States did not change much in the early 1980's (Figure 1.11). This is because the crowding out of capital by the budget deficit was to a large extent compensated for by an inflow of capital from abroad to finance the deficit in the U.S. current account.*

Because of the strong dollar, the US has been running record merchandise trade deficits—$107 billion in 1984, up from previous records of $61 billion in 1983 and $36 billion in 1982. The United States normally runs a deficit in merchandise trade, which is offset in the current ac-

---

*The current account is the net balance in the trade of goods and services, increased by the income earned on investments abroad and diminished by the income from foreign investments in the country.

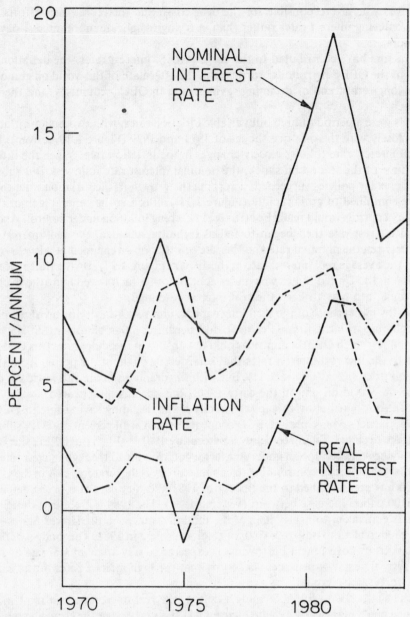

Figure 1.11. Nominal and real interest rates in the United States.

The nominal interest rate shown is the annual average prime interest rate charged by banks—the rate charged to their best customers. The inflation rate is the annual rate of change in the U.S. GNP deflator. The real interest rate is the nominal rate corrected for inflation.

Source: Council of Economic Advisers, 1985.

count by earnings on overseas investments and by a surplus in the export of services. However, the recent merchandise trade deficit is so large and foreign investments so substantial that the United States current account has registered record deficits of $42 billion in 1983 and $102 billion in 1984 (the previous record of $15 billion was set in 1978).

While the inflow of capital has kept U.S. interest rates from rising, the flow of capital into the United States makes capital scarcer and interest rates higher than they otherwise would be elsewhere in the world.

Since it is not likely that interest rates will decline much in the near-term future, the required servicing rates for developing country debt are not likely to drop dramatically either. Therefore, to understand better the options available for ameliorating the debt crisis, it is important to look beyond interest rates and try to understand how the enormous debts that gave rise to the crisis arose in the first place.

After the first oil price shock of 1973-74, many oil-importing countries found it virtually impossible to continue development programmes as well as pay for oil imports at the new oil prices. At the same time, because of the enormous OPEC surpluses available in banks as recycled petrodollars and the high inflation rates, real interest rates were low and sometimes negative—a situation that lasted throughout the decade. Under such circumstances, borrowing was very attractive, and the banks made massive loans available to those oil importing countries judged to be good credit risks, e.g., Argentina and Brazil, and also to some oil-exporting countries, e.g., Mexico, Nigeria, and Venezuela, that went in for borrowing in a mood of confidence emanating from their oil revenues. The lending seemed sensible at the time, as debt-servicing did not look difficult. In fact, considering the three largest debtors—Argentina, Brazil, and Mexico—the overall ratio of debt to exports did not rise in the entire 1975-80 period, a good indication of "credit-worthiness".[19]

The crisis could have been less severe had debtors more effectively managed the problems that gave rise to the need for debt in the first place. In particular, only modest efforts were made in oil-importing developing countries to deal with the root cause of the problem, namely, growing oil imports. Oil imports were allowed to increase at an average annual rate of 6.3 percent between 1970 and 1980—from 3.2 to 5.9 million barrels per day, a rate in excess of the average GDP growth rate during this period.[20]

The Importance of Economic Efficiency: While the solution to the problem of re-ordering the world economy so that development goals are achieved in the South and the countries of the North are adequately prepared to deal with the new era of increasingly scarce resources, will obviously involve many elements, one essential element of the solution is the pursuit of economic efficiency.

Particular attention must be given to the allocation of capital resources. Since the oil exporters' surplus has disappeared and since the United States will probably run a large budget deficit for several years, even if new taxes are raised and the defence budget is cut, it is likely that, unlike the situation in the 1970's, capital resources will be scarce and the real cost of capital will be high worldwide for years to come. Thus more efficient allocation of capital resources is crucial.

A Role for Better Management of Energy Investments: The promotion of economic efficiency in energy production and use is one of the most significant steps that can be taken, in addition to a reduction of the U.S. budget deficit, to put the global economy back on track.

Consider first the situation in the industrialized countries, using the United States as an example. Because of the rapidly increasing costs of new energy supplies, energy supply investments have accounted for an ever larger percentage of all new plant and equipment expenditures in the United States—up from 25 percent in the 1970's to 40 percent in 1982 or, expressed as a percentage of GNP, an increase of from 2½ to 4 percent.

In this context, efforts to improve the efficiency of energy use in the OECD could help reduce interest rates. Over a wide range of opportunities, investments to improve energy efficiency are less costly and involve less capital overall than equivalent investments in the expansion of energy supply. The net capital savings from emphasizing energy-efficiency investments would thus help lower interest rotes by reducing the overall demand for capital from what it would have otherwise been.

The effort to increase energy supplies is also taking a heavy economic toll in developing countries. In 1980, oil imports by developing countries (@ $34 per barrel) cost some $70 billion, which was approximately equal to their balance of payments deficit.[18] Moreover, the share of domestic investments committed to energy supply expansion in developing countries increased from 1-2 percent of GDP in 1970 to 2-3 percent in 1980.[20] The foreign-exchange requirements for energy investment in developing countries totalled some $25 billion in 1982—*about one-third of the foreign exchange required for all kinds of investments.*[20] Costs could be even higher in the future if energy supply expansion continues to be the primary approach to meeting needs for energy services. The World Bank in 1983 estimated that, in order to achieve a targeted increase in per capita use of commercial energy from 0.54 to 0.78 kW between 1980 and 1995, the investments in energy supplies for all developing countries would have to average some $130 billion a year (in 1982 dollars) between 1982 and 1992, or 4 percent of their GDP—the same percentage as in the United States in 1980! And half of this investment would have to come from foreign exchange earnings. This would require an average increase of 15 per cent per year in real foreign exchange allocations to energy supply expansion. Still, despite this big investment in energy supply expansion, oil imports by oil-importing developing countries would increase by nearly one-third to nearly 8 million barrels per day by 1995, the World Bank projected.

As in the case of industrialized countries, the economic goals of developing countries can be met with less capital resources if the criterion of economic efficiency is emphasized when it comes to deciding whether to invest in the expansion of energy supplies or in improvements in the efficiency with which energy is used.

Finally, if opportunities to use oil more efficiently are capitalized on, holding down the overall demand for oil, the world oil price will most likely be lower than it would have otherwise been, as suggested by the historical experience.[21] Figure 1.12 shows OPEC pricing behaviour as a function of the demand for OPEC oil. The resulting reduction in expenditures of export earnings to pay for oil imports from these lower prices could be at least as important as the savings associated with a reduced level of imports. (While the direct saving from reduced imports is proportional to the reduction in imports, the saving from a lower oil price is proportional to all imports.)

1.3.4 *Poverty of the Majorities in Developing Countries—The Importance of Energy for Basic Human Needs*

The Dimensions of Poverty: The fundamental reality of developing countries is the poverty of the majority of human beings who live in them. While an exact measurement of poverty poses conceptual and methodological problems, its pervasiveness, as revealed by the extent to which elementary minimum needs are not satisfied, is undeniable. Whether it is food, shelter, health, education, or employment that is considered, the living standards of the majority in most developing countries are pathetically low. It is a sub-standard life involving a full-time struggle for survival—a type of existence largely unknown and perhaps unimaginable in industrialized countries.

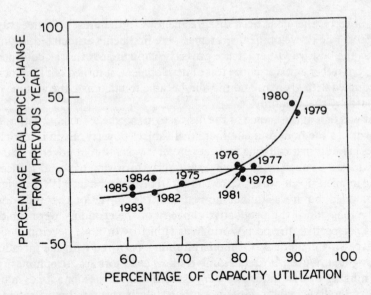

Figure 1.12. The percentage change in the real world oil price versus the OPEC capacity utilization factor (the percentage of productive capacity that is actually producing one year earlier).
Source: Energy Information Administration, May 1984.

The single most important aspect of poverty is hunger, the inability of large fractions of the populations of developing countries to obtain an adequate diet. The inadequacy of food is revealed (Note 1.3-B) in the average per capita food consumption of less than 2200 kcal per day (96 percent of the daily minimum requirement) compared to the average diet in industrialized countries of 3350 kcal per day (129 percent of the daily requirement). The averages do not reflect the fact that the impact of malnutrition is not uniform even among the poor—women and children are particularly subject to malnutrition. Nor do they indicate the poor quality of the diet in developing countries, especially the low levels of protein consumption.

The houses and settlements in which the ill-nourished poor in developing countries live make their existence all the more precarious. Rarely constructed of materials that provide permanent or satisfactory protection against the sun, rain, and winds, most homes have neither water supply nor sanitation. These conditions are ideal for the spread of diseases associated with poor living conditions and lead to high infant mortality and low expectancy.

"There are still countries in Africa where one child in four does not survive until its first birthday... Between 20 and 25 million children below the age of five die every year in developing countries, and a third of these are from diarrhoea caught from polluted water".

And, even when children survive, their life, particularly in rural areas, is burdened with sharing the drudgery of their elders—the drudgery of gathering fuelwood, fetching drinking water and tending livestock. The chances of enrolling and staying in school—especially for girls—are bleak. It is no wonder that about 50 percent of the adults in developing countries are illiterate.

A History of Strategies for Dealing with the Poverty Problem:[22-25] Ever since the process of decolonization began, following World War II, there has been debate on how to eliminate poverty and deprivation in the developing countries. Although there has been broad agreement among development analysts and progressive Third World leaders concerning the importance of eradicating poverty, there has been disagreement as to the most effective way of achieving it.

In the 1950's, when development strategies were first being articulated, economic growth was advocated as the best way to eradicate poverty. Though growth was considered to be, not an end in itself, but rather a performance test of development, it turned out that increasingly development was equated with growth, and the aim became to maximize the volume of goods and services produced by the economy.

The belief was that at least some of the increases in income associated with rapid growth would "trickle down" to the poor and eliminate their abject poverty. Even if a trickle-down process would not occur automatically via market forces, it was felt that governments could spread the benefits downward through progressive taxation, the provision of social services, and other such actions. The implicit slogan was: "Growth first, redistribution later!" This viewpoint may seem to be cold-blooded, but it was in fact motivated by the conviction that the poor would be better served in the long run if the productive capacity of the economy were rapidly built up first.

From this perspective, the post-World War II history of most developing countries must be judged a success. A great deal of "modernisation" occurred with the introduction of western food, clothing and houses, hospitals, universities, cars, airlines, telephones, radio, television, etc. A vast number of new industries were established. At the same time, many of these countries sought to modernize their agriculture, particularly through the so-called "Green Revolution" based on high-yielding varieties and inputs of fertilizer, pesticides, water, etc. The Gross Domestic Product (a measure of the amount of goods and services produced by the country) has shown impressive increases in these countries since the end of World War II.

But the question is: Who has benefited from these material gains which have been achieved? In most developing countries, unfortunately, the answer is that the multinational corporations who have sold them everything from bulldozers, high-rise elevators and irrigation-sprinklers to engineeing consulting services certainly have benefited and the elites within these countries—the industrial, landed, bureaucratic and commercial interests—have benefited as well, but the poor masses are hardly any better off than they were before, and in fact in certain parts of the Third World—sub-Saharan Africa, the barrios and remote rural areas of Latin America, for example—the poor are actually worse off in many respects than they were 30 years ago.

The experience of the past 30 years clearly shows that growth does not necessarily equal development. Far more pertinent than the sheer magnitude of growth is the structure and content of growth, and the distribution of its benefits.

Once a particular pattern of growth takes place, neither its structure and content, nor its benefits, can be easily altered. When the growth process has been for the benefit of the elite, then the material output of that process, e.g., processed and packaged foods, expensive cloth, luxury houses, universities, modern airports, and private cars, cannot be easily transformed into benefits for the masses, i.e., cheap food and cloth, low-cost housing, mass health care, mass education, and public transportation.

And, Gross Domestic Product only measures aggregate growth. It does not distinguish private cars from buses, air conditioners from woodstoves; it does not tell who is receiving the goods and services being produced. Hence, GDP is, by itself, a terribly inadequate indicator of development success or failure.

Because of the failure of economic growth strategies in eliminating poverty, greater attention began to be paid during the 1970's to alternative approaches to the problem of development.

One such alternative approach has been to emphasize employment generation in the belief that the employment thus generated would lead to the creation of the purchasing power needed to lift the poor out of the pit of poverty. It soon became apparent, though, that the roots of poverty in developing countries spring not only from unemployment and underemployment but also from the *low-productivity* of employment.

With this insight, the development debate turned next to the question of income distribution and how income redistribution would affect economic efficiency and growth. To redress gross inequalities in the distribution of income, resources, and power, it was proposed that incremental income be taxed and channelled into public services intended to raise the productivity of the poor. However, it was soon discovered that a policy of redistribution with growth is a very inefficient and slow way to improve the lot of the poor, particularly in very poor countries. It would take many decades for this mechanism to result in an appreciable increase of the income of the poor. At this point in the debate, attention turned to the basic human needs approach to development.

The Need to Target the Satisfaction of Basic Human Needs: The starting point of the basic human needs approach is the pervasive and abysmal character of poverty and the existence of glaring inequalities. The crux of poverty is the fact that the basic human needs of the bulk of the population have not been satisfied. Even elementary minimum standards of food, clothing, shelter, health and education have not been achieved. Therefore the satisfaction of basic human needs must be made a direct and immediate economic objective in development strategy.

The distribution of the goods and services necessary for the satisfaction of these needs cannot be a matter of charity, however, for that would be incompatible with the dignity of the recipients. Human beings need to earn, in the noblest sense of the term, their food, clothing, shelter, health and education. Employment also can be considered a basic human need—and in this way, employment generation can be included in a basic needs strategy. However, as already pointed out, if the employment (provided as a basic need) is the low-productivity variety, then the ensuing production system is unlikely to yield enough goods and services to sustain a high level of satisfaction of the other basic needs.

Poverty and the lack of basic needs are only one side of the coil; the other side is the existence of inequalities in the control of resources and in purchasing power. Inequality in the distribution of purchasing power leads to a skewed demand structure, which in turn makes industry, agriculture and the services produce a pattern of goods and services—the so-called product-mix of the economy—that caters to the demands of elites rather than the needs of the poor. Thus, the elites get richer and the poor poorer, i.e., comparatively speaking, poverty increases.

Under a basic human needs strategy, when a choice has to be made between production for the needs of the poor and the demands of the affluent, then industry, agriculture and the services must assign priority to satisfying the basic needs of the poorest. Hence, inequalities are likely to diminish, and poverty to decline.

Though the satisfaction of basic needs and the reduction of inequalities are clearly linked to each other, the focus of development has narrowed in the basic needs approach from the general goal of reducing inequality to particular aspects of the goal, such as eradication of undernutrition, delivery of adequate health care, education of children, etc. This tighter focus is based, in part, on the realization that the scarce resources available through a redistribution policy can go much further in eliminating human suffering if targeted on meeting a limited set of fairly well defined basic human needs rather than if the revenues from redistributional tax revenues are used instead for more general welfare programmes aimed at reducing inequality.

Also, meeting basic human needs can be considered to be morally more compelling than reducing inequality. The idea that the basic needs of all should be satisfied before the less essential wants of a few, goes back to the founders of the world's great religions, while the goal of reducing inequality does not seem to be of great importance to most people, other than idealogies and utilitarian philosophers.

Furthermore, while reducing inequality is a complex, abstract objective, open to many different interpretations and therefore operationally ambiguous, meeting the basic human needs of the poor is a concrete goal that is amenable to programmatic action.

The basic human needs approach is not without critics, although the main objections represent contradictory points of view:

- It is "an ideological concept that conceals a call to revolution," or
- It is a "minimum welfare job to keep the poor quiet."

Whether these criticisms cancel each other or not, the fact is that programes for the satisfaction of basic human needs have been implemented by a wide variety of political regimes in areas of different sizes, histories, and traditions, including market economies such as Taiwan and South Korea, mixed economies such as Sri Lanka, centrally-planned economies such as China and Cuba, and decentralized planned economies such as Yugoslavia. What these countries have in common is a fairly equal distribution of land, a degree of decentralization of decision-making with adequate central support, and attention to what goes on within the household, particularly to the role of women.

The basic human needs approach has also been attacked on the grounds that it conflicts with economic growth. This point was made by both the superpowers, for example, in response to the International Labour Organisation (ILO) Director-General's Report presented at the ILO Employment Conference in Geneva, June 1976.[23] This report stressed the desirability of making the satisfaction of the basic needs of the poor the primary focus for national and international development efforts.

The impact of a basic human needs approach to development on economic growth is obviously an extremely important, but quite complex, question. Despite the above criticisms, cross-country multiple regression studies indicate that those countries which do well in providing for basic needs have better than average economic growth. This suggests that a basic needs emphasis on development, far from reducing the rate of growth, can be instrumental in increasing it.[24]

Even at quite low income levels, it seems possible to meet the more pressing basic needs without sacrificing economic growth. For example, 69-year life expectancy was achieved in Sri Lanka at a per capita income of $200 (1977) and at an annual per capita growth rate of income of 2 percent between 1960 and 1977.

A mathematical case has been presented to show that the provision of basic needs can result in a higher growth rate (than if these needs are ignored) because the improved nutrition, health, education, etc., that result have the effect of enhancing the productivity of an otherwise underfed, sickly and illiterate worker.[25] In addition to raising labour productivity, it also seems that better education, nutrition, and health are beneficial in reducing fertility, enhancing people's adaptability and capacity for change, and creating an overall political climate favourable for development.

While the basic needs approach to development is by no means universally accepted, international agencies including the U.N. system and the World Bank have used it as a basis for programme formulation.

Among economists and planners, technical questions regarding the definition and quantification of basic human needs have been subject to considerable debate. Nevertheless, the following list of needs and typical indicators illustrates how basic needs can be considered.[22]

Basic Needs	Typical Indicator
Food Consumption	Per capita daily calorie intake
Shelter	Per capita dwelling area
	Fraction of households with access to potable water
	Fraction of households with access to sewage system/disposal
Health	Life expectancy at birth
Education	Literate population (%)
Employment	Fraction of the work-force employed

Energy and Basic Human Needs: Energy is not explicitly mentioned in the above list of basic human needs, but in fact meeting all basic human needs will require energy inputs. Energy strategies, therefore, must *inter alia* be focused on: the growth, preservation, processing, and cooking of food; the construction and maintenance of reliable housing with all necessary services (illumination, sanitation, etc.); the production of adequate clothing that is appropriate for the weather; the provision of sufficient and safe drinking water and sanitation; and the maintenance of a healthy environment, the arrangement of transport facilities, and the establishment of employment-generating industries in areas, countries and regions where unemployment is widespread. In other words, energy-utilizing technologies necessary to meet basic human needs must be provided and energy supplies must be directed to these technologies.

Poverty and Conventional Energy Strategies: Most conventional energy strategies fail to help meet the basic human needs of the poor majorities in developing countries. Our analysis indicates that energy becomes an instrument for the eradication of poverty only when it is directed deliberately and specifically towards the needs of the poor.

Such a needs-based approach is not compatible with the widely-held energy trickle-down strategy (an offshoot of the trickle-down theory of economic development discussed above), wherein the sole preoccupation is with the *supply* of energy, under the assumption that once energy is pumped into society it will automatically trickle down to the end-uses relevant to the satisfaction of basic needs. Just as the economic trickle-down theory has been shown to be ineffective in addressing poverty, so is energy trickle-down an ineffective means of coping with the energy needs of the poor.

If the focus of energy planning is merely on the supply of energy, without scrutinizing the structure of demand, the end-uses of energy and the beneficiaries of energy supply, then whether energy ever reaches the poor to perform the services they need will be largely a matter of chance. Conventional supply-oriented energy strategies are unlikely to make a dent in the poverty of the majorities in developing countries. In fact, such policies may even accentuate inequalities. This is because, in a situation of inequality, the pattern of demand is skewed, with the energy demands of the affluent exerting far greater influence over the nature of energy supply than the needs of the poor. And when energy supply is shaped by such skewed demand patterns, it reinforces privilege and aggravates poverty, leading to an even worse skewing of the pattern of demand.

What is required is to shift energy planners' and policymakers' preoccupation with energy supply to the mundane but vital end-uses of energy such as cooking, which rank high in the priorities of people, and to the energy sources such as fuelwood which are used for these high-priority purposes. Only if energy is deployed as an essential component of a programme to satisfy the basic needs of the population, with special emphasis on the needs of poverty-stricken sectors, is it likely that there will be an improvement in the conditions of the poor as a consequence of actions relating to energy planning.

1.3.5 *Poverty of the Minorities in Industrialized Countries—The Importance of Spreading Energy-Efficiency Improvements*

Quite distinct from the absolute poverty of the majorities in developing countries is the *relative* poverty in industrialized countries. In the poor countries, there is generally not enough to distribute,* but in the rich countries, the problem is mainly the skewed distribution of income.

Almost every industrialized country has its poor and disadvantaged. For example, in 1983 (Note 1.3-C) an estimated 35 million people or 15 percent of U.S. citizens were living at or below the Census Bureau's poverty income level ($10,200 for a family of four).

In most instances, the poor in industrialized countries tend to be inhabitants of economically-depressed regions or ethnic or racial minorities. For example, a disproportionately large fraction of the poor in the United States are black—even though blacks are only 12 percent of the population, they constitute 28 percent of the poor. But even when an industrialized country initially lacks minorities and backward regions, it will often import workers for menial jobs shunned by its own more affluent citizens, thereby creating low-income ethnic minorities. Several affluent countries of Northern Europe, for example, have imported workers from former colonies or from poorer Mediterranean countries for such purposes.

The poor have limited means to acquire the abundant and constantly advertised goods and services of these societies, but at the same time have no choice except to live cheek by jowl with its unpleasant features such as pollution and crime—a situation which tends to aggravate their feelings of being dispossessed. The ensuing social stresses and tensions often constitute a breeding ground for violence.

A major cause of poverty is unemployment, which has taken an upturn in recent years. From an average rate of 3 percent of the labour force in 1964-1973, the average unemployment rate for OECD countries climbed to 5 percent in 1974-1979, and to a peak level of 9 percent in the recession year of 1983, when the aggregate OECD unemployment level reached over 32 million (Note 1.3-D). The industrialized market economies have been increasingly unable to create enough jobs, especially for the large numbers of youth and women who have entered the labour force over the last decade or so.

The energy aspects of the poverty problem for industrialized countries are radically different from those for developing countries, as indicated by the following analysis for the U.S. situation.

In 1980, the poorest 1/8 of U.S. households consumed fuel and electricity for the home and gasoline for automobiles at an average rate of 3.2 kW per capita. The corresponding value for the average US household was 3.6 kW (Note 1.3-E). Thus, the poor in the United States are not "energy poor" in an absolute sense. Indeed, the direct use of energy of the U.S. poor for homes and automobiles is 3.5 times the average use rate in developing countries *for all purposes,* including the indirect use of energy for industrial and commercial purposes and public transportation.

The poor in industrialized countries consume so much more energy than their counterparts in developing countries because of the much wider use of energy-intensive technologies. In the United States (Note 1.3-F) essentially all poor households have electric lights, either gas or electric

---

* Most developing countries also have a skewed income distribution. Nevertheless, the solution for developing countries does not lie entirely in a reduction in inequality; growth is essential.

stoves for cooking, gas or electric water heaters, refrigerators, and television. Moreover, 40 percent have air conditioning, more than 25 percent have freezers, 36 percent have clothes dryers, and 50 percent have automobiles.

Despite this apparent "energy affluence" of the U.S. poor, their *economic* plight has been greatly aggravated by the energy crises of the 1970's. While, on a per capita basis, the poor consumed 5/6 as much energy directly as the average U.S. household, their per capita income was only 1/5 as large. Hence, the poorest 1/8 of households spent 24 percent of their income on fuel and electricity for the home and 15 percent of income on gasoline in 1980. The average households spent only 4 percent of their income on fuel and electricity for the home and 5 percent for gasoline. The poor are clearly the main victims of the energy price shocks of the 1970's.

The high energy expenditures of the poor are also *not* an indicator of affluence. These expenditures are required to meet what are basic needs in the U.S. context. Take space heating and the automobile, which together account for about 3/4 of direct energy use, for instance. About 60 percent of household energy use in the United States is for space heating, which is needed for warmth in winter. On a per capita basis the poor actually use *more* energy to heat their homes than Americans in any other income category (Figure 1.13). This is largely because the homes of the poor tend to be older, less well-insulated buildings with inefficient heating equipment. Moreover, while many better-off Americans have been able to add insulation and storm windows and instal more efficient heating equipment in the aftermath of the energy price shocks of the 1970's, the poor have been unable to afford such investments.

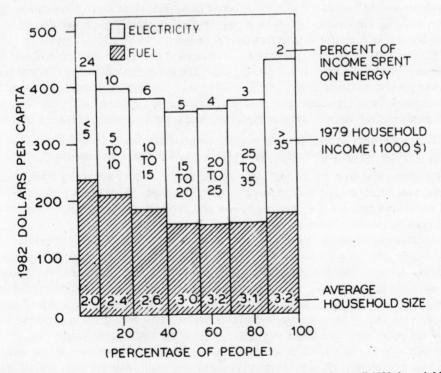

Figure 1.13. Per capita expenditures on household fuel and electricity in the U.S., April 1980 through March 1981. *Source:* Table 2 in Robert H. Williams, Gautam S. Dutt, and Howard S. Geller, 1983.

Similarly, the bygone era of cheap oil fostered in the United States the development of the urban/suburban sprawl which features widely dispersed workplaces, shopping centres, and residential areas that are poorly connected with mass transportation services. This design of human settlements makes the automobile a *necessity* of life for getting to work, to school, and to shopping facilities. And once again, while better-off Americans were able to adjust to the gasoline price increases of the 1970's by buying newer, more fuel-efficient cars, the poor had to make do with driving the behemoths discarded by the more affluent.

Energy clearly exacerbates poverty in the industrialized countries. This linkage is not taken into account in conventional energy planning and policymaking. Rather, conventional energy strategies adopt the ''energy trickle-down'' approach to social welfare and implicitly assume that if energy supplies are increased, these problems will take care of themselves.

The above examples show clearly why the trickle-down approach definitely has not worked for the poor in the US, just as it has not helped to meet the basic human needs of the poor in developing countries. In the industrialized countries, the energy problem of the poor is not that they do not have enough energy to satisfy their needs, but that they consume too much energy and therefore spend too large a fraction of their income on energy.

What is needed is an alternative energy strategy that reckons with the energy-poverty link in industrialized countries and makes the poor less vulnerable to the high costs of energy. The most promising way to do this is by making available to the poor the much more energy-efficient technologies described in Chapter 2 for space heating, household appliances, and the automobile.

1.3.6 *Environmental Degradation in Industrialized Countries: Acid Deposition and Radioactive Wastes—The Importance of Minimizing Emissions from Fossil-fuel Combustion and Solving the Radioactive Waste Disposal Problem*

Since the late 1960's, the environmental side-effects of energy production and use have attracted widespread public attention. Over the years the issues have included thermal pollution from power plants, destruction of surface lands in the strip mining of coal, acid runoff from coal mines, air pollution, nuclear reactor safety, and radioactive waste disposal. Two of these environmental issues which have aroused especially intense public concern in the last few years are:

- ''acid pollution'' arising from fossil-fuel combustion, and
- the disposal of the radioactive wastes from nuclear power production.

Acid Pollution: The term ''acid rain'' was coined over a century ago when it was discovered that air pollution in and around the industrial town of Manchester bleached the colours of fabrics, attacked metal surfaces, and damaged vegetation and materials.[26] But the above-normal activity of rain began to receive serious attention only in the late 1960's.

Acid pollution can be wet or dry. It involves mainly sulphur dioxide, nitrogen oxide, hydrocarbons and ozone. The air pollutants can be deposited hundreds of kilometres from the source and by acting alone or together may cause ''acidification'' in soils, lakes, and rivers and direct damage to plants and buildings. The SO_2 is produced mainly (Figure 1.14) from the combustion of high-sulphur fuels (coal and heavy oils), while NO_x emissions arise largely (Figure 1.15) from motor vehicles and fossil-fuel power plants. Ozone, now looked upon as one of the most important air pollutants, is formed in reactions involving hydrocarbons and NO_x.

Acid pollution is not only a major national problem in many countries, but it is also becoming an awkward international issue because the associated pollutants traverse national frontiers to be deposited in other countries. For example, when the sources and sinks for SO_2 emissions

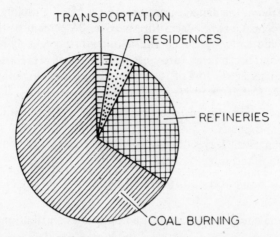

Figure 1.14. Sources of sulphur dioxide emissions in the United States.
Source: R.H. Boyle and A. Boyle, 1982.

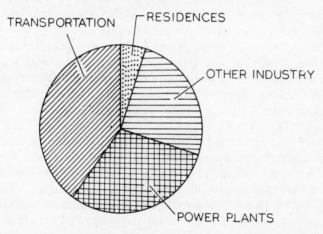

Figure 1.15. Sources of nitrogen oxide emissions in the United States.
Source: R.H. Boyle and A. Boyle, 1982.

in Europe are considered (Note 1.3-G), it turns out that less than 20 percent of the acid deposition on Sweden arises from Swedish sources. Similarly, in North America, more than half of Canada's acid deposition is from U.S. sources—a situation that has already caused discord between Canada and the U.S.

Acid pollution has been held responsible for the corrosion of buildings and statuary, crop damage, lifeless lakes and dead or sterile fish, and damage to forests. It is the threat to forests, which are so vital to the economies of some of the countries subject to acid rain, that is bringing the most pressure on governments to take action on the problem.

The damage to forests appears to be particularly severe in Central and Northern Europe. In West Germany, where forests cover 29 percent of the total land area, the trees on 50 percent of the forest area were found in 1984 to be damaged, and on 1.5 percent of the forest area, trees were seriously damaged or dying, with softwood damage being more serious than hardwood damage.[27]. Furthermore, the damage appears to be spreading rapidly. Severe damage has also been reported in Czechoslovakia, where forests cover 35 percent of the land area and 11 percent of the forest is said to be destroyed and another 11 percent dying[28].

In Scandinavia, spruce and beech forests are most sensitive. Spruce in the southern and western parts of Sweden show a mean needle loss of 10 percent. Three main hypotheses are put forward to explain forest damages in Europe:[29]

- *Acid rain.* Sulphur dioxide and nitrogen oxides form acid rain, which alters the chemicl structure of soils and plants.
- *Ozone.* Ozone damages the leaves directly causing nutrient loss and increase the sensitivity of the plant to fungi and insect attacks.
- *Stress.* The combination effect of different air pollutants and climate factors causes toxic reactions.

In Scandinavia ammonia could in addition be an important contribution to forest damage.[29].

With a business-as-usual energy future, the acid pollution problem can be expected to grow much worse in the years ahead. While total annual SO_2 emissions in Europe were essentially constant between 1972-1982, there would be roughly a 30 percent increase between 1982 to 2002 if energy demand evolved as projected in the IIASA global energy study with a major shift to coal and if no new efforts were launched to control emissions.

Measures to counter acidification at the point of impact (e.g., the liming of surface waters of acidified lakes) are generally not very effective or realistic long-range solutions. Solutions to the problem of acid deposition must be found in the *sources* of pollution—i.e., the emissions must be controlled. Fortunately, there are many options for effectively controlling the emissions that give rise to acid rain (Note 1.3-H). A difficult economic question arises with regard to these control options, though. How much are controls worth? Unfortunately, it is very difficult to quantify the full extent of the damage being done by acid pollution; in fact, even the precise mechanism by which acid pollution is damaging forests is not yet known. This does not mean, however, that nothing should be done until further research clarifies these matters.

The problem with a waiting strategy is that the longer corrective action is delayed, the greater the ultimate damage will be. Hence, it is worthwhile taking aggressive corrective action despite the present uncertainties. A further justification for acting now is that policies designed to limit pollutant emissions and to ameliorate the acid-rain problems could also be effective in promoting the commercialization of a range of technologies needed to make a transition away from oil.

For example, one important technology for controlling acid pollution cost-effectively (Note 1.3-H) is fluidized bed combustion technology, which offers the added benefit of flexibility in use of fuel—e.g., heavy oil, coal of all grades, urban refuse or other organic wastes, wood, etc. Similarly, a switch to methanol as a fuel for motor vehicles would not only lead to a large reduction in NOx emissions from automobiles but also would give rise to other desirable benefits that would smooth the transition away from oil (Note 1.3-I). Likewise, there are many benefits besides the reduction of acid pollution to be derived from a strong commitment to energy efficiency improvements—reduced dependence on oil, reduced total capital expenditures on energy, etc. (Note 1.3-H).

This approach has general applicability in the field of environmental protection. For many environmental issues, it is difficult if not impossible to quantify adequately all the benefits and costs of adopting pollution control technologies. But, despite this uncertainty, it is often desirable to forge ahead and initiate policies to promote the adoption of effective innovative controls, especially if the innovation will serve multiple purposes. In fact, energy planners should try to identify such synergistic solutions and formulate environmental control rules in ways that would help foster the exploitation of such solutions. Unfortunately, this approach to environmental problems is seldom pursued in conventional energy planning.

Radioactive Waste Disposal: The radioactive wastes generated in nuclear-power production must be isolated from the biosphere for thousands of years or perhaps even tens or hundreds of thousands of years. Hence, these wastes must be disposed of in such a way that even without continuous safeguarding there is a negligible probability that the wastes will be released to the environment over the millenia.

Despite the fact that nuclear power has become a significant energy source in many countries, an effective long-term waste-storage scheme has yet to be developed. Over the past few decades, considerable research and development has gone into the radioactive waste disposal problem, and several proposals for disposal have been advanced. Upon close inspection, however, all such proposals have been found to be inadequate.

The failure of these efforts has led many people to believe that the dangers of nuclear power outweigh the benefits. Yet, the volume of radioactive wastes which must be isolated and stored is relatively small, even for a large nuclear power programme. And, there are no obvious insurmountable technical obstacles to isolating these wastes from the environment. Therefore, the radioactive waste disposal problem should be regarded as technically solvable, at least in principle.

Despite the prospect that a satisfactory solution to the radioactive waste disposal problem is likely to be found in due course, society may be making a mistake in expanding the production of nuclear power before such a solution is at hand. From this view, however, arise two oft-asked questions:

- Can society afford to wait?
- Isn't rapid expansion of nuclear power necessary to help bring about a transition away from oil?

It will be shown in subsequent chapters that the pursuit of cost-effective energy-efficiency improvements leads to greatly reduced demands for new energy supplies, including electricity, and to considerable flexibility in the choice of energy supplies—hence, the more risky supply options can be avoided. Radioactive waste disposal is one of the problems that the exploitation of energy-efficiency opportunities buys time to solve.

1.3.7. *Environmental Degradation in the Developing Countries: Deforestation, Soil Erosion and Desertification—The Importance of Energy for Basic Needs*

In contrast to the main environmental problems of the North, which are to a considerable extent the byproducts of affluence and the wasteful use of resources, the main environmental problems of the South involve loss of productive lands via deforestation, erosion, and desertification, and are mainly due to the poor surviving at the expense of nature.

Deforestation: In the mid-1950's, forests covered over one-fourth of the world's land surface; now they cover one-fifth.[30] By extrapolating past trends, the *Global 2000 Report* has estimated that by the year 2000 the world's forest area would be reduced to one-sixth of the world's land area.[30]

Though there is some damage to the forests of the North by acid rain, forest and open woodland areas are in fact expected to increase by 5 percent in Europe from 1970 to the year 2000, and to be reduced only marginally in North America by the year 2000. Most of the ongoing global deforestation (in the literal sense of *net* felling and physical removal of trees) is taking place in the South. According to the International Union for Conservation of Nature and Natural Resources (IUCN), tropical forests are being felled at the rate of 110,000 km² a year; if this rate persists, tropical forests would be eliminated in 85 years.[31] Of the various types of tropical forest, IUCN reports that the most valuable—the lowland rain forests—are being destroyed faster than any. Many such forests, like the lowland forests of Southeast Asia, may not survive to the end of this century."[31] In the Ivory Coast, the forest cover has shrunk by 30 percent over the last decade, and it is estimated that only one-third of the forested area at the beginning of the century is still under forest cover. The coniferous forests of south-eastern Brazil are projected to decline from 5.8 million hectares in 1975 to 0.8 million hectares in the year 2000.[31]

The increasing rate of deforestation is primarily due to the felling of trees for logging and fuel and to the clearing of forests for the purpose of opening up areas for settlements, farming, ranching and industry. The cause of deforestation which is most directly related to energy is the need of over 2 billion people in developing countries for wood as *cooking fuel*. But many of the other factors responsible for deforestation are also energy-related. For instance, an estimated 190 million hectares of cleared tropical forest lands are used for "shifting agriculture" in which the land is burned after rudimentary clearing, crops are grown for a year or two, and then fallowed for about a decade to reestablish soil fertility. As population pressures mount, however, the fallow periods are shortened and the capacities for forest regeneration and soil fertility restoration are lost.[31] The land is then abandoned and often the forest is not reestablished. This type of agriculture is practised by people who do not have access to the energy-intensive agricultural inputs needed to sustain stationary agriculture.

The ongoing process of deforestation must not only be halted but reversed as the need for forest products and benefits grows with increasing population and industrialization. Forests in developing countries are valuable because of their role in the satisfaction of the basic needs of the population. Forest products include fuel (for cooking food and heating homes) and fodder; wood for furniture; lumber for the construction of houses and other structures; timber for railway sleepers, and for electric, telegraph and telephone poles; raw materials for the synthetic fabric (rayon) and paper industries and for medicinal products, etc. And forest management provides employment for the planting, tending, and harvesting of trees. Moreover, even if there were no utilitarian products from forests, their ecological role would still be so fundamental that forests would be invaluable if only for the "services" which they perform for nature. According to the IUCN, forests are " an important renewable resource acting as a reservoir of genetic diversity, and they help regenerate soils and protect them from erosion, they protect areas downstream from floods and siltation, they moderate variations in climate, and they provide recreation and tourism".[31]

Soil Erosion: Soil is an essential life-support system. When soil is not regenerated at the same rate as it is removed, it is eroded and lost virtually forever because soil generation is a very slow process—some 100 to 400 years are required to generate a centimetre of top soil.

The clearing of forests removes the protection of soil cover and is thus a leading cause of soil erosion. And the soil which washes off deforested land is often deposited in the reservoirs behind hydroelectric power stations and irrigation dams, and as their reservoirs fill up with eroded soil, their economic utility declines drastically. Where such man-made water-trapping systems are absent, the removal of watershed forests results in major damage due to flooding.[32] The best-known example is the flooding of the Gangetic plains and delta due to deforestation of the foothills of the Himalayas.

The mismanagement of grazing lands is another major cause of soil erosion. Pasture lands, accounting for about 23 percent of the earth's land area, support about 3 billion domesticated grazing animals, and therefore the bulk of the world's production of meat and milk.[31] Overstocking of animals leads to overgrazing, which lays bare the land to soil erosion.

Land degradation, particularly through soil erosion, is proceeding at an alarming rate—and if current rates continue, as much as one-third of the world's arable land will be lost by the year 2000.[31]

Desertification: Drylands with low rainfall and high rates of evaporation and transpiration are particularly vulnerable, and their overexploitation by humans can turn them into desert-like wastelands. When this occurs, it is known as "desertification." Since drylands cover about a third of the earth's land surface and human pressures on them are increasing, desertification has become a serious global problem (Figure 1.16). The IUCN reports that it threatens the future of some 628 million people, of whom about 78 million are now directly affected by a drop in productivity of the land on which they depend. According to the IUCN, 95 percent of the arid and semi-arid lands are facing desertification, compared to 28 percent of subhumid lands."[31]

The process of desertification can be traced to a number of different primary causes: overgrazing; the ploughing up of dry, high-erodible soils for rainfed agriculture; the poor management of irrigated agriculture, which leads to both the waterlogging and salinization of the soil; and the overcollecting of wood for direct use as fuel or for charcoal production.[31] In some cases, desertification is the result of development projects in which the environmental consequences were poorly understood. In many other cases, though, it is the result of poor people overexploiting dry land resources (soil, vegetation and water) because they have to survive and they have no alternatives.

The Role of Energy in Environmental Protection: Energy strategies can contribute to reduced deforestation, erosion, and desertification, directly or indirectly, in various ways.

The root solution to these problems is a pattern of development that allocates resources to the supply of food, fodder, fuel, and jobs—and energy plays a role in all of these. In particular, attention must be directed to two of the principal causes of environmental degradation in developing countries:

* the use of firewood for cooking, and
* a pattern of pastoral life that leads inevitably to overgrazing of pastureland.

Solutions to these problems require:

* technical alternatives to the low-efficiency firewood stoves widely used for cooking today, and
* a shift to less land-intensive methods of livestock management production coupled with the provision of income-generating alternatives to a pastoral life.

These solutions have energy implications—the former directly through the provision of low-cost alternative fuels for cooking and energy-efficient woodstoves, and the latter indirectly through energy inputs for the production of fodder enabling livestock rearing in stables and through the provision of energy for employment-generating industries. Unless energy strategies for

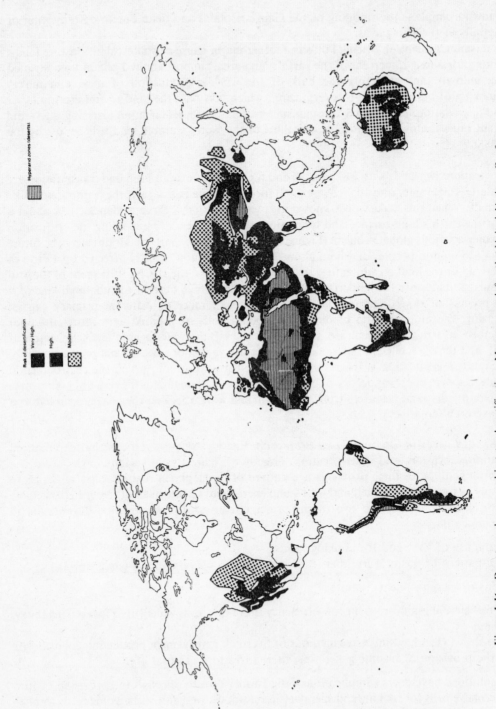

Figure 1.16. Deserts and areas subject to desertification. Map prepared for the United Nations Conference on Desertification.

Source: International Union for Conservation of Nature and Natural Resources, 1980.

developing countries assign very high priority to these needs, the problems of deforestation, soil erosion, and desertification will only worsen.

1.3.8 *Atmospheric CO_2—The Importance of Minimizing Fossil-Fuel Consumption*

The earth is about 4.5 billion years old and has had an atmosphere for most of this time. In the course of just a few decades, a mere instant in geological time, man's activities on earth are altering the atmosphere significantly and thereby bringing about what are likely to be important changes in the world's climate. Specifically, man's activities are resulting in a build-up of carbon dioxide (CO_2) in the atmosphere. Before the Industrial Revolution, the CO_2 concentration in the atmosphere was between 270 and 290 parts per million (ppm). By 1984, it was 340 ppm, that is, 17 percent to 26 percent higher, corresponding to an atmospheric carbon inventory of over 700 billion tonnes. Moreover, the rate of build-up has been increasing—about half of the excess CO_2 has accumulated since the 1950's. The current rate of accumulation is about 1.2 ppm per year.

The CO_2 build-up can lead to climatic change because as the CO_2 concentration increases, more infrared radiation is trapped in the lower atmosphere. This results in a heating of the earth's surface in order to maintain the earth's radiative balance between the outgoing earthshine (infrared radiation) and incoming sunlight (solar radiation) absorbed by the earth and its atmosphere.

There is little doubt that the increased CO_2 level would result in a global average temperature rise of $3 \pm 1.5°C$. To a layman, this may not seem like much. But the warming could take place in a matter of decades and would lead to temperatures higher than those in some of the warmest periods in geological history—the altithermal era 6000 years ago, and the previous (Eemian) interglacial period—125,000 years ago—and approaching those of the Mesozoic era, the age of dinosaurs.[34]

The heating induced by a doubling of atmospheric CO_2 would not be uniform over the earth's surface; it may be two to three times as large at the poles—8 to 10°C. The extra heating at the poles may directly affect high latitude weather and climate (e.g., the warming could lead to the disappearance of the Arctic sea ice). This non-uniform heating may also affect the climate of the entire globe. Global circulation is driven by the "atmospheric heat engine" which operates on the equatorial/polar temperature difference (some 60°C on the average), and it could be significantly altered by the reduction of the temperature difference which the CO_2 build-up will cause.

One of the more noteworthy consequences of the CO_2 build-up would probably be shifts in precipitation associated with a global warming. Present-day atmospheric models are not yet good enough to predict reliably the winners and losers from such shifts, but climatic records from earlier warmer periods leave us some clues. Figure 1.17 shows that in previous warm eras, regions such as North Africa, Southeast Asia and Central America were wetter, but the mid North American continent, Argentina, and much of the U.S.S.R. were drier. Thus, if the geologic record proves to be a meaningful guide, the present "bread-baskets" of the world may be particularly vulnerable to a CO_2 build-up, and major shifts in the geographical distribution of agriculture may result from a CO_2-induced warming.

The precise effect on agriculture of a global heating and of shifts in precipitation patterns would vary from crop to crop and from region to region.[34] One benefit of the CO_2 build-up would be longer growing seasons at high latitudes (e.g., in Canada and northern Siberia) that may open up lands that are now only marginal for growing crops. But in existing agricultural areas, crop production patterns could be significantly changed for better or worse.

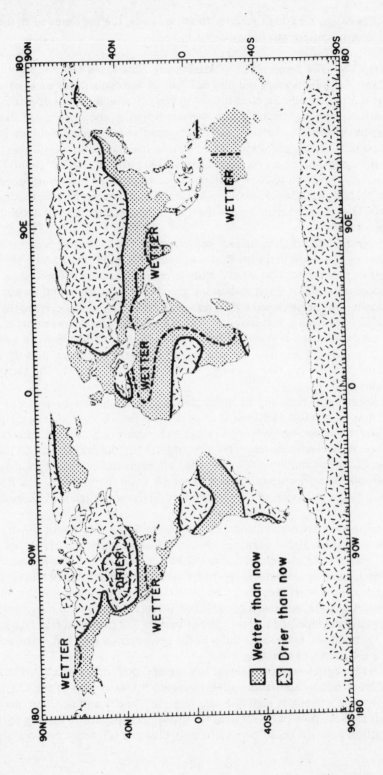

Figure 1.17. Scenario of possible soil moisture patterns on a warmer Earth. It is based on paleoclimatic reconstructions of the Altithermal Period (4500 to 8000 years ago), comparisons of recent warm and cold years in the Northern hemisphere, and a climate model experiment.

Source: W. Kellogg and R. Schware, 1981.

One possible impact of a doubling of CO_2 in the atmosphere, the complete disappearance of Arctic Sea ice, could result in a displacement of climatic zones by 400-800 kilometres or more. Another, the partial melting or disintegration of the entire West Antarctic ice sheet, would lead to a 5 to 6 metre rise in the global sea level which in turn would mean the inundation of many low-lying coastal areas, including cities.

Just how soon could a doubling of the atmospheric CO_2 concentration occur? That depends in part on the origin of the increase in CO_2 levels in the atmosphere. The current increase is primarily attributable to the burning of fossil-fuels, and fossil fuel emissions are expected to dominate the CO_2 build-up in the future as well, but the net reduction in the global biomass (plants absorb CO_2) also probably has been an important factor over the past 100 years (Note 1.3-J). The rate of increase in the atmospheric CO_2 level accounts for about half of the cumulative CO_2 emissions from fossil-fuel combustion. Most of the remainder has probably been absorbed by the oceans.

Assuming that the mix of fossil fuels remains about the same as at present and that half of the CO_2 stays in the atmosphere, a doubling of the atmospheric CO_2 level would occur by 2030 if fossil-fuel use grew at the historical rate—4 percent per year—or 30 years later if the growth rate were cut in half. However, under these growth rates in fossil-fuel consumption, the doubling may actually occur earlier than these dates. One reason is that as petroleum and natural gas supplies are depleted, the fossil fuel mix may shift to coal, coal synthetics, and oil from shale, i.e.,fossil fuels that release considerably more CO_2 per unit of useful energy than petroleum and natural gas (Note 1.3-K); together petroleum and natural gas accounted for two-thirds of fossil-fuel use in 1978. A second reason for the accelerated CO_2 doubling date is the likelihood that as the atmospheric CO_2 level rises, the oceans will become a less and less effective sink for excess CO_2; as a result, the fraction of released CO_2 accumulating in the atmosphere will grow.[35] Moreover, CO_2 is not the only warming agent at work; hence the earth may actually warm up faster than is indicated by consideration of CO_2 effects alone. Human activities are adding other radiatively active trace substances (RATS) or "greenhouse gases" to the atmosphere besides CO_2—chlorofluorocarbons, nitrous oxide, methane, etc. While the impact of any one of these gases is much less than that of CO_2, the collective impact of all these gases could be comparable to or even greater than that of CO_2, thereby increasing the urgency of dealing with the prospect of a global warming.[36]

The CO_2 problem must be dealt with in the formulation of any long-term energy strategy. What are the options? The situation is quite *unlike* that for acid rain, where there are many promising technologies for controlling emissions. While a number of proposals have been made for removing CO_2 from stack gases or directly from the atmosphere, none are practical (Note 1.3-L). The only option for controlling CO_2 emissions is to reduce fossil-fuel combustion. Therefore, we either have to move away from dependence on fossil fuels or learn to live with the consequences of a further build-up of CO_2 in the earth's atmosphere.

In late 1983, major reports on the CO_2 problem were released by the U.S. National Academy of Sciences and Environmental Protection Agency.[37-38] Both of these reports affirmed the seriousness of the CO_2 problem, but they failed to suggest that immediate steps be taken to reduce dependence on fossil fuels even though they clearly identify the combustion of fossil fuels as the major source of CO_2 buildup. Instead, both reports argued for a strategy of adaptation to a high-temperature world. Accommodation to the CO_2 problem is likewise implicit in conventional supply-oriented global energy strategies. Even with the "slow growth" in fossil use envisioned by the WBC and IIASA for the next 40 to 50 years (1 to 2 percent per year), and zero growth in fossil-fuel use thereafter, a doubling of the pre-industrial atmospheric CO_2 level and its attendant consequences would take place late in the 21st century (Note 1.3-M).

Our approach to the CO_2 problem is quite different. It is based on exploring the potential for reducing dependence on fossil fuels via increased efficiency of energy use and a shift to other energy resources.

To what extent should fossil-fuel use be restricted? What level of CO_2 build-up is "acceptable"? A rigorous cost/benefit analysis to determine the answers to these questions may not be possible—not only because complete information on the nature of the risks and benefits involved may never be available but also because many of the risks and benefits are inherently non-quantifiable. Ultimately, an "acceptable" CO_2 level will have to be determined through the political policy-making process on the basis of *indicative but incomplete* scientific information.

The type of information which can be made available to policymakers is an estimate of the percentage reduction in fossil-fuel use that would be necessary if the ultimate acceptable level of CO_2 is to be a particular multiple of the present level, how much reduction might be achieved, and the cost of achieving it.

As will be shown in Chapter 4, technically and economically feasible energy-efficiency improvements and fuel switching can be effective in limiting the atmospheric build-up of CO_2 to levels far below those implicit in conventional energy projections, while at the same time achieving major increases in global economic output.

1.3.9 *Global Insecurity and the Danger of Nuclear War—The Importance of Reducing Dependence on Oil Imports*

The United States and the Soviet Union have about 50,000 nuclear warheads in their arsenals, more than 10,000 of which have explosive yields of the order of one million tons of TNT equivalent. There are so many nuclear weapons in these arsenals, the numbers of civilian deaths which would result from even their "limited" use is so enormous, and the likelihood of being able to control their use after even a few of them had been exploded is so small that all rationalizations for the arms race seem meaningless.[39]

Yet, the arms race continues and even appears to be accelerating. Consequently, the danger of nuclear war increases. As the nuclear weaponry of both superpowers grows more and more formidable, one of the great dangers is that in a confrontation between the two, one or the other becomes convinced that it has to either "use or lose" its nuclear attack force and launches a "pre-emptive strike" against the other side's nuclear attack force, thereby precipitating what will surely become a nuclear holocaust. Moreover, if either side develops a hair-trigger "launch-on-warning" response, in its fear of a possible pre-emptive attack, the probability of an accidental nuclear war is increased.

Each of the superpowers could re-create Hiroshima on a very much larger scale in each of hundreds of cities on the other side, deploying only a few percent of its nuclear warheads. Furthermore, the devastation would spread to vast territories beyond the superpowers. Thus, insecurity is not confined to the two most powerful nations—what now exists is *global insecurity*. The future of life on earth itself is threatened by the prospect that in the aftermath of a large scale nuclear exchange the earth might be enshrouded in darkness and cold in a "nuclear winter"[40, 41]

The direct and immediate causes of global insecurity are superpower interests beyond their own borders. For NATO countries, the interest that represents the most dangerous combination of importance and vulnerability is their oil supply. About 60 percent of the combined oil consumption of the United States and Western Europe was imported from developing countries in 1979. Japan, which is allied through the U.S. to NATO, imports all of its oil. One-half of the total energy consumption of the NATO countries and Japan is in the form of oil, and their transportation sectors are almost totally dependent upon oil.

This dependence on imports would not be a serious concern if the sources of oil supply for the NATO countries and Japan were widely distributed in the developing countries, because then the danger of a major supply disruption would be low. The fact is, however, that two-thirds of the oil imports of the NATO countries and Japan in 1979 came from the Persian Gulf. The dangers arising from a potential disruption of this oil supply are well understood by those in political power. U.S. Defense Secretary Weinberger said:

> "The umbilical cord of the industrialized free world runs through the Strait of Hormuz and into the Persian Gulf and the nations which surround it...."

The dependence on the Persian Gulf is especially troubling to many because of the political volatility of the region. Many countries around the Gulf are subject to revolutions and coups, are often at war with each other, or are involved in confrontations with Israel. The situation is not likely to improve so long as the industrialized market economies depend on oil imports, as more than 80 percent of the reserves of the petroleum exporting nations are located in the Persian Gulf (Figure 1.6).

The sensitivity of the world oil market to events in the Middle East has already been dramatized by its reaction to two relatively minor disruptions:

- the attempts by the Arab nations for a few months after the October 1973 war to cut off oil exports to Israel's NATO allies, and
- the sudden decline in Iran's oil exports as a result of that nation's revolution.

In neither case did the oil imports of the NATO countries and Japan decline by more than a few percent, but in both cases panic buying drove the price of oil up by a large factor (see Figure 1.3).

That conflict in the Middle East can draw in the superpowers and threaten nuclear war is indicated by the experience of October 1973, when the Soviet Union threatened to intervene in the Arab-Israeli War, and the United States, in response, raised the alert status of its nuclear forces.[42]

U.S. anxiety over the possible future course of events around the Persian Gulf has led it to organize the infrastructure of a Rapid Deployment Force which could occupy strategic areas in the Persian Gulf region and could also confront any expeditionary force which the Soviet Union might be tempted to introduce in case of a revolution or war in that region.[43] The presence of mobile Soviet forces in the region (e.g., in Afghanistan) and repeated Soviet warnings that they would not stand idly by if the U.S. intervened militarily in the Gulf add tinder to the Middle East tinderbox.

Since the Rapid Deployment Force would be in a relatively weak position so far from home and without strong reliable allies, the U.S. Government has tried to enhance the credibility of its threat to intervene by making clear that its conventional forces would be backed up by tactical nuclear weapons if necessary.[44]

Thus, the dependence of the industrialized market economies on Persian Gulf oil is a major source of global insecurity and a potential trigger of nuclear war.

The problem of global insecurity has not been dealt with effectively in long-term energy strategies which concentrate on the expansion of energy supplies. For example, in the IIASA world liquid fuel supply/demand projections, it is envisaged that in order to meet rising world oil demand, production in the Middle East/North African region would have to increase more than two-fold by 2030, to its estimated maximum productive capacity. At which time, the region would account for one-third to one-half of world oil production, up from one-quarter in 1983 (Note 1.3-N). In other words, a high level of global dependence on oil from the Middle East, which has been identified as a major source of global insecurity, would be a significant feature of the world half a century from now if the IIASA energy strategy were adopted.

An alternative global strategy that would stress efficiency improvements and the development of indigenous energy resources would enhance global security by making the world much less dependent on oil imports.

1.3.10 *Nuclear Weapons Proliferation—The Importance of Avoiding Nuclear Technologies Involving Fuel Reprocessing and of Pursuing Alternatives to Nuclear Power.*

Developing countries are, by and large, spectators in the nuclear arms race arena, as the superpowers accumulate more and more nuclear weaponry in what has come to be known as the "vertical proliferation" of nuclear weapons. But there is another dimension to the nuclear weapons proliferation problem in which developing countries are closely involved. This is the possible proliferation of nuclear weapons capability to a large number of heretofore non-nuclear-weapons states. The resulting "horizontal proliferation" would increase the probability of nuclear weapons being used in regional disputes. The nuclear weapons used might be "small" by modern superpower standards, but they could have a destructive force comparable to that unleashed on Hiroshima and Nagasaki.

Unfortunately, the existence of nuclear *power* programmes to produce electricity in non-nuclear-weapons countries makes possible the development, construction, and storing of nuclear weapons. Nuclear power programmes require national cadres of nuclear scientists and technicians, a network of research facilities, research reactors, and laboratories, and (for most reactors now in operation) access to enriched uranium. Also, the "spent" fuel from the reactors of these programmes contains large quantities of plutonium, a material usable in nuclear weapons. These aspects of nuclear power create an indissoluble link between nuclear-generated electricity and nuclear weapons proliferation.

In reactors of the types now deployed, plutonium is discharged at a rate of about 15-300 kilograms of plutonium per year for a typical reactor of 1 GW(e) capacity.[45] By 1990, the discharge of plutonium worldwide from reactors already constructed or under construction is likely to exceed 60,000 kilograms *per year*.[46] It only took 6 kilograms to make the Nagasaki bomb.

Although at present the nuclear facilities and spent-fuel repositories located in almost all non-nuclear weapons countries are under international safeguards, this provides only a partial and transitory barrier to proliferation. Several nations that are capable of developing nuclear weapons—including Argentina, Brazil, India, Israel, Pakistan, and South Africa—have not signed the Non-Proliferation Treaty and are, or soon will be, capable of producing weapons-usable material free from international safeguards. Several other states have recently sought greater fuel-cycle independence and may under some circumstances be tempted to give the statutory three months' notice of withdrawal from treaty commitments for reasons of "extraordinary events" that jeopardize their "supreme interests."[47]

If, in the long run, nuclear power comes to produce a significant fraction of world energy—as projected in conventional energy strategies—the safeguarding of weapons-usable materials will become even more precarious. The proliferation risk would escalate not only because a large number of countries would gain access to nuclear weapons-usable material, but also because there would be an important qualitative change in the technology that would greatly facilitate access to weapons-usable material. Because low-cost uranium resources are limited, much of the world's nuclear power would eventually have to be based on plutonium breeder reactors and other systems which require the reprocessing, separation, and recycle of weapons-usable material. Whereas in today's "once-through" nuclear fuel cycle, weapons-usable material is never isolated from radioactive fission products, a 1-GW(e) capacity plutonium breeder reactor would each year discharge and require the reprocessing of about 1000 kilograms of plutonium. Reprocessing and recycling would for the first time give nations virtually instantaneous access to weapons-usable material. Since separated plutonium can be transformed into nuclear weapons in a matter of days, a world with a "plutonium energy economy" would have to treat its commerce in plutonium from reactors as carefully as it treats avowed bomb material.

Several countries are developing breeder reactor technology and are already operating nuclear fuel reprocessing plants to separate plutonium from spent reactor fuel. Although these activities are now concentrated mostly in the industrialized countries, which either already have nuclear weapons or have signed the Non-Proliferation Treaty and accepted international safeguards, it is clear that developing countries would eventually want the same advanced nuclear technologies. As these technologies evolve, there inevitably would be intense efforts by developing countries to develop such technologies themselves and aggressive efforts by industrialized countries to export the technologies. With this ill-starred convergence of "interests", the principal technical barrier to proliferation would be shattered in industrialized and developing countries alike.

Today the close connection between nuclear power and the horizontal proliferation of nuclear weapons in a worldwide plutonium energy economy is not yet "real" for most people—it concerns risks in the very distant future. Fuel reprocessing is not yet well-established as a commercial activity, and most existing nuclear power plants are in countries which either already have nuclear weapons or are regarded as "unlikely proliferators" (Note 1.3-0).

If, however, nuclear power should develop on the scale envisaged in supply-oriented global energy strategies, problems of relatively minor concern today would assume major proportions forty or fifty years from now. If nuclear power were to grow as projected in the IIASA study and the 1978 WEC study, for example, dozens of countries would have nuclear power plants powered by plutonium or other nuclear-weapons-usable fuels. Globally, installed nuclear generating capacity would grow from 121 GW(e) in 1979 to 2600-4400 GW(e) in the period around 2020 or 2030 (Note 1.2-F). At such levels of nuclear power, plutonium would be discharged world-wide from nuclear power plants at rates of 2.6 to 4.2 *million* kilograms per year, and roughly the same quantities of plutonium would be extracted from spent nuclear fuel and refabricated into new fuel each year (Note 1.3-P). This is 400,000 to 700,000 times the amount of material needed to make a nuclear weapon.

Though the IIASA analysts were aware of the proliferation implications of their energy projections, they chose not to come to grips with the issue, saying:

> "The handling of nuclear weapons are the principle (sic) concerns that go along with the use of nuclear fuels...A large number of capable and sizeable groups are already studying these problems, and we did not judge it practical for the relatively small groups of IIASA scientists from many nations, to compete with these efforts... [Also] we regard the problems of nuclear waste handling and proliferation as primarily political ones..."[10]

Of course, the horizontal proliferation of nuclear weapons is a political problem; but it is difficult to imagine a political solution to this problem in a world where nuclear power abounds. It is unlikely that schemes could be devised which would assure that virtually 100 percent of the flows of weapons-usable materials are secured, year in and year out, against diversion to nuclear weapons use. To limit the diversion rate in the IIASA or WEC scenarios for the period 2020 to 2030 to "only" 10 weapon-equivalents per year would require that the international system be 99.996 to 99.999 percent secure. Such a level of worldwide security is inconceivable.

If a global energy strategy is to be compatible with the achievement of global security, it must come to grips with the problem of nuclear weapons proliferation. Because the political feasibility of ever achieving the necessary international controls in the world community of separate and sovereign nations is highly doubtful, the energy strategy advanced here is one which explores the possibilities for making *a transition away from oil without becoming overly dependent on nuclear power technology.*

At the very least this approach requires that nuclear technologies involving reprocessing of spent fuel be avoided. As noted above, the proliferation risk escalates greatly in the shift from "once through" fuel cycles to fuel cycles involving reprocessing.

It is unlikely that some sort of ban on reprocessing technology would *solve* the problem of nuclear weapons proliferation. Any country intent on acquiring nuclear weapons can do so via many routes other than the diversion of plutonium from civilian nuclear power programmes. But avoiding reprocessing would be effective in dealing with perhaps the most politically intractable kind of proliferation, which has been called "latent proliferation".[48]

Latent proliferation refers to the acquisition of nuclear weapons capability without having to make an explicit commitment to do so. It is an important concept because both internal and external forces tend to deter *explicit commitments* to acquire nuclear weapons. Within a governing bureaucracy, it will be difficult to secure a consensus on the desirability of acquiring nuclear weapons, because advocates cannot make a persuasive case that security would be enhanced by having nuclear weaponry. Moreover, an explicit decision to acquire nuclear weapons entails the risks of "getting caught" and subsequently "punished" by other individual countries or by the collective global economic, military, and/or political community—risks that increase, the longer the time required to obtain nuclear weapons.

These forces would not be at work if a nuclear weapons capability were acquired as a consequence of acquiring plutonium recycle technology for the civilian nuclear power programme. With such technology at hand:

> "... countries could proceed with little cost to a point but a step away from the acquisition of nuclear weapons without deciding or announcing in advance their intentions. ... It is useful to remember that India, probably France, and possibly the United Kingdom all proceeded with their nuclear power programs without an explicit decision to divert parts of the program to weapons purposes... By constricting the twilight period between decision and weapon, commercial fuel cycles could provide both a conduit and a mask to a weapons effort that a government might not otherwise be able or willing to undertake."[48]

The avoidance of nuclear fuel reprocessing technologies thus offers an opportunity of dealing with the type of proliferation that is the least amenable to internal political controls. To deter proliferation effectively, such fuel cycles would have to be avoided in all countries, including the nuclear weapons states, as any two-class system that discriminates against countries judged to be proliferation-prone would be ultimately unstable.[49]

We do not mean to suggest, however, that once-through fuel cycles are "proliferation-proof." While the plutonium in spent fuel is "protected" against diversion by the intense radioactivity of the nuclear fission products and is therefore much less accessible than separated plutonium, it is feasible for a country intent on acquiring nuclear weapons to build in a short period, say 60 days, a crude ("quick-and-dirty") reprocessing facility adequate to extract enough plutonium from spent-fuel inventories to make a number of nuclear weapons. The temptation to do this could become great in certain crisis situations.

Because of this proliferation risk inherent even in once-through fuel cycles, nuclear power should be regarded as an *energy source of last resort,* i.e., pursued only in circumstances where viable alternatives do not exist. If emphasis is given in energy planning to energy-efficiency improvements, considerable flexibility is gained in choosing energy supplies, including the flexibility of avoiding proliferation-prone nuclear technologies in a wide range of circumstances.

In the final analysis, though, the threat of nuclear weapons proliferation cannot be averted or significantly reduced solely by avoiding or minimizing dependence on nuclear power. This is necessary but not a sufficient condition for escaping the instabilities of a proliferated world. The horizontal proliferation of nuclear weapons is profoundly influenced by the vertical proliferation of nuclear weapons as manifested in the arms race between the superpowers. As long as the United States and the Soviet Union behave as though their security is enhanced by having nuclear weapons, other countries will want to acquire nuclear weapons as well. The survival of modern civilization and perhaps the human race requires that both horizontal and vertical proliferation be curbed.

1.3.11 *Undernutrition and Food Supplies—The Importance of Energy for the Domestic and Agricultural Sectors*

The United Nations Food and Agriculture Organization (FAO) has estimated that about 450 million people, constituting 23 percent of the 1975 population in a sample of 86 poor countries, are undernourished. The FAO defines an inadequate diet as one providing less than 1600 calories per day, which is only 20 percent above the basal metabolic rate, a criterion of undernourishment which reckons only with the energy required for minimal physical activity.* In contrast, a World Bank estimate,[51] using less stringent assumptions, indicates that 1.1 billion people (about 25 percent of the world's population) are undernourished.

These estimates, however, are only gross averages. In fact, the percentage of undernourished people varies widely. It depends on factors such as:

- the *region* of the world—the countries with the highest percentage of undernourished people are those where the staple foods are either millet and sorghum or roots and tubers;
- the *income group* within the country—the consumption of both protein and the total calories increases with family income (Figure 1.18);
- the *sex* of the population group—due to cultural intra-family factors, women in many developing countries 'eat last and least', even though studies are revealing that they work harder and expend more energy than men, and boys are often given a better diet than girls, [52, 53]; and
- the *season* of the year—food consumption in the Sahelian countries can average 25-30 percent less *before* the harvest than soon after it.[54]

The consequences of undernutrition are severe and far-reaching. Undernutrition in its broadest sense results from a diet inadequate to maintain satisfactory physical and mental development. It impairs a person's capacity to work and lowers resistance to infection. Infection, in turn, increases the food nutrients needed to repair the damage wrought by disease. Undernutrition and chronic infection impair learning ability, further reducing the capacity for effective work.

---

* An industrialized developed country diet averaging about 3350 calories per day is more than double this minimum of 1600 calories per day.

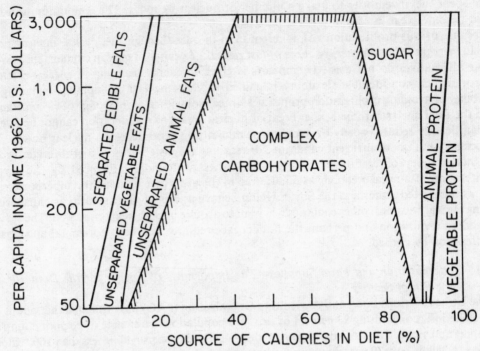

Figure 1.18. Figure showing the increase in protein and calorie intake with income.
Source: Nevin S. Scrimshaw and Lance Taylor, 1980.

Consequently, undernutrition and chronic infection impair learning ability, further reducing the capacity for effective work. Consequently, undernutrition begins a vicious circle broken only by provision of an adequate diet. It debilitates individuals, families, and whole communities. For the poorest, undernutrition is a condition always close to starvation.[55]

With at least 450 million people undernourished, about 10 percent of the world's population, it is evident that undernutrition is a grave problem even at the present population levels. And, by the end of this century, the world may have to feed as many as 2 billion *additional* people. This leads to the question of whether food supplies are and will be adequate for the world's population.[56]

To put the problem into perspective, it is noteworthy that only about 25 million tonnes of cereals would suffice to meet the requirements of the undernourished population today. This 25 million tonnes represents only 1.7 percent of current world production, 15 percent of the world's grain trade, or 10 percent of the cereal now being fed to animals in developing countries. Looked at another way, if the present global production of about 1500 million tonnes of grain were accessible uniformly to the world's population of 4.5 billion, every person in the world would have enough food—about 2570 calories per day.[57]

Hence, the undernutrition of today is clearly *not* because the world does not produce enough food. World agriculture is already producing enough to feed everybody at a Western standard of diet. Inequalities and inefficiencies in food distribution, however, keep the poor and the powerless undernourished.[57]

One important cause of undernutrition is the *non-uniformity* of foodgrain production and food availability in various parts of the world. The developing countries, with three-quarters of the world's population, produce only about 45 percent of the global grain supply. In contrast,

North America, with only about one-tenth of the global population, produces about one-quarter of the world's cereal output. Developing countries that are unable to produce enough food to feed their people can of course enter world markets and buy grain and other food commodities. But paying for these imports can become a terrible drain on their limited export earnings, and reduce further the relatively little capital available for investment in development; moreover, an infrastructure must be created for unloading, storing and distributing these imports. It makes more sense, therefore, to make a serious effort to achieve self-sufficiency at regional and national levels.

Increasing food production and availability in the developing countries is doubly important because it is these countries which will have to absorb the bulk of the 2 billion additional people who may well enter this world by the end of the century. So, the question is whether the developing countries individually and collectively can increase their food production to feed their existing populations and their even larger populations in the years ahead.

The FAO study, *Agriculture: Toward 2000,* calls for a doubling of food production in developing countries by the year 2000 in order to meet their requirements.[58] Achieving this target would require (Figure 1.19):

- bringing new land into agricultural production,
- a greater intensity of cropping, and
- higher crop yields.

The area of land under cultivation is a crucial factor, but it is under severe threat from several pressures, particularly population increases. During 1975, each hectare of cropland in developing countries fed, on average, three people. Because of an estimated population increase of 3 percent per year, each hectare cultivated in 1975 will need to support four people in 1985 and, by the end of the century, close to six people. The FAO calculates that "the total world area available for cropping (counting more than once those areas where multiple crops are grown) is about 4.1 billion hectares". Less than half of this area was cropped in any given year in the 1970's,

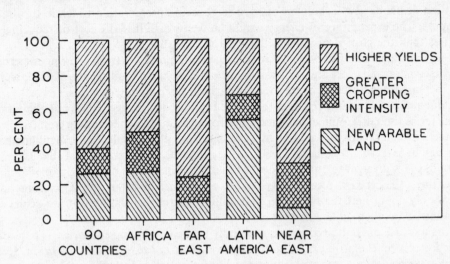

Figure 1.19. Contributions (in percent) to increased food production in developing countries associated with a doubling of food production there by the year 2000, as targeted in the FAO's "Agriculture Toward 2000" study.
Source: Food and Agriculture Organization, 1981.

so there is substantial room for expansion of cropped area, *provided* capital is available for land reclamation and irrigation.''[50]

The bulk of the increase in food production, however, is expected to come from higher yields, as was the case between 1950 and 1980 when 70-90 percent of the increase in output in the developed and developing countries came from gains in per hectare yields of crops such as wheat and rice. Even greater increases in yield may be necessary if the projected 26 percent increase in arable land cannot be realized. The *Global 2000 Report,* perhaps unduly pessimistically, projects that by 2000 land under cultivation will increase by only 4 percent, because "most good land is already being cultivated."[30] Nevertheless, it must be noted that the significant increases achieved in the yields of wheat and rice have not been duplicated in legume yields, even though "the protein in food legumes is nutritionally complementary to that of cereal grains."[55] The nutritionally desirable ratio is one tonne of legume for every two tonnes of cereal, but the Asian legume harvest, for example, is only about one-tenth of that of cereal production.

In most developing countries, agriculture is even now carried out almost wholly in a traditional way with animate sources of energy—human muscles and draught animals—doing all of the work. But often these animate sources cannot meet the power requirements of farming, especially the high-yielding, more intensified farming operations which demand additional energy for pumping irrigation water as well as for planting, tending and harvesting multiple crops. Therefore, inputs of inanimate energy in the form of shaft power are essential for increasing food production. This means that efforts will have to be directed to making these energy inputs available in sufficient quantity at the locations where they are needed to increase output. The FAO projects that a two-fold increase in production will require nearly a five-fold increase in commercial energy use—taking into account not only the direct energy needed for the operation of farm machinery and irrigation pumps but also the energy required to make fertilizers, pesticides, and farm machinery.

Energy is also important for the storage and preservation of food. Food losses to rodents, insects and moulds in developing countries can be as high as 20 percent for cereals and legumes. The situation demands new techniques of food storage and preservation which in turn require energy inputs. The expenditure of energy would be well worth it if the cost of reducing food losses is less than the cost of producing the food which is lost.

The energy available in all steps of the food production process—from growing the crops to storage, preservation and processing—has an important bearing on food supply. There is a critical need for increased quantities of energy, particularly in developing countries.

The actual quantity of energy required to double food production by 2000 is not all that much in absolute terms though. Commercial energy use for agriculture in developing countries would have to increase at an average annual rate of 8 percent, from the equivalent of 0.74 million barrels of oil per day in 1980 to 3.5 million barrels per day in 2000. For comparison, this 3.5 million barrels per day is equivalent to only 6 percent of the 1982 world oil consumption of 58.5 million barrels per day. Also, the extra energy needed for agriculture by 2000, i.e., 2.8 million barrels per day, is only about half the *reduction* (5.3 million barrels per day) in oil use achieved by OECD countries between 1979 and 1981.

While the increase of food production, especially in the developing countries, is a necessary condition for solving the problem of undernutrition, *it is not a sufficient condition.* Even if food supplies were adequate, the problem of distribution remains, and this problem is grounded in income inequalities and the inadequate purchasing power of the poorest households.

The fundamental cause of undernutrition is *poverty*. Even when food is available, the poor, particularly but not solely* in developing countries, cannot afford to buy adequate food for themselves. They lack purchasing power essentially because they do not have employment or because of the low productivity of their labour. An immediate solution to this problem is for the government, through some income transfer mechanism, to supplement the poor's insufficient purchasing power by giving the money, directly or by supplying them with free or very low-cost food. The difficulty with this approach is that it does not get at the roots of the problem. You can give a man food but if he has gastro-intestinal parasites, they will consume as much as 15 percent of what he eats, and he will still be undernourished. The poor need adequate health care and safe water and decent sanitation facilities. They need gainful employment so that they can earn a living rather than have it given to them. They need access to resources—land, capital, energy, information, etc—that will enable them to improve their own productivity. Provide a poor subsistence farmer, for example, with access to fertilizer, irrigation water, and a small tractor or mechanical tiller, and he will, in a very short while, increase his output of rice or wheat or millet or beans enough that he will be able to feed his family adequately for the first time and have something leftover. And by selling his surplus, he will generate income, i.e., increase his family's purchasing power, and make more food available for someone else.

Also, it is worth noting that the incidence of undernutrition tends to vary even within families. Particularly vulnerable are young children (0-5 years) and expectant or nursing mothers. These groups may need supplementary nutritional programmes. And the nutritional status of women generally is worse than for men because, for cultural reasons, they eat less and tend to expend more energy in work than men. Their greater workload may be due to the domestic chores such as gathering firewood, fetching drinking water, etc. These chores may be rendered unnecessary through the provision of cooking fuel and efficient stoves and of water for domestic purposes.

In summary, apart from increasing the production and supply of food in developing countries, many other measures are necessary such as the raising of incomes through employment generation, the provision of a healthy environment, and the making available of cooking fuel and efficient stoves. Several of these measures are strongly energy-related. If energy is to contribute to the solution of the problem of undernutrition, therefore, the energy components of these measures must be built into energy strategies. Such considerations, however, have rarely entered conventional energy strategies.

1.3.12 *Population Growth*[59-63]—*The Importance of Energy for the Needs of Women, Households and a Healthy Environment*

In the final analysis, almost every one of the global problems described so far depends upon the availability and use of natural resources—factors that must be seen in relation to the size of the human population exerting pressure on those resources. Population pressure has been rapidly escalating. The population of the world has increased explosively during this century, particularly since World War II, leading to talk of "a population bomb" inexorably bringing humanity to its doom on this planet. However, these predictions have generally assumed the persistence of the very high population growth rates of the 1950's, which correspond to a doubling of the world's population every 23 or so years.

It is now clear that the recent tremendous increase in the world's population (Figure 1.20) is associated with what is known as a *demographic transition*. In such a transition, the population moves from an old balance of high mortality and high fertility to a new balance of low mortality

---

*Many recent reports have focussed attention to the problem of hunger in the US.

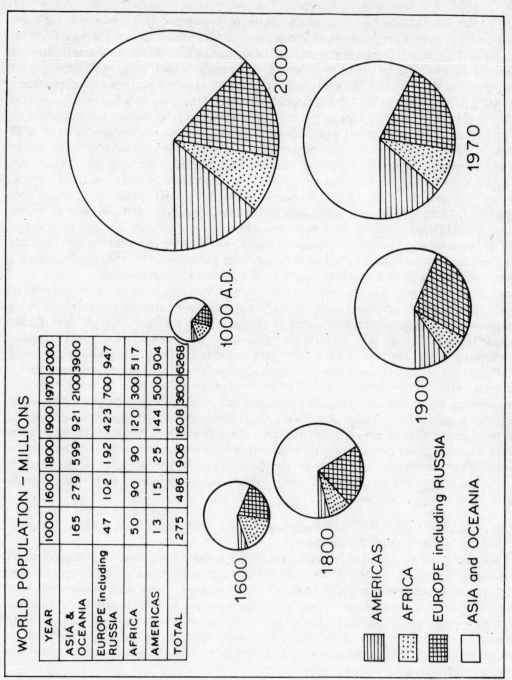

Figure 1.20. One thousand years of population growth.
Source: Population Reference Bureau, "Man's Population Predicament."

WORLD POPULATION – MILLIONS

YEAR	1000	1600	1800	1900	1970	2000
ASIA & OCEANIA	165	279	599	921	2100	3900
EUROPE including RUSSIA	47	102	192	423	700	947
AFRICA	50	90	90	120	300	517
AMERICAS	13	15	25	144	500	904
TOTAL	275	486	906	1608	3600	6268

AMERICAS
AFRICA
EUROPE including RUSSIA
ASIA and OCEANIA

and low fertility (Figure 1.21). Such demographic transitions have occurred in the past—in Western Europe in the 19th century, and in Southern and Eastern Europe in the first quarter of this century.

The demographic transition currently taking place in the developing countries has been initiated by the rapid fall in mortality in these countries. For example, an increase in life expectancy from 40 to 50 years was accomplished in the developing countries in only 15 years from 1950 to 1965. The rapidity of this rise in life expectancy can be appreciated by noting that a similar increase required 70 years (from 1830 to 1900) in Western Europe and 25 years (from 1900 to 1925) in Southern and Eastern Europe (Figure 1.22).

If a large reduction of mortality is not accompanied by a fall in fertility, the population will increase indefinitely. What has happened though in the now industrialized countries is that fertility has also fallen and the low mortality rate is balanced by a new low rate of fertility. Thus, the growth rate of population is low both before and after the demographic transition, but during the transition, the population grows rapidly (Figure 1.23).

The crucial question, therefore, is whether the reduction in mortality which took place in the developing countries between 1950 and 1965 has been followed by a fall in fertility. The evidence seems clear. Until the mid-60's, there was no sign of this fertility decline, but since then fertility has begun to fall in almost all the developing countries except those in sub-Saharan Africa. By the late 1970's, the average "total fertility rate"* in developing countries had fallen to about 4.7, about 20 percent less than it had been 15 years earlier.[59]

It seems therefore that the demographic transition *is* underway and that the populations of the developing countries and of the world as a whole are likely to stabilize eventually. However,

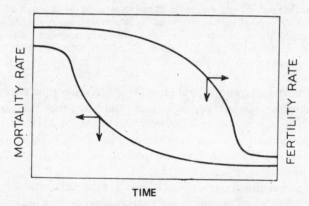

TIME

Figure 1.21. The demographic transition, often considered the central event of the human population, is for the most part complete in the industrialized countries but is in mid-course in the developing ones, where its outcome is as yet uncertain. The effect of the transition on mortality and fertility is shown schematically. Before the transition a high birth rate is approximately in equilibrium with a high death rate. When the transition is complete, a low birth rate is balanced by a low death rate. The decrease in mortality generally precedes the decrease in fertility; during the interval the population can grow rapidly. Mortality declines in the developing countries have been particularly rapid since World War II. Wide spread declines in fertility began only in the 1960's; so far fertility has been reduced by about 20 percent.

Source: D.R. Gwatkin and S.K. Brandel, 1982.

---

* A measure of the average number of children a woman will bear throughout her child-bearing years if at each age she has the average fertility corresponding to that age group.

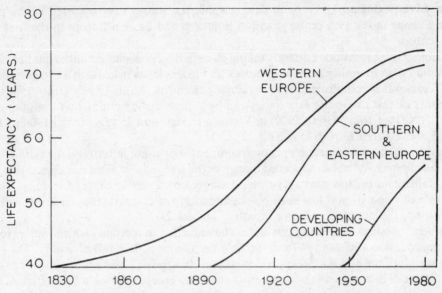

Figure 1.22. An increase in life expectancy from 40 to 50 years occurs early in the demographic transition. The increase in average longevity was achieved in Western Europe between about 1830 and 1900. A second wave of mortality declines in Southern and Eastern Europe began later but proceeded faster. The increase from 40 to 50 years took place in the first 25 years or so of the 20th century. By the 1970's the life expectancy was more than 70 years in almost all of Europe. The third wave of mortality declines, in the developing countries, was the last to begin with but in its early stages was the fastest of all. In 1950 life expectancy in the developing countries was about 40 years. By about 1965 it had risen to 50 accomplishing in 15 years what had required 70 years or so in Western Europe. Life expectancy in developing countries is now about 55 years, between 15 and 20 years less than in the industrialized countries.
Source: D.R. Gwatkin and S.K. Brandel, 1982.

the world is sure to have a growing population for quite some time because of "population momentum." The United Nations medium variant estimate is that the global population will reach 6.1 billion by 2000, 7.8 billion by 2020, and 10.5 billion by 2110, i.e., about 1.4, 1.8 and 2.4 times the 1980 population of about 4.4 billion respectively.[59] The bulk of the population increase is expected to take place in the developing countries, where a *a tripling of the population* to around 9 billion by 2110 is expected.

The enormous increase in global population before it is expected to stabilize around 2100 means that the *per capita* availability of resources will continuously decline in the near future. In fact, *per capita* estimates conceal the differential growth that is likely to take place in the industrialized and developing countries. This differential growth will exacerbate the already serious disparities between North and South.

Against a background of rapidly rising populations in the developing countries, the task of tackling the poverty of the majority of their people becomes an even tougher undertaking. The resources required to tackle the problems associated with today's level of population are enormous, and the task of satisfying a continuously rising population is still more daunting.

The tremendous increases in population imply a scramble for resources which may have a number of dangerous consequences. To the extent that the powerful industrialized countries are intimately involved and even active participants in this scramble, the situation is certainly

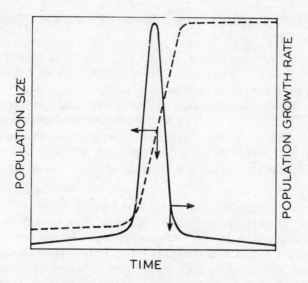

Figure 1.23. The size of the population and its growth rate show the influence of the demographic transition in this schematic representation. Before the transition the population is small and stable. When mortality decreases, the rate of growth increases. The population continues to grow rapidly for a considerable period after the reduction of fertility has caused the growth rate to begin to fall. At some time after the average woman is having about 2.1 children, the population stabilizes again at a much larger size. In Europe in the 19th century the annual rate of population growth rarely exceeded 1 percent. In developing countries, where fertility was very high when mortality began to decline, annual growth rates as high as 4 percent have been recorded. The overall rate is now 2.1 percent. Current projections suggest that by the time stability is achieved, the population of developing countries, now about 3 billion, will be about 9 billion.
Source: D.R. Gwatkin and S.K. Brandel, 1982.

not conducive to world peace. Population growth and the resulting stress from the intensified competition for limited resources are not conducive to global security.

Any use of natural resources is associated with their depletion, the release of emissions, and other environmental impacts. Thus, a rising population invariably implies increased environmental impacts (stepped-up rates of CO_2 release into the atmosphere, increased deforestation and desertification, etc.), unless special measures are taken along the lines suggested in previous sections.

Finally, population levels determine energy demand in a straightforward way—the larger the population, the more the total energy required, with the magnitude of this total energy depending on the per capita energy consumption. For example, the difference between the 1981 United Nations high and low variant estimates (7.1 and 5.7 billion) of the developing country population in 2020 corresponds, assuming a per capita consumption level of say 1 kW (roughly the present average level), to a difference in global energy demand of 1.4 TW, which is about one-third of the world's total oil use today.

This exogenous impact of population on energy is the conventional and obvious aspect of the population-energy connection. But there can be another connection in which energy strategies contribute to a reduction in the intensity of the population problem. To forge this link, one of the most important global energy objectives must be to accelerate the demographic transition, particularly by achieving dramatic reductions in fertility to stabilize global population as quickly as possible and at as low a level as possible.

From this perspective, it becomes imperative to consider the following three general preconditions for a decline of fertility:

- "Fertility must be within the calculus of conscious choice. Potential parents must consider it an acceptable mode of thought and form of behaviour to balance advantages and disadvantages before deciding to have another child."
- "Reduced fertility must be advantageous. Perceived social and economic circumstances must make reduced fertility seem an advantage to individual couples".
- "Effective techniques of fertility reduction must be available. Procedures that will in fact prevent births must be known, and there must be sufficient communication between spouses and sufficient sustained will, in both, to employ them successfully".[61]

The exercise of choice in matters of fertility is a culture-dependent issue, and especially important is the status of women. Do they in fact have a say in the matter? Do they have access to education and gainful employment outside the home? The awareness and availability of fertility-reduction techniques depends upon how successfully information about such techniques has been promulgated and their actual availability to couples of child-bearing age. The relative benefits of reduced fertility depend on such socioeconomic factors as infant mortality and the probability of offspring surviving—the lower the probability, the larger the number of children aspired for and the greater the exposure of the mother to the possibility of additional pregnancies. Similarly, the more the "free" manual labour of children is needed to do essential household tasks such as gathering firewood, fetching drinking water, tending livestock, helping with the weeding and harvesting in crop production, etc., the greater the likelihood of large families. Also, in many developing societies, children become wage-earners at quite a young age—hence they are viewed as economic assets.

These are only a few of the factors which enter people's estimation of the advantages and disadvantages of fertility and family size. It is clear that the reduction of fertility and therefore the acceleration of the demographic transition depends upon the extent to which crucial development goals such as increased life expectancy, improvement of the environment (drinking water, sanitation, housing, etc.), and education of women are achieved.

Almost every one of the socio-economic preconditions for fertility decline depends upon energy-utilizing technologies. Infant mortality has much to do with adequate and safe supplies of domestic water and with a clean environment. The prospects for women's education improve as the drudgery of their household chores is reduced with the availability of efficient energy sources and devices for cooking and of energy-utilizing technologies for the supply of water for domestic uses. The deployment of energy for industries which generate employment and income for women can also help delay the marriage age, another important determinant of fertility. If the use of energy results in child-labour becoming unnecessary for crucial household tasks, an important rationale for large families is eliminated. Thus, energy can contribute to a reduction in the rate of population growth if it is directed preferentially towards the needs of women, households, and a healthy environment.

Conventional energy strategies seldom bother with such seemingly mundane details as cooking stoves and piped water for the households of the poor in the developing countries. But our analysis indicates that these are in fact momentous matters that will greatly affect how we meet some of the most crucial development challenges of our time—halting environmental degradation, eradicating poverty and undernutrition, and accelerating the demographic transition to population stability.

1.3.13 *The Necessity of a Global Energy Strategy that is Compatible with the Solutions to the Other Global Problems*

Most studies of the global energy problem are preoccupied with the supply aspects of energy, tending to overestimate the growth of energy demand and failing to deal adequately with the end-uses of energy and with the opportunities for improvements in end-use efficiency (Section 1.2). There is another serious shortcoming. By approaching the energy problem primarily as the engineering challenge to produce more energy, these analyses tend to overlook connections to the other major global problems and usually fail to deal adequately with the relationships between the energy problem and these other problems. In fact, the energy-supply oriented energy strategies advanced in these studies aggravate other important global problems. Consequently, even if these strategies could yield an energy system that is sustainable, the world of which the energy system is a part would become unsustainable.

Anticipating this criticism, the IIASA analysts state:

"... Whether one considers the accepted fact that dozens of TW yr/yr be supplied only by hard technologies as an attractive or a frightening perspective is not of prime importance here...An evaluation and final decision about what is to be considered attractive or frightening is not be performed, however, by scientists; this is within the domain of politics..."[10]

We agree. What constitutes a desirable or unacceptable energy future has to be determined, not by analysts, but through the political process. At the same time, though, it is the responsibility of analysts to try and spell out the possible consequences—economic, environmental, political, etc.—of taking one or another of the energy paths that lie ahead, because it makes a difference which one our respective governments chose. It is not enough for an energy strategy to provide a solution to the energy problem. An energy strategy must also contribute to the solutions of other global problems, or at least be compatible with their solutions.

In weighing different energy strategies, therefore, their effect on North-South disparities, problems of the global economy, poverty in developing and industrialized countries, global insecurity and the danger of a nuclear holocaust, undernutrition and food supplies, population, and the state of the global and local environment have to be considered. This broader perspective enables us to look beyond the many "technical solutions" to the energy problem which aggravate one or more of the global problems and try to identify alternative energy strategies that do not.

1.4 Social Goals for Global Energy Strategies

What links our interest in solving the energy problem to the other global problems discussed in the previous section is a commitment to certain basic social goals—equity, economic efficiency, environmental soundness, long-term viability, self-reliance, and peace. These are essential elements in the development of developing countries and the continued well-being of industrialized countries. The most important test of any energy strategy, or broad plan for energy, is how it serves these basic social goals.

Equity means several things. It means closing the gap between developing countries and industrialized countries and eradicating poverty in both. It means that in all countries, people, including children, have the necessities of human life—nutritious food, drinkable water, decent shelter and clothing, and basic health care. It means universal education. It means gainful and productive employment for all those men and women who seek it. When evaluating energy strategies, particularly in the development context, concern about equity always causes us to ask: Who benefits?

Economic efficiency has to do with the scarcity of resources. All of human society's most valuable resources—arable land, oil and natural gas, capital, etc.—are in limited supply and therefore they should not be wasted. They should be used as effectively as possible to satisfy basic human needs and to increase the well-being of human society.

Economic efficiency is especially important in the energy sector, in light of the enormous capital requirements for energy supply expansion. If economic efficiency is stressed in allocating resources among various energy supply options and between investment in energy efficiency and energy supply expansion, economic resources will be freed up for other purposes, such as meeting basic human needs or industrial modernization.

In pursuing the goal of *environmental soundness,* we try to ensure the integrity of the life-support system of this earth—the air we breathe, the water we drink, the rain which falls on our fields and forests, the soil which nourishes our crops, etc. And we have seen that the degradation of these natural resources not only diminishes the quality of human life, it also endangers our very survival. We have learned from studying such environmental problems as soil erosion, deforestation and desertification, that if not treated, environmental problems soon become *economic* problems.

Concern for *long-term viability* is simply another way of saying that we must not just be concerned about tomorrow but the next decade and the next century. Systems put in place now will greatly affect the human condition in the future. We owe it to ourselves and future generations to look beyond the short-term, e.g., next year's budget, profit-and-loss statement, or economic progress report.

Energy strategies initiated now which would make oil importers decades from now more dependent on the Middle East, thus jeopardizing future global security, for example, fail the long-term viability test. So do strategies which would trigger a radical change in our climate. So do strategies which would lead to such large-scale quantities of plutonium around the world that nuclear weapons' proliferation cannot be controlled.

Another important aspect of long-term viability is the generation of knowledge and the development of new technologies, especially in terms of reliance on resources which are relatively more abundant and renewable.

Self-reliance refers to the ability of individual people or whole nations to support themselves. It also has to do with their independence, whether it be in everyday life or foreign policy. As it is used here, self-reliance is *not* synonymous with "self-sufficiency". The latter implies that an individual or a village or a nation uses only its own resources to meet its needs. Self-reliance, on the other hand, takes into account the uneven distribution and development of resources in the world and encourages human exchanges in the form, for example, of trade or aid so long as dependence is avoided. Dependence inhibits a people's ability to make independent decisions and to act in their own best interests. Dependence, whether it be dependence on a welfare state or an international bank or on another country's oil, leads to external controls.

Peace, of course in its most immediate sense, means the avoidance of war—foremost the avoidance of nuclear war, because nuclear war could make all of the other societal goals discussed here irrelevant. Peace also means the avoidance of conventional wars because they are such a terrible waste of human life and resources. Important in addition, though, is peace in the sense of a lessening of tensions between nations. For with that, some of the enormous sums now invested in the military, surely the most grossly inefficient allocation of the world's scarce economic resources, could be re-allocated to solving pressing economic problems. Indeed, if converted into investment capital, these re-allocated funds could completely revitalize debt-slowed development efforts in the South and help the North eliminate its chronic unemployment by generating jobs through the modernization of older industries and the more rapid expansion of newer ones.

1.5 Towards Energy Strategies for a Sustainable World

Clearly, the goals discussed above have strong energy links worldwide. Accordingly, global energy strategies need to be formulated in ways that are supportive of, or at least do not conflict with, the achievement of these goals or of the goal which embodies them all—the goal of a sustainable world. This objective for energy is fundamentally different from the conventional one of seeking a sustainable *energy system*, because an energy system can be kept running, in an engineering sense, over the long term even while it is eroding the long-term sustainability of the world.

The preceding analysis (Section 1.3) shows that supply-oriented energy strategies are not supportive of the goal of a sustainable world. The huge capital requirements to expand supply systems make fulfilment of the equity and economic efficiency goals impossible. Failure to reduce oil imports makes self-reliance impossible and peace highly uncertain. Proposed expansion of nuclear power clashes with the long-term viability and peace goals because the nuclear proliferation problem becomes unmanageable if nuclear power grows too much. And large-scale expansion of fossil-fuel use is incompatible with both the environmental soundness and the long-term viability goals.

Because of the failings of supply-oriented energy strategies, we have extended the scope of our analysis to allow consideration of energy strategies that involve changes in the energy-using system as well as the supply system. We have found that the range of outcomes for the global energy system is greatly extended by exploring how energy is being used at present and how resources might be redirected to meeting human needs for energy services more effectively. In fact, the prospects for identifying energy strategies consistent with the achievement of a sustainable world are so hopeful with changes at the point of end-use that we have made the end-use approach to energy the foundation of our analysis.

Energy strategies based on this end-use approach are described in Chapters 2 and 3 for industrialized and developing countries, respectively. These analyses and the global integrating analysis in Chapter 4 show that these strategies are supportive of the achievement of a sustainable world. What follows here are discussions of features of energy strategies required for the achievement of the basic social goals discussed in the previous section.

(1) *Satisfying Basic Human Needs.* For an energy strategy to promote the goal of equity in developing countries and be an instrument for development, then one of its crucial features must be the provision of the energy services needed for the satisfaction of basic human needs. This approach, in contrast to an "energy trickle-down" approach, directs scarce economic resources to pressing needs that would otherwise be neglected—e.g., to mundane energy end-uses such as cooking and energy sources such as domestic fuelwood, and to decentralized energy supplies for agriculture and for rural industries that would create productive employment for the rural poor.

To be compatible with overall development objectives, however, the satisfaction of basic human needs has to be accomplished in ways that enhance rather than diminish the self-reliance of people and in programmes which are economically efficient, environmentally sound, and viable over the long term.

(2) *Meeting the Energy Needs of the Poor and Disadvantaged.* Special attention has to be paid to the energy needs of vulnerable groups in society. In developing countries, these are the urban and rural poor, especially women and children. In the industrial countries, they are poor minorities and the poor in economically-depressed areas. If the needs of these groups are not emphasized, they will be overlooked in the energy planning process.

(3) *Creating Fair Economic Competition between All Energy Sources and End-Uses.* In general, alternative energy technologies do not compete on equal terms in the market. A vast number of handicaps and favours are conferred on specific technologies or groups of technologies. On the supply side, for instance, there are overt and covert subsidies, publicized and disguised preferences. underwriting of research and development expenditures, etc. On the demand side, consumers' lack of information and capital to invest in energy-efficiency improvements and their preoccupation with short-term considerations often prevent them from making economically efficient choices of energy technologies. The result of these various biases is an irrational and wasteful use of resources.

If, however, fair competition for all energy sources and tasks is an integral component of energy strategies, then the social costs of buttressing economically inefficient energy technologies can be reduced and eventually eliminated.

(4) *Promoting Energy-Efficiency Improvements.* As our analysis in Chapters 2 and 3 will show, there are abundant opportunities in both industrialized and developing countries for using energy more efficiently. We have found that it is usually cheaper, at prevailing energy prices, to save an extra unit of energy than to generate it.

We will also show that the energy-efficiency improvements can often be carried out in ways that not only save energy cost-effectively but also encourage broadly desirable technological innovations in consumer and industrial products and in industrial processes. Some examples: in industrialized countries it is feasible to build thermally tight houses that are more comfortable than conventional houses and which cost little or no more than conventional houses; in developing countries it is feasible to produce energy-efficient cookstoves that are more controllable and less polluting than conventional stoves; it is feasible to introduce more energy-efficient industrial processes that lead to marked reductions in overall production costs, thereby promoting industrial renewal in the North and rapid industrial growth in the South.

We also found that if such efficiencies are realized, then the other problems associated with the production and consumption of energy—the equity, environmental, long-term viability and potential international conflict problems—are markedly reduced.

Despite these clear-cut advantages, there are some institutional obstacles to be overcome before the full potential for cost-effective energy efficiency improvement can be realized (Chapter 5).

(5) *Beginning a Transition to Renewable Energy Resources.* In the long run the achievement of a sustainable world requires a shift to renewable energy resources. Our analysis indicates that the prospects for renewable energy are quite favourable for the long run (Appendix A).

For the near term the most promising opportunities for expanding the role of renewable energy lie in modernizing biomass energy resources. Bioenergy already accounts for over 40 percent of energy use in developing countries. But bioenergy is used today very inefficiently, largely for cooking in the domestic sectors of developing countries. By using more efficient stoves and by efficiently converting biomass feedstocks into modern solid, liquid, gaseous, and electrical energy carriers that are suitable for use in energy-efficient end-use devices, even present levels of biomass use could support marked improved living standards in developing countries.

Hydroelectric power is another presently available renewable energy source, for which there remain some unexploited opportunities in industrialized countries, while less than 10 percent of the potential has been exploited in developing countries. Wind power is also potentially important, especially at windy sites in some industrialized countries.

Though not yet commercial, perhaps the most promising renewable energy options for the long run are offered by photovoltaic technologies that directly convert sunlight into electricity (or hydrogen)—especially amorphous silicon technology, for which progress is being made at a breathtaking rate (Appendix A).

With the possible exception of photovoltaic technologies, however, energy strategies emphasizing renewable energy are constrained by land use availability (especially for biomass) or environmental concerns (e.g., for some hydroelectric facilities) when the level of demand is sufficiently high.

Our analysis indicates that renewable energy can come to play an important role in meeting energy needs without running up against such constraints, and thus in ways that are consistent with the goals of economic efficiency and environmental soundness, *if* the pursuit of energy-efficiency improvements keeps overall energy demand from getting too large.

(6) *Generating New Knowledge and Technological Advances.* On the basis of present knowledge it is feasible to identify energy strategies for the next several decades that are compatible with the achievement of a sustainable world. But facilitating the pursuit of energy strategies that serve the long-term viability goal for the indefinite future requires a continuing flow to the market of innovations relating to both energy end-uses and renewable energy. Our analysis indicates that a rich variety of opportunities awaits exploitation in these areas. The rate of exploiting these opportunities would be speeded up if there were a shift of emphasis in the allocation of scarce research and development resources, away from conventional energy technologies, especially coal and nuclear power, in favour of end-use and renewable technologies.

(7) *Promoting National Self-Reliance.* Because low-cost energy resources are non-uniformly distributed around the world it does not make economic sense for many countries to strive for self-sufficiency in energy. Trade in energy is unavoidable and in fact desirable, to the extent that it promotes economic efficiency.

But because low-cost energy resources are *extremely* unevenly distributed throughout the world, there is no global free market in energy, so that classical economic arguments in favour of complete free trade do not hold up for energy, especially oil. A country should seek to restrict trade in energy to sufficiently modest levels that this trade does not reduce a country's ability to pursue economic goals (e.g., by forcing a pattern of export-oriented industrial development to pay the oil import bill, at the expense of industrial development aimed at mass domestic markets) or reduce its ability to pursue an independent foreign policy.

The pursuit of energy-efficiency improvements and renewable alternatives to imported oil, both with presently available technologies and with advanced technologies that could be brought to commercialization with appropriate R&D, would be effective in promoting national self-reliance, for both industrialized and developing countries.

(8) *Compatibility of Energy Strategies with the Solutions to Other Global Problems.* The most important test of any energy strategy, in view of the interrelatedness of the energy problem and other global problems, is that it not aggravate other global problems. If an energy strategy contributes to the solution of another global problem, all the better. At the very least, though, it must not make matters worse.

References

Section 1.1
1. "A Talk with the Shah of Iran," *Time,* April 1, 1974.
2. The World Bank, *World Development Report*, 1985.
3. R.P. Moss and W.B. Morgan, "Fuelwood and Rural Energy Production and Supply in the Humid Tropics", Natural Resources and Environment Series, Vol. 4, United Nations University Press, Tokyo, 1981.
4. D.E. Earl, *Forest Energy and Economic Development,* Clarendon Press, Oxford, 1975.
5. D.O. Hall, G.W. Barnard and P.A. Moss, *Biomass for Energy in Developing Countries,* Pergamon Press, Oxford, 1982.

Section 1.2
6. Jose Goldemberg, Thomas B. Johansson, Amulya K.N. Reddy, and Robert H. Williams, "An End-Use Oriented Global Energy Strategy," *Annual Review of Energy,* Vol. 10, pp. 613-688, 1985.
7. Carroll L. Wilson, Project Director, *Energy: Global Prospects 1985-2000*, Report of the Workshop on Alternative Energy Strategies, 291 pp. McGraw-Hill, New York, 1977.
8. I.J. Bloodworth, E. Bossanyi, D.S. Bowers, E.A.C. Crouch, R.J. Eden, C.W. Hope, W.S. Humphrey, J.V. Mitchell, D.J. Pullin, J.A. Stanislaw, *World Energy Demand to 2020*: Technical Report, a discussion paper prepared by the Energy Research Group, Cambridge Laboratory, Cambridge University, for the World Energy Conference in London, August 15, 1977, and published by the Conservation Commission of the World Energy Conference, London, 1982.
9. World Energy Conference, *Energy 2000-2020: World Prospects and Regional Stresses,* J.R. Frisch, ed., Graham and Trotman, 259 pp., 1983.
10. Wolf Haefele, Project Leader, *Energy in a Finite World: A Global Systems Analysis*. Report by the Energy Systems Program Group of the International Institute for Applied Systems Analysis, 837 pp., Ballinger, Cambridge, 1981. Summarized in Jeanne Anderer, with Alan McDonald and Nebojsa Nakicenovic. *Energy in a Finite World: Paths to a Sustainable Future,* Report by the Energy Systems Program Group of International Institute of Applied Systems Analysis, 225 pp., Ballinger, Cambridge, 1981.
11. Wolfgang Sassin, "Energy", *Scientific American,* pp. 119-132, September 1980.
12. Wolf Haefele, "A Global and Long Range Picture of Energy Developments", *Science,* pp. 174-182, July 4, 1980.
13. D.L. Meadows, "A Critique of the IIASA Energy Models", *The Energy Journal,* Vol. 2, No. 3 pages 17-28, 1981.
14. Bill Keepin, "A Critical Appraisal of the IIASA Energy Scenarios", *Policy Sciences,* Vol. 17, pp. 199-275, 1984.
15. Bill Keepin and Brian Wynne, "Technical Analysis of IIASA Energy Scenarios," *Nature,* Vol. 312, pp. 691-695, 20/27 December 1984.

Section 1.3.2
16. Willy Brandt *et al., North-South: A Programme for Survival,* The Report of the Independent Commission on International Development Issues under the Chairmanship of Willy Brandt, The MIT Press, 1980.
17. Paul Streeten, "New Approaches to the New International Economic Order" *World Development,* Vol. 10, No. 1, pp. 1-17, 1982.

Section 1.3.3
18. The World Bank, *World Development Report 1984,* Oxford University Press, July 1984.
19. Chapter 2, "The United States in the World Economy," in Council of Economic Advisers, *Economic Report of the President,* February 1984.
20. The World Bank, *The Energy Transition in Developing Countries,* Washington D.C., 1983.
21. Energy Information Administration, U.S. Department of Energy, *Annual Energy Outlook 1983*, with Projections to 1995, [Report DOE/EIA-0383 (83), U.S. Government Printing Office, Washington D.C., May 1984].
22. Paul Streeten with Shahid Javed Burki, Mahbub ul Haq, Norman Hicks and Frances Stewart, *First Things First—Meeting Basic Needs in Developing Countries,* World Bank, Washington, D.C., 1981.
23. International Labor Organization, *Employment, Growth, and Basic Needs: a One World Problem,* Geneva, 1976.
24. Norman Hicks, "Is there a trade off between growth and basic needs?", *Poverty and Basic Needs,* World Bank, Washington, D.C., 1980.
25. M.G. Quibria, "An Analytical Defence of Basic Needs: The Optimal Savings Perspective", *World Development,* Vol. 10, No. 4, pp. 285-291, 1982.

Section 1.3.6

26. R.A. Smith, *Air and Rain: The Beginnings of Chemical Climatology*, 1872.

27. L.W. Blank, "A New Type of Forest Decline in Germany", *Nature*, Vol. 314, pp. 311-314, 1985

28. "Pollution, Pathogens, and Pests" *Nature*, Vol. 307, p. 97, 1984.

29. B. Nihlgard, "The Ammonium Hypothesis—an additional explanation to the forest dieback in Europe", *Ambio*, Vol. 14(1), pp. 2-8, 1985.

Section 1.3.7

30. Gerald O. Barney *et al.*, *The Global 2000 Report to the President: Entering the Twenty-First Century*, Volume 1, report prepared by the Council on Environmental Quality and the Department of State, 1980.

31. International Union for Conservation of Nature and Natural Resources, *World Conservation Strategy: Living Resource Conservation for Sustainable Development*, Gland, Switzerland, 1980.

32. John Gribbin, "The Other Face of Development," *New Scientist*, pp. 489-495, November 25, 1982.

Section 1.3.8

33. J. Hansen et al., "Climate Impact of Increasing Atmospheric Carbon Dioxide." *Science*, pp. 957-966, August 28, 1981

34. W.W. Kellogg and Robert Schware, *Climate Change and Society: Consequences of Increasing Atmospheric Carbon Dioxide*, Westview Press, Boulder, Colorado, 1981.

35. V. Siegenthaler and H. Oeschger, "Predicting Future Atmospheric Carbon Dioxide Levels," *Science*, pp. 388-395, January 27, 1978.

36. V. Ramanathan, R.J. Cicerone, H.B. Singh, and J. Kiehl, "Trace Gas Trends and Their Potential Role in Climatic Change," *Journal of Atmospheric Research*, June 1985.

37. Carbon Dioxide Assessment Committee, U.S. National Academy of Sciences, *Changing Climate*, 496 pp., Washington D.C., 1983.

38. Stephen Seidel and Dale Keyes, *Can We Delay a Greenhouse Warming? The Effectiveness and Feasibility of Options to Slow a Build-Up of Carbon Dioxide in the Atmosphere*, a Report of the Strategic Studies Staff of the Office of Policy and Resources Management of the U.S. Environmental Protection Agency, September 1983.

Section 1.3.9

39. Desmond Ball, *Can Nuclear War be Controlled?* Adelphi: Paper #169, International Institute for Strategic Studies, London, 1981.

40. R.P. Turco *et al.*, "Nuclear Winter: Global Consequences of Multiple Nuclear Explosions," *Science*, pp. 1283-1292, December 23, 1983.

41. Paul R. Ehrlich *et al.*, "Long-Term Biological Consequences of Nuclear War," *Science*, pp. 1293-1300, December 23, 1983.

42. Barry M. Blechman and Douglas M. Hart, "The Political Utility of Nuclear Weapons—the 1973 Middle East Crisis," *International Security*, Vol. 7, No. 1, pp. 132-156, Summer 1982.

43. U.S. Department of Defence, *Annual Report, Fiscal Year 1981*, p. 41., and *Annual Report to the Congress, 1983*, pp. III-91 through III-110.

44. Pentagon press briefing quoted in *The Wall Street Journal*, Feb. 1, 1980, *The New York Times*, Feb. 2, 1980 and *The Washington Post*. Feb. 5, 1980.

Section 1.3.10

45. H.A. Feiveson, R.H. Williams and F. von Hippel, "Fission Power: An Evolutionary Strategy", *Science*, Vol. 203, pp. 330-337, January 26, 1979.

46. *Nuclear Proliferation Handbook*, pp. 137, September 1980.

47. *Treaty on the Non-Proliferation of Nuclear Fuel Cycles*, Article X.

49. H.A. Feiveson, "Proliferation Resistant Nuclear Fuel Cycles," *Annual Review of Energy*, pp. 357-394, 1978. H.A. Feiveson and Jose Goldemberg, "Denuclearization," *Economic and Political Weekly*, vol. XV, pp. 1546-1548, 1980.

Section 1.3.11

50. Food and Agriculture Organization, "Fighting World Hunger", 76 pp., Rome, Italy, 1979.

51. World Bank Sector paper on Malnutrition.

52. ASTRA, "Rural Energy Consumption Patterns—A Field Study", *Biomass*, Vol. 2, No. 4, September, 1982.

53. Srilatha Batliwala, "Rural Energy Scarcity and Undernutrition", *Economic and Political Weekly,* Vol. XVII, No. 9, February 27, 1982.
54. "Nutritional Status of the Rural Population of Sahel", International Development Centre, Ottawa, Canada, 1980.
55. Joseph H. Hulse, "Food Science and Nutrition: The Gulf between Rich and Poor", *Science,* Vol. 216, p. 1291-1294, June 18, 1982.
56. Nevin S. Scrimshaw and Lance Taylor, "Food", *Scientific American,* Vol. 243, No. 3, p. 74-84, 1980.
57. D.O. Hall, "Biomass: Fuel Versus Food, a World Problem?" Chapter 25 in *Economics of Ecosystem Management,* D.O. Hall, N. Myers, and N.S. Margaris, eds., Dr. W. Junk Publishers, Dordrecht, the Netherlands, 1985.
58. FAO, *Agriculture: Toward 2000,* Rome, Italy, 1981.

Section 1.3.12
59. Davidson R. Gwatkin and Sarah K. Brandel, "Life Expectancy and Population Growth in the Third World", *Scientific American,* Vol. 246, No. 5, p. 57-65, May 1982.
60. Department of International Economic and Social Affairs, United Nations. *World Population Prospects as Assessed in 1980,* Population Studies Report No. 78, New York, 1981.
61. Ansley J. Coale, "The Demographic Transition Reconsidered", International Population Conference, Liege, 1973.
62. Ansley J. Coale, "Recent Trends in Fertility in Less Developed Countries", *Science,* Vol. 221, p. 828-832, 26 August 1983.
63. Rafael M. Salas, "Beyond 2000: State of the World Population", *Populi, Vol. 8, No. 2, p. 3-10, 1981.*

Figure Captions

R.H. Boyle and Alexander Boyle, *Acid Rain,* Schocken Books/Nick Lyons Books, New York, 1982.

Willy Brandt *et al., North-South, a Program for Survival,* the Report of the Independent Commission of Development Issues, MIT Press, Cambridge, Mass., 1980.

The British Petroleum Company, *BP Statistical Review of World Energy,* 1985.

U. Colombo and O. Bernardini, "A Low Energy Growth Scenario and the Perspectives for Western Europe," report prepared for the Commission of the European Communities' Panel on Low Energy Growth, 1979.

Council of Economic Advisers, *Economic Report of the President,* U.S. Government Printing Office, Washington, D.C., February 1985.

J. Edmonds and J. Reilly, "Global Energy Production and Use to the Year 2050," *Energy, the International Journal,* Vol. 8, pp. 419-432, 1983.

J. Edmonds, J. Reilly, J.R. Trabalka, and D.E. Reichle, "An Analysis of Possible Future Atmospheric Retention of Fossil Fuel CO_2," DOE Rep. DOE/OR/21400-1, Washington, D.C., 1984.

Energy Information Administration, U.S. Department of Energy, *1983 Annual Energy Review,* April 1984.

Energy Information Administration, U.S. Department of Energy, *Monthly Energy Review,* September 1985.

Energy Information Administration, U.S. Department of Energy, *1983 Annual Energy Outlook 1983, with Projections to 1995,* Report DOE/EIA-0383(83), U.S. Government Printing Office, Washington D.C., 1984.

Executive Office of the President, "The National Energy Plan—Overview and Summary," U.S. Government Printing Office, Washington D.C., April 20, 1977.

Food and Agriculture Organization, *Agriculture Toward 2000,* Rome, Italy, 1981.

Jose Goldemberg *et al.,* "An End-Use Oriented Energy Strategy for Brazil," presented at the Second Global Workshop on End-Use Oriented Energy Strategies," Sao Paulo, Brazil, June 4-15, 1984.

Jose Goldemberg, Thomas B. Johansson, Amulya K.N. Reddy and Robert H. Williams, "An End-Use Oriented Global Energy Strategy," *Annual Review of Energy,* vol. 10, pp. 613-688, 1985.

Davidson R. Gwatkin and Sarah K. Brandel, "Life Expectancy and Population Growth in the Third World", *Scientific American,* Vol. 246, No. 5, p. 57-65, May, 1982.

Wolf Haefele, Project Leader, *Energy in a Finite World: A Global Systems Analysis,* Report by the Energy Systems Program Group of the International Institute for Applied Systems Analysis, 837 pp., Ballinger, Cambridge, 1981.

International Union for Conservation of Nature and Natural Resources, *World Conservation Strategy: Living Resources Conservation for Sustainable Development,* Gland, Switzerland, 1980.

A. Jarnegren, F. Ventura and O. Warneryd, *Society Development and Energy Supply,* Report 52-80, (in Swedish), Swedish Council for Building Research, Stockholm 1980, the period 1850-1935; Government Official Report (SOU) 1951:32, "Fuel and Energy", (in Swedish), Stockholm 1951, the period 1936-1950; Government Bill 1975:30, "Energy Economizing, etc.", (in Swedish), Liber, Stockholm 1975, the period 1950-1975; Statistical Abstracts of Sweden 1977 and 1986, (in Swedish), Stockholm 1977 and 1986, the period 1976-1984.

W.W. Kellogg and Robert Schware, *Climate Change and Society: Consequences of Increasing Atmospheric Carbon Dioxide,* Westview Press, Boulder, Colorado, 1981.

A.B. Lovins, L.H. Lovins, F. Krause, and W. Bach, "Energy Strategy for Low Climatic Risk," report prepared for the German Federal Environmental Agency, 1981.

W.D. Nordhaus and G.W. Yohe, "Future Paths of Energy and Carbon Dioxide Emissions," in *Changing Climate,* Report of the Carbon Dioxide Assessment Committee of the U.S. National Academy of Sciences, Washington, D.C., 496 pages, 1983.

Population Reference Bureau, *Man's Population Predicament.*

D.J. Rose, M.M. Miller, and C. Agnew, "Global Energy Futures and CO_2- Induced Climate Change," MITEL Report No. 83-015, Massachusetts Institute of Technology Energy Laboratory, MIT, Cambridge, Massachusetts, 1983.

Frank von Hippel, "Global Risks of Energy Consumption," in *Health Risks of Energy Technologies,* Curtis C. Travis and Elizabeth L. Etnier, eds., pp. 209-227, published for the AAAS Selected Symposium 82 by Westview Press, Boulder, Colorado, 1983.

R.H. Williams, G.S. Dutt, and H.S. Geller, "Future Energy Savings in U.S. Housing," *Annual Review of Energy,* Vol. 8, pp. 269-332, 1983.

The World Bank, *The World Development Report,* Oxford University Press, 1983.

The World Bank, *The World Development Report,* Oxford Univesity Press, 1984.

Notes

T.V. Amentano and J. Hett, eds. "The Role of Temperate Zone Forests in the World Carbon Cycle," Carbon Dioxide Effects Research and Assessment Program, U.S. Department of Energy Report, CONF-7903105, UC-11, February 1980.

C.F. Baes *et al.,* "The Collection, Disposal, and Storage of Carbon Dioxide, pp., 495-519, in *Interactions of Energy and Climate,* Wilfred Bach, Jurgen Pankrath, and Jill Williams, eds., Proceedings of an International Workshop, Munster, Germany, March 3-6, 1980, D. Reidel Publishing Company, Dordrecht, Holland, 1980.

I.J. Bloodworth, E. Bossanyi, D.S. Bowers, E.A.C. Crouch, R.J. Eden, C.W. Hope, W.S. Humphrey, J.V. Mitchell, D.J. Pullin, J.A. Stanislaw, *World Energy Demand to 2020: Technical Report,* a discussion paper prepared by the Energy Research Group, Cambridge Laboratory, Cambridge University, for the World Energy Conference in London, August 15, 1977, and published by the Conservation Commission of the World Energy Conference, London, 1978.

The British Petroleum Company, *BP Statistical Review of World Energy,* 1981.

The British Petroleum Company, *BP Statistical Review of World Energy,* 1984.

W.S. Broecker et al., "Fate of Fossil Fuel Carbon Dioxide and the Global Carbon Budget," *Science,* pp. 409-418, October 26, 1979.

S. Brown and A.E. Lugo, "Preliminary Estimate of the Storage of Organic Carbon in Tropical Forest Ecosystems," pp. 65-117, in *The Role of Tropical Forests on the World Carbon Cycle,* S. Brown, A.E. Lugo, and B. Liegel, eds., Carbon Dioxide Effects Research and Assessment Program, U.S. Department of Energy Report, CONF-800350, UC-11, August 1980.

Sandra Brown and Ariel E. Lugo, "Biomass of Tropical Forests: A New Estimate Based on Forest Volumes", *Science,* pp. 1290-1293, March 23, 1984.

Council of Economic Advisers, *Economic Report of the President,* U.S. Government Printing Office, Washington D.C., February 1985.

Joel Darmstadter, *Energy in the World Economy: A Statistical Review of Trends in Output, Trade and Consumption Since 1925,* published for Resources for the Future by the Johns Hopkins Press, Baltimore and London, 1971.

O.W. Dykema [Energy Systems Group, Rockwell International], "SOx and NOx Control in Combustion", paper presented at the Canadian Electrical Association Seminar on Flue Gas Desulphurization, Ottawa, September, 1983.

Freeman Dyson, "Can We Control Carbon Dioxide in the Atmosphere?" *Energy,* Vol. 2, pp. 287-291, 1977.

D.E. Earl, *Forest Energy and Economic Development,* Clarendon Press, Oxford, 1975.

Richard J. Eden, Project Leader, "World Energy Demand to 2020: Technical Report", a discussion paper prepared by the Energy Research Group, Cambridge Laboratory, for the Conservation Commission of the World Energy Conference, August 15, 1977, published by the World Energy Conference, 1978.

Energy Information Administration, U.S. Department of Energy, *1980 Annual Report to Congress, Volume Two: Data,* April 13, 1981.

Energy Information Administration, U.S. Department of Energy, "Residential Energy Consumption Survey: Housing Characteristics, 1980," DOE/EIA-0314, June 1982.

Energy Information Administration, U.S. Department of Energy, *Monthly Energy Review,* June 1982.

Energy Information Administration, U.S. Department of Energy, "Residential Energy Consumption Survey: Consumption and Expenditures, April 1980 Through March 1981," DOE/EIA-0321/1, September 1982.

Energy Information Administration, U.S. Department of Energy, "Residential Energy Consumption Survey: Consumption Patterns of Household Vehicles, Supplement: January 1981 to September 1981," DOE/EIA-0328, February 1983.

Energy Information Administration, U.S. Department of Energy, *Montly Energy Review,* April 1985.

Exxon, *How Much Oil and Gas?, May 1982.*

H.A. Feiveson, Frank von Hippel, and Robert H. Williams, "Fission Power: An Evolutionary Strategy," *Science,* Vol. 203, pp. 330-337, January 26, 1979.

Timothy Greening, "Oil: The Formidable Barriers to Additional Supplies", *The Energy Economist,* Issue 11, pp. 2-4, September 1982.

Wolf Haefele, Project Leader, *Energy in a Finite World: A Global Systems Analysis,* Report by the Energy Systems Program Group of the International Institute for Applied Systems Analysis, 837 pp., Ballinger, Cambridge, 1981.

D.O. Hall, G.W. Barnard and P.A. Moss, *Biomass for Energy in Developing Countries,* Pergamon Press, Oxford, 1982.

N.H. Highton and M.J. Chadwick [Beijer Institute], "The Effects of Energy Shifts on Sulphur Oxide Levels in Europe: 1972-2002", in National Swedish Environment Protection Board, Report 1637: *Strategies and Methods to Control Emissions of Sulphur and Nitrogen Oxides,* pp. 165-172, Stockholm 1983.

M. King Hubbert, "The Energy Resources of the Earth," *Scientific American.* Vol. 224, pp. 60-70, 1973.

Richard A. Kerr, "Carbon Budget Not So Out of Whack," *Science,* pp. 1358-1359, June 20, 1980.

E.N. Lorenz, *Nature and Theory of the General Circulation of the Atmosphere,* World Meteorological Organization, Geneva, 1967.

Ariel E. Lugo, "Are Tropical Forest Ecosystems Sources or Sinks of Carbon?" pp. 1-20, in *The Role of Tropical Forests on the World Carbon Cycle,* S. Brown, A.E. Lugo, and B. Liegel, eds., Carbon Dioxide Effects Research and Assessment Program, August 1980.

Cesare Marchetti, "On Geoengineering and the CO_2 Problem," *Climate Change,* Volume 1, pp. 59-68, 1977.

Richard Nehring, *Giant Oil Fields and World Resources,* RAND Corporation report, June 1978.

Office of Mobile Source Air Pollution Control, U.S. Environmental Protection Agency, *Preliminary Perspective on Pure Methanol Fuel for Transportation,* September 1982.

Office of Technology Assessment, U.S. Congress, *Energy from Biological Processes,* Washington D.C., 1980.

J.S. Olson et al., "Changes in the Global Carbon Cycle and the Biosphere," Oak Ridge National Laboratory Report, ORNL/EIS-109, 169 pp., 1978.

Organization for Economic Cooperation and Development, *OECD Economic Outlook,* Vol. 36, Paris, December 1984.

The Scientific Committee on Problems of the Environment (SCOPE) of the International Council of Scientific Unions, *SCOPE 13: The Global Carbon Cycle,* B. Bolin, E.T. Degens, S. Kempe, and P. Ketner, eds., 491 pp., John Wiley & Sons, New York, 1979.

W. Seiler and P.J. Crutzen, "Estimates of Gross and Net Fluxes of Carbon Between the Biosphere and the Atmosphere from Biomass Burning," *Climate Change,* pp. 207-247, Vol. 2, 1980.

Minze Stuiver, "Atmospheric Carbon Dioxide and Carbon Reservoir Changes," *Science,* pp. 253-258, January 20, 1978.

E.T. Sundquist and G.A. Miller, "Oil Shales and Carbon Dioxide," *Science,* Vol. 208, pp. 740-741, May 16, 1981.

United Nations, *World Energy Supplies 1973-1978,* 1979.

United Nations, *1979 Yearbook of Energy Statistics,* 1981.

United Nations, *1979 Yearbook of International Trade Statistics,* Special Table B, pp. 1076-1123, 1980.

United States Census Bureau, *Statistical Abstract of the United States 1985,* U.S. Government Printing Office, Washington D.C., December 1984.

Ralph Whitaker, "Trade-offs in NOx Control", pp. 18-25, *EPRI Journal,* January/February 1982.

Robert H. Whitaker and Gene E. Likens, "The Biosphere and Man," in *Primary Productivity of the Biosphere,* Helmut Lieth and Robert H. Whitaker, eds., Springer-Verlag, New York, 1975.

Mason Willrich and Theodore B. Taylor, *Nuclear Theft: Risks and Safeguards,* a report to the Energy Policy Project of the Ford Foundation, Ballinger, Cambridge, 1974.

G.M. Woodwell *et al.,* "The Biota and the World Carbon Budget," *Science,* pp. 141-146, January 13, 1978.

G.M. Woodwell et al., "Measurement of Changes in Terrestrial Carbon Using Remote Sensing," Carbon Dioxide Effects Research and Assessment Program, U.S. Department of Energy Report, CONF-7905176, UC-11, May 1979.

G.M. Woodwell, J.E. Hobbie, R.A. Houghton, J.M. Melillo, B. Moore, B.J. Peterson, and G.R. Shaver, "Global Deforestation: Contribution to Atmospheric Carbon Dioxide," *Science,* Vol. 222, pp. 1081-1086, December 9, 1983.

World Energy Conference, *Energy 2000-2020: World Prospects and Regional Stresses,* J.R. Frisch, ed., Graham & Trotman, 1983.

World Meteorological Organization, *Meteorological Aspects of the Utilization of Solar Energy as an Energy Source,* Technical Note No. 172, 1981.

2. An End-Use Oriented Energy Strategy for Industrialized Countries

In this chapter we describe a long-term energy strategy for industrialized countries which takes into account both the implications of ongoing structural changes for the future demand for energy services and cost-effective opportunities for energy-efficiency improvements.

The implications of pursuing an end-use oriented energy strategy are illustrated by detailed long-range projections of energy demand for two countries: Sweden and the United States. The results in both cases show that the end-use approach to the energy problem can lead to future levels of energy demand far below present levels—even with large increases in economic output in these countries. These low levels of energy demand make the energy supply problem far more manageable than it would otherwise be. As long as the level of energy demand is not too large, it is no longer necessary to push all energy supply options "to the limit." It becomes possible instead to avoid or minimize dependence on the more costly and troublesome options.

Much of what has been learned from analyses of the Swedish and US situations is applicable to most other industrialized countries as well—especially most countries in the Organization for

Economic Cooperation and Development (OECD).* In view of the scarcity of data available about patterns of energy use in centrally planned industrialized countries, the extent to which our findings are relevant to the Council of Mutual Economic Assistance (CMEA) countries* is less certain. The fact that the average level of per capita energy use in CMEA countries is comparable to that for OECD countries—although the average level of amenities is lower in CMEA countries—suggests that there are large opportunities for energy-efficiency improvement in these countries as well. What can be achieved in industrialized countries such as the United States and Sweden is, we think, strongly indicative of what can be achieved in any industrialized country.

2.1 The Energy Situation in Industrialized Countries

2.1.1 *Patterns of Energy Use*

In 1980, per capita primary energy use averaged about 6 kW in both OECD and CMEA countries—more than six times the average level of energy use in the rest of the world (Notes 2.1-A and 2.1-B). There is a wide range, however, in per capita energy use among industrialized countries of the West, as indicated in Figure 2.1. Also, while this figure indicates that per capita energy use tends to increase with per capita GDP, there are noteworthy variations in the energy/GDP relationship. Japan, France, and Italy, for example, are modern industrial economies which require only about one-half as much energy to generate a dollar of GDP as does the United States. Both differences in the efficiencies of energy-using technologies and in the mix of economic product account for the wide variation.

While the share of final energy demand accounted for by the residential/commercial sector varies relatively little (30 ± 6 percent) among major regions, there is considerable variation in the energy consumption attributable to industry and transportation. The industrial share varies from a low of 36% percent for the United States to a high of 58% percent for CMEA countries, while the transportation share varies from a low of 17% percent for CMEA countries to a high of 33% percent for the United States (Notes 2.1-A and 2.1-B). These variations reflect, in part, the greater emphasis given to heavy industry in CMEA economies, compared to the broader-based, less energy-intensive economic activities of the United States, and the much greater use of personal automobiles in the United States—one automobile for every two persons in the United States versus one automobile for every 40 persons in CMEA countries.

2.1.2 *OECD Oil Dependency and the Oil Crises of the 1970s*

The oil crises of the 1970s were brought on by the dependency of the OECD countries on foreign oil—a situation that grew out of a rapid shift to oil in the decades following World War II and the inadequacy of domestic oil resources. While a major shift to oil also took place in CMEA countries in this period, internal production in these countries was able to keep up with the growth in demand, so that the oil demand/supply situation in CMEA countries has remained essentially independent from that of the rest of the world.

In 1950 oil accounted for 29 percent of total energy use in OECD countries and, of this, 29 percent or 2 million barrels per day was imported. By 1973 the situation had changed dramatically. Oil use at the time of the first oil shock had soared to 53 percent of total energy use in OECD countries, 67 percent of which, or 26 million barrels per day, was imported.

---

* Countries of Western Europe and North America, plus Japan, Australia, and New Zealand. See Appendix B.

* Here we refer to the Soviet Union and countries of Eastern Europe, although Vietnam, Mongolia, and Cuba are also CMEA countries. See Appendix B.

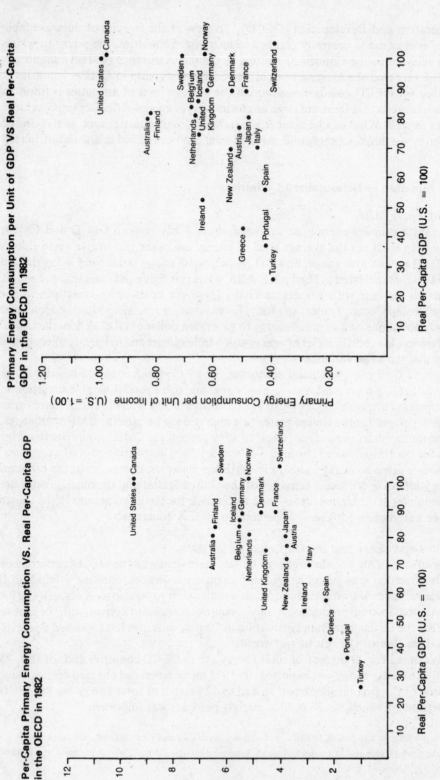

Figure 2.1. Primary energy consumption per capita vs. real per capita GDP (left) and primary energy consumption per dollar of GDP versus real per capita GDP (right) for OECD countries in 1970.

In these figures the values shown for GDP per capita are not simply based on exchange rates, which do not adequately reflect differences in purchasing power among countries. Rather the values shown for per capita GDP are adjusted for differences in the purchasing power of currencies for the year 1970 (I.B. Kravis *et al.*, 1978).

Sources: Total primary energy: United Nations, 1983;
Population: United Nations, 1981;
Real per capita GDP: column 5, Table 4 of I.B. Kravis *et al.*, 1978.

The three-fold real increase in world oil price in 1973 (Figure 1.3), brought on in large part by the dependency on imported oil, initially caused oil use to fall in all OECD regions. But after a modest softening in the world oil price, oil demand growth resumed again, preparing the way for the second oil shock of 1979 (Figure 1.3).

The industrialized countries have paid a heavy price for their oil habit. The International Energy Agency (IEA) concluded that the major slowdown in economic activity of the OECD countries in the 1980-81 period can be attributed in large measure to the direct and indirect effects of the second oil price shock: The IEA calculated that the deterioration of terms of trade vis-a-vis OPEC countries and the tightening of monetary and fiscal policies in reaction to the oil induced thrust to inflation cost OECD countries an income loss in excess of one trillion dollars in this period.[1]

The sharp increases in the world oil price have contributed to real price increases of the order of 50 to 100 percent for other energy forms as well in OECD countries (Note 2.1-C). Relative price increases have tended to be greater in the United States than elsewhere, especially in the transportation and industrial sectors, where U.S. energy prices prior to the first oil price shock were especially low (Note 2.1-D).

The energy price increases of the 1970's have had the effect of decoupling energy consumption and economic growth in the OECD countries—prior to 1973 they had moved in lock-step. Between 1973 and 1984, energy use in these countries increased only 2 percent while GDP increased 29 percent (Figure 2.2 and Note 2.1-Ea). An especially dramatic effect of the oil price shocks is

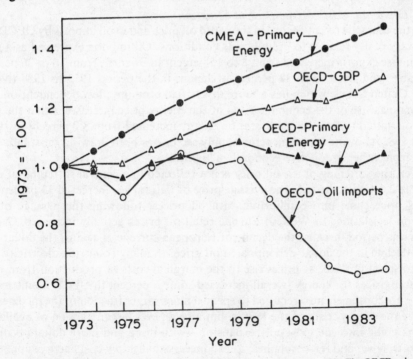

Fig. 2.2.　Primary energy consumption, net oil imports, and GDP, 1973-1984 (1973 = 100). The OECD values are aggregat values for the countries of the Organization for Economic Cooperation and Development. The CMEA curv is for aggregate primary energy use in the Soviety Union and Eastern Europe.

Source: Energy data are from British Petroleum Company, 1985. GDP data are from OECD, 1984b and OECD, 1985

that between 1979 and 1984 OECD countries collectively reduced their oil use by 7 million barrels per day and their oil imports by 10 million barrels of oil per day. This reduction in imports was more than the total oil consumption by all oil-importing developing countries in 1981!

In CMEA countries, however, total energy use grew between 1973 and 1984 at an average rate 3.5 percent per year (essentially the same rate of growth as in the period 1965-1973) and the use of oil at 2.9 percent per year (Figure 2.2 and Note 2.1-Eb).

2.1.3 *The Unviability of the Present Situation*

The world oil price fell 32 percent in constant U.S. dollars, 1980-84 (Figure 1.3). Because world oil production capacity is expected to be considerably in excess of oil demand for several years, the world oil price is expected to remain considerably below the 1980 price throughout the rest of the 1980's, barring a major political crisis in the Middle East.

The expectation of a "soft" world oil market in the near term has led to complacency about the "energy problem"—especially in the United States, where energy policy all but disappeared from the political agenda after the election of President Ronald Reagan.

Despite the absence of energy crisis conditions, however, major new policies at charting a new energy course are needed because: (1) the high economic costs of present patterns of energy production and use are a burden on the economies of the industrialized world; (2) the continuing high high dependence on scarce fluid fossil fuel is laying the basis for yet another oil crisis; and (3) there are energy-related environment problems that need to be solved.

The High Economic Costs of the Present Situation

Despite the dramatic recent decline in the world oil price and in oil imports by OECD, imported oil is still very costly relative to pre-oil crisis conditions. Oil imports expressed as a percentage of merchandise exports increased from 8 to 15 percent in Sweden, from 10 to 26 percent in the United States, and from 18 to 32 percent in Japan, in the period 1973 to 1984 (Note 2.1-F).

For the United States, which has a more diversified economy, less dependent on imports, a more telling measure of the economic costs of the energy price increases is that the percentage of GNP committed to retail expenditures on energy increased from 8.2 percent in 1972 to 13.7 percent in 1982, showing that the energy supply sector has been drawing substantial economic resources from other productive economic activities.[2]

Moreover, the softening of the oil price is not reflected in a general softening of real energy prices (Note 2.1-G). In the U.S. the average price of natural gas increased 33 percent, 1980-84, as the gas price came into equilibrium with oil prices, following the passage of gas price deregulation legislation. In Western Europe retail oil prices actually increased 12 percent on average in this period, because the decline of European currencies against the dollar more than offset the decline in the dollar-denominated oil price. In many countries, electricity prices also continue to spiral upward, as prices rise to the marginal costs of production from new power plants; Retail prices for energy overall increased some 5 percent in OECD countries, 1980-84.

Not only are consumers now required to spend a much larger fraction of the incomes on energy than before the energy crises of the 1970's, but also an ever larger fraction of available capital resources is going into energy supply expansion—with more and more dollars being required to bring forth fewer and fewer supplies, as producers exhaust low-cost energy supplies and turn to higher cost options. Figure 2-3 shows that in the United States the percentage of new plant and equipment expenditures committed to energy supply increased from about 25 percent in 1972 to over 40 percent in 1982. This shift of capital resources to the energy supply sector has contributed to the general problem of capital scarcity which has inhibited the renovation of the industrial bases of the industrialized world.

The Dependence on Oil

Despite the 17 percent reduction in oil use between 1979 and 1984, oil still accounted for 44 percent of all energy used in OECD countries in 1984, and nearly half of this oil was imported. Thus, OECD countries are still critically dependent on a scarce commodity, for which most of the remaining reserves lie in the Persian Gulf.

A shift away from oil could be made to synthetic liquids based on abundant resources of coal, oil shale, and tar sands. But this would be quite costly. A 1983 estimate by the U.S. Department of Energy puts the production cost of synthetic crude oil from coal or oil shale in the range of $50-$80 per barrel (1982$)—or two to three times the world crude oil price in 1982 ($29 per barrel).[3]

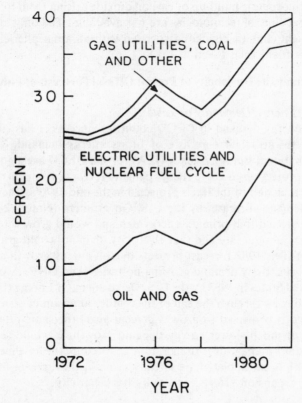

Figure 2.3. Capital investment in energy supply in the United States, as a percentage of total U.S. investment in new plant and equipment.

Capital expenditures on energy supply are compiled from various sources. Those for the oil and gas industry are from February issues of *The Oil and Gas Journal* (various years), the trade journal of the oil and gas industry. Those for the electric utility industry are from the Annual Statistical Report of *Electrical World,* the trade journal of the U.S. electric utility industry. Those of the gas utility industry are from the American Gas Association report *Gas Facts: A Statistical Record of the Gas Utility Industry* (various years).

The total investment in new plant and equipment in the U.S. is from "Economic Indicators," published for the Joint Economic Committee of the U.S. Congress by the President's Council of Economic Advisors (various issues).

The Need for an Environmentally Sound Energy Strategy

A major shift from dependence on petroleum, a relatively "clean" fossil fuel, to synthetic fuels based on the much more abundant but inherently much "dirtier" coal, oil shale, and tar sands fossil fuels would also lead to major environmental problems associated with the mining of these fuels, their conversion to more convenient energy carriers and their final use.

Many of these environmental problems can, in principle at least, be controlled to an acceptable degree by employing appropriate environmental controls. It will be a major challenge to the political process, however, to reach a consensus on how much should be spent controlling the environmental damage caused by energy-producing technologies and then enforce those controls. So far, for example, most governments have failed to take any effective action to reduce acid precipitation, which is caused, to a considerable extent, by the sulphur dioxide and nitrogen oxides emitted during coal combustion, although it is doing significant environmental damage (Section 1.3.6).

In the case of the atmospheric build-up of carbon dioxide arising from fossil-fuel combustion and use, environmental control technologies are not available. The only apparent options are to accept the consequent climatic changes associated with an atmospheric CO_2 build up or to reduce dependence on fossil fuels (Section 1.3.8).

2.2 Looking to the Future: The Unviability of Present Official Forecasts of Future Energy Demand

2.2.1 *The History of Energy Demand Forecasts*

The stagnation of energy demand in OECD countries since 1973 has caused forecasters to continually revise downward their forecasts of future energy demand. The energy demand forecasts of the International Energy Agency (IEA) for the OECD are a good example (Figure 2.4). The IEA's 1982 projections involve energy demand levels near the turn of the century that are only slightly more than half of the levels projected in the mid-1970's. There has been a similar trend in the energy demand forecasts by the U.S. Government (Note 2.2-A).

In 1977 the IEA projected that primary energy demand would grow through the end of the century at a 3.7 percent annual rate and that electricity demand would grow about 40 percent faster than GNP. But their 1982 forecast projects overall energy growth at only half the rate projected in 1977 and electricity demand growing no faster than GNP. A similar situation was envisaged for the United States in 1983 by the U.S. Department Of Energy (DOE): overall energy demand is projected to grow through the end of the century at an annual rate of only 1.3 percent and electricity demand is projected to grow 3 percent year, the same rate as the GNP.

With the prospect of much slower energy demand growth, it would seem that the energy problem has become quite manageable. In fact, though, because the baseline consumption levels from which this growth is extrapolated are already so high, these recent forecasts will require heroic energy supply expansion efforts, especially for electricity.

2.2.2 *The Overemphasis on Electricity*

Up until the energy crises of the 1970's electricity demand grew twice as fast as total energy use and about twice as fast as the economy in many parts of the world.

This rapid growth was expected to continue well into the 21st century, eventually transforming the fossil-fuel-intensive energy economies of the industrialized world into virtual "all-electric" energy systems based largely on nuclear power. This vision appeared to be quite plausible because electricity prices were declining dramatically in many parts of the world (see, for example, Figure 2-5).

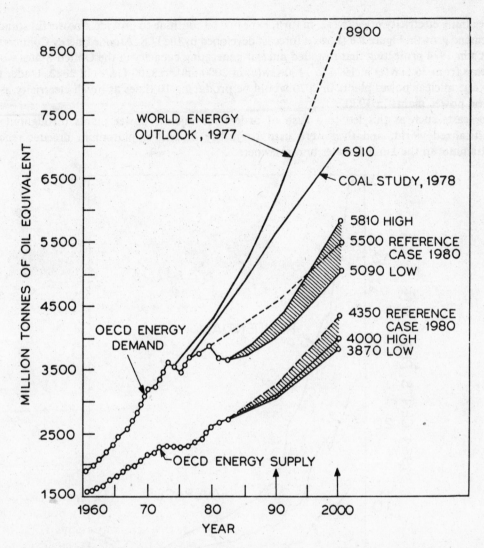

Figure 2.4. OECD Primary energy demand and supply (in million tonnes of oil equivalent)—history and projections. The alternative projections were made in 1977, 1978 and 1980 by International Energy Agency.
Source: International Energy Agency, 1982.

Declining prices were made possible by substantial improvements in the efficiency of central station power generation (Figure 2.6) and in the technology of power transmission and distribution, and by a dramatic increase in the scale of operations (Figure 2.7) which made possible the exploitation of scale economies. A shift to nuclear power was expected to cut costs even further. Large nuclear power plants were being ordered for very low prices in the 1960's, and it was predicted in 1969 that the real price of electricity in the United States would again be cut in half by the year 2000, due to a major shift to extremely low-cost electricity from the breeder reactor.[4]

Declining electricity prices were, in turn, expected to continue to provide a powerful stimulus to demand growth. Figure 2.8 shows a forecast developed by the U.S. Atomic Energy Commission (AEC) in 1974 projecting that installed nuclear generating capacity in the United States would increase from 36 GW(e) in 1974 to 1100 Gw(e) in 2000 and to 3300 Gw(e) in 2020. Under this forecast, nuclear power plants in 2020 would be producing 10 times as much electricity as all electric power plants in 1980.

Forecasts such as this led to a rush of orders for nuclear power plants throughout the industrialized world, and they were used to justify ambitious plutonium breeder reactor programmes in the United States and elsewhere.

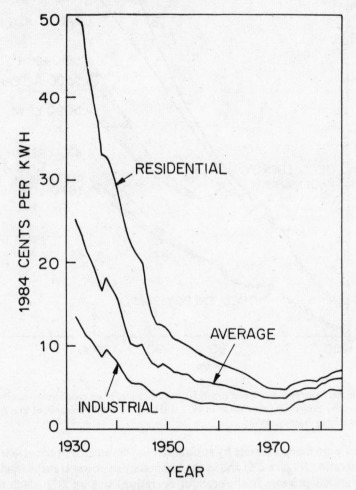

Figure 2.5. The history of electricity prices in the United States, in 1984 cents per kWh. Prices were converted to constant dollars using the GNP deflator.

But these official forecasts soon proved to be completely unrealistic. Even before the oil price shock of 1973, electricity prices in many areas had begun to rise, and they have continued rising ever since. Electricity prices have been rising largely because capital costs for new nuclear and coal-fired central station power plants have been escalating dramatically for a decade in many

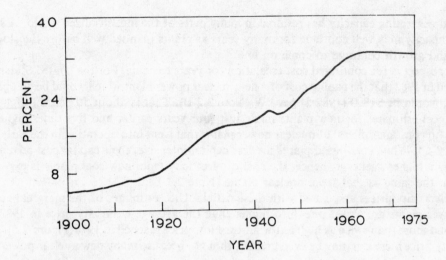

Figure 2.6. The average efficiency of electricity production at steam-electric thermal power plants in the United States. *Source:* R.H. Williams, 1978.

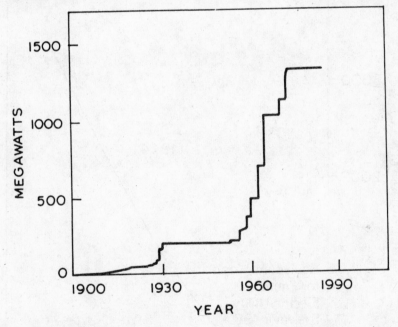

Figure 2.7. The largest steam-electric turbine generator in service in the United States. *Source:* R.H. Williams, 1978.

countries. These cost escalations have been due to problems of quality control, changing regulatory requirements, and to construction bottlenecks and delays that have developed because each big project has been in many ways unusual. The capital cost-induced electricity price escalation, in turn, helped to cause a dramatic slowdown in the growth of electricity demand. Many plants were built in anticipation of more rapid electricity demand growth, and so considerable excess electrical generating capacity has resulted in many parts of the industrialized world. A state of excess capacity may well continue for many years as plants planned well before the slowdown in demand growth continue to come on line.

The prospect is for continued cost escalation for years to come. For the United States, DOE estimated in 1982 that the capital cost of a new nuclear power plant which would begin operation in 1995, would be $1800 (1982 $) per kW (Note 2.2-B). That is about 50 percent higher than official cost estimates for new plants made just four years earlier and five times higher than the cost (in constant dollars) of nuclear power plants that went into operation in the early 1970's (Note 2.2-C). While coal-based plants are less capital-intensive, their capital cost advantage is offset by a higher fuel cost; hence, the cost of electricity from new coal plants is expected to be about the same as that from nuclear plants (Note 2.2-B).

The electricity prices associated with these official U.S. estimates of the marginal costs for new power plants are about one-third higher than the average electricity price in 1982 (Note 2.2-B) and more than twice as high as the all-time low price reached in 1970 (Figure 2.5). Actual electricity price increases may be even greater, though, because many new nuclear power plants

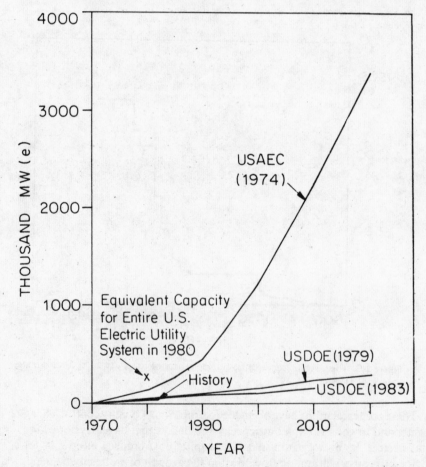

Figure 2.8. U.S. government projections of future nuclear power generating capacity made in 1974, 1979, and 1983.

are proving to be far more costly than official estimates (Note 2.2-D). The average capital cost (in real terms) for all nuclear power plants under construction in the United States is about 30 percent higher than the 1982 DOE estimate (Note 2.2-E).

Energy planners now generally realize that a future with electricity which is essentially "too cheap to meter" is a fantasy, and they have drastically reduced the levels of electricity demand envisaged for the future. The U.S. Department of Energy's 1983 projection for nuclear power generation by 2010, for instance, is just 1/16 of what was projected by the AEC in 1974 (Figure 2.8). But even the "slow" growth that is now being projected for electricity demand would involve a very intensive construction effort. Under the 1983 DOE forecast, for example, electric utilities would have to build some 440 large (1000 Mw) electric power plants between 1983 and 2000. The total investment cost? More than $1 trillion.[5]

Such projections cannot be taken too seriously by the investment community, however, because there is little evidence that official or trade association projections of future electricity demand have stabilized. Figure 2.9 shows the continuing downward trend in recent Swedish government forecasts. Similarly, between 1974 and 1983 the annual projections made by *Electrical World* (the trade association of the U.S. electric utility industry) for installed electrical generating capacity needed in the United States by 1995 went down, on average, by 56 Gw(e) *each year* (Note 2.2-F)— an amount adequate to provide 1/7 of total U.S. electricity requirements. Such continual

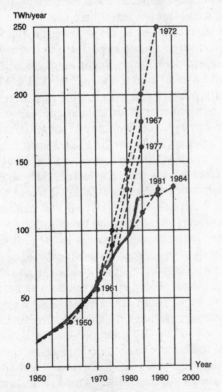

Figure 2.9. Official electricity demand forecasts for Sweden, in TWh/year. The years shown on the graph are the years the forecasts were made.
Source: T.B. Johansson and P. Steen, 1985; Swedish National Energy Administration (STEV), 1984.

adjustments in forecasts make investment in new electrical generating capacity a very risky financial undertaking.

What are the realistic prospects for future electricity demand growth? Electricity would seem to be an attractive energy carrier for the post-petroleum era. It is a high quality energy carrier that can be derived from a variety of alternative abundant energy sources and can be readily used in diverse applications.

The potentially large new markets for intensive use of electricity, markets presently dominated by oil, natural gas and coal, include: electric cars, process heat applications in industry, and space heating applications in buildings. In all these applications, however, the prospects for significant growth in electric sales are rather poor, owing largely to the high cost of electricity versus the available alternatives for providing the same energy services (Box 2.1: The Prospects for Energy-Intensive Markets for Electricity).

There are other new markets for electricity which appear more promising: computers, robotics, and a wide range of new information and electronic control technologies. But the intensity of electricity use in these activities will tend to be low, so volume of electricity demand will not be high.

BOX 2.1: The Prospects for Electricity-Intensive Markets for Electricity

The prospects for energy intensive applications of electricity—for automobiles, industrial process heat, and space heating—are reviewed here.

The automobile might seem to be an especially promising application in which electricity, based on abundant energy resources, could substitute for increasingly scarce petroleum. A typical electric car with a lead-acid battery might weigh about 1400 kg (including some 400 kg for the battery) and have a "fuel economy" of 0.21 kWh per km[6]. At the U.S. average residential electricity price (7 cents per kWh), this corresponds to an electricity cost of 1.5 cents per km. In terms of energy costs, the electric car would seem to be attractive relative to a typical new car, having a fuel economy of some 10 litres per 100 km (24 mpg). At the 1982 U.S. gasoline price of $0.34 per litre ($1.30 per gallon), the operating cost would be more than twice as much as that for the electric car. But the cars being compared here are radically different from one another. The electric car would have much less interior space—less in fact that a gasoline car weighing 1000 kg; it would have poor acceleration— typically 0 - 40 km/hour in 9 seconds, compared to 0 - 80 km/hour in the same amount of time for a gasoline-powered car; its range would be only about 50 km between recharges, compared to about 400 km for the gasoline car; its initial cost (exclusive of the cost of the battery) would be higher than the cost of a gasoline car of comparable performance, owing to the need for more structural strength and a larger power plant; and the battery replacements over the life of the car would add up to about as much expense as the initial cost itself.[6] The potential car buyer today could buy instead a small gasoline car of high fuel economy like the 4-passenger Honda City Car.[7] With a fuel economy of 5 litres per 100 km (47 mpg), gasoline would cost 1.7 cents per km for this car; its first cost would be not only less than that of the electric car but perhaps 25 percent less than that of other gasoline cars; it has an acceleration of 0 - 72 km/hour in 9 seconds; it has a range of 750 km on a single tank of gasoline; and of course there would be no costly battery replacements. The energy-efficient gasoline car thus provides unbeatable competition for the electric car. Unless there are radical advances in battery technology, the prospects for the electric car are dim.

What about industrial process heat markets? For low temperature (< about 300 degrees C) industrial process heat applications, which accounted for about 60 percent of industrial process heat use in the United States in 1977,[8] the competitive prospects for electricity are not bright, because electricity is some 6-8 times more expensive than coal or biomass energy alternatives. For high-temperature process heat applications, the prospects are brighter,[9] but because of the marked price disadvantage, electricity would have to compete here entirely on the basis of the special advantages offered by electricity over fossil fuels for direct heating applications—the possibilities for rapid heating, for more concentrated heat treatments of materials, etc. While there are a number of promising applications of this type, the potential volume of electricity sales associated with these applications is probably rather limited.

For space heating applications in buildings, there may be significant markets for electricity, if sufficiently efficient heat pumps can be utilized to overcome the 2 ½ - fold (or more) price advantage versus fossil fuels. However, the potential size of this market will be significantly limited by competition with opportunities for energy demand reduction through building shell improvements. As will be shown in Section 2.5.3.1, there are opportunities to meet space heating needs more cost-effectively by tightening up the building shell to the point where most space heating needs are provided by the heat given off by lights, by applicances, by the building occupants, and by sunlight coming in the windows. With such shell improvements the residual heating requirements are usually sufficiently low that it is difficult to justify the installation of capital-intensive heating systems like heat pumps.

In addition to the factors limiting new markets for electricity, there are many promising new opportunities for reducing the use of electricity in existing markets through the use of more energy-efficient electrical devices that are becoming available (Section 2.5.3).

All these factors considered together indicate that the levels of future electricity demand implicit in official forecasts are both unrealistically high and involve a great waste of economic resources. This conclusion is supported by quantitative analyses for Sweden and the United States (Section 2.6). Our long-range energy demand scenarios for these countries do involve a continuation of the historical trend towards electrification of their economies, but levels of future electricity demand are far lower than those envisaged in any official forecasts.

2.2.3 What's Wrong With Energy Demand Forecasts?

A fundamental problem with forecasting models is that they represent the energy future as an extension of the past, basing future demand projections upon historical data that correlate energy demand and demographic variables, GNP, energy prices, etc. Such models fail to take into account changes that are occurring in industrialized countries.

As will be shown in Section 2.4, there are many new economic activities which are not energy-intensive. For such activities, which are growing in importance, GNP is a very weak variable upon which to base energy-demand projections. Moreover, over the last several years, in the aftermath of the six-fold increase in the world oil price, a wide range of technological choices have emerged for providing energy services with much less energy inputs and at lower total cost. Energy forecasts extrapolated from past trends do not adequately reflect these new opportunities.

Nevertheless, these forecasts could become self-fulfilling prophesies because they provide a rationale for diverting still more capital and other economic resources to energy producers from other important economic activities; they could be used politically to show the "inevitability of demand growth," thereby justifying the need for tax relief and/or new subsidies for energy suppliers so that they can meet "the growing need for energy."

In the rest of this chapter, the conventional energy modelling techniques are abandoned. Instead, we explore the energy future of industrialized countries from an end-use perspective, examining both ongoing structural changes in the energy economy and opportunities for more efficient energy use. And, instead of attempting to forecast future energy demand, our analysis will show what it *could be*, consistent with constraints of technical and economic feasibility and the solutions of the other major global problems discussed in Section 1.3.

We think it more useful to policy-makers for us to illuminate future choices rather than construct forecasts. Of course, there are often institutional obstacles to the pursuit of the most economical solutions to a problem such as the energy problem, but highlighting the technological opportunities and their potential impacts on the overall energy supply/demand situation could provide impetus to the formulation of policy initiatives aimed at overcoming such obstacles.

It is a mistake to view trend as destiny. There is such a wide range of future energy paths that the energy future of industrialized countries is much more a matter of choice than prediction.

2.3 Rethinking on Energy Demand

The sharp changes in the economic environment relating to energy demand over the last decade along with the dwindling global supplies of fluid fossil fuels and the global security and environmental risks associated with present patterns of energy production and use are powerful incentives for searching for alternative, more hopeful futures.

Conventional energy demand models which attempt to predict future energy demand on the basis of historically determined responses to energy prices and incomes have continually failed in attempts to predict the energy future. Moreover, and perhaps more important, because they ignore the details in the patterns of energy consumption and the technological opportunities for changing these patterns to adjust to the new constraints on energy, such modelling efforts provide no guidance to policy-makers as to the technological alternatives available.

An alternative approach to analyzing energy demand, employed pursued in this book, involves disaggregating energy demand by energy-using activity so as to highlight the technical, economic, and institutional issues that bear on energy demand. Aggregate energy demand is then obtained by identifying an energy intensity for each activity, multiplying this with the activity level and summing up over all activities in society.

As noted above, we are particularly interested here in examining the effect on future energy demand of both ongoing shifts in the mix of energy-using activities and the introduction of cost-effective alternative technologies that require less inputs of energy to provide the energy services needed in a modern industrialized society.

The "energy saving" structural changes that we identify and describe are not the short-term adverse changes induced by the oil price shocks or by the adverse economic conditions of recent years, e.g., factory closings, increased unemployment, and increased poverty. Nor do we consider energy-saving structural changes that involve "lifestyle changes"—such as savings in space conditioning energy via reduced indoor temperatures in winter or increased indoor temperatures in summer, or savings in automobile fuel use via a shift to bicycles or walking, and the like. Rather, what are examined here (Section 2.4) are the long-term structural shifts associated with the transition to "post-industrial" economies, a transition which began well before the onset of the energy crises of the 1970's.

This transition involves both a continuing growing share for services in the mix of goods and services in the economies of industrialized countries and, within the goods-producing sector, a shift in emphasis from the processing of basic materials to fabrication and finishing activities. These ongoing shifts are of major importance to future energy demand because the new activities are in general much less energy intensive than the old.

The potential impact of introducing more energy-efficient technology, as is indicated by the examples shown in Table 2.1, is very large. For most energy-intensive activities, there is a substantial difference between the energy performances of typical devices now in wide use and the most efficient devices that are now available, suggesting a continual shift towards higher average energy productivities as more investment is made in new technology. The numbers in the last column in Table 2.1 illustrate the general phenomenon that there are many advanced technologies still under development that could reduce energy requirements even further over the long run.

Rethinking energy demand must not stop at simply noting opportunities for technological change. Economic issues are also important. It makes economic sense, for example, to compare the costs of energy-efficiency improvements with the costs of equivalent amounts of *new* supplies

Table 2.1. Alternative Energy Intensities for Illustrative Activities

Energy-Using Activity	Energy-Intensity Measure	Alternative Energy Intensities		
		Present Average (a)	Presently Known Best Technology (b)	Advanced Technology (c)
Automobile	litres of gasoline equivalent per 100 km	10 - 15 (d)	5 (e)	2.3 - 3.6 (f)
Residential Space Heating	kJ of heat output required of the heating system, per square metre of floor area per degree day	120 - 160 (g)	50 (h)	15 - 18 (i)
Refrigerator/Freezer	kWh per litre per year	3.4 (j)	1.3 - 1.7 (k)	0.7 (l)
Steel Production	GJ of final energy per tonne of raw steel (m)	22 - 27 (n)	15 (o)	9 - 12 (p)

(a) Refers to present averages of stocks in the indicated industrialized communities.
(b) Refers to examples of presently commercially available technology, as specified for each item.
(c) Refers to data for advanced prototypes, as specified for each item.
(d) The lower value is the West European average for the existing fleets of passenger automobiles in 1980 and the higher is the U.S. average.
(e) This is the fuel economy of the 4-passenger Volkswagen Diesel Rabbit and gasoline-fueled-Honda City Car.
(f) The higher value refers to the Volvo LCP 2000 prototype VW. The lower value refers to the 4-5 passenger PERTRAN car proposed by Battelle Memorial Institute researchers (Section 2.5.3.3).
(g) The average for the existing housing stock in the U.S. and Sweden.
(h) The average for a number of better designed houses in the U.S. and Sweden (Section 2.5.3.1).
(i) The energy performance of the best super-insulated house built in the U.S. and Sweden (Section 2.5.3.1).
(j) The average energy performance of the existing stock in the U.S.
(k) The lower value is for a Toshiba 411-litre unit and the higher for an Amana 515 litre unit (Section 2.5.3.1).
(l) A 368 litre prototype built by L. Schlussler.
(m) For steel production based on a 50 percent scrap steel charge.
(n) The lower value refers to the 1976 average values for Sweden and the higher to the 1979 average for the U.S.
(o) Refers to new steel technology being commercialized in Sweden (Section 2.5.3.4).
(p) Refers to advanced Plasmasmelt and Elred technologies under development in Sweden (Section 2.5.3.4).

and to pursue investments in energy efficiency to the point where the cost of the last "saved joule" is comparable to the cost of the last "produced joule." However, this is almost never done. Indeed, comparisons of energy-efficiency investment costs are usually not even made with the *average present* costs of energy supply.

Consumers are often reluctant to make energy-efficiency investments because much of the savings they would realize in terms of lower energy bills are often offset to a large extent by the required additional investment. In fact, a common feature of some energy-using situations is that the total cost of providing a given energy service varies little over a wide range of energy consumption levels associated with different end-use technologies. This phenomenon is illustrated in Figure 2-10, which shows the total cost of owning and operating an automobile in cents per

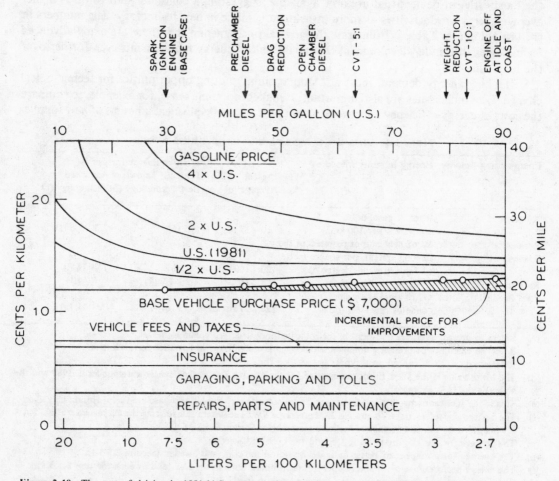

Figure 2.10. The cost of driving in 1981 U.S. cents per km and per mile, versus automotive fuel economy.

The energy performance indicated here is based on computer simulations of an automobile having various fuel-economy improvements added in the sequence indicated at the top of the graph (see also Figure 2.27 and Note 2.5.3.3-F). The base case car is a 1981 Volkswagen Rabbit (gasoline version).

The estimated costs of the various fuel-economy improvements are given in Note 2.5.3.3-L. The figure shows that the reduced operating costs associated with the various fuel-economy improvements are roughly offset by the increased capital costs of these improvements, over a wide range of fuel-economy improvements.

Source: Frank von Hippel and Barbara G. Levi, 1983.

kilometre, as a function of the automotive fuel economy. The cost curve was constructed by taking into account both the estimated extra capital costs associated with fuel-economy improvements and the fuel savings that would be associated with these improvements, at alternative gasoline prices. What is remarkable about this curve is that the total cost per km varies only slightly with fuel economy, over a wide range of fuel economies—from about 6 to below 3 litres per 100 km. The reduced operating costs associated with various fuel-economy improvements are roughly offset by the increased capital costs of those improvements.

Since making capital investments in energy-efficiency improvement typically involves a good deal of "hassle" for consumers, they are unlikely to make investments to the extent justified by considerations of cost, especially if the overall economic benefit to them is only marginal. But while energy efficiency may be a matter of indifference to the individual consumer, society as a whole is usually better off if capital investments in energy-efficiency improvements are made, because of the reduced pressure on insecure or depletable resources or the reduced environmental impacts that result. This suggests that government intervention is needed to make full use of technical and economic opportunities to use energy more efficiently (Chapter 5).

The "broad minimum" phenomenon illustrated in Figure 2.10 must be put into perspective, however. When energy efficiency improvements are introduced in such examples, it is assumed that "all other factors are equal". But in reality it may often be the case that all other factors are not equal. It will be shown in Section 2.5.3.1, for example, that new "super-insulated" houses can be built today which cost little if any more than conventional houses, because the added costs of insulation, multiple pane windows, etc., are offset by compensating capital cost savings.

Technological innovations are continually being developed. What determines whether they are adopted or not? Assuming the availability of capital, a buyer will usually choose to replace an existing technology with a new one only if the new technology has several clear advantages over the old. Berg and Solow found that in industry, for example, a technological innovation which improves several factors of production similtaneously—the productivity of labour *and* capital *and* energy—is more likely to be adopted than one which, say, just improves energy-efficiency.[11-12] In other words, energy efficiency is one but by no means the only factor which drives the technological adaption process.

This phenomenon has been so powerful throughout the history of modern technology that improvements in energy productivity have sometimes occurred even in periods of declining energy prices, as is illustrated by the history of modern lighting technology.

Figure 2.11 shows that since the time of Edison's initial research in the late 19th century, there has been almost a hundred-fold improvement in lighting efficacy (measured in lumens per Watt). The continuing improvement in lighting technology took place at the same time as electricity prices were tumbling (Figure 2.5) and was motivated primarily by considerations of durability, light quality, etc., but incidentally led to energy-efficiency improvements as well. The now rising electricity prices provide a powerful incentive to make even more improvements, and the opportunities for further improvements are substantial. Fluorescent lights have about five times the efficacy of the incandescent lights which dominate residential lighting use, and the theoretical maximum efficacy of white light (220 lumens per watt) is about three times that of typical fluorescent bulbs.

The point to be made here is that the need to reexamine the technology of energy use, brought on at this time by the energy price shocks of the 1970's, can be a spur to technological progress generally. Throughout this chapter, we shall see many examples of how this can happen.

Accordingly, policies to promote energy-efficiency improvement must not be too narrowly focused on forcing the deployment of particular technologies. Rather they should encourage

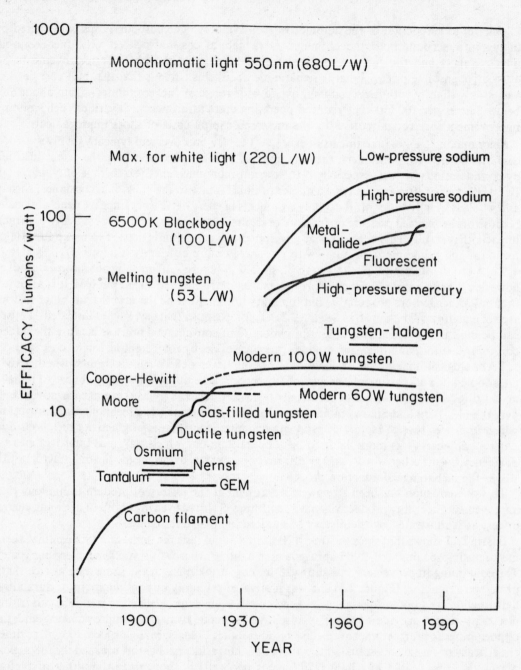

Figure 2.11. Approximate efficacies of various light sources plotted as a function of time, showing the evolution within each technology as well as the progress from one technology to another. The efficacies for several standard sources are indicated for comparison.

Source: J.M. Anderson and J.S. Saby, 1979.

improved energy efficiency in sufficiently flexible ways that overall economic productivity is enhanced as well.

In the long run, the energy price shocks of the 1970's may come to be viewed as powerful stimuli for quickening the pace of technical change, which had lagged in the United States and many other industrialized countries since the late 1960's but which historically has been so important to economic progress—accounting for nearly 90 percent of the productivity gains of the U.S. economy between 1909 and 1949.[12]

2.4 Structural Changes Affecting Energy Demand

- The patterns of consumer and industrial activities in industrialized countries are undergoing major changes that will affect the future demand for energy services in ways that cannot be discerned simply by looking at the future as an extension of the past.

This phenomenon will be illustrated here, first by showing the major shifts in these activities that can be expected in the Swedish context, followed by a more general discussion of the ongoing saturation of many energy-intensive activities and the associated shifts in the mix of economic output in many industrialized countries.

2.4.1 *Three Decades of Energy Growth—The Case of Sweden*

In the period 1950 to 1979 final energy use increased in Sweden by almost a factor of three. The increase is related to four roughly equal components: population growth and increased energy use per capita for industry, transportation, and housing (Figure 2.12).

In the time period considered, population increased by 19 percent, from 7.0 to 8.3 million. But the population of Sweden has stopped growing and in fact is expected to go into a slow decline—to 8.1 million by 2000 and to 7.7 million by 2020.

The paper and pulp industry and the primary metals industry accounted for 55 percent of final industrial energy use in Sweden in 1979. These industries also accounted for ¾ of the increase in industrial energy use in the period 1950 to 1979. Now, however, both of these sectors of industry face structural changes.

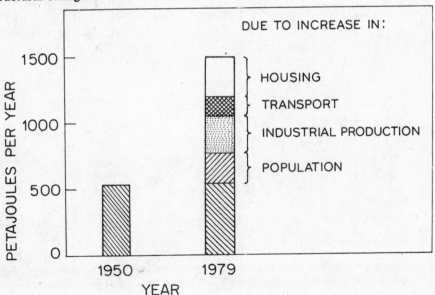

Figure 2.12. Factors accounting for the increase in energy use in Sweden, 1950-1979.

The future growth of the pulp and paper industry is constrained by competition for fertile land from timber production, biomass fuel production, and agriculture. The total annual increment of coniferous roundwood, the traditional raw material for paper production in Sweden, is already largely committed to the pulp and timber industries. In the future, one might consider short-rotation plantations for producing the needed feedstock. However, in order for the produced paper to be able to compete in world markets, the short-rotation or other deciduous wood feedstock would have to be priced so low that the alternative use of the wood as fuel would be very competitive. For this reason, the production of pulp can be expected to increase marginally or not at all.

The steel industry, which dominates the primary metals industry in Sweden, faces considerable competition on the world market. It is likely that the steel industry will not be able to expand physical steel production—measured in tonnes—and may even have difficulties in maintaining the current production volume.

Nevertheless, both the paper and steel industries may expand output—measured in value-added—by giving more emphasis to specialty products of higher value-added. Moreover, industrial production generally might shift in the future away from the steel and paper industries to activities that are not fundamentally constrained—such as the production of chemicals or the fabrication and finishing of basic materials. Such a shift would have a profound impact on energy demand, because these other industries are much less energy-intensive. Specifically, the chemical industry in Sweden requires only half as much energy per dollar of value added as does the primary metals and pulp and paper industry; and the fabrication and finishing industry requires only 1/15 as much (Figure 2.13).

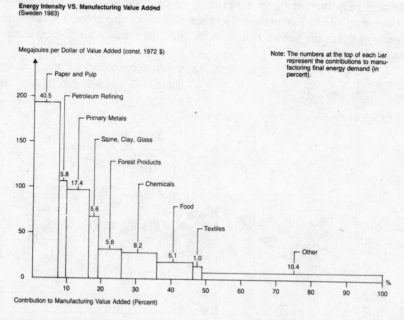

Energy Intensity VS. Manufacturing Value Added
(Sweden 1983)

Megajoules per Dollar of Value Added (const. 1972 $)

Note: The numbers at the top of each bar represent the contributions to manufacturing final energy demand (in percent).

Paper and Pulp — 40.5
Petroleum Refining — 5.8
Primary Metals — 17.4
Stone, Clay, Glass — 5.6
Forest Products — 5.8
Chemicals — 8.2
Food — 5.1
Textiles — 1.0
Other — 10.4

Contribution to Manufacturing Value Added (Percent)

Figure 2 13. Final energy intensity versus manufacturing value-added for Swedish manufacturing industries in 1983. The number at the top of each bar is that sector's contribution to total final energy use in Swedish manufacturing, in percent.

Personal transportation by automobiles in Sweden increased by a factor of seven to eight between 1950 and 1979 and now accounts for about 60 percent of the energy use in the transportation sector. Specific fuel demand, i.e., the number of litres of gasoline per 100 km, increased little between 1950 and 1979, and the distance driven per vehicle was roughly constant. This means that automotive fuel use increased in this period essentially in proportion to the number of vehicles.

While, in general, the number of vehicles is related to the overall level of economic activity and the size of the population, there are indications of saturation in the number of cars in Sweden. There are also indications of a saturation in time spent in automobiles by each person (Section 2.4.2).

Energy for space heating accounted for about one-third of final energy use in Sweden in 1979. Between 1955 and 1979 there was a threefold increase in energy use for space heating, because of more and larger dwellings and higher indoor temperatures.

Further changes in these factors appear to be limited in the future. Indoor temperatures will not be increased much because indoor temperatures above some optimal level are uncomfortable. The number of dwellings is related to the household size and population. These factors will not drive up demand in the future because the size of the population has now stabilized, and the household size is already relatively small, averaging 2.4 persons per household. The size of the dwellings, however, is more directly related to incomes and the relative costs of houses.

In summary, the factors which have driven the growth in final energy use in Sweden over the last 30 years are unlikely to continue to do so in the future.

These observations for Sweden apply in part to other highly industrialized countries as well. The population of most Western European countries has stabilized. (It is expected to continue growing only in Mediterranean countries). While the limitations on the future output of the paper industry are specific for Sweden, the steel industry in most Western industrialized countries faces problems similar to those in Sweden. In addition, the factors affecting space heating and automotive transportation are more or less similar in other Western industrialized countries.

2.4.2 *Saturation in Energy-Intensive Activities—A More General Discussion*

Energy consumption is not an end in itself. Rather it is used to provide energy services such as space conditioning, hot water for bathing, lighting, refrigeration for food storage, information transmission, mobility, mechanical power to do work, etc. Energy consumption associated with such activities has increased dramatically in industrialized countries in recent decades, as these countries have become more prosperous. But in looking to the future, sharp breaks with historical trends can be expected as the levels of consumer services associated with especially energy-intensive activities approach saturation.

Time Limits on Transportation

Time is an ultimate constraint on a person's activities. Time budgets have proven to be a valuable tool to gain better understanding of the constraints on an individual who must find time for eating, sleeping, working, shopping, transportation, leisure, housekeeping, education, and so on. A Norwegian study has found that the average person in Norway spends

- 10.6 hours/day for personal needs,
- 3.6 for paid work,
- 4.0 for other work,
- 5.2 for off-time (including travel to and from work)
- 0.6 for education and miscellaneous.

Similar results have been obtained for Sweden.[13]

Transportation is a major energy-using activity that demands a great deal of the individual's time in a modern society. The average American now spends about one hour per day in an automobile, and the average Swede about the same time in transportation. These numbers suggest that there may be no major increase in the time spent on transportation. In any case, it is difficult to see how spending more time in transportation activities could be construed as an improvement in the quality of life.

Of course, the time constraint is not a serious one at this time for air passenger transportation, which can be expected to continue rising with incomes. However, even with substantial future growth, air travel will continue to make only a relatively minor contribution to overall energy use for many years; in the United States, where the level of air travel is already relatively high (about 1800 km per capita in 1980), air passenger travel accounted for only 2 percent of total energy use.

Saturation in Energy-Intensive Household Activities

The most significant energy-using household activities in temperate-climate industrialized countries are space heating and water heating, which together account for 80 to 90 percent of final energy use in households (Note 2.4-A). While these have been major growth activities in the past, they will not be in the future, as most houses are now provided with domestic hot water, and most houses in climates where space heating is important now have central heating (Note 2.4-B).

The heaviest energy users among electrical devices in the home, accounting for one-third of residential electricity use in the United States, are lights and refrigerators, which are present in essentially all homes (Note 2.4.-B).

Air conditioning is an activity that is far from saturation in Europe, but in much of Europe air conditioning is not needed. And where it may be needed, the future growth of this activity generally will not affect residential energy budgets much. Even in the United States, where over half of all households had air conditioning in 1980, air conditioning accounted for only 3 percent of residential energy use.

There may well be considerable future growth in the use of labour-saving and time-saving appliances such as dishwashers, clothes dryers, and food freezers. Once again, however, such growth would not have major consequences for energy. Automatic dishwashers use about the same amount of energy as washing dishes by hand (Note 2.4-C). Clothes dryers and freezers are not really in the "energy big leagues"; even in the United States, where the saturation levels for these appliances are relatively high (40 percent and 60 percent respectively), these appliances account for only 4% of residential energy.

As incomes continue to rise, the energy intensity of consumer expenditures can generally be expected to decline, at the margin, because of these saturation effects. The marginal home equipment investment dollar is shifting to home products whose use will not lead to significantly greater residential energy use. Such products include high-power equipment of very low duty factor (e.g., hair dryers and power tools for the home workshop, yard, and garden), and low-power electronic equipment, such as video recorders and home computers.

Saturation in the Use of Basic Materials

The new products that consumers are investing in are also characterized by a high ratio of value-added to material content. This tendency towards lower material intensity is apparent even in the replacement market for conventional appliances. As more affluent consumers buy replacements for conventional appliances, the fancier, more expensive units tend to be less material-intensive and thus require less energy to manufacture per dollar than their lower priced counterparts (Figure 2.14).

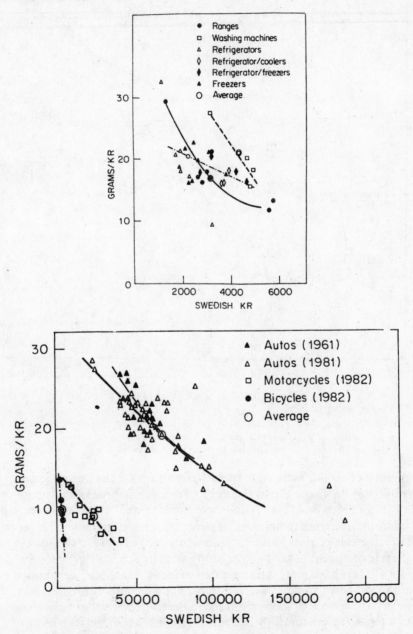

Figure 2.14. Material intensity (in grams per Kr) of various household appliances (top) and transport vehicles (bottom) in Sweden, as a function of their sales prices (in Swedish Kr).

Source: T.B. Johansson and P. Steen, *et al.*, 1985.

The reduced material intensity of consumer products is reflected in a reduced material intensity of the economy as a whole, as affluence increases. As Figure 2.15 shows, the amount of steel and cement consumed per dollar of GDP has been declining sharply in Sweden since per capita GDP reached $6000 to $7000 (1975 U.S. $) in the mid-1960's.

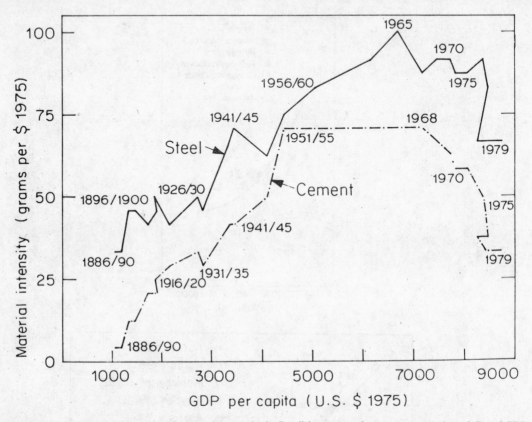

Figure 2.15. The material intensity of steel and cement in the Swedish economy (in grams consumed per dollar of GDP) versus standard of living (in GDP per capita) over time in Sweden.
Source: T.B. Johansson, P. Steen, E. Bogren, and R. Fredericksson, 1983.

This saturation effect is illustrated in two different ways for the United States in Figure 2.16. The bottom part of the figure shows the trend over time in apparent consumption (in kg) per dollar of GNP for seven basic materials—three "traditional" materials (steel, cement, paper) and five "modern" materials (ammonia, ethylene, chlorine, aluminium). This shows that in the United States, the role of steel in the U.S. economy peaked in 1920 (versus 1965 for Sweden) and that the role of cement peaked before 1930 (compared to the 1960's for Sweden). Further, the peaks for all these basic materials, even the "modern" materials, were passed by the 1970's, indicating a broad-based trend away from basic materials use in the economy.

The top part of Figure 2.16 shows the trend in the United States of basic materials over time in per capita terms (kg consumed per capita). Here it is seen that there is strong evidence of the onset of saturation, e.g., a cessation of growth in kg per capita consumed, for most materials. Indeed, detailed case studies of these materials indicate that the prospects for bulk demand growth

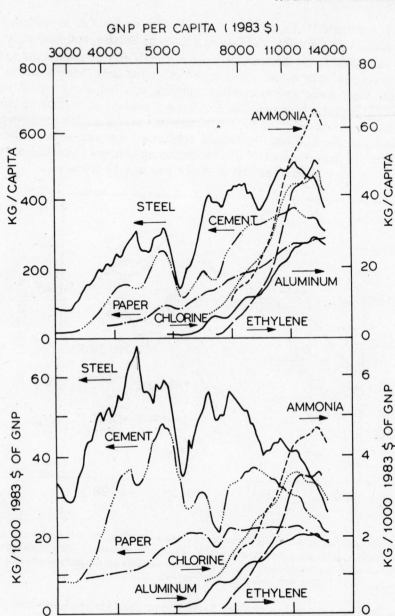

Figure 2.16. Basic materials use in the US, in kg per dollar of GNP (bottom) and kg per capita (top). The data are for apparent consumption (production plus net imports, adjusted for stock changes). To help clarify long-term trends, annual consumption data are averaged over running 5-year periods and plotted against the mid-year in these periods.

Source: Eric D. Larson, Robert H. Williams, and Drew Bienkowski, 1984.

for these materials are poor[14]. Figure 2.17 shows similar evidence of saturation for steel, cement, and aluminium in Germany, the United Kingdom, Sweden, and France.

The onset of saturation in the demand for basic materials may be understood in terms of the history of the economies of industrialized countries. In the early "industrializing" stages, large investments were made in infratructure building and industrial capital equipment formation—material-intensive activities. The economies continued to be material-intensive as their industrial bases approached maturity, but the production of materials-intensive consumer durables (e.g., automobiles and large household appliances) was emphasized. The onset of saturation in the use of basic materials marks the beginning of a new "post-industrial" phase of economic growth, some general features of which will now be discussed.

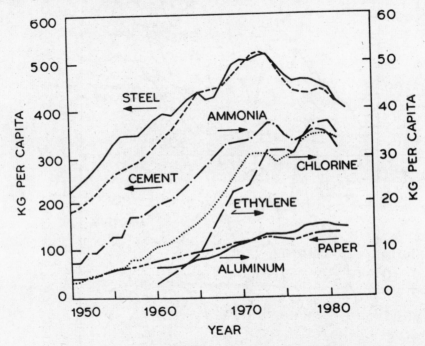

Figure 2.17.   The per capita use of some basic materials in European countries. The data are for aggregated apparent consumption in Germany, France, and the U.K., averaged over running 3-year periods and plotted against the mid-year. The ammonia curve is for an aggregate of data for Germany and France only.

Source: Marc Ross, Eric D. Larson, and Robert H. Williams, 1985.

2.4.3 *The Production Mix: Towards More Services and Less Material-Intensive Products*

Future energy use must be examined in the context of expectations of future levels of social welfare or well-being. While there is no simple relationship between energy use and social welfare or even between true social welfare and various existing measures of material well-being, it is nonetheless useful to discuss the future of energy demand in the context of some readily available and understandable measure of well-being. The most common measure of social welfare is the Gross Domestic Product (GDP), which is the total dollar value of the consumption of goods and services plus gross investment. Despite its many shortcomings, we shall occasionally use the GDP as a measure of social welfare, essentially because it is the only measure available.* The ongoing transition in industrialized countries to post-industrial economies means that the mix of economic activities contributing to GDP can be expected to differ in important ways in the future from what it has been in the past.

The Growing Importance of Services

In part, the reduced emphasis on the use of basic materials in modern economies is associated with the ongoing shift to a service economy—where considerable emphasis is given to mass services: finance, insurance and marketing services, information services, (e.g., computer software services), medical services, recreational services, education, communications, etc.

The increase in services production relative to goods production has been going on for decades in industrialized countries**, and is evident from the long-term trends in employment in Sweden and the United States (Figure 2.18). In the early years of industrialization, the shares of employment accounted for by both manufacturing and services grew while employment in agriculture declined; modernization of agriculture made it possible to produce more and more food with fewer and fewer people. Over the last couple of decades, though, services have grown relative to manufacturing, mining, and construction as well.

The shift is also reflected in the economic output of the goods and services producing sectors. In the United States, the goods-producing sector grew just 0.83 times as fast as GNP in the period 1970 to 1980 (Note 2.4-D); approximately the same relationship held for the period 1960 to 1980.

The Growing Importance of Fabrication and Finishing

Besides the ongoing shift to services, the onset of saturation in the use of basic materials is reflected in a marked shift *within* the goods-producing sector away from the processing of basic materials and towards the fabrication and finishing of these basic materials—value-added intensive processes, involving extensive use of computers, robotics, and high technology generally, which demand much less inputs of energy than the processing of basic materials.

Consider the situation (1978) in the United States and Sweden, with the industrial sector disaggregated into mining, agriculture, and construction (MAC), the basic materials processing subsector of manufacturing (BMP), and "other" manufacturing, which involves the fabrication and finishing of basic materials (Note 2.4.-E, Figure 2.13, Figure 2.19):

---

* The GDP does not measure vital non-economic aspects of human welfare; it does not reflect the value of leisure to well-being; it does not reflect the adverse impacts on quality of life of pollution and other "externalities" associated with production and consumption; it does not reflect the distribution of economic benefits among the people; and, as a practical matter, it does not measure the increasingly important economic activities associated with the "hidden" or "informal" or "irregular" economy. This "other" economy involves a wide range of unreported productive activities ranging from "household industries"[15, 16] such as the growing of vegetables in backyard gardens and cabinet-making in basement workshops to the bartering of professional services (a ploy on the part of a growing number of middle class people to avoid taxes), and to overtly illegal activities such as prostitution and drug dealing. It has been suggested in the case of Sweden that this "other" economy accounts for about half of all hours worked and a quarter of all value produced;[13] for the United States the "irregular" economy has been estimated to contribute from 10 percent - 25 percent of total economic output.[17]

** For the purposes of the present discussion, the goods-producing sector in the United States is taken to be agriculture, mining, manufacturing, and construction. The services-producing sector is taken to be everything else: finance, insurance, real estate, services, communications, public utilities, wholesale and retail trade, and transportation, and government.

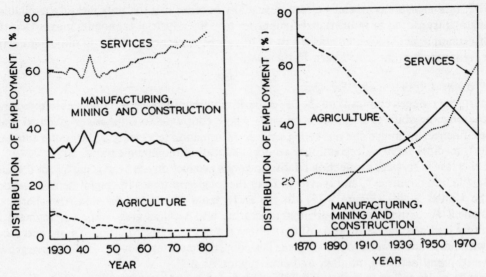

Figure 2.18. Sectoral distribution of employment in the United States (left) and Sweden (right). For the United States the employment measure is the number of full-time equivalent employees. For Sweden it is the number of employees working more than half-time.

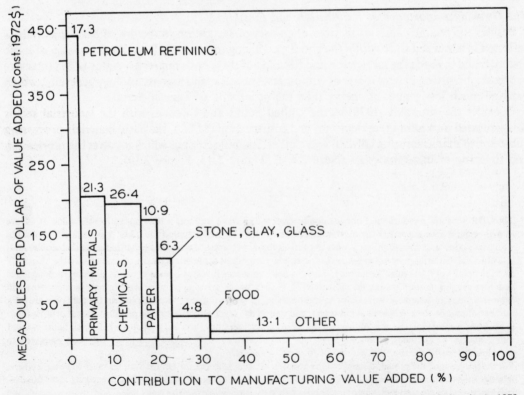

Figure 2.19. Final energy intensity versus manufacturing value-added for U.S. manufacturing industries in 1978. The number at the top of each bar is that sector's contribution to total final energy in U.S. manufacturing, in percent.

	MAC		BMP		Other	
	U.S.	Sweden	U.S.	Sweden	U.S.	Sweden
% of final energy use	25	35	25	37	50	28
% of final energy was	15	10	73	82	11	8
Units of energy required per dollar of output	3	1	14	7.5	1	1

Thus, while the fabrication and finishing of basic materials is economically very important in both countries, the energy use in this subsector is low; the BMP subsector of manufacturing dominates energy use.

Shifts in output among these subsectors towards less material-intensive activities have been pronounced. In the 1970's the rate of growth of industrial output for fabrication and finishing activities in the United States was 4.3 percent per year on average, compared to 3.0 percent per year for the BMP, and 1.2 percent per year for MAC. Similarly, in Sweden there is strong evidence of the declining importance of the production of basic materials as contributors to economic growth: fabrication and finishing activities grew in this same period at an annual average rate of 2.0 percent per year, compared to 1.1 percent per year for industry as a whole, 1.2 percent per year for the primary metals sector, and a 1.4 percent per year rate of decline for the cement industry.

These ongoing shifts to fabrication and finishing, which typically require an order of magnitude less energy per unit of output than the processing of basic materials, can have a profound effect on industrial energy use. For the United States the shift away from the processing of materials accounted for an annual rate of reduction of 2 percent in the ratio of industrial energy use to GNP in the period 1973 to 1984.[18]

Actually, the shift away from basic materials processing industries began well before the energy crises in the United States and other advanced industrial economies. Figure 2.20a shows the history of this shift in the United States from the end of World War II through 1982. Figure 2.20b shows that the onset of the decline and the rate of decline has varied markedly, though, from industry to industry within the BMP subsector. In the U.S. the decline has been particularly pronounced for primary metals and stone, clay and glass—especially energy-intensive industries which have been in relative decline for most of the post-war era. Paper is an energy-intensive industry which grew about as fast as manufacturing overall until the first oil crisis—after which it too began a slow relative decline (Figure 2.20b). The United States chemical industry stands out among BMP industries as an especially energy-intensive industry which actually grew twice as much in the period 1947-1976 as manufacturing overall. But even this industry has apparently reached a turning point, subsequently growing no faster than all of manufacturing. Moreover, there are strong signs that the chemical industry too is poised to go into relative decline. The once robust petrochemical subsector in particular is having to make rapid adjustment to middle age. There are no prospects for significant overall growth in domestic markets, which are approaching saturation, and the competitive position of the United States industry on the world market has been eroded by stiff new competition from countries enjoying the advantage of cheaper raw materials. The industry is adjusting by shifting its output to lower volume but higher value-added specialty chemicals for which it is believed the technology is still beyond the reach of many developing countries.[18,19]

The ongoing shift from material intensity in manufacturing is actually greater than these data would suggest, because even within subsectors major shifts are occurring. As already noted,

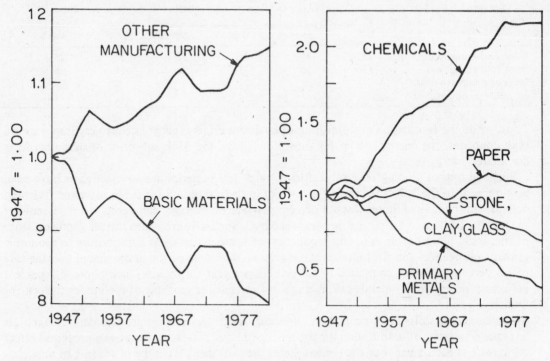

Figure 2.20. The ratio of manufacturing value-added for various subsectors of manufacturing to total manufacturing value-added over time in the United States, relative to the value of this ratio in 1947.

The figure on the left shows the declining importance of the basic materials processing subsector overall since the end of World War II, while the figure on the right shows that the trend away from basic materials started at different times and proceeded at different rates for different industries.

The value-added measure used here is the quantity Gross Product Originating, compiled by the Bureau of Economic Analysis of the United States Department of Commerce.

a shift is underway within the chemical industry toward higher value-added and hence less material-intensive production. Scrutiny of data for the U.S. paper industry provides some insights into such intra-subsectoral shifts. *Economic output* (gross product originating) for this industry grew at an average rate of 3.1 percent per year during the period 1970 to 1980. During this same period, the *physical output* of paper (measured in tonnes per year) grew at an average rate of just 1.8 percent per year. The difference between these two growth rates is attributable to the increased emphasis being given to high value-added finishing activities. The coating, cutting, and packaging of papers and the making of bags and boxes, etc., are becoming more important in the paper industry relative to the much more energy-intensive basic pulp and paper production activities[20]:

A visit to a typical paper and pulp mill quickly reveals this structure: very few employees are in evidence in the vast wood handling, pulping, bleaching and paper-making facilities. In the relatively small facilities where the product is cut up and packaged, many workers are seen...*

---

* High value-added is closely associated with the number of workers. In general, value-added is the sum of employee compensation plus profits plus indirect business taxes plus capital consumption allowances. In 1982 employee compensation accounted for ¾ of value-added in the paper industry, which is about the same as for all of manufacturing.

Just as shifts from material-intensive subsectors to fabrication and finishing-intensive subsectors lead to reduced energy requirements per dollar of output, so do such shifts within subsectors. In the paper industry, for example, energy use tends to be concentrated in the bulk processing operations rather than in the finishing operations; thus the tonnage of paper produced is a much better index of energy use than the more rapidly growing output measured in value-added terms.

The ongoing trends towards reduced material intensity (and hence energy intensity) of the economies of highly industrialized countries can be expected not only to continue but perhaps even to accelerate, as an economically rational response to the sharp increases in energy prices that have taken place.

2.5 Technical Opportunities Affecting Energy Demand
2.5.1 *Measuring the Productivity of Energy Use*

How much energy is really needed to run the economy and support a comfortable standard of living? To address this question meaningfully, it would be useful to have indices of energy performance for major energy-using activities that indicate the potential for improving the productivity of energy use. Unfortunately, the energy efficiency concept commonly used is an inadequate indicator of the long-term potential for improving the productivity of energy use. A couple of examples illustrate this point.

Household furnaces are typically described as being 60 to 80 percent efficient, which means that 60 to 80 percent of the heat released in fuel combustion is delivered as useful heat to the living space. This measure suggests that a 100 percent efficient furnace would be the best you can do, but this is incorrect because devices exist for actually delivering more heat. A heat pump is such a device. A heat pump extracts heat from the local environment (outdoor air, indoor exhaust air or groundwater) and delivers it plus the energy to run the heat pump as heat at a temperature appropriate for space heating. In other words, a heat pump delivers as heat more than 100 percent of the energy needed to run it. In fact, heat pumps available today can provide 2 to 3 units of useful heat for each unit of electricity consumed.

Air conditioners are rated by a coefficient of performance (COP), which is the ratio of the heat extracted from a cooled space to the electrical energy consumed. A typical air conditioner might have a COP of 2. Unfortunately, this measure provides no hint as to how this performance compares to the maximum possible, which is a COP much greater than 2.

The efficiency used in the examples is defined as:

$$\frac{\text{Energy transferred to the purpose of the system}}{\text{Energy input to the system}}$$

This efficiency is often called the "first law efficiency" because it is based on the first law of thermodynamics which holds that energy is neither created nor destroyed. This efficiency concept enables one to keep track of energy flows and hence is useful in comparing devices of a particular type. It is wholly inadequate, however, as an indicator of the general potential for energy savings.

A much more meaningful efficiency is one which measures performance by comparing the actual energy use to the theoretical minimum amount of energy required to perform a given task. For example, in heating a house, the task of the heating system could be defined as providing warm air to maintain rooms on a certain temperature schedule for a season, given particular heat losses from the house. The task of the engine and transmission of a car could be defined as maintaining a 90 km per hour speed for 1 km, given the car's drag and tyre losses. For a given task the theoretical minimum energy requirements can be determined without reference to the actual equipment used, that is, without reference to the use of a furnace for heating or use of an internal combustion engine for a car. An efficiency that measures energy performance, defined as:

$$\frac{\text{Theoretical minimum energy consumption for a particular task}}{\text{Actual energy consumption for a particular task}}$$

is called the second law efficiency, because the minimum energy required to perform a task is determined by the second law of thermodynamics.

In using the second-law efficiency concept, energy consumption must be measured in units that reflect the quality of the energy involved. The quality of a unit of energy is high if a large fraction of that energy can be converted into useful work. Highly organized energy (such as electrical energy, chemical or fuel energy, or the energy of motion of a falling weight or falling water) is high quality energy, because all or nearly all of it can be converted into useful work. In contrast, thermal energy, or heat, i.e., the energy of the random motion of molecules, is disorganized energy which can be converted only partially into useful work. The measure of energy use required for second-law efficiency calculations which takes into account energy quality differences is "available work.[21] For electricity, which in principle can be converted 100 percent into useful work, the available work is equal to the more familiar heating value. For fuels it is approximately equal to the heating value.* For thermal heat sources, on the other hand, the available work is much less than the heat content. For example, the available work contained in 1 Joule of heat in water heated to 60 degrees C from an ambient temperature of 13 degrees C is only about 0.1 Joule, reflecting the fact that *in principle* one cannot extract much useful work from warm water (Note 2.5.1-A).

The difference between the first-law and second-law efficiencies can be illustrated by a concrete example. Consider space heating, in which an amount of heat Q is delivered to a building at 30 degrees C by a gas furnace with a first-law efficiency of 60 percent. The minimum amount of available work W required to provide this heat is that which is required to run an ideal heat pump. If the outdoor temperature is 4 degrees C, then an ideal heat pump would provide (Q = 12) units of heat for each unit of electricity consumed (Note 2.5.1-B). If F is the amount of fuel actual consumed by the furnace, then the second-law efficiency is

$$W/F = (W/Q) \times (Q/F) = (1/12) \times (0.6) = 0.05$$

Thus, while the first law efficiency for a gas furnace (60 percent) gives the misleading impression that only a modest improvement is possible, the second-law efficiency (5 percent) correctly indicates a twenty-fold maximum potential gain in theory (Figure 2.21).

This example shows that the provision of low-temperature heat via fuel combustion in furnaces is quite inefficient in relation to the performance of technologies that might eventually be used. Table 2-2 similarly shows that for a wide range of energy-intensive activities, the second-law efficiencies are low.

Just how far can we expect to go towards achieving the theoretical maximum of 100 percent second-law efficiency? In practice, 100 percent efficiency is never achieved, because of limits of available technology and cost. At some point, the energy savings associated with a further efficiency gain are not worth the additional capital cost. Reasonable long-term goals might be efficiencies in the range of 20 to 50 percent for typical practical systems.[22] The values at the

---

* For example, for methane gas the available work is 1.008 times the lower heating value.

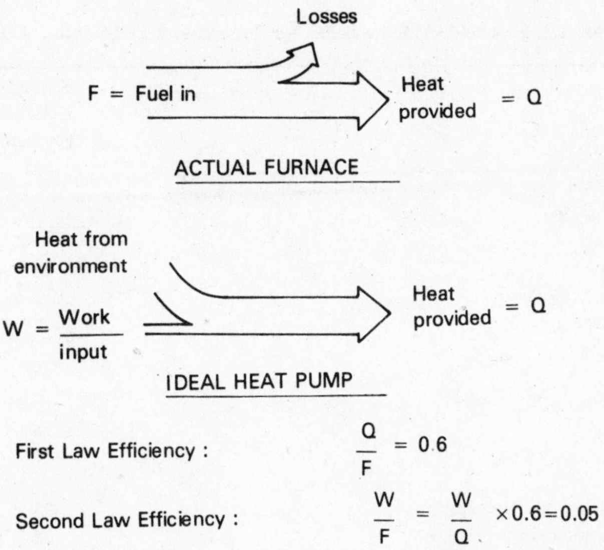

Figure 2.21. Schematic of energy flows for efficiency evaluation of a typical household furnace system in use in the United States. The first-law efficiency is the ratio of furnace output to input. The second-law efficiency compares the work input for a thermodynamically ideal heat pump to the fuel input for the actual furnace.
Source: M.H. Ross and R.H. Williams, 1981.

high end of this range would be characteristic of highly engineered devices designed for specialized tasks (mainly in industry), and the values at the low end would be representative of what could be achieved with more flexible, less sophisticated devices suitable for wide applications in homes, commercial buildings, and transportation.

While the second-law efficiency measure suggests large opportunities for energy savings, it does not tell the whole story, because the efficiency is for a given task, which can often the modified without adversely affecting the utility of the goods or services being provided. For example, in the food processing industries one way of defining the task of concentrating fruit juices might be the evaporation of the water from the juice—a very energy-intensive process. More generally, the task might be defined as water removal, which could be accomplished with far less energy via a reverse osmosis membrane technology. As another example, the second-law eficiency for gas furnaces indicated above does not reflect the energy savings possible from improvements aimed at reducing the thermal leakiness of the building shell—by means of increased insulation, reduced air infiltration, use of storm windows, etc. Improving furnace efficiency and reducing the thermal leakiness of buildings are complementary approaches to improving energy productivity.

As this brief discussion suggests, the *technical* potential for energy productivity improvement is large. At any given time, though the actual potential is limited by available technology and by economic considerations.

2.5.2 *Criteria for Judging the Economic Viability of Energy*

Productivity Improvements

While an improvement in the energy productivity of an energy using system often leads to savings in operating costs, the realization of these savings often requires an increased initial

Table 2.2. Second-Law Efficiencies for Typical Energy-Intensive Activities

		Second-Law Efficiency (percent)
Sector		
1.	Residential/commercial:	
	Space heating:	
	Furnace	5
	Electric stove	2 ½
	Air conditioning	
	Water heating:	
	Gas	3
	Electric	1 ½
	Refrigeration	4
2.	Transportation: Automobile	12
3.	Industry:	
	Electric-power generation	35
	Process-steam production	28
	Steel production	23
	Aluminium production	13

investment. How much extra investment can be economically justified *to save energy* depends on the costs of energy supply alternatives and on how future costs and benefits are related to present costs. To provide a basis for making economic judgments about investments in energy productivity improvement, we now turn to discussions of alternative economic figures of merit and discount rates.

Economic Figures of Merit

There are several figures of merit that can be used to measure the cost-effectiveness of energy efficiency investments: the simple payback, the cost of saved energy (CSE), the internal rate of return (IRR), and the lifecycle cost (LCC) or annualized lifecycle cost (ALC).

The most easily understood and most widely used measure is the simple payback, the ratio of the initial investment to the first year's savings. Despite the simplicity of this index, it is the least desirable index that one could use. Because it contains no information regarding the expected lifecycle of the investment and the time value of money, it cannot be used to rank investments by their cost-effectiveness.

The alternative indices—the CSE, the IRR, and the LCC or ALC—, which do provide a basis for ranking (Box 2.2: Alternative Economic Figures of Merit) will be used, as appropriate, in the present analysis to describe the economics of energy efficiency improvement. While one cannot expect the typical energy consumer to use such sophisticated concepts in making economic judgments about energy using devices, indices such as these should be used in policy analysis to help determine the most efficient allocation of scarce economic resources.

The Discount Rate

Whatever the methodology used to weigh the benefits of expected future energy savings against the cost of the extra initial investment in energy efficiency improvements, some measure of the time value of money, i.e., discount rate, is needed to permit quantitative economic evaluations.

Box 2.2: Alternative Economic Figures of Merit

The Cost of Saved Energy: The cost of saved energy (CSE) is an index which permits a ready comparison between investments in energy efficient and energy supply alternatives. Simply stated, the CSE is the annual repayment cost for a hypothetical loan taken out to pay for the investment in energy-efficiency improvement, divided by the expected annual energy savings. The decision rule that is used with the CSE concept is that investments ought to be pursued in order of increasing CSE up to the point where the CSE is equal to the cost of the energy supply alternative being considered. An attraction of the CSE concept is that it is an index which can be specified independently of the price of energy. The index does require the specification of an effective cost of capital, i.e., an interest or discount rate, a measure of the time value of money, however.

The Internal Rate of Return: An alternative approach to the economics of energy efficiency is to treat the alternative options as investment opportunities and try to get the consumer to make economic decisions the way corporate investors do: by calculating the internal rate of return (IRR) associated with the extra investment required to improve energy efficiency. Simply stated, the IRR is the real rate of return realized from the dollar value of the energy savings resulting from an energy-efficient investment. The decision rule relating to the IRR is that investments are cost-justified if the IRR is in excess of some threshold "hurdle rate," and those alternatives with the greatest IRR in excess of this threshold would be favoured. An attraction of the IRR concept is that a discount rate does not have to be specified, but expected future energy prices must be.

The Lifecycle Cost: The Lifecycle (LCC) is the total discounted present value of all future costs associated with providing a particular energy service over a specified period. The annualized lifecycle cost (ALC) is the total annualized cost associated with the provision of this energy service: the operating cost (mainly fuel) plus the annualized cost of the initial investment, as defined for the CSE concept. While the LCC (or ALC) concept requires the specification of both a discount rate and the energy price, a comparison of the LCC (or ALC) for different energy systems provides a simple means of identifying the least costly energy strategy.

Unfortunately, there is no "correct" discount rate that can be derived from economic theory (Box 2.3: Alternative Approaches to the Discount Rate).

One way to determine the discount rate is to look at the actual discount rates implicit in decisions relating to investments in energy efficiency. Casual observations suggest that such discount rates are often very high—especially for investments made by individual consumers. Consumers often think in terms of "simple payback" and tend to make investments in energy efficiency when the payback is less than 2 to 4 years, corresponding to discount rates of about 50 and 25 percent respectively. Empirical studies of the discount rates implicit in consumer purchases of energy-using equipment support such casual observations and indicate typical discount rates of 30 to 100 percent and even higher.[25-27] A consumer discount rate of 50 percent implies a dollar saved ten years from now is not worth 2 cents today (only 1.7 cents).

---

* Principal plus interest charges for a loan of term equal to the expected lifetime of the investment.

Box 2.3: Alternative Approaches to the Discount Rate

One approach to the discount rate for conservation investments would be to adopt the "textbook" or flexible approach to capital budgeting for business. With this approach, a wide range of possible projects that might be undertaken are examined, and investments are made in those that offer real internal rates of return in excess of the corporation's real cost of capital—typically 10 to 15 percent per year for projects that are financed with equity capital.

If capital resources are especially scarce, however, "hurdle rates" required to justify investments may be higher than this. When capital is so scarce that it must be rationed, available capital funds are often allocated to projects in descending order of IRR; if the total capital cost for all projects exceeds the available funds then some projects will not be funded and the "hurdle rate" becomes the IRR for the last project funded.

Much lower discount rates are often typical of electric utilities. For privately owned utilities a large fraction of the capital requirements are financed with debt (borrowed money), which is less costly than equity (capital raised by issuing new stock). In the U.S., the Electric Power Research Institute (EPRI) has recommended that a 4 percent real discount rate be used for the evaluation of projects by privately owned utilities, a value which EPRI believes reflects the long-term trend for the average cost of capital to utilities.[23] For publicly owned utilities, which use no equity financing, the cost of capital would tend to be less than for privately owned utilities.

If the individual consumers were to adopt the "textbook" approach to capital budgeting for business, their cost of capital would be the appropriate discount rate. If the consumer has a good enough credit rating to borrow, the effective cost of capital would depend on the source. Examples of relatively low-cost sources of capital in the U.S. include home mortgages and credit union loans, for which real interest rates tend to be in the 4 to 10 percent range. Credit card borrowing, characterized by a real interest rate of 12-14 percent, is an example of a relatively high-cost source of capital.

Still another way to establish an appropriate discount rate is to set it equal to the consumer's "opportunity cost" of capital—i.e., the rate of return that would be realized with the best alternative investment opportunity available.

Among alternative investment opportunities available to the consumer in the U.S., relatively secure investments such as savings certificates and money market funds typically result in real rates of return in the range of 2 to 5 percent per year or less. For riskier investments, the return would be somewhat higher. The real after-corporate-tax rate of return for stockholders' equity for non-financial corporations in the United States was 6.5 percent for the 25-year period 1955 to 1979.[25]

Investments in energy-productivity improvements are not strictly comparable to ordinary financial investment opportunities available to consumers, however. One factor that would lead to assigning a higher discount rate to home investments in energy-productivity improvement is the fact that once the investment is made, the principal cannot be withdrawn for other purposes—making the investment opportunity less flexible. Offsetting this disadvantage, however, is the advantage, in countries where the interest on financial investments is subject to income-tax, that the returns on investments in energy-productivity improvements are tax-free energy savings.

Such short paybacks and high discount rates are not, however, appropriate indices for measuring cost-effectiveness for society as a whole. One major objective of public policy should be to promote economic efficiency so that society can extract the most benefits out of its scarce resources. But high discount rates, if applied to efficiency investments, undervalue "saved energy" and encourage continued inefficient use of scarce energy resources.

For our analysis, we use discount rates of 5 to 10 percent.* Five is close to what typical consumers might realize in alternative investments, to the average cost of capital for private utility investments, and to the discount rate (6 percent) recommended by the Swedish Energy Commission for evaluating alternative energy technologies. Ten percent is typical of moderately priced capital sources available to middle-income consumers and is often used as the hurdle rate in assessing long-term projects in the private sector when capital resources do not have to be rationed.

The Limitations of Energy Cost/Benefit Analysis

While economic assessments along the lines indicated above can be very useful in helping decide the most efficient allocation of economic resources between energy supply and energy-productivity improvement, there are limitations to relying exclusively on such analyses.

First, future energy prices are quite unpredictable. The world oil price is especially volatile and can change rapidly in response to political turmoil in the Middle East. And, besides exogenous factors like this one, future prices can also be affected by energy policies. Where prices are determined by marginal costs which are rising, major investments in energy-productivity improvement can lead to lower future energy prices. For example, in a country like Sweden, where the electricity supply system has been based primarily on low-cost hydro-electricity but future hydro supplies are limited, investments in electricity-productivity improvements could dampen electricity price rises by obviating the need to build new, more costly thermal electric power plants.

In addition, there are many external social costs that are not reflected in market-determined prices. "Social premiums" might be added to market-determined prices to reflect these costs, but determining the appropriate magnitude of these premiums is extremely difficult. For example, how much of a social premium should be added to the world market price for oil to account for the increased risk of nuclear war arising from dependency on insecure Middle East oil sources? Or, how much should fossil fuels be taxed to steer the world away from a course leading to global climatic change arising from the atmospheric build-up of CO_2? Attempting to answer such questions is complicated by the fact that some of the external costs associated with conventional energy production modes (fossil fuel-based or nuclear) are inherently unquantifiable.

Still another practical problem is that many of the most attractive investments in energy-productivity improvement tend to serve multiple purposes, and it is often difficult to quantify these benefits. For example, homeowners save energy by having their houses well-insulated, but an added benefit is that such houses are less "drafty," and the people who live in them less subject to cold-aggravated illnesses. Introduction of the direct-injection stratified charge engine for motor vehicles would allow wide flexibility in the choice of fuels as well as high fuel economy. And, as already noted (Section 2.3), major innovations in industry tend to be those that simultaneously improve many or all factors of production.

---

* The discount rates used in this book are "real", i.e., corrected for general inflation. Thus a 12 percent discount rate in current dollars at a time of 7 percent inflation corresponds to a 5 percent real discount rate.

Despite these limitations, it is nevertheless usually worthwhile to carry out cost/benefit analyses along the lines described above. As we will show by example, the economics of investing in energy-productivity improvements are distinctly favourable in a large number of cases. Considerations of externalities and distributed costs and benefits should be taken into account, though, in those important cases where the results of the cost/benefit analysis are more ambiguous, e.g., where there is a "broad minimum" in the lifecycle cost curve, as illustrated in Figure 2.10.

2.5.3 *Opportunities for Energy-Productivity Improvement by Sector*

In the following subsections, the technological and economic aspects of opportunities for major improvements in energy productivity are discussed for residential buildings, commercial buildings, transportation, and industry.

2.5.3.1 *Residential Buildings*
Space Heating

Space heating accounts for between 60 and 80 percent of final energy use in residential buildings in industrialized countries (Note 2.4-A). Fortunately, the prospects for improving the productivity of energy use here are good. As a result, the energy requirements for space heating can be reduced· from a major to a minor fraction of household energy use.

There are two complementary ways to improve the energy productivity of space heating: by modifying the building shell so as to reduce heat losses and make optimal use of the available sunshine;* and by improving the heating system.

Shell modifications in new houses: Heat losses can be reduced by increasing the insulation, by adding more glazing (extra panes of glass) to the windows, and by reducing natural air infiltration. Good indoor air quality can be maintained in a "tight" house by substituting forced ventilation with heat recovery for natural ventilation; a heat exchanger can be used to transfer heat from the stale exhaust air to the incoming fresh air or to other useful purposes.

A good measure of the energy performance of a house is the energy output required of the heating system, adjusted for floor area and climate variations (Note 2.5.3.1-A). Table 2-3 lists the energy performance of various groups of new houses in the United States and Sweden that incorporate major energy saving features. This table shows that enormous improvements are possible relative to both the existing housing stock and typical new construction in both countries.

In Sweden, there has been much more experience in building thermally tight houses than elsewhere.[28] For example, the energy performance (adjusted for climate and floor area) of houses built to conform to the 1975 Swedish building standard would be comparable to that of the Minnesota and Oregon houses shown in Table 2-3—houses which have energy performances far better than is typical for new construction in the United States. Houses that perform considerably better than these standards are being built routinely today in Sweden, as indicated by the Skane examples in Table 2-3.

Furthermore, still better energy performance is now being realized by builders of so-called "super-insulated" houses. "Super-insulated" housing designs such as the Northern Energy Home (NEH) being sold in New England in the United States and the Wolgast house in Sweden are

---

*. We concentrate here on single family dwellings, for which the most data on energy performance are available. Potential improvements for multi-family dwellings are comparable to or even greater than for single family units. In principle, space heating requirements shoule be less for multi-family units, owing to the lower ratio of building shell area to enclosed space.

characterized by: low air infiltration rates plus mechanical ventilation used with devices to recover heat from the warm exhaust air; triple or even quadruple glazing; indoor shutters to reduce window heat losses at night; and more insulation than is used in conventional construction (Table 2.3). With super-insulated designs such as the NEH, space heating requirements could be reduced to very low levels in moderate climates. For example, an NEH with a floor space of 120 square metres located near New York City (having an average United States climate) could be heated with electric-resistive heaters with just 1400 kWh per year, roughly the electricity requirements of a typical new refrigerator/freezer (Table 2.4). In slightly warmer climates, such houses could be nearly entirely heated by the heat given off by appliances, body heat from its inhabitants, and sunshine coming in the south-facing windows.

Table 2.3. Space Heat Requirements in Single-Family Dwellings
(Kilojoules per square metre per degree day)[a]

United States	
Average, housing stock[b]	160
New (1980) construction in U.S.[c]	100
Mean measured value for 97 houses in Minnesota's Energy-Efficient Housing Demonstration Program[b]	51
Mean measured value for 9 houses built in Eugene, Oregon, U.S.[d]	48
Calculated value for a Northern Energy Home, New York Area[a]	15
Sweden	
Average, housing stock[f]	135
Homes built to conform to the 1975 Swedish Building Code[g]	65
Mean measured value for 39 houses built in Skane, Sweden[h]	36
House of Mats Wolgast, in Sweden[i]	18
Calculated value for alternative versions of the prefabricated house sold by Faluhus[i]	83
Version #1	83
Version #2	17

(a)   The required output of the space heating system (i.e., heat losses less internal heat gains less solar gains) per unit floor area per heating degree day. The number of heating degree days (HDD) is a measure of the severity of the winter heating season.

(b)   See R.H. Williams, G.S. Dutt, and H.S. Geller, 1983.

(c)   As reported by the National Association of Home Builders (J.C. Ribot, A.H. Rosenfeld, F. Flouquet, and W. Luhrsen, 1983).

(d)   These are one-storey houses, with an average floor area of 103 sq. metres, 15 sq. metres of double-paned windows, 15 cm (30 cm) of fibreglass insulation in the walls and floor (ceiling), and an average air infiltration rate of 0.25 air changes per hour. The energy performances of the houses were adjusted to standardized conditions: an internal heat load of 1.0 kW and an indoor temperature of 20 degrees C. See J.C. Ribot, A.H. Rosenfeld, F. Flouquet, and W. Luhrsen, 1983.

(e)   The Northern Energy Home (NEH) is a super-insulated home design which is sold in New England. The NEH design is based on modular construction techniques. The house is constructed of factory-built wall and ceiling sections (120 × 240 cm × 23 cm) which are mounted on a post and beam frame. The calculations presented here were carried out by Dan McMillan of the American Council for an Energy-Efficient Economy using the Computerized Instrumented Residential Audit computer program (CIRA) for a house with the following features: 120 sq. metres of floor area; 12% of wall area (14 sq. metres) in windows, with 60% on the south side; triple glazed windows with night shutters; 20 cm of polystyrene insulation in walls, 23 cm in ceiling; 0.15 ACH natural ventilation plus 0.35 ACH forced ventilation plus 70% efficient air-to-air heat exchanger; internal heat load of 0.65 kW, corresponding to the most energy-efficient appliances available in 1982 plus 3.06 occupants on average. The indoor temperature is assumed to be 21 degrees C in the daytime, setback to 18 degrees at night. The New York climate is characterized by 2700 degree days.

(f)   In 1980 the average fuel consumption for space heating, floor area, and number of heating degree days were 98.5 GJ, 120 sq. metres, and 4474 DD respectively, for oil-heated single-family dwellings (Lee Schipper, 1982). Here a 66% average furnace efficiency is assumed.

(g)   A single-storey house with 130 square metres floor area, no basement, electric resistance heat, an indoor temperature of 21 degrees, and 4010 degree days should consume this much for space heating (Lee Schipper, 1982).

(h)   The average for 39 identical, 4 bedroom, semi-detached houses (112 square metres of floor area; 3300 degree days).

(i)   The Wolgast house has 130 sq. metres of heated floor space, 27 and 45 cm of mineral wool insulation in the walls and ceiling respectively, quadruple glazing, low natural ventilation plus forced ventilation via air preheated in ground channels. Heat from the exhaust air is recovered via a heat exchanger. The local climate is characterized by 3800 degree days. See P. Steen, T.B. Johansson, and R. Fredriksson, and E. Bogren, 1981.

(j)   The Faluhus has a floor area of 112 square metres. The more energy-efficient Version #2 (with extra insulation and heat recuperation) costs 3970 SEK (U.S. $516) per square metre compared to 3750 SEK (U.S. $488) per square metre for Version #1.

What are the costs of improved energy performance in new construction? The costs tend to vary quite a bit with the builder—and also with the experience of the builder, but the cost of saved energy is generally less than the price of heating fuel or electricity (Note 2.5.3.1-B).

Although one would expect super-insulated houses to be very expensive, there is a growing body of evidence suggesting that the *net extra cost* of super-insulated houses may not be very large in comparison to the cost of conventional houses—see Note 2.5.3.1-C for the cost of the NEH and Note 2.5.3.1-D for the cost of the Swedish Wolgast house. A particularly good piece of data on the cost of super-insulated houses is provided by the two different versions of prefabricated houses offered in Sweden by Faluhus (Table 2.3). These two versions are identical except for their energy performance characteristics. While the more energy efficient version is one of the most energy efficient houses available, the associated cost of saved energy is still less than the present Swedish price of electricity, even though present Swedish electrical rates are exceedingly low and far below marginal costs (Note 2.5.3.1-E).

Heating systems for new houses: The remarkable achievements in reducing heat loss through super-insulated house designs has been matched in recent years by spectacular improvements in the design of furnaces and heat pumps.

Several manufacturers in both Europe and North America have recently started marketing high-efficiency natural gas and oil furnaces that use one-third less fuel than conventional furnaces. Even though these furnaces typically cost $600 to $700 more than other new furnaces, they would be cost-effective in the United States as long as the heating requirements are greater than half the average for the existing housing stock (Note 2.5.3.1-F).

For super-insulated houses, however, these furnaces at present prices would be cost-effective only in very cold climates, e.g., Canada, Sweden, or in New England or the upper Midwest of the United States. For super-insulated designs in moderate climates, where space heating is needed only on the very coldest days, a different strategy becomes possible.

With super-insulated designs, it is often possible to offset much of the extra expense of thermal tightening by not having to install a costly heat distribution system and by installing one or more small wall-mounted space heaters instead of a much more costly central furnace. With a super-insulated design, there would be less internal temperature variation with such a heating system than in ordinary houses with conventional heating systems. While, at present, small space heaters are less efficient than centralized condensing furnaces, potentially low-cost space heaters with efficiencies of over 90 percent are in an advanced state of development and may be commercially available by the late 1980's (Note 2.5.3.1-G).

Substantial energy savings are also possible in electrically-heated houses, through the use of heat pumps. Recent development work has resulted in dramatic improvements in the performance of electric heat pumps. The most efficient unit available on the market in the United States in 1982 provided 2.6 units of heat per unit of electricity consumed—i.e., it had a coefficient of performance (COP) of 2.6. This unit requires one-third less electricity than typical heat pumps in use, which have a COP of about 1.7.

Most heat pumps sold in the United States extract heat from outside air. A problem with heat pumps that use outside air as their "heat source" is that in very cold weather they perform no better than resistance heaters. Good performance is possible in cold weather, however, with heat pumps that extract heat instead from other warmer sources.

An ingenious heat pump concept employed in new Swedish housing involves extracting heat from the warm exhaust air in air-tight, mechanically ventilated houses and using it mainly to preheat domestic hot water. A very high COP of 3 is achieved with such systems, which are cost-effective even in well-insulated houses in Sweden (Note 2.5.3.1-H).

The economics of heat pumps for space heating become less favourable for super-insulated houses located in moderate climates. Indeed, an NEH house (Table 2.3) located in the New York City area would have an annual space heating cost of jut $120, if electric-resistive heaters were used and electricity were priced at marginal cost ($0.085 per kWh). It would be hard to justify the extra expense of a heat pump on the basis of the expected space heating savings when the heating requirements are so low—about 1/10 of the average for United States houses. But even in such instances heat pumps may make economic sense if air conditioning is also needed, because a heat pump can operate "in reverse" as an air conditioner in summer.

Another clever Swedish heat pump concept is the "grey water" heat pump—a large centralized device (10 to 20 MW) which uses sewage water as the heat source and provides hot water @ 70 - 80 degrees C for a district heating system. With the economies of scale provided by centralized operation, this system provides exceedingly low-cost heat (Note 2.5.3.1.-I).

Shell modifications for existing houses: Improving the thermal performance of existing houses is crucial since houses already built will dominate the housing stocks of industrialized countries for many decades—both because of the long life of existing structures and because of the expected slow net growth of the housing stock. Thermal design improvements are generally more costly for existing than for new houses due to the fact that once built, many house features cannot be readily changed.

Nevertheless, there are a number of cost-effective conservation measures widely available to homeowners today. These familiar options include the addition of insulation, storm windows, clock thermostats, and a number of "retrofits" to the heating system.

There is now good evidence, however, that the full potential for energy conservation is not realized when only conventional measures such as these are performed. Detailed measurements in the late 1970's revealed that houses have obscure defects in their thermal envelopes, leading to heat losses far in excess of what would be predicted by traditional heat-loss models (Note 2.5.3.1-J).

Fortunately, though, instrumented analysis procedures developed over the last few years permit these defects to be located quickly.[29] The instruments include a house pressurization device which increases air leaks, permitting easy detection and quantification of air infiltration heat losses, and an infrared viewer which can be used to find irregularities in the thermal envelope. For instance, an infrared inspection of an insulated attic floor while the house is slightly pressurized can reveal air leaks and other defects in only a few minutes. Without the diagnostic equipment, the defects would have to be located by lifting up the insulation—a process that takes hours and yet may miss important defects.

Audits of houses to ascertain needed thermal improvements based on the use of such instrumentation are expensive relative to walk-through, non-instrumented, "pencil and paper" energy audits which are offered by many electric and gas utilities in the United States. However, if some of the obscure defects discovered with the aid of the instruments are corrected on the spot, substantial energy savings can be realized even in houses which already have attic and wall insulation and storm windows. Many of the obscure defects can be corrected with small quantities of low-cost materials, so that an expensive instrumented energy audit can be transformed into a cost-effective "housedoctor" visit, where a pair of technicians spend a few hours in a house, diagnosing its defects and fixing many of them on the spot.[30]

A test of the house-doctor concept was carried out in the Modular Retrofit Experiment (MRE) by gas utilities in New Jersey and New York in collaboration with the Buildings Energy Research Group (BERG) at Princeton University.[31] In the MRE, the one-day, two-person house-doctor visit saved, on average, 19 percent of the gas use associated with space heating. Subsequent, more conventional shell modification retrofits brought the total fuel savings to an average of 30 percent, for an average total investment of about $1300; the associated real internal rate of return in fuel savings was nearly 20 percent (Note 2.5.3.1-K).

The achievements commercially demonstrated in the MRE do not represent the limit of what can be realized with shell improvements of existing dwellings. One important experiment exploiting unconventional retrofit opportunities resulted in an energy savings of two-thirds in a U.S. house which prior to being modified was regarded as "thermally tight" by U.S. standards (Note 2.5.3.1-L).

Heating systems for existing houses: As in the case of new houses, the energy savings that are possible with shell modifications can be complemented by improvements in the efficiency of the heating system. The economics of efficient furnaces or heat pumps in the "retrofit market" tend to be much better than in the new housing market, however, because of the relatively large heat loads involved. Even after major shell improvements the residual heat loads are much larger than they are in new super-insulated houses.

For example, if new condensing gas furnaces were introduced in the MRE houses *after* the shell improvements were made, space heating fuel use could be reduced from 70 to 44 percent for an extra investment of $1000, corresponding to a real rate of return of 15 percent (Note 2.5.3.1-K).

Other Residential End-Uses

Besides space heating, other important end-uses in the residential sector for which major energy savings are possible include water heating—which accounts for between 10 and 35 percent of residential energy use in nine large industrialized countries (Note 2.4-A); refrigeration—which accounts in the United States and Sweden for over ¼ of residential electricity use; and lighting—which typically accounts for about half as much electricity as refrigeration.

Recently, gas-fired water heaters which use only half as much fuel as conventional units have become available (Note 2.5.3.1-M); so have heat-pump water heaters which use only one-third as much electricity as conventional electric-resistive units (Note 2.5.3.1-N).

Major innovations in refrigeration technology have led to the recent introduction of refrigerators, refrigerator/freezers, and freezers that use far less electricity than units now in wide use (Note 2.5.3.1-O). In the case of refrigerator/freezers, for example, the most energy-efficient units on the market in Europe, Japan, and the United States require annually only 1.3 to 1.6 kWh per litre of cooled volume, compared to 3.5 kWh per litre for the average units (450 to 500 litres) in wide use in the United States, and far more efficient units are under development (Table 2.4).

The cost of saved energy for presently available more efficient units is typically far less than the price of electricity (Note 2.5.3.1-P), and the first costs of the more efficient units may go down as more efficient refrigeration technology becomes commonplace. As in the case of new houses, important synergisms may work to keep the first cost down, e.g., investments to reduce heat losses may be offset by the reduced costs of lower-capacity motors and compressors. In fact, a survey of new refrigerators and freezers in Sweden indicates little correlation between energy performance and first cost (Figure 2.22).

Table 2.4. Energy Performance of Typical and Energy-Efficient Refrigerators and Refrigerator/Freezers

Brand (Origin)	Model	Type[a]	Capacity (litres)	Specific Electricity Use[b] (kWh per litre per year)	Electricity Use[b] (kWh per year)
Average, 45 US Models	(1983)	SD	363	1.94	703
Hitachi (Japan)	617A	SD	169	1.36	230
Gram (Europe)	K215	SD	215	1.26	270
Kenmore (US)	564.86111	SD	311	1.19	370
National (Japan)	211	SD	207	0.99	205
Laden (Europe)	40.830	SD	305	0.95	290
Bosch (Europe)	KS2680SR	SD	255	0.86	220
Gram (Europe)	K395	SD	395	0.80	315
Gram (Europe)	prototype[c]	SD	200	0.52	104
Average, 488 US Models	(1983)	TD	518	2.46	1275
Bosch (Europe)	KS3180ZL	TD	310	1.77	550
Amana (US)	TSC18E	TD	510	1.71	870
National (Japan)	291(HV/T)	TD	290	1.65	480
Amana (US)	ESR14E	TD	402	1.58	635
Electrolux (Europe)	TR1120C	TD	315	1.51	475
Amana (US)	prototype	TD	510	1.43	730
Kelvinator (US)	prototype	TD	510	1.39	710
Toshiba (Japan)	GR411	TD	411	1.31	540
Amana/Kelvinator	conceptual[d]	TD	510	1.14	580
Schlussler (US)	prototype[e]	TD	368	0.68	252

(a)   SD = single door; TD = two door.

(b)   Consumption values for different countries may not be directly comparable, as they are based on standardized tests which may vary from country to country.

(c)   Refrigerator with automatic defrost but no freezer. This prototype was designed and analyzed at the Physics Laboratory III, Technical University of Denmark, in cooperation with the refrigerator manufacturer Bdr. Gram A/S, Vojens, Denmark, and the compressor and control systems manufacturer Danfoss A/S, Norborg, Denmark (Jorgen S. Norgard *et al.*, April 12, 1983).

(d)   This is the estimated consumption that would result from combining the efficient compressor utilized in the Kelvinator prototype with the other energy-saving features of the Amana prototype (Howard S. Geller, 1983).

(e)   A horizontal refrigerator/freezer designed by Larry Schlussler at the University of California at Santa Barbara (Larry Schlussler, June 1978).

The wide use of energy-efficient refrigeration technology would have a major impact on electricity supply planning. For example, if the existing stock of refrigerators and freezers in the United States were replaced by the most efficient models commercially available in 1982,

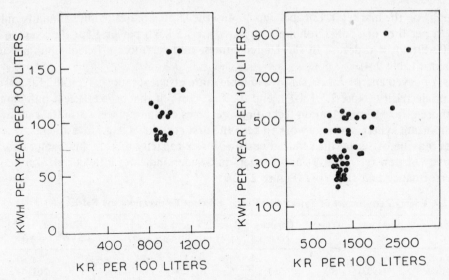

Figure 2.22. Specific electricity requirement (kWh per litre of inner volume for new refrigerators (left) and freezers (right) sold in Sweden, as a function of the initial sales price per unit volume.
Source: T.B. Johansson, P. Steen, *et al.*, 1985.

the savings would be equivalent to the output of 18 Gw(e) of baseload electrical generating capacity—i.e., about 18 large coal or nuclear power plants.

The incandescent bulb has dominated residential lighting since it was first introduced in 1879, despite the five-fold efficiency improvement that is possible with fluorescent bulbs. The more efficient fluorescent lighting has not caught on largely because most people feel that the soft yellow light of the incandescent is preferable to the harsher white light of fluorescents for the living environment of the home.

Over the last several years, though, a great deal of effort has gone into the development of high-efficiency lightbulbs that could replace incandescents. At this time several candidate bulbs are emerging (Note 2.5.3.1-Q). Not only are these bulbs 3 to 5 times as efficient as incandescents, they also some have the desired colour characteristics of the incandescent, they are compact in size like the incandescent, they can be screwed into incandescent sockets, and they last several times as long. While they cost far more than incandescents, these more efficient bulbs are cost-effective in many circumstances (Notes 2.5.3.1-R).

Conclusion

For the major energy-using activities of the home, there are substantial opportunities to improve energy productivity cost-effectively. An indication of the overall potential for energy savings is presented in Table 2.5. This table shows per capita residential energy use, first for average households in the United States and Sweden at present and then for hypothetical all-electric households having all major energy-using amenities, based on the most efficient technologies commercially available in 1982. While these hypothetical households enjoy a much higher level of amenities than average households today, they would use only about 300 Watts per capita, which is not only much less than present levels of household energy use in the United States and Sweden but also far less than the 1100 to 1400 Watts per capita used in most Northern European countries at present to support much lower levels of amenities (Note 2.4-A). Energy use could be reduced even further with advanced technologies still under development.

2.5.3.2 *Commercial Buildings*

Commercial buildings comprise a heterogeneous mix of office buildings, retail stores, warehouses, schools, hotels, theatres, churches, etc.

A remarkable characteristic of the commercial building sector is its high energy intensity. In the United States the average energy intensity [measured in Gigajoules (GJ) of final energy per square metre of floor area] is almost 50 percent higher than it is for the residential sector. Moreover, for the three activities which account for about 90 percent of United States commercial energy use (Note 2.5.3.2-A):

- Space heating is on average over 50 percent more energy-intensive than in the residential sector;.
- Air conditioning is nearly 550 percent more energy-intensive; and
- Lighting is nearly 580 percent more energy-intensive.

These numbers are particularly surprising when it comes to space heating and lighting. One would think that the space heating energy requirements of commercial buildings would be much lower than for residential buildings, owing to the much lower ratio of "building skin" to enclosed space; for commercial buildings heat losses through the building walls are relatively minor, whereas they are critical for small, single-family residential buildings. The much more intensive use of electricity for lighting is puzzling when one considers that lights are used mainly in daytime in many commercial buildings but at night in houses, and that commercial buildings use primarily fluorescent lights, which are about five times more energy-efficient than the incandescent lights which predominate in home lighting.

*Space Conditioning in Existing Buildings**

An insight into the puzzle of the intensity of energy use of commercial buildings in the United States can be gleaned from looking at the details of energy use of a hypothetical Kansas City office building typical of office buildings constructed in the United States before the oil crisis of 1973.** Located in a typical United States climate, this building (Note 2.5.3.2-B) has three storeys with the perimeter of each floor lined by individual offices and the remaining "core" area undivided (Figure 2.23).

While the energy requirements per square metre of floor space for this building are about twice the average for the U.S. commercial building stock in 1980, there are many commercial buildings in the United States with comparable or worse energy performance (Figure 2.4).

Figure 2.25 shows the energy performance of the Kansas City building for heating and cooling by month of the year. Remarkably, the building "requires" air conditioning even in the dead of winter and space heating even in mid-summer.

The first piece of this puzzle has to do with the fact that commercial buildings have a tendency to become overheated by internal heat sources (from people, lights, and other equipment). Because of the small ratio of building shell to building volume, there is little opportunity for this heat

---

* This section is based largely on reference 33.

** Office buildings are emphasized here in part because of their relative importance—they accounted for 1/6 of all commercial building space in the US in 1980. Also, they have been given more analytical attention than other commercial building types. In addition, they are characterized by a total energy intensity and energy intensities for space heating, air conditioning, and lighting which are very close to the estimated "average" energy intensities for all commercial buildings.[34]

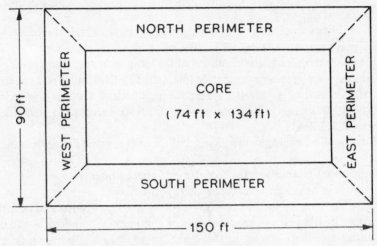

Figure 2.23. Layout of the hypothetical U.S. office building in Kansas City discussed in the text. *Source:* W.S. Johnson and F.E. Pierce, 1980.

Table 2.5. Final Energy Use in the Residential Sector (Watts per capita)

End Use	Average Households at Present		All Elect., 4-Person Hshlds. with the Most Efficient Technology Available in '82/83[d]	
	U.S. '80[a]	Sweden '78/82[b,c]	U.S.	Sweden[b]
Space Heat	890	900	60[e]	65[f]
Air Cond.	46	—	65[g]	—
Hot Water	280	180	43[h]	110[i]
Refrigerator	79	17	25	8
Freezer	23	26	21	17
Stove	62	26	21	16
Lighting	41	30	18[j]	9[j]
Other	80	63	75	41
Total	1501	1242	328	266

(a)   The total consists of 360 W of electricity and 1140 W of fuel.

(b)   For details, see T.B. Johansson *et al.*, 1983b.

(c)   This total consists of 350 W of electricity and 890 W of fuel. 50% of the electricity is for appliances and 50% is for heating purposes.

(d)   With 100% saturation for the indicated appliances, plus dishwasher, clothes washer and clothes dryer.

(e)   For an average-sized, detached, single-family house (150m² of floor space); ave. US climate (2600 degree C days); a net heating requirement of 50 KJ/m²/DD(Table 2.3); a heat pump with a seasonal average COP = 2.6 (the highest efficiency for new air-to-air units)

(f)   For a Faluhus (Table 2-3) in a Stockholm climate (3810 degree C days). This house uses a heat exchanger to transfer heat from the exhaust air stream to the incoming fresh air.

(g)   For the average cooling load in air-conditioned US houses (27 GJ per year) and a COP = 3.3 (the COP on the cooling cycle for the most efficient heat pump available in 1982).

(h)   For 59 l/capita/day of hot water (@ 49°C) (or 910 k Wh/year/capita) and the most efficient (COP = 2.2) HP water heater available, 1982.

(i)   For 1000 kWh/year/capita hot water energy use via resistive heat. Ambient air-to-water heat pumps are not competitive at the low Swedish electricity prices.

(j)   Savings achieved by replacing incandescents with compact fluorescents.

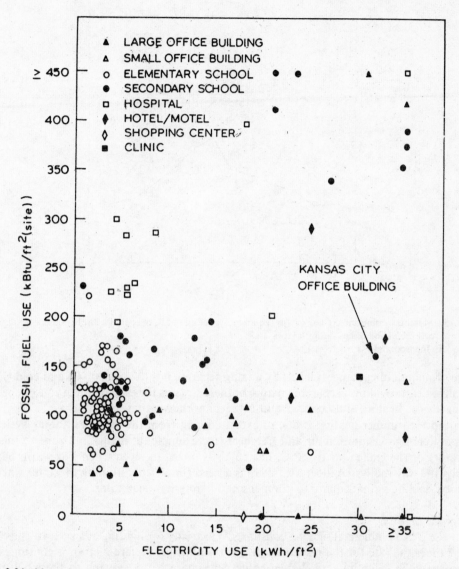

Figure 2.24. Metered energy performance data for 223 commercial buildings, mainly in the United States, together with a data point showing the energy performance of the hypothetical Kansas City office building discussed in the text.
Source: Howard Ross and Sue Whalen, 1983.

to be conducted to the out-of-doors through the building shell; hence this excess heat must be carried out of the building by the ventilation system.

The second piece of the puzzle has to do with the fact that, unlike the situation is single-family residences, where the indoor air mixes well and evens out the temperature differences in different parts of the building, the different zones of commercial buildings are subject to different heat gains and losses and thus have quite different heating and cooling requirements. For the prototypical building examined here, this problem is dealt with by following the widespread

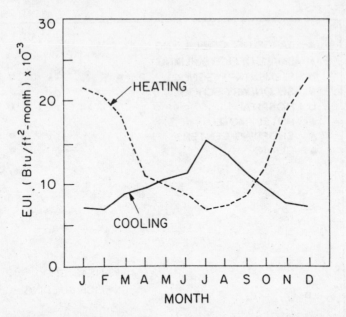

Figure 2.25. Monthly heating and cooling energy requirements per unit of floor space for the hypothetical Kansas City office building discussed in the text.
Source: W.S. Johnson and F.E. Pierce, 1980.

practice of maintaining comfort levels by mixing separate hot (32° to 50° C) and cold (13° C) air streams, i.e., providing a separate mixture of these streams for each building zone, depending on its needs for heating and cooling, and of keeping the ventilation rate fixed.

The need for summer heating is due to excess cooling from the air-conditioning system. If there is not enough ventilation air and free heat (for example, at night) in a zone to raise the temperature of the chilled air to 22°C, heated air is mixed in, even when it is warm outside! Similarly, the way the system operates, there is a need for air conditioning in winter whenever ventilation cooling is inadequate to compensate for internal heat gains.

*Commercial Lighting**

While the U.S. commercial sector "consumes" electricity for lighting at a rate which is about almost 7 times the rate for the residential sector (in kWh per square metre), it consumes over 30 times as much "lighting" (in lumen-hours per square metre), owing to the much higher efficiency of fluorescent bulbs. How can one account for such a high lighting consumption level? Essentially, three factors are involved: an emphasis on uniform lighting (instead of task lighting); relatively low efficiencies of light transmission to working surfaces; and generally high lighting standards.

In modern homes, people generally feel that lighting contrasts are pleasing—so that rooms are often lighted non-uniformly. But in the workplace the emphasis has been on uniform lighting. Although a desk worker can get the same amount of illumination on his or her work from an 18-Watt fluorescent bulb on the desk (equivalent in output to a 60-Watt incandescent) as from

---

* This section is based largely on reference 33.

160 Watts of bulbs in the ceiling, designers of commercial buildings have not been much interested in "task lighting."

In addition, not all the light emitted by fluorescent bulbs is useful light. Bulbs emit light in all directions. Much of the light therefore must be reflected at least once, with attendant losses, before it reaches working surfaces. Fluorescent bulbs are also ordinarily enclosed in "diffusers," with the result that even light initially emitted in a useful direction must penetrate a translucent cover or be reflected from a louver grid—again with attendant losses—before it reaches a working surface. As a result, typically only 20-50 percent of the light emitted by the fluorescent bulbs actually reaches working surfaces.

And finally, there is a good deal of evidence that existing illumination levels are much higher than needed, suggesting that a fundamental reappraisal of lighting standards is in order (Note 2.5.3.2-C).

New Commercial Construction

Commercial buildings being built today are far better energy performers than the typical pre-1973 vintage office building described earlier. In fact, newly constructed buildings in both the United States and Sweden are only about half as energy-intensive on average as the existing commercial building stock (Table 2.6).

But there are still major opportunities for further improvement. An exercise conducted in the United States in the late 1970's suggests that, given a quick course in energy conservation techniques, ordinary teams of United States architects and engineers can design buildings that have energy requirements which are much lower than even typical new buildings.[35] The purpose of the exercise was to lay the basis for the federal Building Energy Performance Standards being considered at that time. A sample of 1661 commercial buildings was chosen as representative of U.S. design practices a few years after the 1973 oil embargo. For a random sample of 168 of these buildings, the original design teams were contacted and asked to redesign the same building with improvements in energy efficiency, without violating any of the requirements of their original clients and without changing the overall first costs of the buildings significantly. Only "off the shelf" technologies could be used. Since few of the original design teams were expert in energy efficiency, the Research Corporation of the American Institute of Architects (AIA/RC) assisted them by providing a three-day training session and a workbook with practical design ideas and simplified analytical tools. The results of this AIA/RC "redesign" effort are summarized in Table 2.6: the average energy intensity for the redesigned buildings was about 3/5 as large as that of typical new buildings. This improved performance was achieved at very little additional cost and is far from the economic optimum.

Subsequently, the AIA/RC chose three buildings for a much more detailed analysis and selected architectural teams noted for their expertise in designing energy-efficient buildings.[35] These teams were asked to redesign the original buildings in such a way as to minimize the life-cycle cost (LCC)—specifically to save energy to the extent of giving the building owner at least a 10 percent (real) rate of return on his investment. The redesigns involved careful lighting techniques, use of state-of-the-art control technology, and hydronic heat pumps (or the equivalent), and resulted in buildings requiring only about 40 percent as much energy as typical new buildings (Table 2.6).

Perhaps the most energy-efficient commercial building constructed in the late 1970's is the Folksam Building in Farsta, Sweden, near Stockholm. The Folksam Building is a 7-floor office building (75m × 17m) whose perimeter is lined by offices (2.4m × 4m) surrounding an

Table 2.6. Site Energy-Intensity Factors for Commercial Buildings

	(Gigajoules/m^2 per year)		
	Fuel	Electricity	Total
United States			
Average 1979 Building Stock[a]	0.82	0.49	1.31
Current U.S. Practice[b]	0.16	0.57	0.73
AIA/RC Redesigns[b]	0.07	0.40	0.47
AIA Lifecycle Cost Minimum Designs[b]	0.04	0.28	0.32
Enerplex South, Princeton, NJ[c]	—	0.31	0.31
Sweden			
Average 1982 Building Stock[d]	0.66	0.38	1.04
Swedish Norm for New Construction[b]	0.57	0.19	0.76
Folksam Building, Farsta[e]	0.07	0.39	0.46
Harnosand Building, Harnosand[f]	0.12	0.13	0.25

(a)   For an average of 2700 heating degree C days (Energy Information Administration, December 1983).
(b)   Table 1.12 (p. 39) and Figure 1.61 (p. 165) in Solar Energy Research Institute, 1981.
(c)   For 2700 heating degree C days. Calculated, not measured values (L.K. Norford, June 1984).
(d)   Consumption corrected to normal weather (4010 heating degree C days) (Lars-Goran Carlsson, March 1984).
(e)   Measured values for the representative period December 1978 to December 1979 (3810 heating degree C days). "Fuel consumption" is the energy actually delivered by the district heating system (K. Welmer, 1981).
(f)   Measured values for 4600 heating degree C days (K-Konsult, February 1984).

unoccupied core area. Despite the cold climate (3810 Degree Days), the offices in this building still get overheated in the daytime, even on cold winter days. With an ordinary design, the building would require cooling in the daytime and heating at night. The Folksam Building designers dealt with this problem instead by storing the excess heat produced in the daytime for use at night (or for heating up the building in the morning). Storage is accomplished in this building by means of the "Thermodeck" concept, which involves passing the office ventilation air through long tubular cores in the massive but hollow concrete floor slabs on its way to the offices. As a result of this storage scheme, the air temperature rise in the offices during the day is only about 2 degrees, so that cooling is unnecessary. In the summer the day/night temperature difference is exploited—excess heat is stored in the slabs during the day, as in winter, but at night the slabs are cooled with outside air.

At the time of this writing, the most energy-efficient commercial building in Sweden was the Harnosand Building in northern Sweden. Utilizing the Thermodeck principle, the pre-heating of ventilation air with solar panels, and microprocessor controls for a better matching of energy supply and demand, the building uses only about half as much energy as the Folksam Building (Table 2.6).

Existing Commercial Buildings

How much energy savings can be achieved by means of retrofits in existing buildings? In Sweden, improved energy management, involving little or no capital investment, e.g., night setbacks of thermostats, adjustments in ventilation to better match needs, etc., typically leads to savings of 20 to 30 percent. One U.S. study measured the energy savings in commercial buildings which before retrofit had the energy consumption levels as shown in Figure 2.24; after retrofit, the average measured savings in 184 buildings was 23 percent, and the corresponding cost of saved energy for 56 of these buildings where cost data were available was $2.8 per GJ (1982 dollars), assuming a 10-year retrofit life and a 10 percent real discount rate.[36] These savings fall far short of the economic potential, however, because there is probably much more that can be done at an average cost of saved energy less than average price of energy in the commercial sector—which averaged $10 per GJ in the US in 1982 (Note 2.1-D).

One estimate of the overall potential for energy savings by means of retrofits in the U.S. commercial sector comes from a survey of 14 experienced architects and engineers which indicates that a reasonable target for the United States would be a 50 percent reduction by the year 2000,[35] which would cost, on average, about $22 per square metre of floor area, or $5 per GJ of saved energy and is thus clearly economically efficient.

The rate of capturing potential energy savings should *in principle* be faster for retrofits of commercial buildings than for existing residential buildings. This judgment arises from the fact that energy use in the commercial sector is dominated much less by building shell losses than by the performance of the energy-converting equipment in the building—equipment for which the lifetimes are far shorter than the life of the building itself.

2.5.3.3 *Transportation**

In 1979 transportation accounted for 45 percent of all liquid fuel consumption in the OECD nations.[37] Within the transportation sector, passenger transportation accounted for over 70 percent and passenger cars and light trucks alone for over 60 percent of oil consumption (Figure 2.26 and Notes 2.5.3.3-A and 2.5.3.3-B). Trucks and ships accounted for most energy consumed in the transport of freight.

Light Vehicles

The single largest class of consumers of energy in the transportation sector of the OECD nations is that of light vehicles, which include both passenger cars and light trucks, i.e., pickups, vans, etc. Because passenger cars constitute more than 80 percent of all OECD light vehicles (Note 2.5.3.3-C), the potential for passenger car fuel-economy improvements is particularly important. However, since light trucks differ little from passenger cars, either in terms of usage or technology, most of this analysis would apply to them as well.[38]

Technological potential: Passenger cars in North America are both larger and consume more fuel per kilometre of travel than those elsewhere in the OECD. Large cars became less popular in North America, however, after the 1979-80 gasoline price increase. Therefore, a typical European car, the gasoline-engine-powered VW Rabbit (Golf), is taken as a reference OECD passenger car in discussing the potential for technological improvement through improvements in the conversion efficiency of the drive train, reduction in vehicle weight, and reductions in aerodynamic and rolling resistances.

To drive a model year 1981 gasoline-powered VW Rabbit with manual transmission through an EPA combined driving cycle (Note 2.5.3.3-D) requires delivering to the wheels the mechanical energy equivalent of approximately 1.1 litres of gasoline per 100 km (1/100 km) (Note 2.5.3.3-E). The delivery of this much energy to the wheels involves fuel consumption at a rate of about 7.9 1/100 km (30 mpg), corresponding to an average efficiency of only 13.5 percent for the drive train in converting fuel to mechanical energy at the wheels—suggesting a very large potential for improving the efficiency of the drive train.

---

* Franke von Hippel contributed to the writing of this section.

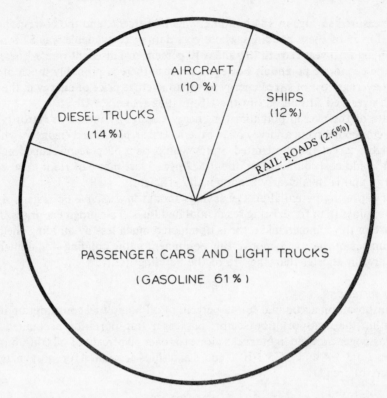

Figure 2.26. The distribution of oil consumption for transportation in the OECD countries in 1979 by end-use activity. Transport oil consumption in the OECD countries totalled 16 million barrels per day in 1979.
Source: Frank von Hippel and Barbara Levi, 1983.

One possibility for improving the efficiency of the drive train is to shift from a gasoline to a diesel engine. This has been done in the diesel version of the Rabbit, which has an energy performance of 5.3 1/100 km (45 mpg)*, and which for several years was the most fuel-efficient car available in the United States.

Since the Volkswagen diesel Rabbit first came on the market, several other high fuel-economy cars have been introduced (Table 2.7). As of 1985 the most fuel-efficient cars available in the U.S. were two gasoline-fuelled models: the Sprint, a 4-passenger vehicle marketed by Chevrolet and manufactured by Suzuki, with an estimated on-the-road fuel economy of 4.11/100 km (57 mpg), and the sporty 2-passenger Honda CRX, with a fuel economy of 4.3 1/100 km (54 mpg).

Impressive as these new models are, they exploit only a fraction of presently available technology that can be applied to the automobile. Most of the improvements already achieved are from weight reduction, reduced rolling resistance from the use of radial tyres, reduced aerodynamic drag, and the use of pre-chamber diesel engines or lean-burn gasoline engines. Much more can be accomplished with present technology by shifting to direct-injection diesel engines (the kind used in trucks) or spark-ignited, direct-injection diesel engines that have multi-fuel

---

* While diesel fuel has a heating value about 11 percent higher than that of gasoline, gasoline production requires more processing energy to refine. The result is that the primary energy required per litre of refined fuel is about the same for the two fuels.

capability, by introducing the continuously variable transmission and the feature of engine-off during idle and coast, by using lightweight materials more extensively, and by further reductions in aerodynamic drag.

Frank von Hippel and Barbara Levi of Princeton University have carried out a computer simulation of the following sequence of fuel-economy changes to give an indication of what can be accomplished with modifications of the VW Rabbit that go beyond the use of a diesel engine (Note 2.5.3.3-F):

Table 2.7. Fuel Economy for Passenger Automobiles[a]

Car	Fuel	Fuel Economy [1/100 km (mpg)]		Maximum power	Curb weight	Passenger capacity
		Urban	Highway	(kW)	(kg)	
Commercial						
1985 VW Golf, Jetta	diesel	5.7 (41)	4.0 (59)	39	1029	5
1985 VW Jetta (Turbocharged)	diesel	5.7 (41)	4.4 (54)	51	1029	5
1985 Honda City Car (Jazz)	gasoline	4.5 (53)[b]	—	50	655	4
1985 Honda CRX	gasoline	4.3 (54)	3.4 (69)	45	779	2
1986 Honda CRX	gasoline	4.1 (58)	3.2 (73)			2
1985 Chevrolet/ Suzuki Sprint	gasoline	4.5 (52)	3.5 (68)	36	676	4
1986 Chevrolet/ Suzuki Sprint	gasoline	3.8 (61)	3.1 (77)			4
1985 Nissan Sentra	diesel	4.7 (50)	3.7 (64)	41	850	5
1985 Ford Escort/ Mercury Linx	diesel	4.9 (48)	3.5 (67)	39	945	5
1985 Toyota Starlet	gasoline	3.9 (61)[b]	—			4
1985 Daihatsu Charade	diesel	4.2 (56)[b]	—			
1985 Peugeot 205	diesel	5.2 (45)[c]	4.3 (55)[c]			5
Prototype[d]						
VW Auto 2000[e]	diesel	3.7 (63)	3.3 (71)	39	780	4-5
VW-E80[f]	diesel	3.2 (74)	2.4 (99)	29/38	700	4
Volvo LCP 2000[g]	multifuel	4.2 (56)	2.9 (81)	39/66	707	2-4
British Leyland ECV-3[h]	gasoline	5.1 (46)	.5 (67)	54	664	4-5
Renault EVE +[i]	diesel	3.7 (63)	2.9 (81)	37	855	4-5
Renault Vesta[j]	gasoline	2.3 (104)	1.9 (124)	24	511	2-4
Peugeot Vera 2[j]	diesel	4.6 (51)	3.5 (67)	46	791	4-5
Peugeot ECO 2000[k]	gasoline	3.0 (78)	2.4 (99)	21	450	4
Toyota Lightweight Compact (1)	diesel	2.9 (80)[c]	—			5-6
Designs						
Volvo LCP (potential)[m]	multifuel	3.2 (74)	2.35 (100)			
Cummins/NASA Lewis Car[n]	multifuel	3.3 (72)	2.44 (96)	52	1364	5-6
Pertran Car[o]	diesel	2.2-2.4	(100-105)	—	545	5-6

(a)   As determined in the US Environmental Protection Agency (EPA) laboratory test. The urban and highway values are often presented for a combined driving cycle involving 55% urban driving and 45% highway driving. To better correlate with actual road performance, a new US rating involves multiplying the urban lab test value (in mpg) by 0.9 and the highway lab test value by 0.78. This correction is important for gasoline engines but not for diesels, for which the lab test is a fairly good measure of on-the-road performance.

(b)   Measurements were made for the Japanese driving cycle, which is an urban driving cycle. To convert this EPA lab test value back to the original Japanese driving cycle value, divide by 1.12.

(c)   Measurements were made for the European driving cycle. To convert the EPA lab test value for the urban cycle back to the original European driving cycle value, divide by 1.12. The EPA lab test value for the highway cycle is equivalent to the European 90 km per hour highway test.

(d)   Prototype data are from Deborah Bleviss, 1985.

(e)   Three-cylinder, direct-injection, turbocharged diesel engine; more interior space than the Rabbit; engine off during idle and coast; use of plastics and aluminium; a drag coefficient of 0.26 (Hans-Wilhelm Grove and Christian Voy, 1985).

(f)   Three-cylinder, direct-injection diesel engine; Polo size; flywheel start/stop. The Higher power version has a supercharger.

(g)   The 39 kW car has a 3-cylinder, direct-injection, turbo-charged, diesel engine. The 66 kW car has a 3-cylinder, heat-insulated, turbo-charged, intercooled diesel engine with multi-fuel capability. Extensive use of aluminium, magnesium, and plastics; a drag coefficient between 0.25 and 0.28 (Rollfe Mellde, 1985a).

(h)   Uses plastic panels.

(i)   Supercharged, direct-injection diesel engine; engine off during idle and coast; low aerodynamic drag.

(j)   Lightweight materials; low aerodynamic drag.

(k)   Two-cylinder gasoline engine, lightweight materials, very low aerodynamic drag.

(l)   Three-cylinder, direct-injection diesel; continuously variable transmission (CVT); lightweight materials; large interior volume.

(m)   The Volvo LCP 2000 plus CVT and engine-off during idle and coast features (Rollfe Mellde, 1985b).

(n)   4-5 passenger; 4-cylinder, direct-injection, spark-assisted, multi-fuel capable, adiabatic diesel with turbo-compounding; CVT; 1984 model Ford Tempo body (R.R. Sekar, R. Kamo, and J.C. Wood, 1984).

(o)   Pre-chamber diesel engine with supercharger; CVT; lightweight materials; flywheel for energy storage in breaking (S.L. Fawcett and J.C. Swain, 1983).

- Reduction of aerodynamic drag and rolling resistance,
- A shift from a pre-chamber to an open chamber (direct injection) diesel,
- The introduction of a continuously variable transmission (CVT) and an associated reduction in peak engine power,
- A (10 percent) reduction in weight,
- An expansion of the CVT range,
- An engine-off feature during coast and idle.

Incorporating all of these features would improve the Rabbit Diesel's fuel economy to 2.6 litres/100 km (89 mpg) (Figure 2.27).[39]

All these innovations are based on proven technology and are represented in prototypes that have already been built (Table 2.7). Several prototype cars have been built with lower aerodynamic drag than was assumed here. The open chamber (direct-injection) diesel is the variety of diesel engine used in trucks; it was incorporated in the VW Auto 2000 and the Volvo LCP 2000 prototypes. The CVT was incorporated in the Toyota Lightweight Compact prototype, and several manufacturers are planning to introduce CVTs in production vehicles. Various manufacturers have produced much lighter prototype 4-passenger cars than the one considered here. And the VW Auto 2000 has incorporated the engine-off feature for idle and coast.

Much more can be done to reduce vehicle weight—through the use of high strength steels, aluminium, plastics and ceramics. For example, researchers at the Aluminum Company of America have shown that the weight of GM's X-Body car can be reduced from 1230 kg to 700 kg by shifting to an aluminium-intensive design,* involving the use of 270 kg of aluminium.[40] Volvo's prototype LCP 2000 is a lightweight (700 kg) car owing to the extensive use of aluminum, magnesium, and plastics (Table 2-7). Japan is expected to make a major leap forward by shifting to lighter weight, nonmetallic cars based on the use of ceramics and plastics in the 1990's.[41] A particularly ambitious weight-reduction target has been established for the PERTRAN car, a design proposed by automotive researchers at the Battelle Memorial Institute in Columbus, Ohio (Table 2.7). Weighing 545 kg, the diesel PERTRAN would have an estimated U.S. EPA Urban Driving Cycle fuel economy of 2.3 litres/100 km (100 mpg).[42]

Still another opportunity for seeking additional fuel-economy gains is through the use of even more efficient engines. Among advanced designs the so-called adiabatic diesel is especially promising (Note 2.5.3.3-G).

There are thus many very promising options, based on both existing technology and new technology expected to be available within a decade, for dramatically decreasing the fuel

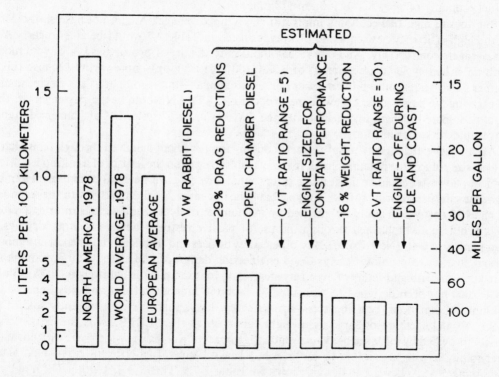

Figure 2.27. The potential for improved automotive fuel-economy.

Taking the VW diesel Rabbit as a base case, the added fuel-economy improvements include aerodynamic drag reduction, a shift from a pre-chamber to an open-chamber diesel engine, a shift to a continuously variable transmission, vehicle weight reduction, and an automatic engine shut-off when power is not required at the wheels.

The fuel-economy values are based on the United States Environmental Protection Agency's composite (55% urban, 45% highway) driving cycle.

Source: Frank von Hippel and Barbara Levi, 1983.

---

* Even though aluminium production is much more energy-intensive than steel production, the resulting fuel savings more than offset the increased manufacturing energy requirements.[38]

consumption of the ordinary automobile—perhaps to 2.3 1/100 km (100 mpg) in the decades immediately ahead.

Trade-offs: A number of trade-offs must be taken into consideration in developing fuel-economy targets for light vehicles. Among these are those relating fuel economy to performance, consumer comfort, safety, and air pollution.

Does fuel-stinginess imply a sluggish vehicle that will make highway entry or passing difficult or dangerous? Not necessarily. A shift from the gasoline to the diesel version of the VW Rabbit does involve a 30 percent reduction in peak engine power. If desired, however, this loss could be regained by super-charging—with an estimated additional fuel-economy benefit of 10 percent but at an estimated additional cost of $250-$300 (Note 2.5.3.3-H). Evidence that a super-efficient car can be a good performer is provided by Volvo LCP 2000. Despite a fuel economy of 3.6 1/100 km (65 mpg), this prototype requires only 11 seconds to accelerate from 0 to 60 mph (96 km/hour), compared to 17 seconds for the popular automatic Chevrolet Cavalier, which has a fuel economy of only 8.4 1/100 km (28 mpg).[43]

But doesn't high fuel economy imply a tiny, cramped vehicle? Again, not necessarily. Most of the high-efficiency new cars and prototypes listed in Table 2.7 would be able to carry 4 or 5 passengers comfortably. And one recent addition to the list of prototypes is a Toyota model with an urban cycle fuel economy* of 2.9 1/100 km (80 mpg)—noteworthy because this is neither a two-seater nor a minicompact nor a subscompact—but a compact-size vehicle capable of carrying 5-6 passengers. Likewise the proposed PERTRAN would be a 5-passenger car that would compare in length and height with the GM X-Car or Chrysler K-Car (current designs), but would be somewhat narrower.

Aren't fuel-efficient cars unsafe? Fuel-economy improvement achieved through reduction in vehicle size reduces the potential crush space available to cushion the impact of a collision. Current vehicles, however, could be made much safer so that, given a governmental commitment to more rigorous safety standards, safety could probably be increased simultaneously with fuel economy at an extra cost which could be justified by the number of deaths and serious injuries averted. This could be accomplished through the use of passive restraint devices such as air bags or automatic seat belts (Note 2.5.3.3-I) or other safety devices and through fundamental automotive design changes to promote safety. Good engineering design can be effective in improving the structural strength and safety of even very small cars; for example, the lightweight (707 kg) Volvo LCP 2000 has been designed to withstand 35 mph (56 km/hour) front and side impacts and 30 mph (48 km/hour) rear impacts—stricter safety levels than for cars currently sold in the US.[43] A safety feature of large lightweight cars such as the PERTRAN is the extra "crush space" provided by the large body to cushion the impact of a collision. Moreover, with advanced technology it will be feasible to build even "heavy" super fuel-efficient cars—such as the Cummins/NASA Lewis car design shown in Table 2-7.[44] This 3.0 1/100 km (79 mpg) car, equipped with a CVT and an advanced, multi-fuel capable, direct-injection, adiabatic diesel engine and turbo-compounding, would weigh 1360 kg, approximately the average weight of new cars being sold in the US today.

What about air pollution? A problem posed by efforts to improve fuel economy by shifting to diesel engines is that diesels emit on average about 100 times the weight of particulates produced by gasoline engines of comparable performance. But again there are technical fixes that can be brought to bear on this problem. Mercedes Benz and Volkswagen have now developed emission

---

* The Japanese fuel-economy test does not involve highway conditions.

control devices that make it possible for their diesel cars to meet the strict 1986 California particulate emissions limit of 0.2 grams per mile.[43] Moreover, the use of diesel fuel is not necessary to the realization of high fuel-economy. Several of the high fuel economy cars listed in Table 2.7 have gasoline engines. Moreover, the high energy-efficiency of diesel-firing can be achieved with spark-assisted, direct-injection diesel engines—engines such as those of the Volvo LCP 2000 and the proposed Cummins/NASA Lewis car, which could be fired with gasoline or alcohol, fuels that would lead to low particulate emissions.

The cost of super-efficient automobiles: To what extent are high fuel-economy design changes cost-effective? von Hippel and Levi made cost estimates of the various improvements they examined. They showed that over the full range of fuel-economy improvements considered, from about 6 1/100 km (40 mpg) to under 3 1/100 km (80 mpg), the total cost of owning and operating a car would be approximately constant, with the extra costs of fuel-economy improvement roughly offsetting the value of the fuel savings (Figure 2.10).

Looking more closely at the von Hippel/Levi analysis suggests, however, that at the margin there is a considerable range in the cost-effectiveness of the options considered.

In 1981 the diesel version of the Rabbit, with a fuel economy of 5.3. 1/100 km (45 mpg), cost some $525 more than the gasoline-powered Rabbit in the United States. Despite this marked increase in the first cost, the added investment was cost-effective from the perspective of the consumer, even in the United States, where the retail price of gasoline averaged $0.36 per litre ($1.35 per gallon) in 1981; with a .10 percent real discount rate, the cost of saved energy would be just $0.20 per litre ($0.75 per gallon) of gasoline equivalent (Note 2.5.3.3-J).

von Hippel and Levi estimated that the improved aerodynamic design, reduced rolling resistance, the use of a direct injection diesel engine, and the introduction of the CVT—which would collectively improve performance of the diesel Rabbit from 5.3 1/100 km (45 mpg) to 3.3 1/100 km (71 mpg) (Note 2.5.3.3-F)—would involve an increased first cost of perhaps $500 (Note 2.5.3.3-L); the associated cost of saved energy would be $0.25 per litre ($0.92 per gallon), which again is certainly cost-effective.

The costs of the remaining measures considered by von Hippel and Levi (weight reduction, expansion of the CVT range, and the idle-off feature) are more difficult to estimate, but may be as high as another $700 (Note 2.5.3.3-L). If so, the cost of saved energy associated with improving performance from 3.3 1/100 km (71 mpg) to 2.6 1/100 km (89 mpg) would be about $1.00 per litre ($3.70 per gallon), which is not cost-effective. But several considerations suggest that the cost estimates by von Hippel and Levi may be too high.

One reason these estimates of the costs of fuel-economy improvement may be too high is that most of the fuel-saving measures being explored by manufacturers would offer consumer benefits other than just fuel savings, so that charging the extra costs exclusively against fuel economy is inappropriate.

Consider first the multiple benefits of the ongoing trend towards the use of more plastics in cars. Analysis by John Tumazos of Oppenheimer & Company in the U.S. indicates that emphasis on plastics means that the cost of owning and operating a car in the U.S. will be reduced $150 to $250 per year, not only because of higher fuel economy but also because of longer product life due to corrosion resistance, cheaper repairs, and lower insurance rates.[45] Also, according to Russell Shew of the Washington D.C.-based Center for Auto Safety, plastic cars may offer safety benefits too. U.S. government tests on the Pontiac Fiero, which has an all-plastic shell, showed the car's crash performance to be better than for any other vehicle in its weight category.[45]

Similarly, the engine-off during idle and coast feature may be able to extend engine life; and the continuously variable transmission would eliminate the noticeable jerkiness in shifting with automatic transmissions.[43]

Another reason why the von Hippel/Levi cost estimates may be too high is that the costs of improvements are probably not simply additive. Some extra costs may be offset by savings that arise in the synergistic processes of technological innovation. For example, the extensive use of plastics in auto bodies can lead to production cost savings in both fabrication and assembly. Such production cost savings opportunities have led the developers of the Volvo LCP 2000 to conclude that in mass production this car would cost the same as today's average subcompact.[43]

The extent to which first costs will increase with fuel economy is still uncertain, although arguments such as those given here suggest that the net increased cost may prove to be modest. But there is little doubt that high levels of fuel economy can be achieved without increasing much the total cost of owning and operating a car (Figure 2.10).

Heavy Trucks

Diesel fuel for highway use accounts for 14 percent of all petroleum use for transportation in the OECD countries (Notes 2.5.3.3-A and 2.5.3.3-M). Virtually all of this diesel fuel consumption is by trucks—mostly by heavy trucks, i.e., trucks with more than about 9 metric tonnes gross vehicle weight with maximum rated load (Note 2.5.3.3-N).

Since the engines of diesel trucks are already relatively efficient, the opportunities for fuel-economy improvements may be more limited than is the case for light vehicles. In the United States, where the average fuel intensity of large combination trucks was 4.7 1/100 t-km in 1978 (Table 2.8), at least a 50 percent reduction in fuel intensity would seem feasible for new trucks over the next two decades. This could be accomplished through engine improvements (reduction of engine cooling losses by lining the combustion chamber with heat-resistant ceramic material, turbo-compounding, and use of a freon vapour "bottoming cycle" to extract more power out of exhaust gases); aerodynamic drag reductions; and improved tyres (Table 2.8).

That this may actually be a relatively modest goal is indicated by recent experience with a particular new combination truck in the United States. Operating under typical U.S. freight haulage conditions, this truck would have a fuel intensity 40 percent below the U.S. average in 1978 (Note 2.5.3.3-S), without the benefit of a bottoming cycle.

Passenger Aircraft

Air travel, which accounted for 10 percent of OECD energy consumed in transportation (Note 2.5.3.3-A) and 80 percent of total international passenger travel (Note 2.5.3.3-T) in 1979 is so energy-intensive that fuel costs accounted for fully 30 percent of total operating costs of United States commercial airlines in 1980, even though their fuel efficiency had improved markedly between 1973-80. During that period, while passenger-km flown on U.S. carriers increased by 57 percent, fuel consumption increased only 3 percent. By 1980 the average fuel intensity of the fleet had dropped to 5.3 1/100 seat-km (44 seat miles per gallon) or 9.4 1/100 p-km (25 passenger miles per gallon).[46] Thus, by 1980, the efficiency of flying in the United States was on average about equal to that of driving with two persons in a passenger car.

A 50 percent reduction in fuel intensity over the next couple of decades does not seem an overly ambitious target. New intermediate-sized (two engine) wide-body short-haul aircraft which were flight tested in 1981 (the Boeing 757 and 767) are expected to consume 25 percent less fuel per seat-km than the 1980 fleet average. Non-evolutionary design changes such as the introduction of modern turbo-props and twin-bodied aircraft could reduce fuel consumption still further (Table 2.9).

Ships

About 70 percent of the fuel used for shipping by the OECD nations in 1979 was used in international commerce (Note 2.5.3.3-U). At $30 per barrel ($220 per tonne), the petroleum consumed in OECD international shipping in 1979 would have been worth about $15 billion, or about $4.00 per tonne of cargo shipped (imports plus exports), which is about half the average total shipping cost per tonne of U.S. Imports that year (Note 2.5.3.3-V). Obviously, there is an incentive to decrease the energy intensity of shipping.

Table 2.8. Opportunities for Reductions in Heavy Truck Fuel Consumption

	Fuel Consumption (litres per 100 t-km)
1978 Average for U.S.[a]	4.67
Engine Improvements[b]	3.04
Rolling Resistance Reductions[c]	2.63
Aerodynamic Drag Reductions[d]	2.0

(a) For 1.01 billion tonne-km of freight hauled by intercity trucks, which consumed 47.3 billion litres of fuel in 1978 (Note 2.5.3.3-0).

(b) An estimated 23.5 percent savings would be achieved by shifting to an adiabatic turbo-compound diesel engine; a further 15 percent savings would be achieved by using the waste heat of the exhaust to drive a freon vapour engine (Note 2.5.3.3-P).

(c) The targeted savings would be the result of introducing "advanced technology" radial tyres and the replacement of some pairs of tyres by "wide singles", thus cutting in half the losses in tyre sidewalls (Note 2.5.3.3-Q).

(d) The aerodynamic drag coefficient is assumed to be reduced from 0.75 to 0.45, a level which Renault expects to achieve for its new trucks by 1986 (Note 2.5.3.3-R).

Table 2.9. Opportunities for Energy-Efficiency Improvement in Passenger Aircraft

	Litres/100 seat-km	Seat mpg	Percent Savings
US Average-1980[a]	53	44	
Narrow Body	65	37	
Wide Body	47	50	
New Wide Bodies (Boeing 757,767)[a]	28-32	75-85	
Future Evolutionary Improvements[c]			
Large	20	120	
Large Doubledeck (1137 seats)	13	175	
Future Non-evolutionary Improvements			
Twinbodies[d]			30
Turbo-props[e]			20

(a) *Source:* J.B. Smith, 1981.

(b) Based on test flight results for the Boeing 757-200, assuming Pratt and Whitney PW 2037 engines, 196 passengers, and a 500-1000 nautical mile "block" (R.R. Ropelewski, August 30, 1982).

(c) *Source:* D.J. Maglieri and S.M. Dollyhigh, February 1982.

(d) *Source:* J.C. Houbalt, April 1982.

(e) *Source:* L.J. Williams and P.J. Galloway, February 1981.

Ship engines are relatively efficient today. The average fuel-to-mechanical energy conversion efficiency of the best low-and medium-speed diesels, is of the order of 35 percent.[47] However, the peak efficiencies of 55 percent which could be achieved in heavy trucks with adiabatic high-speed diesels equipped with bottoming cycles (Note 2.5.3.3-P) should be achievable with this technology in ships as well. Hence, a 40 percent reduction in fuel intensity for shipping is a reasonable target over the next two decades.

Savings from improved efficiency would be augmented by reductions in the tonnage of petroleum shipped in international commerce. Almost one half the tonnage shipped in 1977 was petroleum and its refined products (Note 2.5.3.3-W). Petroleum probably accounted for a similar fraction of the tonnage moved in interior and coastal shipping (Note 2.5.3.3-X).*

Conclusion

The significance of the technological opportunities for improved energy efficiency discussed here can be appreciated by noting that if the Cummins ½ NASA Lewis car had been the norm for the automobile in OECD countries in 1979, if the energy intensities of truck freight and air passenger travel had been half as great as they actually were, and if the energy intensity of shipping had been 0.6 times as large, then oil use in transportation would have been about 450 million tonnes lower than it actually was—lower by an amount equal to about 60 percent of OECD oil imports in 1979.

2.5.3.4 *Industry*

The energy price shocks of the 1970's resulted in relative price increases for industry much larger than those for other energy-consuming sectors (Section 2.1)**. Within the industrial sector, the energy-intensive basic material processing industries, which accounted for 70 percent of industrial energy use in OECD countries in 1979 (Note 2.5.3.4-A), have been especially hard hit. Many energy-intensive industries, which in the past have had the benefit of energy prices that were much lower than the average for all industry (see, for example, Figure 2.28), are experiencing much larger relative energy price increases than the average for all industry. The data shown in Table 2.10 provide a measure of the relative economic impacts of high energy prices on different manufacturing activities in the United States in 1980: the ratio of energy costs to value-added ranged from 15 to 76 percent for various basic material processing sectors and sub-sectors, but averaged only 3 percent for other manufacturing activities.

Economic conditions thus provide a powerful incentive for seeking improvements in energy productivity. Fortunately, as in other sectors, there is a wide range of technical opportunities for making such improvements. It is useful to classify these opportunities into four categories:

- good housekeeping measures
- fundamental process changes
- product changes
- new energy conversion technologies.

---

* The fraction of shipping energy used in the transport of petroleum is considerably smaller than its fraction of the tonnage shipped, however, since the average volume of tankers is much larger than that of ships in general and the propulsive energy required per tonne decreases approximately with the surface-to-volume ratio of the ship and the density of the cargo (Note 2.5.3.3-Y).

** Because the cost of providing them energy involves much less unit transport and marketing cost, industrial users are more sensitive than other energy consumers to cost increases at or near the point of energy production.

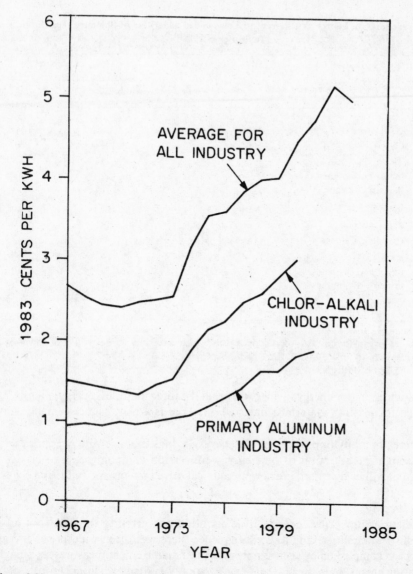

Figure 2.28. Industrial electricity prices in the United States, in constant (1983) cents per kWh.
Sources: Average industrial prices: Energy Information Administration, March 1984, and Energy Information Administration, April 1984.

Specific industry prices: Bureau of the Census, *Fuels and Electric Energy Consumed,* various years.

Good-housekeeping measures are those that can be carried out in the very near term, requiring little in the way of capital investment.

Industrial process change refers to new ways to make given products that not only require less energy but also generally lead to improved overall productivity as well; here the possibilities will be illustrated with discussions of significant changes, now going on and in the offing, in the steel, paper, chemical, and cement industries.

Table 2.10. Expenditures on Energy as a Percent of Value-added and Value of Shipments for U.S. Manufacturing Industries in 1980[a]

Standard Industrial Classification		Compared to	
		Value-Added	Value of Shipments
No.	Name		
26	Paper & Allied Products[b]	16	6
261-3	Pulp & Paper Mills[b]	31	11
28	Chemicals & Allied Products[c]	23	
281	Industrial Inorganics	27	
286	Industrial Organics[c]	76	
29	Petroleum Refining	35	4
32	Stone, Clay, & Glass	15	8
3241	Hydraulic Cement	45	
33	Primary Metals	23	8
331	Basic Steel	32	
3334	Primary Aluminium	46	
	Other Manufacturing	3	

(a) *Source:* Marc H. Ross, 1983.
(b) This does not include the costs of wood-derived fuels, which accounted for nearly half of all energy consumed by the paper industry in 1980 (G.J. Hane *et al.*, September 1983).
(c) Includes the cost of feedstocks.

In some instances, major energy savings will result if entirely new, less energy-intensive materials are substituted for existing materials; one such material is a new "super-cement" now under development.

Finally, there are many opportunities for energy-savings that are widely applicable in many industries through the adoption of new energy-conversion technologies; two classes of such technologies that will be discussed are new mechanical drive systems and industrial cogeneration.

Good Housekeeping

During the era of low priced energy, little attention was given to the details of how energy is used in industrial facilities, and as a consequence there was often widespread energy waste.

Much can be accomplished by such simple improved management measures as: plugging leaks in and insulating steam lines, turning off energy supply systems when not in use, etc. Simple "good housekeeping" measures such as these can be promoted effectively by a variety of measures: the direct metering of major energy-using sections of industrial facilities; charging energy costs to production departments instead of to general overhead; using sophisticated inspection and maintenance equipment such as infrared scanners; establishing training programmes in energy-conserving techniques for the operation of energy-intensive equipment; and using automatic control systems for energy-intensive equipment.

The marketing to industrial (and commercial) customers of energy use monitoring and control technologies and of the associated management or consultant services has in fact evolved into a major industry in its own right, as is evident from the emergence and rapid growth of various new publications that are helping to meet the energy management information needs of large corporate energy users (Note 2.5.3.4-B).

Between 1973 and 1981, primary energy use by industry in OECD countries declined 6 percent or by the equivalent of 2 million barrels of oil per day (100 million tonnes of oil per year), while at the same time GDP increased by 20 percent.[48] Most of this improved energy performance is due to shifts in the industrial mix towards less energy-intensive products (Section 2.4 and Note 2.5.3.4-C) and to the adoption of "good housekeeping" measures such as those described above. Despite this progress, it is generally thought that the opportunities for improving energy management are still not exhausted, although further improvements may require more ingenuity and effort.[49]

Nevertheless, the greatest opportunities for further gains in industrial energy productivity involve going beyond simple good management and making major new capital investments.

Process Innovation

The history of modern industry tells us that new processes for providing familiar products are most likely to overcome resistance to technical change and displace existing processes if they offer opportunities for simultaneous improvements in several factors of production—e.g., reduced labour, capital, materials, and energy requirements.[11,12] As already noted (Section 2.3), this has been such a powerful phenomenon that energy requirements have often been reduced in the process of technological innovation even during periods of declining energy prices.

One example of the importance of process change is illustrated in Figure 2.29, which shows that the energy required to produce a tonne of ammonia has been reduced by about a factor of five since the turn of the century. Another example is the recent introduction of the float-glass process, that has captured most of the market for flat glass-making. Previously, the manufacture of high-quality flat glass involved extensive grinding and polishing, which consumed 10 to 20 percent of the glass. The elimination of these steps not only reduced costs but saved both direct energy needs for power and indirectly the energy needed to make the glass that was ground into waste.

New processes are being continually developed. Important R&D areas from which industrial process innovations are likely to continue emerging include: powder metallurgy, plasma

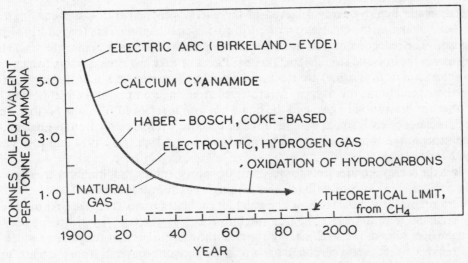

Figure 2.29. The energy requirements for producing ammonia as alternative processes were introduced over time, in tonnes of oil-equivalent per tonne of ammonia.

metallurgy, computer-assisted design and manufacturing, laser processing of chemicals, bio-technology, membrane separation technology, and the use of microwaves for localized rather than volumetric heating. Improvements in all such areas will make it possible to do more with less—that is, produce higher value-added with less inputs of the various production factors, including energy.

Process innovations are not specifically targeted at energy demand reductions—i.e., the goal of process innovation is not to minimize the cost of providing energy services. Rather, it is a more general one—to minimize the total cost of production. Nevertheless, the energy demand reduction that is achievable through process innovation is often much greater than what can be achieved by simply "retrofitting" an existing process with an "energy conservation" device such as a heat recuperator.

An example illustrates the point. Consider the painting or coating of beverage cans. The conventional method of curing the coating is to evaporate the solvent in an oven. Curing the coating this way requires considerable time and energy because the entire volume of the can must be heated. Some energy could be saved by putting a heat recuperator on the oven, but this approach would not reduce the curing time. An alternative curing process which has been developed involves applying ultraviolet (UV) radiation to the surface to be cured. The UV radiation strikes photo-initiators in the coating solvent, and these photo-initiators incite the liquid coating to polymerize and harden. Because the energy is applied to the surface and not the volume of the cans, large reductions in both curing times and energy use are achieved simultaneously—and the reduction in energy use is far greater than what could be achieved through the use of recuperators.[50]

Process change has been and can be expected to continue to be a major factor in improving energy productivity in industry. The scope for change is suggested by some of the possibilities in the steel, paper, chemical, and cement industries discussed below.

Steel: About 5/6 of all steel production takes place in industrialized countries (Note 2.5.3.4-D), where it accounts for a major fraction of all manufacturing energy use: e.g., 1/6 in Sweden and 1/7 in the United States.

The dramatic improvements achieved in the energy productivity of steel-making can be appreciated by noting the changes over time in coking coal requirements for reducing iron ore to pig iron. The reduction of iron ore takes place in a blast furnace, which is the single largest energy user in the iron-and steel-making process. Pieces of coke and chunks of ore (which contain iron in the form of iron oxide) are stacked in the blast furnace and hot air is blown through. Under these conditions, the oxygen is transferred from the iron ore to the carbon and leaves the furnace as blast furnace gas, and the liquid pig iron is removed from the bottom. In 1804 some 5.5 tonnes of coking coal were required to produce a tonne of pig iron in England.[51] The requirements in the United States had been reduced to 1.6 tonnes by 1913 and to 0.9 tonnes by 1972.[52]

Besides the energy requirements for the blast furnace operation, additional energy is required for the various steps in conventional steel-making steps shown in Figure 2.30:

- the preparation of the feedstock material for the blast furnace (to make coke out of coking coal and to agglomerate the iron ore into sinter or pellets);
- the operation of the steel-making furnace, the most important function of which is the removal of the excess carbon remaining in the pig iron recovered from the blast furnace.
- the reheating of the steel in "soaking pits" so that it can be transferred to primary rolling

mills to be formed into slabs, blooms, or billets, after it has been poured from the steel-making furnace into rough shapes and cooled in ingots;

- the subsequent reheating at secondary rolling mills for further shaping into plates, bars, wires, tubing, etc.

At each step in the steel-making process, there are opportunities to improve energy and total productivity. The theoretical minimum amount of energy required to produce a tonne of steel is 7 GJ from iron ore and 0.7 GJ from scrap.[53] At present, steel-making in Sweden and the United States is based on a 50/50 mix of iron ore and scrap, so that the theoretical minimum is about 3.9 GJ per tonne of raw steel. By comparison, the actual energy used to produce raw steel was 27 GJ per tonne in the United States in 1979 and 22 GJ per tonne in Sweden in 1976 (Table 2.11).

Table 2.11. Unit Energy Requirements (in Gigajoules per tonne) for Raw Steel Production in Sweden, With a Comparison to Present U.S. Practice[a]

	I[c]		II[d]	III[e]	IV[f]	
					Alternative New Technologies[b]	
	Sweden Ave., 1976[b]	Modern Technology	Maximum Energy Recovery	Elred	Plasma-Smelt	U.S. Average, 1980[g]
Electricity[b]	2.9	1.8	1.8	1.3	4.2	2.0
Oil and gas	7.6	4.3	2.2	1.3	1.3	7.5
Coal	11.9	9.0	9.0	9.4	3.3	17.5
TOTAL	22.3	15.1	13.0	11.9	8.7	27.0

(a)   Here the mix of iron ore and scrap feedstocks is assumed to be 50/50, which is approximately the present average for both Sweden and the US.
(b)   *Source:* T.B. Johansson, P. Steen, E. Bogren, and R. Fredriksson, January 1983a..
(c)   This energy structure should result from changes planned for the mid-1980s by the Swedish steel industry.
(d)   Same as "Modern Technology", except that the potential for energy recovery with presently available commercial technology has been fully exploited.
(e)   Same as "Modern Technology", except that the blast furnaces are replaced by a process called Elred, which is under development by Stora Kopparberg AB (S. Eketorp *et al.,* 1980).
(f)   Same as "Modern Technology", except that the blast furnaces are replaced by a new process called Plasmasmelt, which is under development by SKF Steel AB (S. Eketorp *et al.,* 1980).
(g)   For 108 million tonnes of produced raw steel. The energy consumption data by energy carrier are from Energy and Environmental Analysis, February 1983.
(h)   Here electricity is evaluated at 3.6 MJ per kWh (i.e., losses in generation, transmission, and distribution are not included).

The markedly better energy performance of the Swedish steel industry arises because the relatively small industry of that country has had to be innovative to secure its niche in the global steel market as a seller of speciality steels. The Swedish steel industry is not the most energy efficient, however. The Japanese are probably the leaders in this regard—a tonne of their steel in 1978 required only about 17 GJ of energy inputs.*[54]

---

* The performance of the Japanese industry relative to Sweden's is better than indicated by these numbers, because the ore fraction in Japanese steel was 75 percent.

The potential for practical energy productivity increases in steel production is illustrated with four alternative technological structures for the Swedish steel industry in Table 2.11. Structure I is based on current plans of Swedish industry for the 1980's. In Structure II, presently available commercial technology for heat recovery has been fully exploited, mostly by using combustible gases in cogeneration (the combined production of heat and electricity—see below). Structures II and IV are based on iron-making processes now under development in Sweden—the Elred process (Note 2.5.3.4-E) and the Plasmasmelt process (Note 2.5.3.4-F). In both cases, the objective is to reduce overall costs and reduce environmental problems by:

- being able to use powdered ores (concentrates) directly, without agglomeration of the ore into sinter or pellets,*
- being able to use ordinary steam coal instead of having to process the much more costly metallurgical coal into coke,
- integrating various individual operations.

Of these alternative processes, the Plasmasmelt process would be especially appealing in coal-poor, hydro-rich countries, while in countries where electricity prices are high, e.g., the United States, it may be preferable to focus on less electricity-intensive processes like Elred or iron-making processes that instead of producing hot molten metal produce solid direct reduced iron (DRI). Direct reduction processes convert iron ore in various forms into sponge iron at temperatures much below the melting point, using a wide variety of reductants other then metallurgical coke (Note 2.5.3.4-G).

The various new iron-making processes at or near commercialization are by no means the ultimate in what might eventually be achieved in terms of energy and total productivity improvements in the steel industry. Other promising advanced processes that attempt to integrate now separate operations to save on capital, labour, and energy costs include direct casting (Note 2.5.3.4-H), direct steel-making (Figure 2.30a and Note 2.5.3.4-I), and dry steel-making (Figure 2.30b and Note 2.5.3.4-J).

The dry steel-making process, which ends up with the final product in powder form, i.e., there would be no melting, holds forth the promise of very low capital costs, suitability for small-scale operations, and a potential 40 percent energy savings relative to conventional processes.[50]

If the dry steel-making technique is mastered, it will likely become the industry norm because finished products of exceptionally uniform quality can be achieved with powder metallurgy.[55] Moreover, large productivity gains would be possible in fabrication and finishing if powder is pressed into the final shapes instead of having to form the final product out of conventional work-stock (bar, plate, etc.), which results in the production of large quantities of scrap which must be reprocessed to be used.[11]

Clearly, the opportunities for process improvement are by no means even close to being exhausted in the steel industry. But, while high energy prices encourage the pursuit of these opportunities, this incentive is offset by the stagnation in the demand for steel in industrialized countries, which creates a poor climate for technological innovation.

Paper: The United States, Japan, and Canada account for about one-half of all the paper produced in the world (Note 2.5.3.4-D), with the United States alone accounting for one-third of the world total. In terms of per capita production, however, Sweden, a major paper exporting

---

* Because lower and lower quality ores are being exploited, the required preliminary processing leaves the ore concentrated in powdered form.

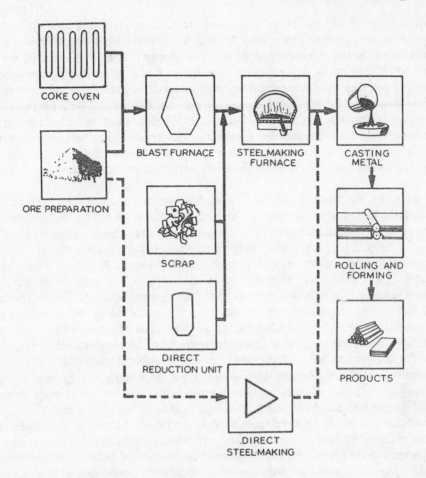

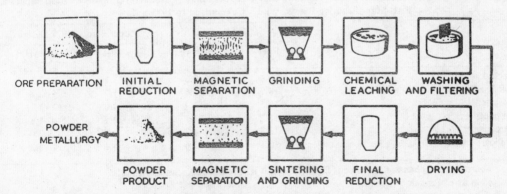

Figure 2.30. The top figure shows alternative processes for steel-making. The conventional approach is indicated with solid lines, direct steel-making with dashed lines. The lower figure is a schematic representation of a dry steel-making process.

Source: G.J. Hane *et al.*, 1983.

country accounting for 3½ percent of total world production, is in the lead—producing each year about 700 kg per capita or nearly three times as much as the United States (Note 2.5.3.4-D). The importance of the paper industry to the Swedish energy profile is reflected by the fact that it accounts for 40 percent of manufacturing energy use, compared to 10 percent in the US.

While there have been significant improvements in the energy efficiency of paper production in the United States, and further improvements are expected (Note 2.5.3.4-K), the most significant innovations are taking place in Sweden. Because of the dominant role of the paper industry in the Swedish economy and the fact that the potential for expanding paper production in Sweden is limited by constraints on the supply of low-cost wood feedstock (Section 2.4.1), it is not surprising that Sweden is a world leader in energy-efficient process technology for the paper industry.

Several years ago, the Swedish Pulp and Paper Association commissioned the Swedish Steam Users Association to design new pulp and paper-making plants optimized to the technology and the energy prices prevailing at that time. The results of that effort, completed in 1977, are summarized in Table 2.12. Six products were examined in detail. Overall, the analysis indicated that a 50 percent average reduction in unit fuel requirements and a 20 percent average reduction in unit electricity use, relative to the energy performance of 1973 vintage plants, were possible. The measures suggested were characterized as good management practices. Today's higher energy prices and further technological developments have made additional improvements in energy efficiency cost-effective. While a systematic, detailed analysis of these further opportunities is not yet available, the scope for further change is suggested by an enumeration of measures being considered for the various parts of the paper-making process (Figure 2.31).

Some readily available additional opportunities for improving energy productivity, with papermaking process changes, include counter-pressure drying and use of pressurized ground pulping and oxygen bleaching (Note 2.5.3.4-L). In addition, there are a number of promising advanced technologies still under development (Note 2.5.3.4-M). Moreover, the large steam loads of the paper industry make it an ideal candidate for the extensive use of modern cogeneration technologies such as the gas turbine (see *Industrial Cogeneration,* below), and the extensive use of mechanical drives makes the industry a good candidate for installation of new motor control

Table 2.12. Unit Fuel and Electricity Requirements (in Gigajoules per tonne of product) for Model Swedish Paper and Pulp Manufacturing Plants, With a Comparison to 1973 Values[a]

	1973 Averages		Model Plants (1977 Designs)	
	Fuel	Electricity	Fuel	Electricity
Bleached sulfate[b]	20.4	2.8	13.1	2.3
Kraft liner[c]	18.7	3.6	11.7	2.8
Newsprint[d]	10.5	7.4	4.0	7.2
Fine paper[e]	11.2	3.1	4.9	1.8
Soft paper[f]	12.7	4.3	4.8	3.0
Cardboard paper[g]	10.1	2.6	5.1	2.2

(a)   *Source:* T.B. Johansson, P. Steen, E. Bogren, and R. Fredriksson, January 1983.
(b)   Pulp dried for sale
(c)   Sulphate pulp integrated with a kraft liner mill.
(d)   Newsprint mill with mechanical pulp production of its own.
(e)   Fine paper mill without pulp production of its own.
(f)   Soft paper mill without pulp production of its own.
(g)   Cardboard mill based primarily on waste paper.

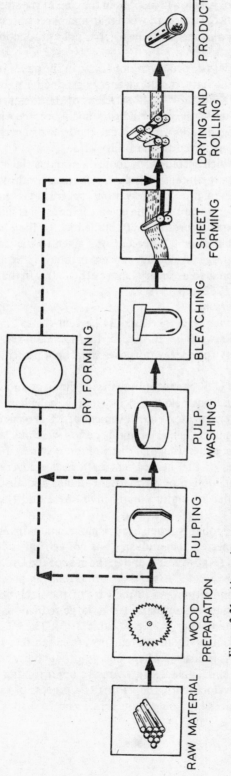

Figure 2.31. Alternative paper-making processes. The conventional approach is indicated with solid lines, the dry process with dashed lines.

Conventional paper-making from wood, the primary feedstock, consists of the following steps:

- *Feedstock preparation:* The bark, which is low in fibre content, is removed mechanically.
- *Pulp production:* The mass of fibres or "pulp" required for paper-making is produced from the wood by some combination of chemical, mechanical, and/or thermal means. (The pulp is essentially pure cellulose with chemical pulping, which removes the lignin from the wood.)
- *Pulp processing:* The wood pulp is then washed, screened, thickened, and often bleached.
- *Paper-making:* In the paper machine the watery pulp is poured onto a fast-moving wire or plastic screen. In the filtering, pressing, drying, and rolling that follows the fibres bond to each other to form paper, which is continuously fed into rolls.

Of these operations, most of the energy is required for pulping, bleaching, and paper-making. Energy is required largely as low quality heat (used mostly to evaporate water) and electricity (used mostly to run mechanical drives).
Source: G.J. Hane *et al.*, September 1983.

devices (see *Motor Control Technology*, below). Also, as in the case of the steel industry, process integration generally leads to lower energy use, e.g., when pulp and paper operations are combined, the pulp does not have to be dried, and the waste heat from pulp production can be used for drying.

Thus, there are major opportunities for process innovations in the paper industry that could lead to major improvements in productivity, including energy productivity. By effectively using residual fuels—the organic-rich "black liquor" waste streams of the chemical pulping process, bark, and perhaps forest residues as well—the paper industry might become wholly self-sufficient in energy, or, with emphasis on cogeneration, even a net exporter of energy, with electricity produced in excess of onsite needs exported to the utility grid.

Chemicals: The chemical industry, which accounts for about ¼ of industrial energy use in OECD countries (Note 2.5.3.4-A), is an energy-intensive industry characterized by a broad diversity of products and continuing fundamental process innovations, spurred in recent decades by rapid demand growth. The importance of innovation in the chemical industry is indicated by its R&D effort. In the United States R&D expenditures in 1980 totalled $4.5 billion, nearly all of which was for "in-house" efforts ($30 million or 0.7 percent was given to university research).

Expenditures on energy are such a large production expense in today's chemical industry (Table 2.10) that major efforts have been made to use energy more efficiently. In the case of the United States chemical industry, between 1972 and 1979:
- value-added increased 37 percent,
- the physical production of chemicals increased 34 percent (Note 2.5.3.4-N),
- the use of fuels as feedstocks went up 26 percent (Note 2.5.3.4-O), and
- the use of fuels and electricity for energy purposes went up 4 percent (Note 2.5.3.4-O).

Thus, the energy-intensity of the U.S. chemical industry declined at an average annual rate of 3.8 percent, 1972-1979; even when feedstocks are included in the calculation of energy inputs, the result is an impressive annual average rate of improvement of 2.8 percent during this period.

But the energy-efficiency improvements achieved in this period were due largely to improved "housekeeping" (especially in steam generating and distribution systems), improved operating practices (especially in steam-heated or fuel-heated reactors), heat recovery from exothermic reactions, and more efficient thermal design of distillation columns and absorbers.[56] Very little of the improvement can be attributed to major capital investment and fundamental process change.

While further improvements will probably require much more capital investment, continuing significant improvements can be expected. The industry has not yet fully adjusted to the price increases that have already occurred (Figure 2.28), and in the case of the United States industry, even greater price increases can be expected. Also, there are promising prospects for reducing energy-intensity through process innovation. As in the case of the steel industry, the process innovations in this industry in the past have usually led to large reductions in specific energy use, even in periods of low or declining energy prices (Table 2.13). There is no indication that opportunities for further energy-productivity improvements are exhausted. Even in the case of the "mature" ammonia production industry, where energy-productivity improvements have already been so large (Figure 2.29 and Table 2.13), there are opportunities for further gains. Two of the most recent process developments, the "Purifier" process of C.F. Braun and the Humphrey and Glasgow process, possibly could reduce fuel and feedstock use for ammonia production by 10-40 percent.[57]

While the extreme diversity of the chemical industry makes very difficult a thorough discussion of the technical opportunities for process improvement, some insights of the future possibilities can be gleaned from the generic opportunities in two areas: *reaction chemistry* and *separation and concentration* processes.

Table 2.13. Historical Energy Requirements Per Unit of Output for Selected Chemicals Produced in the U.S.[a]

Soda Ash		Ammonia		Chlorine	
(Solvay Process)		(Haber-Bosch Process)		(Diaphragm Cells)	
Date	Energy	Date	Energy	Date	Electricity
	(GJ/tonne)		(GJ/tonne)		(kWh/tonne)
1868	60	1917	93.0	1916	4400
1894	31	1923-50	81.0	1947-73	3300
1911	28	1965	52.0	1980	2400
1925	17	1972	46.5		
1942	15	1978	41.2		
1970	14				

Ethylene Di-Chloride		Ethylene Oxide		Polyethelene	
Date	Index	Date	Index	Date	Index
1967	100	1970	100	1956	100
1973	15	1973	85	1973	40
		1974	79	1975	18

(a) *Source:* Robert V. Ayres, February, 1983.

To drive a chemical reaction, enough energy must be supplied to break existing chemical bonds and/or to form new ones. Typically, however, far more energy is used than is theoretically required. Much energy is wasted in exciting the wrong bonds, and often reactants must be heated up to very high temperatures before the desired reaction will take place. There is much that can be done to better direct energy inputs to the targeted reactions. Here, opportunities relating to catalysis, laser chemistry, and bio-technology are briefly described.

A *catalyst* is a substance that speeds up desired reactions or forces a series of reactions to dominate, so that a specific product results, while undesired products are suppressed. Catalysts have been used successfully for many years, often to major advantage with regard to energy savings. For example, a three-fold improvement in energy efficiency resulted when the catalyzed ammonia production process replaced the old non-catalyzed process.[58] Much more can be done with catalysis—e.g., with catalysis based on elements such as tungsten, molybdenum, and ruthenium.[50]

Laser chemistry offers another promising approach for selectively targeting energy inputs to obtain the desired products.[59] The advent of the laser may one day obviate the need to search for the optimum combination of temperature, pressure, and catalyst to bring about a particular result. The laser holds forth the promise of selective control over chemical reactions by introducing only the particular energy levels necessary to weaken or break specific chemical bonds as needed to achieve a desired chemical result. The laser may well emerge as a major precision tool of chemical transformations, with attendant major fuel-conserving benefits. Because of the present high cost of laser energy, the most likely near-term application is for catalytic purposes, i.e., a small amount of laser energy would be used to accelerate a reaction that occurs naturally at a low rate. But as laser costs come down it may well become feasible not only to trigger reactions but also to supply all the energy that is needed to drive the reactions thermodynamically.

If *bio-technology* were to play a major role in the chemical industry, it would represent a revitalization of an old industry rather than the creation of a new one—several chemicals were produced from biomass by means of fermentation prior to the discovery of cheap petroleum feedstocks. The major advantages of bio-technology are that:

- Biologically produced chemicals can be produced from renewable feedstocks. At the levels of requirements for chemical feedstocks, biomass supplies are often abundant. Even in the United States, where the petro-chemical industry is quite mature, the annual use of petro-chemical feedstocks is far less than what could be provided with forest or other biomass resources (Note 2.5.3.4-P).
- Biological reactions involving enzymes take place under moderate conditions of temperature, pressure, etc., while chemical synthesis reactions typically take place under extreme conditions which require large amounts of energy to maintain.
- In biological systems, a complex synthesis scheme can often be carried out in a single step, while many steps with intervening separations are required in chemical synthesis.
- With genetically engineered organisms, it may be possible to maximize the yields of the desired products with minimal production of wastes.
- With biological processing, toxic waste disposal problems will often be reduced. Wastes are often bio-degradable and sometimes valuable sources of nutrients.

For typical chemical reactions, 70 percent of the capital costs and 80 percent of the energy consumed are in the *separation* and *concentration* steps.[50] Often these separation processes are very energy-intensive and incomplete, involving "brute force" techniques. Some of the more promising alternative approaches involve membrane separation, super-critical fluid extraction, and freeze crystallization.

Membrane separation processes use differences in the rates of diffusion across a semipermeable membrane to separate substances. Membrane separations are possible for gas-gas, liquid-gas, or liquid-liquid separations (Figure 2.32). The potential untapped opportunities for using membranes in chemical processing are vast.[60]

Membrane separation requires much less energy than "brute force" methods. For example, desalinating sea-water requires some 280 Megajoules (MJ) per tonne of fresh water produced through distillation, but only 60 MJ per tonne by means of reverse osmosis.[21] And even commercially available membrane-separation techniques could be significantly improved, as is suggested, for example, by the fact that the theoretical minimum amount of energy required to desalinate seawater is 3.5 MJ per tonne.[21] A recent example of major savings involving membranes took place in the chlor-alkali industry, which produces chlorine and sodium hydroxide. The introduction of membranes in this industry in the United States has led to a 25 percent reduction in energy use.[54]

At temperatures and pressures above the critical point of a fluid, there is no distinction between the liquid and gaseous phases and in this state fluids are said to be "super-critical". *Super-critical fluids* (SCF) are of interest for chemical separation purposes because large quantities of many substances which are insoluble under normal conditions can be dissolved in SCF. Since the enhanced solubility tends to be selective, SCF can be used for separating components of mixtures. SCF separation techniques often require only about half as much energy as distillation. But because of the high pressures involved, the technique is capital-intensive. Engineering data are needed to provide the basis for system designs based on the use of SCF.[50]

Freezing can result in energy savings of 75 to 90 percent when used as an alternative to distillation in chemical separations. Separation is achieved with this *freeze crystallization* technique

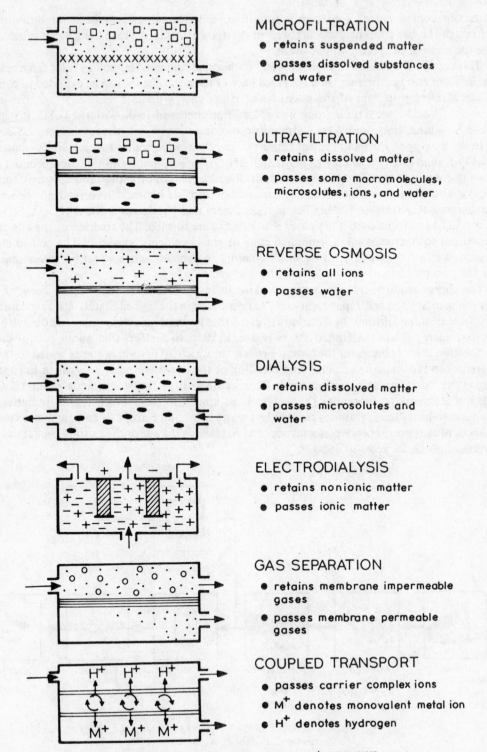

MICROFILTRATION
- retains suspended matter
- passes dissolved substances and water

ULTRAFILTRATION
- retains dissolved matter
- passes some macromolecules, microsolutes, ions, and water

REVERSE OSMOSIS
- retains all ions
- passes water

DIALYSIS
- retains dissolved matter
- passes microsolutes and water

ELECTRODIALYSIS
- retains nonionic matter
- passes ionic matter

GAS SEPARATION
- retains membrane impermeable gases
- passes membrane permeable gases

COUPLED TRANSPORT
- passes carrier complex ions
- M^+ denotes monovalent metal ion
- H^+ denotes hydrogen

Figure 2.32. Types of membrane-separation processes.
Source: G.J. Hane *et al.*, September 1983.

when one component of a mixture is solidified while the others remain in the liquid phase. Although the basic mechanisms are well understood, further development work is needed to use the concept in specific applications.

This cursory discussion shows that present chemical production technology is far from reaching the limits of energy efficiency, and the rapid pace of technical change can be expected to continue if not accelerate, in light of the pressures of rising energy prices.

Cement: World cement production was 852 million tonnes in 1978, with the U.S.S.R., Japan, the U.S., China, Italy, and West Germany accounting for one-half the total (Note 2.5.3.4-D).

In the making of Portland cement (Figure 2.33), the finely ground raw materials—limestone ($CaCO_3$), sand (SiO_2), and clay ($Al_2O_3{:}2SiO_3{:}2H_2O$)—are heated by combustion product gases flowing counter to the flow of raw materials down the length of a long, tilted, slowly rotating tubular kiln. As the raw materials heat up, carbon dioxide is first driven from the limestone, leaving lime (CaO). With further heating, dicalcium and tricalcium silicates ($2CaO{:}SiO_2$ and $3CaO{:}SiO_2$) and tri-calcium aluminate ($3CaO{:}Al_2O_3$) are formed. The rotation of the kiln causes the output to emerge as small, roundish balls of glassy cement "clinker." The cooled clinker is ground to a fine powder, with a small quantity of gypsum added to retard the setting time for the concrete mix.

The energy requirements for cement-making in several countries varies from a low of 4 GJ per tonne in Sweden and Japan to nearly 7 GJ per tonne in the United States. The fuel consumed is used to heat the kiln and the electricity to grind the raw materials and clinker. The much higher level of energy use in the United States is due, in part, to the fact that about half of cement production there is based on the "wet" process (in which 30 percent water is added to the raw materials to facilitate the grinding and blending of the raw materials), whereas in Europe and Japan the "dry" process predominates. This wetting of the feedstocks adds about 1.2 GJ of additional energy to evaporate the water. U.S. kilns also have much higher radiation and convention heat losses, and they recover less waste heat than European and Japanese kilns. At the root of the poorer energy performance of U.S. plants is the age of the existing capital stock—it averaged about 23 years in 1980.[50]

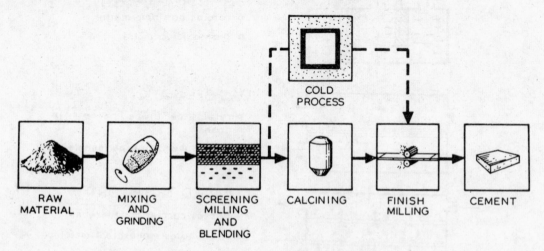

Figure 2.33. Cement making processes.
Source: G.J. Hane *et al.,* September 1983.

Besides shifting to the dry process, energy use can be reduced by using suspension pre-heaters, by flash calcining (Table 2.14), by shifting to cements that are less energy-intensive than Portland cement, or by pursuing the possibility of "cold processing" (Note 2.5.3.4-Q).

Product Change

Not only can much be done to reduce the energy-intensity of familiar products by introducing innovative manufacturing processes, but also there are major opportunities to introduce new products that can accomplish the same tasks more effectively—sometimes with much less energy.

Product design affects the energy required in manufacture and product use through the choice of, the durability of, and the amounts of materials involved. Product design can lead to reduced energy use if it facilitates the potential for materials recycling; this is particularly important for metals, because recycled steel requires only 35 percent as much energy to create finished steel as the use of iron ore, and recycled aluminium less than 10 percent as much as primary aluminium manufacture. Product design can also lead to reduced energy use if it makes long product life attractive—facilitating repair, re-manufacture, and reuse.[61] Product weight reduction through materials substitution can lead to energy savings. In the case of transportation vehicles, energy-saving materials substitution strategies will lead to increased manufacturing energy use, but this is more than offset by reduced energy use in operation, e.g., the aluminium-intensive car (Section 2.5.3.3). In other cases, though, design change may not mean increased energy use in manufacturing. This may very well be the case, for example, with the non-metallic car—a light-weight car manufactured entirely out of ceramics and plastics—expected to be developed in Japan in the 1990's.[41]

Some of the most exciting possibilities for energy-saving materials substitution involve developing entirely new primary materials that may prove to be more appropriate "building blocks" for Post-Petroleum Era societies than many of the primary materials now in wide use. One such class of materials is super cements.

Table 2.14. Energy Requirements for Cement Production (Gigajoules per tonne)

	U.S. Average (1978)[a]	F.R.G. Average (1973)[b]	Japan Average[c]	Suspension Pre-heater[d]	Sweden Average (1983)[e]
Fuel	6.3	4.3	3.6	3.8	3.56
Electricity[f]	0.52	0.3	0.43	0.46	0.40
Total[g]	6.8 (8.0)	4.6 (5.3)	4.0 (5.0)	4.3 (5.3)	3.96 (4.90)

(a)   *Source:* J.T. DiKeo, 1980.
(b)   *Source:* Florentin Kraus, March 1981.
(c)   *Source:* Haruki Tsuchiya, April 1982.
(d)   *Source:* M.H. Chiogioji, 1979.
(e)   There are two Swedish plants, which produced 1.2 million tonnes in 1983, using a combination of suspension pre-heating and flash calcining [C.Wilk (Cementa Co., Skovde, Sweden), personal communication to Per Svenningsson, February 1984].
(f)   Here electricity is counted as 3.6 MJ per kWh (i.e., losses in generation, transmission, and distribution are not included).
(g)   The numbers in parentheses are primary energy requirements which include losses in the generation, transmission, and distribution, under the assumption that power generation is based on thermal plants, so that

$$\text{Primary energy use} = \text{Fuel} + 3.34 \times \text{Electricity}$$

Super cements: Energy requirements for processing of basic materials vary widely. Of the materials we have described, Portland cement is the least energy-intensive—requiring some 30 GJ to produce 1 cubic metre; it takes six times as much energy to produce a cubic metre of polystyrene and 29 times as much to produce a cubic metre of stainless steel. And cement is derived from commonplace materials—limestone, clays, and sands.

Could not a great deal of energy be saved and resource availability constraints be avoided by substituting relatively low energy-intensity cements for metals and plastics? Of course, we do not make greater use of materials like cement today because cements do not meet certain relevant performance criteria, such as: tensile strength (the resistance to pulling) and fracture toughness (the resistance to impact). Inorganic materials like cements do tend to be stiffer than alternative materials, but they tend to have low tensile strength and low fracture toughness. Recent research and development, however, has led to the discovery of ways to improve dramatically cements in these aspects.[62]

A new "super-cement" has been developed called macro-defect-free (MDF) cement, which differs from ordinary cement in that the pores in the cement are "kneaded out"—i.e., the maximum pore diameter is reduced in size from the usual millimetre or so to just a few micrometres. The MDF cements produced this way have dramatically increased tensile strength— comparable to that of aluminium. In fact, it is entirely possible to make a spring out of MDF cement (Figure 2.34). The fracture toughness of MDF cement is also much improved—it is comparable to that of jade, so that, for example, a block of it could be hollowed out into a tube on a lathe. To make the MDF cements highly resistant to impact, they must be reinforced by fibres, the technology for which has recently been developed. Since a cement object is made at low temperature, the reinforcing fibres can be inexpensive organic ones that have low melting temperature. Remarkably, strips of fibre-reinforced MDF cement can be made pliable and bent like strips of metal.

New Energy-Conversion Technologies

Complementing the opportunities for process change and product substitution are generic opportunities for introducing new energy conversion and conservation technologies targeted specifically at energy use. These opportunities typically yield savings of the order of 20 to 50 percent. While such relative savings tend to be less dramatic than what can often be achieved with process innovation or product change (where two-fold or even larger savings are sometimes possible), the savings opportunities associated with many improved energy conversion devices are often widely applicable throughout industry, so that the aggregate savings can be significant. The opportunities here include the use of more insulation on furnaces, the use of radiation reflectors, heat recovery devices, induction heating of metals, microwave heating, better mechanical drive systems, and cogeneration.

To illustrate the possibilities, we will take a closer look at mechanical-drive technology and industrial co-generation.

Mechanical drive: Mechanical drive accounts for a major share of industrial electricity use in industrialized countries. In both the United States and Sweden, for example, industrial motor drives accounted for about ¾ of total industrial electricity use (Note 2.5.3.4-R).

Most industrial motors are fairly efficient, so that only rather modest electricity savings are possible by shifting to high-efficiency motors. A 1976 study estimated that if all new and replacement motors in the United States were high-efficiency devices, electricity savings of about 7 percent could be realized by 1990.[63]

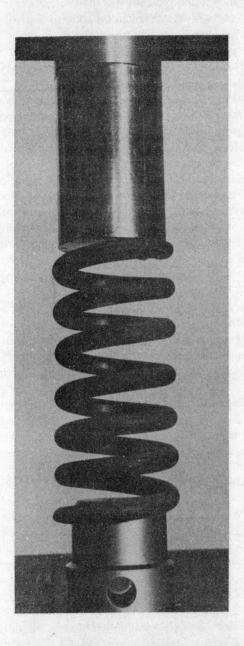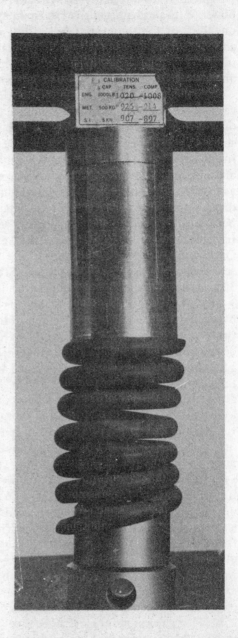

Figure 2.34. A spring made of cement is shown under 300 pounds of tension (left) and under no tension (right). The cement is specially prepared to have no pores or other internal flaws larger than a few micrometres across. As a result, the cement, which is called macro-defect-free (MDF) cement, is highly resistant to fracture; it is as strong as aluminium.

Source: J.D. Birchall and Anthony Kelly, 1983.

Far greater savings are achievable, however, by using better linkage and control technology with motors. A study in British light industry has shown that typically less than half of the input power to the mechanical drive system in a plant is actually delivered to the tool tip, with about one-third of the total input power lost in the gear-boxes and in throttling.[64] One possible alternative to the use of electro-mechanical drive systems that would be economically attractive involves hydraulic power transmission; reportedly it would lead to 20 percent electricity savings and net savings of capital as well.[65]

Another possibility involves the use of new semiconductor motor control technology for variable-load situations involving pumps, compressors, fans, etc. For example, constant speed, overpowered motors are typically used to move gases, and the gas flow is regulated by baffles; similarly, throttling valves are used for controlling liquid flows. An alternative to throttling involves the use of an alternating current variable-speed drive (VSD).[66] With VSD, the voltage and frequency are varied simultaneously maintaining a constant voltage-to-frequency ratio so as to efficiently modulate a standard induction motor. The VSD allows a motor-drive process to be controlled by reducing the energy input to the motor rather than by dissipating the unwanted portion of the motor's output. Energy savings of 20 to 50 percent or more can be realized. Important VSD applications exist not just in industry but for motors throughout the economy. One estimate is that half of alternating current motor usage in the United States could be economically improved by the use of VSD controls by 1990, with an average energy saving of 30 percent for the motors affected.[67]

The compelling attraction of VSD's for both new and retrofit applications is cost. Pay-backs of one to three years are possible in a wide variety of applications. Even in applications where the energy savings are modest (on the order of 20 percent or less), good rates of return can often be achieved (Note 2.5.3.4-S). Due to improvements in solid-state technology, the reliability of VSD devices has improved in recent years, and costs have tumbled and can be expected to continue falling.[68]

Industrial cogeneration: In cogeneration, electricity and useful heat are produced together in a common thermal power plant. The energy in the combusted fuel is, in effect, used twice. The very high temperature heat released in combustion is used first to produce electricity, and then the "waste heat" discharged from the power-generating unit is recovered and used for industrial purposes.

The cost of electricity produced in cogeneration is often cheaper than electricity purchased from utilities. This is perhaps surprising in light of the fact that cogeneration is generally carried out on a small scale. Compared to 500 to 1000 MW(e) for a central station thermal power plant, an industrial cogeneration system typically is in the 3 to 100 MW range, and cogeneration systems for buildings or building complexes can range in size from 10 kW to 10 MW. But the diseconomies of small-scale operation are offset in cogeneration systems by significant opportunities for savings. In cogeneration, the extra fuel required to produce electricity, beyond what would be required to produce the heat alone, typically amounts to only about one-half of the fuel required to produce electricity at a conventional thermal electric power plant.

While cogeneration systems can be used for providing heat to buildings or clusters of buildings through "district heating", the application with the largest fuel-savings potential is cogeneration in the steam-using, basic materials processing industries—such as the pulp and paper industry, the chemical industry, the petroleum refining industry, the food industry. Because these industries typically require large steam loads which are fairly constant, day and night and year round, the electricity they produce tends to be base-load electricity. Thus, industrial cogenerators can produce power which, from a utility's point of view, is similar in quality to the base-load electricity that would be produced by large central-station coal or nuclear power plants.

Historically, however, utilities have discouraged such cogeneration systems, by refusing to provide back-up power, by charging excessive rates for back-up power, and by refusing to purchase electricity produced in excess of on-site needs. But in today's new era of high-cost central station power and uncertain future demand, industrial cogeneration should look attractive to utilities as well.

If utilities were willing and able to rely on cogenerated power for planning purposes, they would not have to base their investment decisions on dubious long-term electricity demand forecasts (Section 2.2.3)—simply because the lead time for cogeneration facilities is typically only about three years, compared to the 8 to 12 years required for conventional power plants. Thus, if utilities were to be involved with cogeneration projects, they could plan their addition of new capacity in small increments as they see how demand is evolving, and avoid building excesses of generating capacity (the situation at present in many industrialized countries—see Section 2.2.2), which are a heavy financial burden.

Is the cogeneration potential large enough for cogeneration to play a major role in utility planning? This is the crucial question. Whether or not the power produced will be significant in a power planning context depends on the cogeneration technology deployed. With the steam turbine system, the most familiar cogeneration technology, the power generating potential of industrial cogeneration is quite limited. However, with alternative "electricity-intensive" cogeneration technologies such as the gas turbine, the gas turbine/steam turbine combined cycle, or the diesel engine, the potential can be quite large (Note 2.5.3.4-T).[69] It has been estimated that when electricity-intensive cogeneration technologies are emphasized, the industrial cogeneration potential in the six basic materials processing industries in the United States, is about one-quarter of total U.S. electricity generation in 1980.[49] Comparably large contributions can be expected in other countries which have large, steam-intensive basic materials processing industries.

Electricity-intensive cogeneration systems are often very efficient with relatively low capital costs. As a result, even small-scale units (e.g., 5 MW gas turbine systems) using costly fuels, e.g., natural gas in the United States, can typically produce electricity at a cost which is lower than the price of electricity purchased from the utility.[70]

Although electricity produced in electricity-intensive cogeneration systems based on costly oil or natural gas is often competitive, it would be desirable if these cogeneration technologies could also be operated on lower quality fuels that would be more appropriate for the Post-Petroleum Era.

There are a number of promising prospects for using low quality fuels such coal or organic residues, e.g., crop residues, forest residues, urban refuse, etc., or other biomass feedstocks from "energy farms" (see Appendix A: Renewable Energy Resources) with electricity-intensive cogeneration technologies. Some are commercially ready today and others are still under development (Note 2.5.3.4-U). These technologies involve either the direct burning of these feedstocks or the burning of a gaseous fuel derived from them.

Biomass-based cogeneration options are particularly interesting.[71] Biomass is a renewable feedstock which, unlike coal, is low in sulfur and toxic trace elements. Moreover, since the costs of producing biomass fuels are likely to be less than present oil and gas prices, the prospects are good that electricity-intensive cogeneration technologies fuelled with biomass will also be competitive in many circumstances with electricity purchased from the utility.

With the deployment of electricity-intensive cogeneration technologies, the total amount of cogenerated electricity is typically much greater than on-site needs; hence, some institutional mechanism is needed for distributing the electricity produced in excess of ôn-site needs to other consumers.

In the United States, an institutional innovation introduced in the Public Utilities Regulatory Policy Act of 1978 (PURPA) encourages such activity.[72] PURPA requires utilities to purchase electricity produced by qualifying cogenerators at a fair price* and to provide back-up power to cogenerators at non-discriminatory rates. After this legislation was passed and rules to administer the law were formulated, the law and rules were challenged in court. Eventually, however, the law and the rules were upheld by decisions of the U.S. Supreme Court in 1982 and 1983. After these Supreme Court decisions, the cogeneration industry in the U.S. began to grow rapidly, reversing a long-term decline. New cogeneration capacity registered in filings with the U.S. Federal Energy Regulatory Commission for qualifying cogeneration status totalled 4,000 MW in 1983, 7,000 MW in 1984, and 10,000 MW in 1985,[73] and rapid growth is expected to continue.

An alternative institutional arrangement for cogeneration that may be even more conducive to the deployment of electricity-intensive cogeneration technologies would be to have the utilities involved in owning (or co-owning) cogeneration facilities.[22] While historically utilities have shunned such activity, some imaginative U.S. utilities envisage aggressive involvement in electricity-intensive cogeneration projects.[74]

2.6 Integrated Examples

In this section, we show what the ongoing shifts to post-industrial economies and the exploitation of opportunities for energy-productivity improvement could mean for Sweden and the United States—two highly industrialized but quite different modern economies.

Sweden is a small, affluent, country with a cold, northern European climate. It has an open economy, with a sizable basic materials processing sector dominated by steel and paper. It is often regarded as a very energy-efficient country.

The United States, on the other hand, is a large, affluent, culturally-diverse country with climates that range from areas as cold as those of the northern-most reaches of Europe to areas that are essentially sub-tropical. The United States has the world's largest economy, which is almost closed. It also has the world's most energy-intensive industrialized economy, measured in terms of either per capita energy use or energy use per dollar of GNP.

Despite the wide differences in these two economies, our analysis indicates that in each case it would be feasible in the decades ahead to reduce per capita energy use by about a factor of two at the same time the output of goods and services continues to expand, if end-use oriented energy strategies are pursued.

Because it has been possible to demonstrate similar results for such different economies and because the technical and economic factors underlying this analysis are widely applicable, it is reasonable to postulate that what can be achieved in these countries might be applicable to other industrialized countries as well.

2.6.1 *An End-Use Oriented Energy Future for Sweden*

In 1984, 50 percent of all energy used in Sweden was imported oil; hydro power supplied 17 percent; biomass, 14 percent; nuclear power, 12 percent; and coal, 7 percent. Primary and final

---

* A utility in the United States is required to pay qualifying cogenerators for the electricity they wish to sell a price equal to the cost which the utility could avoid by not having to provide that power by alternative means.

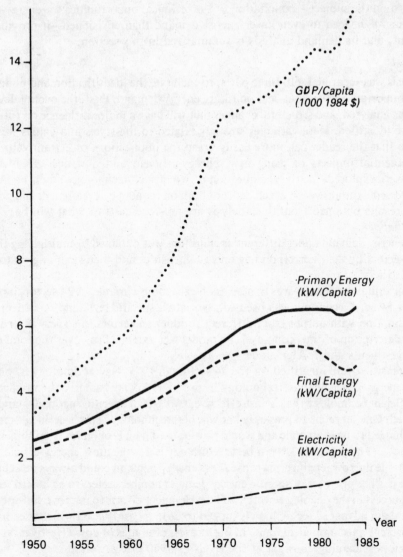

Figure 2.35. Primary energy use, final energy use, and electricity use in kW per capita and GDP (in 1984 dollars) per capita , in Sweden.
Source: Statistics Sweden, 1986.

energy use per capita were 5.6 kW and 4.8 kW, respectively; and per capita GDP was $11,350 (1984$) (Figure 2.35).

A 1981 energy policy decision of the Swedish Parliament calls for reduced dependence on oil and no use of nuclear power after the year 2010. The decision is that the energy supply system should be based on sustainable, domestic, and preferably renewable energy sources—with the least possible environmental impact, and that energy demand should be at the lowest possible level, compatible with economic and social goals.

A study which explores what might be meant by lowest possible energy demand compatible with economic and social goals was published in 1981.[75] While this study does not represent the final word on the subject—exploitation of the technical opportunities for energy-efficiency improvements might lead to even lower energy demand than envisioned—it provides a good starting point, and its demand analysis is summarized in this section.

Methodology

The analysis was separated into three parts, to facilitate the identification and understanding of the most important factors that bear on future energy demand. First, the overall development of society was analysed, and structural changes that will have a major influence on future energy demand were identified. Here variables strongly related to life-styles and values were treated explicitly, so that the reader can more easily grasp the implications of certain value changes. Next, the potential impacts of using new, energy-efficient and now cost-effective end-use technology were explored. Lastly, the question of when such technology might be in common use was addressed—the answer to which depends both on economic development and government policy. The results obtained from this analysis are not forecasts of what will happen but of what *could happen.*

The total energy demand under different assumptions was obtained by multiplying the activity levels in all sectors by their corresponding energy intensities and then summing up to cover all activities in society.

The overall volume of activity was treated as an exogenous variable. A 24-sector, input-output model of the Swedish economy was used to associate a specific level and mix of private and public consumption with output from different production sectors in society in order to give a consistent description of the economy. The model was extended to cover the total economy, including investments and foreign trade.

Energy intensities were identified for the 24 sectors in 1975. New technology is, on average, much more energy-efficient than technology in use in 1975. To quantify the impact of shifts to energy-efficient technology that is cost-effective, two sets of specific energy-demand numbers were identified. One set refers to presently known, best technology. This is defined as technology that is, or is judged to be, economic and would thus be used if an economically optimal investment were made now. (The presently known best technology is not the most energy-efficient existing technology in the narrow, engineering sense. Less energy-demand could always be achieved, but at an extra cost.) The other set of specific energy demand numbers refers to advanced technology, where some success in the ongoing research and development efforts to improve end-use efficiency is assumed. Again advanced technology is judged to be in a cost bracket which is economic and does not represent a technological limit. In both cases, estimates of cost-effectiveness are based on energy prices remaining constant at the levels of 1980 prices.

Because future economic growth is uncertain and outside the scope of this study, the calculations of total energy use were performed for various combinations of end-use technologies and different levels and mixtures of final demand for goods and services. Attention was focused in the analysis on the cases where the overall volume of goods and services is 50 percent and 100 percent higher than in 1975, with a mix of activities similar to that in 1975 except for saturation effects (see notes to Tables 2.15 and 2.16).

Energy-Efficient Technology

In this section, the choices of the energy performance levels used for the major energy end-uses in the Swedish case study are described. The basis for these choices was discussed in section 2.5.3.

Buildings: Residential and commercial buildings accounted for 25 percent and 10 percent of final energy use in Sweden in 1975 (Table 2.15). These sectors are, in other words, major energy users. The potential for more efficient use of energy in them is very large, as we have shown in Sections 2.5.3.1 and 2.5.3.2.

The Government's 10-year plan aimed at reducing energy use in existing buildings by about one-third provides the basis for the choice of the presently known best technology for existing buildings, both residential and commercial. The increase in efficiency is to be achieved through improvements in insulation, windows, and heat recovery. The targeted improvement is not optimal, though, because it is estimated for an energy price about 30 percent below 1981 prices.

The Swedish building codes of 1981 require new building designs to be optimized for an energy price of about $4 per GJ (1980 $), or one-third of the consumer oil price. These building codes are thus not stimulating an optimum level of energy efficiency and cannot be taken as the measure of the presently known best available technology for new buildings.

For our analysis, we used the level of energy use achieved in a single-family home (the house of Mats Wolgast—Table 2.3) outside the city of Uppsala (having a climate with 3800 degree days, about 9 percent more than Stockholm) as an example of the presently known best technology for new residential buildings. This super-insulated house has a floor area of 130 square metres a mechanical ventilation system with heat recovery, and a number of thoughtful design details. The capital investment was slightly below the norm for a house.of comparable size, because the high degree of super-insulation achieved made it possible to avoid the purchase of a costly central oil heating system. The house is heated instead with a few small electric-resistance heaters and a wood stove. The energy input from electricity (60 percent) and wood (40 percent) for heating and hot water is about 110 megajoules per square metre (MJ/m^2) per year, compared with an average of about 800 MJ/m^2 for single-family dwellings in Sweden. Comparable energy performance is now achieved with some commercially available pre-fabricated house, (e.g., Version #2 of the Faluhus in.Table 2.3).

The energy performance of the Folksam building in Stockholm, completed in 1977, is taken to be representative of presently known best technology for office buildings. A total of 133 MJ/m^2 per year is required to heat this building, compared with an average of 900 MJ/m^2 for comparable office buildings in Sweden (Section 2.5.3.2).

For the level of advanced technology, greater heat recovery in ventilation air and more efficient use of hot water was assumed. Indeed, energy performance much better than that of the Folksam building has already been demonstrated for the Harnosand building, constructed in 1981. For this building measured total energy use per square metre of floor space is only about half of that for the Folksam building (Table 2.6).

For appliances, both best available and advanced technology levels were based on the analysis of Norgard.[76] Assuming Norgard's "strong" efficiency improvements, that on average give a 52 per cent specific energy-demand reduction from the 1975 level, and assuming saturation for all appliances, a total of 11 TWh per year was used in all cases.

The results of the analysis for buildings are summarized in Table 2.15. This table shows that final residential energy use would decline to 35 percent of the 1975 level with a 50 percent increase in the consumption of goods and services and to 43 percent of the 1975 level with a 100 percent increase in goods and services consumption, if current best technology were used. If advanced technology were used, residential energy use would instead decline to 25 percent or 30 percent of the 1975 level. Similar reductions would be realized for commercial buildings. Moreover, the buildings sector's share of total energy use would be reduced from 35 percent in 1975 to less than 20 percent in all the alternative energy-efficient scenarios—indicating that on the basis

Table 2.15a. Final Energy Use Scenarios for Sweden (in Petajoules)[a]

| | 1975 | | | Consumption of Goods and Services Up 50% | | | | | | Consumption of Goods and Services Up 100% | | | | | |
| | | | | Present Best Technology | | | Advanced Technology | | | Present Best Technology | | | Advanced Technology | | |
	Fuel	Elect.	Total	Fuel	Elect.	Total	Fuel	Elect.	Total	Fuel	Elect.	Total	Fuel	Elect.	Total
Residential[b]	295	61	356	61	65	126	36	54	90	78	73	151	50	58	108
Commercial[c]	104	36	140	11	39	50	3	37	40	11	41	54	5	38	43
Transportation															
Domestic	223	11	234	183	11	194	137	7	144	210	13	223	160	9	169
Intl. Bunkers	47	–	47	32	–	32	25	–	25	40	–	40	29	–	29
Industry															
Manufacturing[d]	410	137	547	293	153	446	230	141	371	353	191	544	275	175	450
Agric., Forestry and construction	65	14	79	40	21	61	37	13	50	54	29	83	49	19	68
TOTALS	1141	259	1400	622	289	911	467	253	720	751	347	1098	569	299	868

Table 2.15b. Final Per Capita Energy–Use Scenarios for Sweden (in kW per capita)

| | 1975 | With Consumption of Goods and Services | | | |
| | | Up 50% | | Up 100% | |
		Present Best	Advanced	Present Best	Advanced
Residential[b]	1.36	0.48	0.34	0.58	0.41
Commercial[c]	0.53	0.19	0.15	0.21	0.16
Transportation					
Domestic	0.89	0.74	0.55	0.85	0.65
Ints. Bunkers	0.18	0.12	0.10	0.15	0.11
Industry					
Manufacturing[d]	2.09	1.70	1.42	2.08	1.72
Agric., Forestry, and construction	0.30	0.23	0.19	0.31	0.26
TOTALS	5.34	3.48	2.75	4.19	3.31

(a)   The population is assumed to be 8.3 million in all cases. The per capita GDP was 8,320 (1975)$.
(b)   Heated floor space is assumed to increase from 36 square metres per capita in 1975 to 55 (73) square metres per capita with a 50% (100%) increase in per capita consumption of goods and services.
(c)   Commercial buildings floor space is assumed to increase from 12.7 to 16.4 (20.2) square metres per capita for a 50% (100%) increase in per capita consumption of goods and services.
(d)   The mix of industrial output is assumed to shift towards less material-intensive products, as discussed in the text.

Table 2.16. Disaggregated Energy–Use Scenarios for Swedish Transportation and Manufacturing
(in Petajoules per year)

| | 1979 | With Consumption of Goods and Services | | | |
| | | Up 50% | | Up 100% | |
		Present Best	Advanced	Present Best	Advanced
Domestic					
Transportation					
Automobiles[b]	148	86	54	101	65
Trucks[c]	50	43	29	47	32
Rails Freight[d]	4	4	4	7	7
Other	50	61	58	65	65
TOTALS	252	194	145	223	169
	1975				
Manufacturing					
Pulp and Paper[e]	212	148	122	148	122
Iron and Steel[f]	119	76	61	86	72
Fabrication and Finishing	54	61	54	90	76
Cement	40	36	32	47	40
Chemical	32	36	36	50	47
Other	90	90	65	122	94
TOTALS	547	446	371	544	450

(a)   Transportation data disaggregated by end-use are not available for 1975.

(b)   The number of person-km of car travel is assumed to increase from the 1979 level of 41 billion by 25 percent (50 percent), for the case of a 50 percent (100 percent) increase in the per capita consumption of goods and services.

(c)   The volume of truck freight is assumed to increase from the 1979 level of 14 billion tonne-km by 12.5 percent (25 percent), for the case of a 50 percent (100 percent) increase in the per capita consumption of goods and services.

(d)   The volume of rail freight is assumed to increase by 60 percent (120 percent), for the case of a 50 percent (100 percent) increase in the per capita consumption of goods and services.

(e)   Production in the paper and pulp sector was limited to a 50 percent increase over the 1975 level in both cases due to limited raw material supply.

(f)   Because of the considerable competition faced by the export-oriented Swedish steel industry, the output of the steel industry is assumed to be only 23 percent (44 percent) higher than in 1975, if per capita consumption of goods and services increases 50 percent (100 percent).

of present knowledge the relative savings potential is greater for buildings than for any other sector.

Transportation: Transportation accounted for 20 percent of Swedish energy use in 1975, and here too the opportunities for major efficiency improvements are large, as indicated in Section 2.5.3.3.

Consider first the automobile, which accounts for about 60 percent of transportation energy use. The average fuel economy of automobiles on the road in Sweden was 10.9 1/100 km (21 mpg) in 1982, and 1985 models averaged 8.6 1/100 km (27 mpg). There are now automobiles on the market with a fuel economy of about 5.0 litres per 100 kilometres (47 mpg), which was selected as representing the presently known best technology.

Further improvements in fuel economy can be achieved through such modifications as more efficient engines, lighter materials, reduced air drag and rolling resistance, more efficient transmissions, etc. Prototypes are now reaching 3.1 1/100 km (78 mpg), which was adopted as representing advanced technology.

Based on a combination of technical and organizational improvements, the energy intensity for truck freight was assumed to be reduced by 22 and 48 per cent respectively, for best available and advanced technology.

Table 2.15 shows for transportation that, associated with a 50 percent increase in the consumption of goods and services, final energy use would decline to 80 percent of the 1975 level, and with a 100 percent increase in goods and services to 93 percent, if current best technology were used; it would be reduced to 61 percent or 71 percent of the 1975 level if advanced technology were used. Within the transportation sector, the expected relative savings would be considerably greater for automobiles (Table 2.16): final energy use for autos would be 58 percent or 68 percent of the 1975 level if current best technology were used, and 36 percent or 44 percent of the 1975 level if advanced technology were used.

Industry: In 1975 Swedish industry accounted for 39 percent of total Swedish energy use. The largest industrial energy users are the paper and pulp industry and the iron and steel industry, accounting for 39 and 22 percent of the total, respectively.

The analysis of the pulp and paper industry is based on the set of model Swedish plants designed in 1977 and discussed in Section 2.5.3.4. These designs represent what should have been ordered for new plants at that time. Because these plants were designed before the 1979-1980 energy price increases, we assume a further 12 percent reduction in the energy-intensities for paper and pulp production to characterize best available technology. Energy-intensities with advanced technology were taken to be 18 percent less than with best available technology.

A continuation of the ongoing trends towards more paper and less pulp and towards a greater share of pulp provided by mechanical pulping (an electricity-intensive method of grinding wood into pulp) was assumed. In spite of this, it was found that the overall energy-intensity for the pulp and paper sector would be 54 or 62 percent lower than in 1975, if the presently known best technology or advanced technology were introduced, respectively.

These innovations would by no means exhaust the technical possibilities, as the theoretical minimum amount of energy required for paper production is less than 0.11 GJ per tonne of dry wood, which is less than one percent of present energy-intensity.

For the iron and steel industry, the analysis was based on the four alternative industrial structures described for the Swedish steel industry in Section 2.5.3.4 (Table 2.11).

The measure of presently known best technology was taken to be an average of the energy performances for Structure I (based on current steel industry plans for the 1980s) and Structure II (the same as Structure I except that the potential for heat recovery has been fully exploited). For advanced technology, an average of the energy performances of Structure III (Elred technology) and Structure IV (Plasmasmelt technology was used). Energy use in the Swedish steel industry in 1976 would have been lower by about one-third or one-half, if best available or advanced technology had been used, respectively.

Again, even the use of advanced technology would not represent a technological limit, as the theoretical minimum amount of energy required to produce a tonne of steel from a 50/50 mix of iron ore and scrap (the mix appropriate for Sweden) is 3.9 GJ, or about 1/6 of the average level of energy use in the Swedish steel industry in 1976.

Estimates of the overall impact on energy demand of adopting various generic energy-saving technologies—e.g., added insulation, heat recovery, use of heat pumps, new lighting technology, variable-speed motor control devices, etc.—in other manufacturing industries are summarized in Table 2.16.

For the industrial sector as a whole, the net effect of economic expansion, a shift in the mix of output towards less material-intensive products, and the adoption of more energy-efficient best available technology, would be a final energy use level 81 percent of the 1975 level with a 50 percent increase in the consumption of goods and services and a level equal to the 1975 level for a 100 percent increase in good and services consumption, with the adoption of present best technology. With advanced technology, final energy use would be reduced to 67 percent or 88 percent of the 1975 level, with a 50 or 100 percent increase in goods and services consumption. And industry's share of total energy use would increase from 45 percent in 1975 to the range of 55-60 percent.

Total Energy Demand

Alternative future energy-demand scenarios for Sweden are summarized in Table 2.15 and in Figure 2.36, which shows energy-demand level results where the overall volume of goods and services is 1.0, 1.5, and 2.0 times greater than in 1975, compared to the 1975 energy demand level and to a forecast for 1990 made by Sweden's National Industrial Board. The effect of energy-efficiency improvements is dramatic. Compared to a total energy use level of 1400 petajoules (PJ) in 1975, the level of annual energy use with a 50 percent or 100 percent increase in the level of goods and services production would be about 910 PJ or 1100 PJ with best available technology and 720 PJ or 870 PJ with advanced technology. The overall reduction in energy demand would be accompanied by an *increase* in electricity's share of final demand from 19 percent in 1975 to 32-35 percent with these alternative, more energy-efficient futures (Table 2.15a).

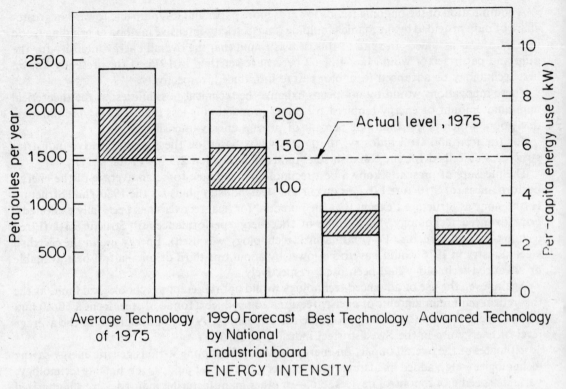

Figure 2.36. Final energy demand in Sweden for four levels of energy intensity at 100, 150, and 200 percent of the 1975 level of consumption of goods and services.

The sensitivity of the level of future energy demand to changes in parameters other than the overall level of goods and services and the choice of end-use technology was also explored. Particularly important for a small country such as Sweden is how different trade conditions might affect energy demand. Figure 2-37 displays the results of this sensitivity analysis, relative to a base case in which a 50 percent increase in the overall level of goods and services and best available technology are assumed.

Uncertainties

Aside from the effects of different levels of economic output and the choice of end-use technology, there are uncertainties relating to key assumptions in this analysis that could possibly change in significant ways the estimates of future energy demand presented here.

A future society with a mix of personal consumption expenditures only modestly changed from that of today was assumed. (It was assumed, however, that the demand for buildings and transport amenities would grow more slowly than the overall volume of goods and services—see notes to Tables 2.15 and 2.16). This assumption thus ignores the possibility that more resource-conserving life-styles may become important in our time.

Also, constant coefficients were assumed for the input-output analysis and only very limited consideration was given to changes in the mix of industrial output—e.g., it was assumed that there would be a shift from pulp only to more pulp/paper production and that output of the pulp/paper industry would increase no more than 50 percent, and that the growth of the steel industry would be constrained by competition (see notes to Table 2.16). Hence, the ongoing

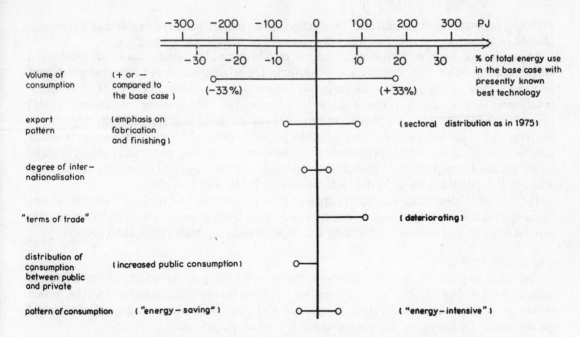

Figure 2.37. Sensitivity analysis showing the changes in final energy demand associated with alternative assumptions, relative to a base case scenario in which the overall level of consumption of goods and services is 1.5 times the 1975 level and presently known best end-use technology is deployed.

1. Sweden has traditionally been a large exporter of basic materials such as steel, iron ore, paper and pulp. In the base case, the contributions to total exports were assumed to change: for fabrication and finishing from 38 percent in 1975 to 53 percent; for iron and steel from 7.4 to 3.4 percent; for paper and pulp from 14 to 11 percent; for transportation from 8 to 5 percent; for fabrication and finishing from 38 percent in 1975 to 53 percent; for iron and steel from 7.4 to 3.4 percent; for paper and pulp from 14 to 11 percent; for transportation from 8 to 5 percent; for chemicals from 5 to 7 percent; and for shipbuilding from 5.5. to 1.5 percent. If, instead, the mix of exports remained the same as in 1975, annual energy use would be about 10 percent higher. Alternatively, if all the increase in exports were shifted to engineering products, i.e., products involving considerable value-added in fabrication and finishing, annual energy demand would be reduced by about 8 percent.

2. The degree of internationalization (total imports divided by total domestic demand) was assumed to be 47 percent in the base case. If this index were as low as 29 percent (the 1975 level), energy demand would be 3 percent lower; energy demand would instead be 3 percent higher than in the base case if the degree of internationalization were 65 percent. This indicates that the energy intensities of exports and imports are roughly equal (excluding energy items).

3. The impact of a deterioration in the terms-of-trade (the ratio of the index of export prices to the index of import prices) by 1.5 percent per year would be to increase energy demand by 10 percent. (In 1979 the Swedish government commission set up to study the consequences of dispensing with nuclear power projected that terms-of-trade would deteriorate by 1.5 percent per year during the 1980's.) This situation would be unacceptable in the long run, however, as it implies that by 2000 exports of 4 dollars value would be required to pay for every 3 dollars of value of imports.

4. A doubling of public expenditures (which requires that private spending be 26 percent higher than in 1975 in order to keep the overall level of economic output 50 percent higher) would reduce energy demand by 6 percent.

5. To illustrate the importance of the mix of household expenditures, these expenditures was disaggregated into 13 categories. In two calculations, the total increase in expenditures were disaggregated into 13 categories. In two calculations, the total increase in expenditures was assigned to the six most and the six least energy-intensive categories. The result was that total energy demand was higher by 9 and lower by 4 percent respectively.

and very important shift in industrial production to less material-intensive (and thus less energy-intensive) activities (Section 2.4) was only partially dealt with in this analysis.

The mix of consumption activities changes not only because of changes in consumer preferences and saturation effects but also because of shifts in the relative prices for different goods and services. In this analysis, it was assumed that the relationships between the costs of different production factors remain constant at the 1975 levels. The base year for this analysis (1975) pre-dated the second oil price shock, and in 1975 the economy had not yet fully adjusted to the oil price increases of 1973/74. Therefore, the projected demand for energy-intensive products may be too high, given today's price structure. This shortcoming may be partially offset, though, by the explicit assumptions that growth in the demand for buildings and transportation services will fall short of the growth in the total volume of goods and services.

These considerations suggest that, in spite of the uncertainties and necessary simplifications in the approach used, the calculated total energy-demand levels associated with each combination of consumption and end-use technology are, if anything, too high rather than too low.

The issue of "When"

The third step of the analysis addressed the issue of the pace at which energy-efficiency improvement will be introduced. *Assuming* that the more energy-efficient options will be chosen if they are found to be cost-effective, the pace at which new technology is adopted will depend on the rates of turnover of the capital stock for major energy-using activities.

Machinery and equipment in industry have economic lifetimes of 10-20 years, depending on the sector of industry. The automobile fleet is renewed approximately every 15 years. The structural components of buildings have very long lifetimes, 50-100 years, but many energy-related installations in buildings have shorter lifetimes—for example, 10-30 years for furnaces, windows, ventilation systems, and facades.

Using these rates of capital turnover, the impact on average specific energy-demand levels was calculated for both presently known best technology and advanced technology. Presently known best technology was assumed to be routinely installed beginning in 1985, and advanced technology beginning in 1995. It was concluded that, if this more efficient technology were introduced at the rates of normal capital turnover, available technology might become average technology around the turn of the century, and advanced technology during the first decade of the next century.

While many of the new, more energy-efficient technologies will be introduced in this manner, as a response to high energy prices, it is unlikely that all the more energy-efficient technologies considered here would be adopted at this pace, solely as a response to market forces, as there are major institutional obstacles inhibiting the most cost-effective technological choices—especially in relation to energy efficiency improvements for buildings and the automobile. *New government policies to overcome these obstacles will be needed to realize the full level of energy savings described here* (Chapter 5).

Energy Supply for an Energy-Efficient Sweden

Sweden has no known domestic resources of oil, gas, or coal. Shifting from the present major dependence on imported oil (Table 2.17), to reduce vulnerability to oil supply disruptions and to improve the balance-of-payments situation, has been an important objective for Swedish energy policy during the last decade. But the choice of replacements for oil in Sweden has been a subject of controversy.

The official perspective up until the late 1970's included a large expansion of nuclear power—

Swedish production of which had become the world's largest on a per capita basis by the early 1970's. Growing public opposition to nuclear power in the 1970's and the accident at Three Mile Island in the United States in 1979 led to a nuclear referendum in Sweden in 1980. Based on the advice of the referendum, the Parliament decided that Sweden should limit the use of nuclear power to the present programme, and that there should be no use of nuclear power at all after the year 2010, a time set to allow for the use of these reactors through their technical life-times.

Table 2.17. Energy Supply for Sweden (in Petajoules per year)[a]

	1984[b]	A Year in the Period (2010-2020)
Primary Energy Source		
Oil	734	130 to 0[c]
Natural Gas[d]	—	—
Coal	104[e]	70
Peat	2[f]	50[g]
Nuclear Power	175	—
Imported Electricity	1	—
Hydropower	241	240
Wind power	—	40
Solar Heating	—	5[h]
Biomass		
Paper and Pulp Industry Wastes	168[i]	140
Forests	41[j]	90[k]
Agricultural Residues, Reeds	—	10[l]
Biomass Plantations	—	275 to 655[m]
TOTAL	1466	1050 to 1300
Conversion, Transmission and Distribution Losses	200	150 to 400
Final Energy Use	1266[n]	900

(a) For a 50 percent increase in the consumption of goods and services relative to 1975 and the deployment of presently known best available energy end-use technology.

(b) Source: Statistics Sweden, 1986.

(c) It is assumed that oil is used as automotive fuel or not at all.

(d) Sweden presently imports some 10 PJ of natural gas each year but has no known domestic resources.

(e) Coal is used mainly for metallurgical purposes.

(f) Source: Bengt Sunfeldt, Swedish National Energy Administration, personal communication.

(g) Present plans are to produce some 22 to 40 PJ per year. The peat production potential is about 100 PJ per year. The plans are kept low because of environmental effects at the production sites and a relatively high sulphur content of many peat bogs.

(h) Solar heating on a large scale would require seasonal storage. This contribution is for domestic hot water and limited amounts of solar space heating. The combination of solar panels and low-temperature heat storage for use with heat pumps is attracting considerable interest at present.

(i) Black liquor, bark, and other wastes.

(j) *Source:* National Swedish Industrial Board (SIND) 1983.

(k) This includes thinnings, residues, etc. The level is less than half of the potential, because we assume no whole tree harvesting due to the large environmental effects of such operations.

(l) Agricultural wastes are produced at a level of about 40 PJ per year. Reeds are from natural stands.

(m) Bioenergy from plantations is the swing energy source. The higher value is for the case where methanol is produced from wood @ a 50% conversion efficiency; the lower for the case where petroleum is used instead for such purposes.

(n) Ref. (b) gives total final energy use as 1243 PJ, but this estimate does not include the use of peat and wood fuel in the residential and commercial sectors. An efficiency of 75 percent is assumed here for such uses.

The 1981 Parliamentary decision calls for developing sustainable, preferably domestic an renewable energy sources, with the least possible environmental impact. Swedish policy to shif from dependence on imported oil and the 1981 Parliamentary decision are in line with the concern articulated in Chapter 1 regarding the global security and environmental risks associated with continued major reliance on oil imports, fossil fuels generally, and nuclear power. But can northern country like Sweden make a major shift to renewable energy resources?

Earlier work has indicated that Sweden would be able to build a fully renewable energ system.[77] In this effort a "Solar Sweden" was designed for an energy-demand level of almos 1800 PJ per year. It is obviously much easier to design a fully renewable system for the muc lower energy-demand levels of the scenarios presented in Table 2-15.

To illustrate the possibilities, supply scenarios have been constructed for total primary energ (Table 2-17) and for electricity (Table 2-18), assuming that the per capita consumption of good and services increases 50 percent and currently best available end-use technology is deployec If renewable supplies can be provided for this demand scenario, it would obviously be possibl to meet the comparable or lower demand levels shown in Table 2.15 for the cases involving th use of advanced end-use technology with either a 50 percent or 100 percent increase in th consumption of goods and services.

Consider first electricity generation. Traditionally, hydropower has dominated electricit generation in Sweden. During the last decade, however, hydropower expansion has essentiall come to a halt, and additional capacity has been obtained from nuclear power sources. By 1984 hydro's share of electricity generation had fallen to 56 percent (still much larger than the 2(percent global average), and nuclear power's share had risen to 41 percent. When the presen

Table 2.18. Electricity Supply for Sweden (in Petajoules per year)[a]

	1984[b]	A Year in the Period (2010-2020)
Hydropower	241[c]	240[d]
Wind power	—	40[e]
Oil	15[f]	—
Nuclear Power	175[c]	—
Imported electricity (net)	1	—
Biomass plantations	—	41[g]
TOTALS	432	32¡[h]

(a) For a 50 percent increase in the consumption of goods and services relative to 1975 and the deployment of presently known best available energy end-use technology.

(b) *Source:* Statistics Sweden, 1986.

(c) This is the net power output from the power stations, i.e., own electricity consumption is not included.

(d) This is the level of hydropower production now decided upon by Parliament.

(e) This would be the production from about 1500 large wind power units. The system effects of this wind power production would be handled by the hydropower system, which has sufficient capacity to deal with the variations in wind-energy production from even a wind-energy system three times as large, although at increasing system costs [Kraftsam (The Swedish Utilities Association), 1984].

(f) This is the electricity production via cogeneration facilities for industry and district heating.

(g) Biomass from plantations is the swing fuel.

(h) Electricity production less 10 percent transmission and distribution losses equals the electricity demand given in Table 2.15.

nuclear stations are taken out of operation, alternative electricity supplies will be needed, even with the energy-efficient scenarios under consideration here.

There are unexploited hydropower resources in Sweden that are technically and economically suitable for expanding hydro production by about 50 percent. Environmental and aboriginal concerns, however, have led Parliament to limit hydropower development to the production level assumed here—a 23 percent increase in hydro production over the 1982 level (Table 2.18).

Wind power is now being considered and may contribute a significant but limited amount of power in the future. Here it is assumed that wind would provide about 1/10 of the total electricity supply (produced, for example, in 1500 large units)—which is far less than what could be regulated with storage in the hydro-electric supply system, to compensate for the variability of the wind (Table 2.18).

Electricity requirements in excess of what could be provided by hydro and wind resources might be provided by biomass-fired thermal plants—for example, in efficient sawdust-fired gas turbines for cogeneration applications or for the production of power only.* Here it is assumed that 13 percent of all electricity is produced using biomass. If this were produced in power-only plants from wood raised in biomass plantations, some 0.4 million hectares of plantations would be required.

Biomass can also be used for a variety of direct combustion purposes, and as a feedstock for making both gaseous and liquid energy carriers.

While essentially all roundwood production from the natural forests in Sweden is already committed to other purposes, a variety of alternative biomass options are available—including wastes from the natural forests, forest product industry wastes, agricultural residues, reeds, and biomass from plantations.

Willows, birch, and alder are considered the most promising species for plantations and may yield 10—25 tonnes of dry matter per hectare per year in large-scale plantations, depending on climate and technology of cultivation. About 1.4 million hectares of forest area and 2.3 million hectares of bogland have suitable soil characteristics and sufficient water supply for energy forest plantations. Presently, however, the most interesting land use category for energy forest plantations in Sweden is surplus agricultural land. These areas are estimated to be 0.4—1 million hectares for 1990. Willow plantations thus have a potential of the order of 1,000 PJ per year or more. For the present analysis, two alternative plantation production levels—275 and 655 PJ per year—are assumed. In the former case, petroleum would be used as automotive fuel while, in the latter, this petroleum would instead be entirely replaced with methanol derived from biomass (@ 50 percent conversion efficiency). Even at the lower level of plantation development, though, oil use would be only 18 percent as large as in 1984. The land requirements for this much plantation energy would be 1 to 3 million hectares. For comparison, forestry now uses over 20 million hectares, and agriculture some 3 million hectares.

Our analysis indicates that, with efficient end-use technology, it would be feasible not only to transform the Swedish energy supply problem into one that is not nearly as formidable as it appears in Sweden today, but also to move Sweden largely or even entirely to dependence on domestically produced, renewable energy sources, without running into serious land use or

---

* Biomass can be converted to electricity at an efficiency of 26 percent in small (3 MW), low-cost, gas turbines. An innovative system so designed has been installed at Red Boiling Springs, Tennessee in the United States, where its owners have a contract to sell the produced electricity to the Tennessee Valley Authority.[78] If the waste heat from the gas turbine exhaust were recovered as steam in a waste heat recovery boiler and the steam injected into the combustor, the power output of the turbine would increase to 5 MW and the efficiency to some 30 percent or more.[70]

other environmental constraints. Of course, at higher demand levels, all domestic supply sources would have to be pressed for larger contributions—at an increasingly higher economic and environmental cost—or fossil fuels or nuclear power would have to be reconsidered.

2.6.2. *An End-Use Oriented Energy Future for the United States*

Per capita primary energy use in the United States averaged 11 kW in 1980, 2 ½ times that for Western Europe and Japan and 10 times that for developing countries. With only 5 percent of the world's population but accounting for ¼ of global energy use, the United States is responsible for a large share of the major global externalities associated with energy production and use described in Chapter 1.

The energy price shocks of the 1970's have sharply curbed the growth in energy use in the United States, however. In 1984 per capita primary energy use was 10 percent lower than in 1973, even though real per capita GNP was 17 percent higher (Figure 2.38). The stagnation in energy demand has had several consequences. Excess electrical generating capacity is growing in most parts of the country, as new plants for which construction was started before the energy crises of the 1970's continue to come on line, despite sluggish electricity growth. There has been a shutdown of excess oil refining capacity, as a necessary response to the 18 percent reduction in oil use between 1978 and 1984; and nearly all the ambitious synthetic fuels programmes concocted in the immediate aftermath of the second oil shock have been abandoned.

Despite this dramatic recent break with long-term historical trends, the official prognosis is that growth in energy use will soon resume. The U.S. Department of Energy (DOE) projected in 1983 that in the period 1980-2010 primary energy use will increase 30 percent, with most of the increment coming from coal—up 2.4 times, and nuclear energy—up 4.4 times.[79]

Because of the expectation of a rising world oil price (associated with slowly rising world oil consumption) and a major shift to more costly domestic energy resources, the DOE projected that this energy path would be associated with a doubling of the real average retail price of energy in this period. Considering the energy price increases that have already occurred, this implies a four-fold overall real energy price increase, 1972-2010. In addition to these direct costs, the major global externalities associated with energy production and use would be exacerbated with this energy course.

This course, however, is not inevitable. Recent history (Figure 2.38) suggests that it is not necessary to increase energy use in order to improve economic well-being. And the potential for decoupling energy and economic growth is actually far greater than even this recent experience suggests because of both the ongoing shift to a post-industrial economy (Section 2.4) and the opportunities for energy-efficiency improvement (Section 2.5).

Methodology

To illustrate the importance of these factors for long-term energy planning, energy demand scenarios to the year 2020 are developed here. In these scenarios, the projected activity levels for major energy end-uses are based on extrapolations of historical data, while the energy intensities assigned to these activities are based on the use of energy-efficient technology.

The resulting scenarios should be regarded neither as indicators of the upper limits of the potential for energy-efficiency improvements nor as forecasts of what will happen. The improved technologies considered are either already commercially proven or are based on proto-types or advanced developments that could lead to commercial products in the near future. In all cases, efficiencies are far below thermodynamic limits and, in most instances, possibilities for further

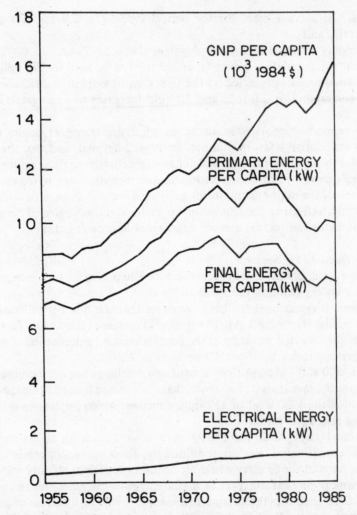

Figure 2.38. Per capita primary energy use, final energy use, electricity use, in kW (actually kW-years per year); and per capita GNP in the United States, in 1984 dollars.

improvement are apparent. At the same time, while the scenarios are based only on end-use technologies which are judged to be cost-effective, the economy will not automatically adopt all of these improvements. The scenarios overstate what can be accomplished through market forces acting alone. New public policy initiatives will be needed to overcome the institutional obstacles that inhibit the introduction of some energy-efficient technologies.

The economic and demographic context: Alternative economic scenarios are considered, involving 50 percent and 100 percent increases in per capita GNP, 1980-2020. In the former, the average annual growth rate for per capita GNP is 1 percent per year, the average for the priod 1973-1983. In the latter, the average growth rate is 1.7 percent per year, corresponding to what would be achieved if:

- The economy continued the ongoing gradual shift from goods to services production,
- Labour productivities of the goods-producing and services-producing sectors grew in the

future at the average rates for the period 1953-1978, 2.1 and 1.3 percent per year, respectively, and

• The economy returned to full employment (Note 2.6.2-A).

Assuming that the population grows from 228 million in 1980 to 296 million in 2020—the "middle population series" projected by the U.S. Census Bureau in 1982 (Note 2.6.2-A)—the two scenarios correspond to 2.0-fold and 2.6-fold increases in aggregate GNP respectively, 1980-2020.

Modelling future energy-using activities: In the residential and transport sectors, the major energy-using activities are relatively few in number, fairly well-defined, and, for the most part, well-established. (Most less well-defined, rapidly changing activities in these sectors contribute little to overall energy demand.) For these sectors, long-term projections based on historical trends are made for each of the major energy-using activities, and the associated energy intensities are indicated explicitly, reflecting the discussion of end-use technologies in Section 2.5. For the commercial and industrial sectors a more aggregated approach is taken.

Future Energy Demand by Sector

With the above assumptions, energy futures can be laid out sector by sector, showing explicitly the potential for energy-demand reduction.

Residential sector: It is assumed for both scenarios that the number of households increases from 82 to 119 million (Note 2.6.2-A). The projected increase reflects two factors: a 30 percent increase in population and the maturing of the population, i.e., a decline in the average household size from 2.8 persons today to about 2.4 persons in 2020.

Since the mid-1970's, the average floor area of new dwellings has not increased (Note 2.6.2-B). Accordingly, for both scenarios it is assumed that the average heated floor space *per household* remains constant at the 1980 level of 139 square metres, which implies an increase by 2020 of 1/6 in the living space *per capita*.

It is also assumed that by 2020 there is 100 percent saturation for all major appliances in the home except air conditioning. For air conditioning, the assumed saturation level is only ⅔, because in most parts of the country where air conditioning is not already widely used, it is not needed. This assumption that the level of major energy-using amenities in the home becomes independent of GNP simplifies the analysis in a manner involving little loss of generality, because of the already high average level of major energy-using amenities in the home. The household amenities which are sensitive to income are characterized either by low power usage, e.g., home entertainment systems and computers, or high power usage but very low load factors, e.g., power equipment for the shop and yard, and are of minor consequence in the overall household energy budget.

For space heating, which accounted for nearly 60 percent of final residential energy use in 1980 (Table 2.19), the problem of retrofitting old "energy-guzzling" houses will persist for decades. Because the housing stock turns over slowly, it is estimated that 3/5 of the houses built before 1981 will be standing in 2020, at which time they will account for 2/5 of all houses (Note 2.6.2-C).

A 30 percent average reduction between 1980-2020 is assumed for the required output of space-heating systems in fuel-heated houses built before 1981. This norm is based on the average, cost-effective savings achieved for the 29 houses retrofitted in the utility-based Modular Retrofit Experiment conducted in New Jersey in 1981 (Note 2.5.3.1-K). No corresponding savings are assumed for existing electrically-heated houses, because these houses involve much lower heat losses than fuel-heated houses (Note 2.6.2-D).

The corresponding norm for new houses built after 1990 is assumed to be the average, climate-adjusted, energy performance achieved for the 97 houses built under Minnesota's Energy Efficient Housing Demonstration Program (EEHDP), a 50 percent reduction relative to typical new construction (Table 2.3 and Note 2.6.2-D). The average heat load for houses built in the period 1981-1990 is assumed to be midway between that of new houses built in the early 1980's and the post-1990 norm.

Besides these heat-load reductions, the norm for heating system performance in 2020 is assumed to be equal to that of the most efficient furnaces and heat pumps commercially available in 1982: an annual fuel utilization efficiency of 95 percent for condensing gas furnaces and a seasonal average coefficient of performance of 2.6 for electric heat pumps (Note 2.6.2-E).

If these heating-load and heating-system efficiency norms were realized, aggregate final energy use for residential space heating in 2020 would be only one-third as large as in 1980 (Table 2.19).

For end-uses other than space heating, energy performance in 2020 is assumed to be that of the most efficient technology commercially available at present (Notes 2.6.2-D, 2.6.2-E, and 2.6.2-F).

For the residential sector as a whole, the resulting final fuel use would be only 40 percent as large in 2020 as in 1980 and electricity use only 70 percent as large (Table 2.19). Final energy use per household would decline from 130 Gigajoules (GJ) in 1980 to 44 GJ in 2020, corresponding to an average rate of decline of 2.7 percent per year, 1980 to 2020. While this is a rapid rate, it is far less than the 7.6 percent per year average rate of decline for the period 1978 to 1982, when final energy use per U.S. household decreased 30 percent.[80]

The energy performance levels indicated for 2020 should not be construed as the limits of what can achieved. Heat-load reductions of ⅔ and more have been demonstrated for existing housing.[81] In new construction, the houses built under the EEHDP programme may be out-performed by super-insulated houses now offered by some builders (Table 2.3). The annual electricity use assumed for refrigerator/freezers (some 580 kWh per year) is more than twice as large as the consumption level that has been measured in a high-efficiency prototype,[82] and there are opportunities for further improvement of other appliances as well.

Some of the projected energy savings would be realized in the normal working of the market. However, the indicated energy-demand levels would most likely not be realized without directed government action, because of numerous institutional obstacles to cost-effective investments in energy-efficiency improvement; these include: inadequate consumer information about energy-saving opportunities and their cost-effectiveness; inadequate availability of capital for investments in energy-efficiency; barriers to life-cycle cost minimization arising from a split of responsibility between those who would make investments and those who would bear operating costs; and, generally, the difficulties associated with implementing energy-efficiency improvements when the commercial infrastructure needed to deliver these improvements to residential users is not yet well-established. Policy choices for dealing with these institutional obstacles are discussed in Chapter 5.

Commercial buildings: The growth of the commercial buildings sector is closely coupled with the expansion of the service-sector. Here the linear relationship established in the period 1970 to 1979 between commercial floor space and service sector employment is assumed to continue, so that in both scenarios the volume of commercial floor space would expand 50 percent, between 1980-2020, in conjunction with a 65 percent increase in service-sector employment in this period

Table 2.19. Final Energy-Use Scenario for the U.S. Residential and Commercial Sectors[a]

I. Residential Sector

End Use	Number of Units in Use (Millions)				Annual Energy Use (Exajoules)[d]			
	1980[b]		2020[c]		1980, actual		2020	
	Elect.	Fuel	Elect.	Fuel	Elect.	Fuel	Elect.	Fuel
Space Heat	14.3	66.7	43.4	75.8	0.29	6.02	0.19	1.93
A/C								
Central	22.2	—	62.4	—	0.24	—	0.37	—
Room (#HH w)	24.8	—	19.0	—	0.09	—	0.04	—
Hot Water	26.1	55.3	43.4	75.8	0.33	1.64	0.14	0.67
Refrig/Freezer	93.0	—	119.2	—	0.56	—	0.25	—
Freezer	33.7	—	119.2	—	0.16	—	0.24	—
Range	44.4	37.1	43.4	75.8	0.12	0.28	0.12	0.40
Dryer	38.3	11.8	43.4	75.8	0.16	0.06	0.14	0.33
Lights	81.6	—	119.2	—	0.29	—	0.13	—
Miscellaneous					0.34	—	0.34	—
TOTALS					2.59	8.00	1.96	3.33

II. Commercial Sector

	Commercial Floor Area (billion square metres)		Annual Energy Use (Exajoules)[f]			
			1980		2020	
	1980	2020	Elect.	Fuel	Elect.	Fuel
TOTALS	4.23	6.37[e]	2.03	4.34	1.80	1.40

(a)   To conform to U.S. convention, final energy use is defined here somewhat differently from the use of this term elsewhere in this book. Here final energy use equals primary energy use less losses in the generation, transmission, and distribution of electricity. Losses associated with petroleum-refining and transport are counted as final use by the industrial and transport sectors, respectively.

(b)   The 1980 distribution of appliances by end uses in 1980 is from Energy Information Administration, June 1982.

(c)   The distribution of appliances in 2020 is based on the following assumptions:
- The number of households, the average size of household, and the distribution of heating systems by energy carrier are as given in Notes 2.6.2-A and 2.6.2-C.
- All electrically heated houses in 2020 will have electric water heaters, dryers, and ranges; all fuel-heated houses will have fuel-heated water heaters, dryers, and ranges.
- All electrically heated houses and ¼ of all fuel-heated houses will have central air conditioning in 2020; ¼ of all fuel-heated houes will have room air conditioners in 2020.
- The saturation level for all major appliances will be 100% in 2020.

(d)   These figures are based on the calculations presented in Notes 2.6.2-D, 2.6.2-E, and 2.6.2-F. The efficiencies of the heating and cooling systems in 2020 are assumed to be equal to those labelled "best 1982 new" in Note 2.6.2-E.

(e)   See Note 2.6.2-G.

(f)   Based on Notes 2.6.2-G and 2.6.2-H.

(Note 2.6.2-G). As in the case of residential buildings, the long life of commercial buildings implies that retrofits of the existing building stock will be important for decades to come. In 2020, it is estimated that 3/5 of the 1980 building stock will still be standing and will at that time account for 2/5 of the total commercial building stock (Note 2.6.2-G).

For retrofits, a 50 percent reduction in final energy use per square metre is assumed relative to 1980 levels, based on the results of a Solar Energy Research Institute survey of experienced architects and engineers which ascertained their judgements of the retrofit potential by the year 2000.[35]

The assumed norm for new buildings constructed after 1990 is the average energy performance of the three buildings designed in an American Institute of Architects Research Corporation exercise aimed at life-cycle cost minimization (Table 2.6). As in the residential case, the assumed norm for buildings constructed in the period 1981-1990 is the average of the performance of buildings being-built at present and in the post-1990 period.

With these assumptions, final fuel use in commercial buildings in 2020 would be about 30 percent of the 1980 level and electricity use would be about 90 percent of the 1980 level (Table 2.19). Again, the energy performance values underlying the commercial sector scenario do not represent the limits of what can be achieved, as our discussion in Section 2.5.3.2 clearly indicates.

As in the residential sector, market forces acting alone cannot be expected to provide adequate incentive to realize the full energy savings indicated here, despite the cost-effectiveness of the savings measured at today's energy prices. An indication of the weakness of market forces is the fact that in 1979 the cost of energy in commercial buildings averaged about $10 per square metre of floor space,[83] only some 2-3 percent of the cost of salaries for the employees who worked in these buildings—the dominant operating cost. The rapidly growing industry marketing ''energy services'' may be able to help achieve some of the energy-saving opportunities in the commercial building sector, especially in large, existing commercial buildings.[84] But new public policy initiatives will be necessary to realize the full cost-effective savings potential (Chapter 5).

Transportation: Light vehicles (automobiles and light trucks), air passenger transport, and truck and rail freight together accounted for more than ¾ of transportation energy use in 1980 (Table 2.20).

Since the average American already spends an hour a day in the car, future levels of automobile use are not likely to increase much with per capita GNP. Thus, it is assumed with both a 50 percent increase in per capita or a 100 percent increase that the number of light vehicles per adult remains constant at the present level of 0.8 and that the average light vehicle is driven the same amount as today (17,000 km/year).

In contrast to the automotive situation, there is *no* evidence that air travel demand is reaching saturation levels. In accordance with a relationship to GNP established in the period 1970-1979, it is assumed that between 1980-2020 air travel per capita increases 2.1 times with a 50 percent increase in per capita GNP or 3.3 times with an increase of 100 percent in per capita GNP (Table 2.20).

The mix of freight between truck and rail is assumed to remain fixed at the 1980 level, and the total volume of freight is assumed to grow slightly more slowly than GNP, as it did in the 1970's—up 1.8 times or 2.3 times with per capita GNP increases of 50 percent or 100 percent (Table 2.20).

It is assumed that the average light vehicle on the road in 2020 would have a fuel economy of 3.1 1/100 km (75 mpg). This is the fuel economy that could be achieved with widespread use of cars that perform like the vehicle modelled by von Hippel and Levi—a VW Rabbit Diesel modified with reduced aerodynamic drag, reduced rolling resistance, an open chamber instead of a pre-chamber diesel, a continuously variable transmission, and a 16 percent reduction in weight (Section 2.5.3.3)[35]—or like the heavy super-mpg car with an adiabatic diesel engine proposed by Cummins Engine Company and NASA Lewis researchers (Section 2.5.3.3)[40]. Again, this target certainly does not represent the limit of what can be achieved. Neither of these cars includes features that would be incorporated in the proposed 100 mpg PERTRAN car—including greatly reduced vehicle weight and regenerative braking.[42]

For both truck and air passenger transport, a 50 percent reduction in energy-intensity is assumed, 1980-2020, along the lines of discussion in Section 2.5.3.3.

Under the above conditions, total transportation energy use in 2020 would be reduced to 60 percent of the 1980 level with a per capita GNP increase of 50 percent or 69 percent of the 1980 level with a 100 percent increase in per capita GNP (Table 2.20). A striking feature of the energy use in transportation in this analysis is the much diminished contribution from light vehicles—24 to 28 percent in 2020, down from 59 percent in 1980.

Again public policies are needed to promote the achievement of the technically feasible, cost-effective savings identified here. The price of energy may be effective in promoting efficiency improvements in areas such as air travel, where the cost of fuel is a major fraction of the total

Table 2.20. Alternative Final Energy-Use Scenarios for the U.S. Transportation Sector
(Exajoules per Year)

Transport Mode	1980		2020			
			GNP/C up 50%		GNP/C up 100%	
	Fuel	Elect.	Fuel	Elect.	Fuel	Elect.
Automobiles and Light Trucks	12.3[a]	—	3.5[b]	—	3.5[b]	—
Commercial Air Passenger Transport	1.6[c]	—	2.2[d]	—	3.5[d]	—
Intercity Truck Freight	1.6[e]	—	1.5[f]	—	1.9[f]	—
Rail Freight	0.6[g]	—	0.5[h]	0.2[h]	0.7[h]	0.2[h]
Other	4.7	—	4.7	–	4.7	—
TOTALS	20.8	—	12.4	0.2	14.3	0.2

(a) *Source:* Energy and Environmental Analysis, April 15, 1982.

(b) Here it is assumed that (1) the number of light vehicles per person aged 16 and over is 0.80 (compared to (0.78 in 1980) and (2) the average light vehicle is driven 17,000 km (10,600 miles) per year, the same amount as in 1980. Following the discussions of the technical possibilities outlined in Section 2.5.3.3, it is assumed that the average fuel economy of light vehicles is increased by 2020 to 3.1 litres per 100 km (75 mpg)—an average for a car population that might be 20% 2-passenger vehicles @ 2.1 litres per 100 km (110 mpg), 50% 4-passenger vehicles @ 3.0 litres per 100 km (78 mpg) and 30% 5/6 passenger cars plus light trucks @ 4.1 litres per 100 km (58 mpg.)

(c) Total air travel [domestic and international travel via certified route air carriers plus travel by charter (supplemental) air carriers] in 1980 amounted to 420.6 billion p-km @ an average fuel intensity of 3.82 MJ per p-km (23.1 passenger miles per gallon). See Bureau of the Census, December 1981.

(d) Here it is assumed that: (1) the historical relationship between revenues R and GNP (each expressed in billion of 1972$) established in the period 1970-1979:
$$R = 9.017 \times 10^{-5} \times (GNP)^{1.64}, r = 0.959,$$
persists in the future; and (2) the revenue per p-km remains constant at the 1980 level ($0.038 per p-km). On the basis of the discussion in Section 2.5.3.3, it is assumed that the average energy-intensity of air travel can be reduced in half, 1980-2020.

(e) Total intercity truck freight in 1980 amounted to 825 billion tonne-km @ an average fuel-intensity of 2.0 MJ per tonne-km. See G. Kulp and M.C. Holcomb, 1982.

(f) Here it is assumed that: (1) the historical relationship between the total volume of truck and rail freight F(t+r) (in billion tonne-km) and GNP (in billion 1972$) established in the period 1971-1980
$$F(t+r) = 4.14 \times (GNP)^{0.86}, r = 0.94,$$
persists in the future; and (2) the truck/rail mix of freight is maintained at the 1980 level. On the basis of the discussion in Section 2.5.3.3, it is assumed that the energy-intensity of truck freight is reduced in half, 1980-2020.

(g) The rail freight volume in 1980 amounted to 1345 billion tonne-km @ an average fuel-intensity of 470 kJ per tonne-km. See G. Kulp and M.C. Holcomb, 1982.

(h) It is assumed here that: (1) by 2020 half of rail freight is electrified; and (2) in terms of final energy, the electricity intensity of electric rail freight is 1/3 that of diesel rail freight.

cost, but in the case of the automobile, the price incentive alone may be inadequate to achieve high fuel economy. Even at high gasoline prices, market forces are particularly weak in promoting fuel economies much better than about 8 1/100 km (30 mpg)—both because the cost of fuel represents a small fraction of the total cost of owning and operating a car and because it appears that over a wide range of fuel economies the total cost per km is quite insensitive to the fuel economy level (Figure 2.10). Since society as a whole would be better off because high fuel economy would reduce its dependence on oil imports, even though the consumer is little motivated to seek high fuel economy, some form of market intervention to promote high fuel economy is desirable (Chapter 5).

The industrial sector: Modelling the likely mix of future industrial activity and the energy demand associated with this activity is an especially difficult task in the U.S. context. The U.S. industrial base, like that of other industrialized countries, is being fundamentally reshaped, as the country continues the transition to a Post-Industrial Economy (Section 2.4.3); hence, the mix of industrial activities several decades in the future can be expected to be quite different from that of today. Also, unlike the situation in Sweden, which has a small open economy, for which most of the industrial energy use is concentrated in just two industries (steel and paper), the U.S. industrial base is quite diverse. Because of these complexities, the industrial sector is modelled here with more aggregated descriptors than those used in other sectors.

Consider first the growth of total industrial output.* If, in the future, industrial output grows 0.83 times as fast as GNP, as it did in the 1970's (Note 2.6.2-A), reflecting the ongoing shift from goods to services production, the industrial sector as a whole would expand 1.8 times with a 50 percent increase in per capita GNP increase and 2.3 times with a 100 percent increase in per capita in GNP, 1980-2020.

Within the industrial sector, the energy-intensive basic materials processing (BMP) and the mining, agriculture, and construction (MAC) subsectors and the value-added-intensive fabrication and finishing activities that make up "other manufacturing" (OMFG) are treated separately here.

The trend away from the processing of basic materials in the United States (Section 2.4.3) is captured by the lower curves in Figure 2.39, which show the trends, from 1960 through 1984, in indices of aggregate materials consumption and production per capita (energy-weighted consumption and production of paper, steel, aluminium, petroleum refinery products, cement plus a combination of 20 large volume industrial chemicals).[18] The curves shown here are consistent with essentially zero average growth in materials use and production per capita since the early 1970's. Market analyses for eight of the most significant basic materials, from an energy use standpoint, indicate very little likelihood of revival from the demand stagnation of the last decade.[14]

In light of this evidence, saturation, i.e., zero growth per capita, is assumed here for the output of the BMP and MAC subsectors of industry, which would thus grow in the future only as fast as population, or 30 percent, 1980-2020. Together with the assumption about total industrial output growth, this implies that the OMFG subsector would grow 2.3 times with a 50 percent increase in per capita GNP and 3.3 times with a 100 percent increase, 1980-2020.

It is also assumed here that the energy-intensity of each subsector is reduced in half, 1980-2020. Such an average improvement in energy productivity seems to be well within the range of practical

---

Here industrial output is Gross Product Originating, the value-added measure compiled by the Bureau of Economic Analysis of the U.S. Department of Commerce.

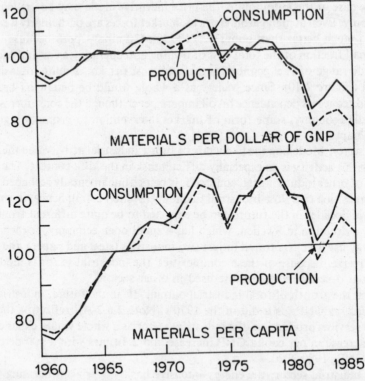

Figure 2.39.   Trends in materials consumption and production per dollar of GNP and per capita in the United States. Here aggregate consumption and production are measured by indices which are energy-weighted sums of physical consumption and production of paper, steel, aluminium, petroleum refinery products, cement plus a combination of 20 large-volume industrial chemicals, normalized to 100 in 1967. The energy weights are energy-of-manufacture intensities from the late 1970's.
Source: Marc Ross, Eric D. Larson, and Robert H. Williams, 1985.

feasibility, in light of the large potential for energy-efficiency improvement in industry (Section 2.5.3.4), very little of which has been captured to date. The economic incentive is also especially strong in the industrial sector to seize these opportunities; the average final energy price to industrial consumers tripled in real terms, 1972-1982, compared to only a doubling of the real energy price in other sectors in this period (Note 2.1-D).

With the above assumptions, final energy demand in industry would decrease by 30 percent between 1980 and 2020 if per capita GNP increases 50 percent or by 24 percent if per capita GNP increases 100 percent (Table 2.21). It is assumed under both economic growth rates that the historical trend towards electrification continues, so that the electrical share of final demand increases from 11 percent in 1980 to 25 percent in 2020.

Table 2.22 shows how trends in important industrial parameters in these projections compare to the changes that actually took place in the historical period 1973-1984. The high growth scenario, for example, involves the same average future GNP growth rate (2.5 percent per year) as in this historical period, yet considerably more modest rates of decline both for final industrial energy use (0.7 percent versus 1.6 percent per year) and for the ratio of industrial energy use to GNP (3.2 percent per year versus 4.1 percent per year). This scenario, however, involves somewhat slower electricity demand growth (1.3 percent versus 1.9 percent per year).

An End-Use Oriented Energy Strategy

While the industrial sector is probably more responsive than any other sector to market forces in making energy-efficiency improvements, there is still need for new policy initiatives to help improve the economic climate for adopting more energy-efficient technology.

It is particularly important that energy-intensive industries have adequate incentives to seek major improvements in industrial productivity by introducing wholly new processes and products—the adoption of which would typically lead to far greater energy savings than

Table 2.21. Alternative Final Energy-Use Scenarios for the U.S. Industrial Sector

	1972	1980	2020	
			GNP/C up 50%	GNP/C up 100%
Industrial Output (billion 1972$)	406.1[a]	464.7[a]	837[a]	1074[a]
Annual Fuel Use (Exajoules)	25.2[b]	23.7[b]	14.1[c]	15.1[c]
Annual Electricity Use (Exajoules)	2.3[b]	3.0[b]	4.7[c,d]	5.1[c,d]
Fuel Intensity (MJ/1972$)	62.1	51.0	16.8	14.1
Electricity Intensity (MJ/1972$)	5.7	6.4	5.6	4.7
Final Energy Intensity (MJ/1972$)	67.8	57.5	22.5	18.8

(a) Here industrial output is gross product originating in industry (manufacturing, construction, mining, and agriculture, forestry, and fisheries) (Note 2.6.2-A).

(b) Historical data on fossil-fuel and electricity use are from Energy Information Administration, April 1983a and April 1983b. Industrial wood consumption data are from Energy Information Administration, August 1982.

(c) Final energy use for the year 2020 is determined by assuming that: (1) the outputs of the basic materials processing (BMP) and mining, agriculture, and construction (MAC) subsectors of industry grow only as fast as the population, 1980-2020, because of demand saturation, and (2) the average energy-intensity of each industrial subsector [BMP, MAC, and other manufacturing (OMFG)] is reduced in half, 1980-2020, via energy-efficiency improvements. Assumption (1) implies that in the period 1980 to 2020 the BMP and MAC subsectors grow 30% and the OMFG subsector grows 2.3-fold and 3.3-fold, associated with a 50% and 100% increase in per capita GNP, respectively.

(d) The electrical fraction of final energy use in industry in 2020 is assumed to be 0.25 (up from 0.11 in 1980). This increase is based on the assumption that the relationship established in the period 1970 to 1980, between Fel, the electrical fraction of final industrial and time, persists in the future:

$$\text{Fel (t)} = -6.738 + 0.00346 \times t, \; r = 0.9836.$$

Table 2.22. Historical and Projected Trends for Selected U.S. Industrial Indicators

	(Average annual growth rates, percent per annum)		
Indicator	Average Historical Rate (1973 - 1984)	Low Growth Scenario (1980 - 2020)	High Growth Scenario (1980 - 2020)
GNP	+ 2.5	+ 1.7	+ 2.5
Final			
Fuel Use[a]	− 2.0	− 1.3	− 1.1
Electricity Use	+ 1.9	+ 1.1	+ 1.3
Energy Use[a]	− 1.6	− 0.9	− 0.7
Final Energy Use/GNP	− 4.1	− 2.6	− 3.2
Electricity Use/GNP	− 0.6	− 0.6	− 1.1

(a) Includes industrial wood use.

modification of existing processes with energy-saving "retrofits".[11] Because the demand for basic materials is growing so slowly (and is actually declining for some basic materials), the basic materials processing industries in the United States, as in other industrialized countries, are rather unpromising contexts for innovation. To help foster innovation in the face of this stagnant demand, policies that encourage industrial R&D and investments in new plant and equipment generally should be given careful consideration.

Integrated Results

The results of the demand analysis are summarized in Table 2.23. By pursuing throughout the economy opportunities for energy-efficiency improvement, U.S. per capita final energy use between 1980-2020 could be reduced, from 9.0 to 4.3 kW with a 50 percent increase in per capita GNP or to 4.6 kW with a 100 percent increase. While the electrical share of final energy use would nearly double, demand for electricity overall would increase only 0.3 percent or 0.5 percent per year.

Perhaps the most uncertain quantities underlying this projection are the energy intensities assumed for commercial buildings and automobiles and the growth of the basic materials processing sector of industry.

For commercial buildings there has been a tendency to underestimate *new* building loads, especially for office computers and the air-conditioning systems required for their support.[85] It is uncertain whether this is a long-term problem or will become unimportant with technological advances that reduce the energy-intensity of computers and other office machinery. However, if the average final energy intensity of commercial buildings were 1.0 GJ per square metre per year, ⅔ of the 1980 average and double what was assumed here, aggregate US energy use in 2020 would be up only 6%.

The 75 mpg fuel economy assumed for the automobile in 2020 could be achieved for large cars only with the commercial success of certain advanced technologies (e.g., the adiabatic diesel engine). With present technology 75 mpg could be achieved only with some down-sizing. In light of uncertainties about future technologies and the American love affair with large cars, the 75 mpg target might not be achieved even with supportive new public policies. However, if the average fuel economy were 50 mpg instead of 75 mpg, US energy use in 2020 would be only 4% higher, for the case where per capita GNP increases 100 percent.

For industry, it was assumed that the BMP and MAC sectors grow in proportion to the population, despite the powerful evidence for essentially zero growth in the production of basic materials to the year 2000[18]. If, instead, the BMP and MAC sectors were not to grow, 1980-2020, the result would be aggregate energy use being 9% lower than that projected.

This sensitivity analysis shows that the principal finding of this scenario analysis, that final energy use can be reduced in half even if per capita GNP doubles, is not very sensitive to changes in some of the boldest assumptions involved.

Many efficiency improvements will be made naturally as the capital stock grows and turns over—since typical new energy-using buildings and equipment tend to be much more energy-efficient than the existing stock. Accordingly, it is very likely that the recent downward trend in per capita energy use (Figure 2.38) will continue.

But to bring about a reduction in per capita energy use of 50 percent or more would require new policy initiatives—some of which would be specific to particular energy-using sectors, while others would be of a broad-gauged nature, intended to help promote a more even-handed treatment in the marketplace of investments in energy supply and in energy-efficiency improvement (Chapter 5).

Table 2.23. Alternative Scenarios for Total and Per Capita Final Energy Use in the U.S.

	1980 (actual)				2020 GNP/C up 50%				2020 GNP/C up 100%			
	Total (Exajoules per year)			Per capita (kW)	Total (Exajoules per year)			Per Capita (kW)	Total (Exajoules per year)			Per capita (kW)
	Fuel	Elect.	Total		Fuel	Elect.	Total		Fuel	Elct.	Total	
Residential[a]	8.0	2.6	10.7	1.5	3.3	2.0	5.3	0.6	3.3	2.0	5.3	0.6
Commercial[a]	4.3	2.0	6.3	0.9	1.4	1.8	3.2	0.3	1.4	1.8	3.2	0.3
Transportation[b]	20.8	—	20.8	2.9	12.4	0.2	12.6	1.3	14.3	0.2	14.5	1.5
Industry[c]	23.7	3.0	26.7	3.7	14.1	4.7	18.8	2.0	15.1	5.1	20.2	2.2
TOTALS	56.8	7.6	64.4	9.0	31.2	8.7	39.9	4.3	34.1	9.1	43.2	4.6

(a) See Table 2.19.
(b) See Table 2.20.
(c) See Table 2.21.

Table 2.24. U.S. Primary Energy Supplies (Exajoules per Year)[a]

	1980	2020	
		GNP/C up 50%	GNP/C up 100%
Coal	16.22	11.4[b]	13.3[b]
Oil and Natural Gas	57.60	25.9[c]	27.8[c]
Nuclear	2.89	7.5[d]	7.5[d]
Hydro	0.99	1.4[d]	1.4[d]
Wind and Photovoltaics	—	0.9[d]	0.9[d]
Biomass	2.34	4.6[e]	4.6[e]
TOTALS	80.04	51.7	55.5

(a)   Here nuclear energy is counted as the thermal energy released in the fission process (33% thermal efficiency); hydro, wind, and photovoltaics as the electricity produced.

(b)   Assumed to be the residual. Coal-fired power plants are assumed to have a 40% thermal efficiency.

(c)   Assumed to be ½ of total primary energy use. Oil and gas-fired central station power plants are assumed to be steam-injected gas turbines with a thermal efficiency of 50% (Marine and Industrial Engine Projects Department, General Electric Company, July 1984). Gas-fired cogeneration systems are assumed to require 1.5 units of fuel, in addition to what would otherwise be required to produce steam alone, for each unit of electricity provided.

(d)   See Table 2.25.

(e)   The level projected for 2010 in Office of Policy, Planning, and Analysis, October 1983. It is assumed that ½ of all cogenerated electricity is based on biomass. Biomass-fired cogeneration systems [e.g., sawdust-fired gas turbines (J.T. Hamrick, February 1984)] are assumed to require 1.5 units of fuel, in addition to what would otherwise be required to produce steam alone, for each unit of electricity provided. Biomass used for other purposes is assumed to be converted to useful solid, liquid, and gaseous energy carriers @ an average conversion efficiency of 70%.

Table 2.25. U.S. Electricity Supplies (Exajoules per Year)

	1980	2020	
		GNP/C up 50%	GNP/C up 100%
Coal	4.18	2.3[a]	2.5[a]
Oil and Natural Gas	2.13	0.9[b]	1.0[b]
Nuclear	0.90	2.5[c]	2.5[c]
Hydro	0.99	1.4[d]	1.4[d]
Wind and Photovoltaics		0.9[d]	0.9[d]
Cogeneration	0.15	1.4[e]	1.5[e]
TOTALS	835	9.4 (f)	9.8 (f)

(a)   Assumed to be the residual.

(b)   Assumed to be 10% of the total.

(c)   It is asumed that in 2020 nuclear generating capacity is 120 GW(e), the level projected for the year 2000 in Office of Policy, Planning, and Analysis, U.S. Department of Energy, October 1983. This implies that only plants constructed after 2000 are to replace retired nuclear units.

(d)   The level projected for 2010 in Office of Policy, Planning, and Analysis, U.S. Department of Energy, October 1983.

(e)   For 1978 it has been estimated that the industrial steam load technically suitable for association with cogeneration is some 13% of total industrial fuel use (M.H. Ross, 1981). Here the same percentage is assumed as a basis for estimating the cogeneration potential for 2020. It is further assumed that the average electricity/heat output ratio is 210 kWh per GJ, which is typical for gas turbine-based systems (R.H. Williams, 1978).

(f)   Electricity production is the electricity demand shown in Table 2.23, divided by 0.93, to account for transmission and distribution losses.

Energy Supply Implications

If total final energy demand were as low as indicated in Table 2.23, the energy supply problem would become quite manageable. Illustrative supply mixes are shown for primary energy and electricity generation in Table 2.24 and 2.25, respectively.

The aggregate numbers shown in Table 2.24 imply that primary energy use per capita would decline from 11.1 kW in 1980 to 5.5 kW with a 50 percent increase in per capita GNP or to 5.9 kW with a 100 percent increase. The fractional reduction in primary energy use is somewhat less than that for final energy, owing to the projected greater role for electricity.

The energy supply mix shown here is compatible with the solutions to the global problems described in Chapter 1. Tables 2.24 and 2.25 show that energy requirements associated with substantial economic growth could be met even if:

- Dependence on oil and gas were reduced to less than 50 percent of the 1980 level;
- The use of coal were reduced to 70-80 percent of the 1980 level; and
- There were no expansion of nuclear power beyond what is planned for the year 2000.

Conclusion

Exploiting the many economic opportunities for more efficient energy use could lead to a dramatic reduction in overall energy use along with continuing economic prosperity.

Pursuit of such a course would free up for other activities economic resources that are now directed to energy. There are two reasons for this. At today's prices and for a wide range of energy-saving options, it is less costly to provide needed energy services by investing at the margin in energy-efficiency improvements instead of in energy supply expansion. And, if energy demand is significantly reduced this way, energy prices would generally be lower because the more costly supply options could be avoided.

In addition to the direct economic benefits, considerable flexibility would be gained in putting together an energy supply mix. In a low energy-demand future, not all energy supply options need be pushed to their limits. It becomes possible to avoid or reduce dependence on the more troublesome supply options and to provide energy supplies in a manner consistent with the solutions of the world's major problems relating to equity, the economy, the environment, and security—all of which have strong links to energy (Chapter 1).

References

Section 2.1

1. International Energy Agency, *World Energy Outlook,* Paris, 1982.
2. Energy Information Administration, U.S. Department of Energy, *State Energy Price and Expenditure Report 1970-1982,* DOE/EIA-0376(82), April 1985, 1984.
3. Office of Policy, Planning, and Analysis, U.S. Department of Energy, *Energy Projections to the Year 2010; a Technical Report in Support of the National Energy Policy Plan,* DOE/PE-0029/2, October, 1983.

Section 2.2

4. Paul Macavoy, *Economic Strategy for Developing Nuclear Breeder Reactors,* MIT Press, Cambridge, 1969.
5. U.S. Department of Energy, "The National Energy Policy Plan," DOE/S-0014/1, October 1983.
6. R.L. Graves, C.D. West, and E.C. Fox, "The Electric Car: Is It Still the Vehicle of the Future?" Oak Ridge National Laboratory Report No. ORNL/TM-7904, August 1981.
7. Charles A. Lave (Department of Economics, University of California at Irvine), "The Potential U.S. Market for Micro/Mini Cars," paper presented at the International Policy Forum, Hakone Prince, Hakone, Japan, May 1982.

8. Solar Energy Research Institure, *A New Prosperity: Building a Sustainable Future, The SERI Solar/Conservation Study,* Brick House Publishing, Andover, Mass., 454 pp., 1981.
9. Philip S. Schmidt, *Electricity and Industrial Productivity,* Pergamon Press, 1984.
10. Marc Ross, "Trends in the Use of Electricity in Manufacturing," *Technology and Society* (the publication of the Institute of Electrical and Electronics Engineers Society on Social Implications of Technology), March 1986.

Section 2.3

11. Charles A. Berg, "Energy Conservation in Industry: the Present Approach, the Future Opportunities," a report prepared for the President's Council on Environmental Quality, May 1979.
12. Robert M. Solow, "Technical Change and the Aggregate Production Function", *The Review of Economics and Statistics,* XXXIX, No. 3, pp. 312-320, August 1957.

Section 2.4

13. L. Ingelstam, "Arbetets varde och tidens bruk—en framtidsstuddie," ("The Value of work and the use of time—a future study"), Liber Forlag, Stockholm, 1980.
14. Eric D. Larson, Robert H. Williams, and Andrew Bienkowski, "Materials Consumption Patterns and Industrial Energy Demand in Industrialized Countries," PU/CEES Report No. 174, Center for Energy and Environmental Studies, Princeton University, Princeton, New Jersey, December 1984.
15. J. Gershuny, *After Industrial Society—The Emerging Self-service Economy,* MacMillan Press, 181 pp., 1978.
16. Scott Burns, *The Household Economy: Its Shape, Origins, & Future,* 252 pp., Beacon Press, Boston, 1975.
17. Edgar L. Feige, "How Big Is the Irregular Economy?" *Challenge,* pp. 5-13, November-December 1979.
18. Marc H. Ross, Eric D. Larson, and Robert H. Williams, "Energy Demand and Materials Flows in the Economy," PU/CEES Report No. 193, Center for Energy and Environmental Studies, Princeton University, Princeton, New Jersey, July 1985.
19. Steven Greenhouse, "Shake-out for Petrochemicals", *The New York Times,* February 20, 1984.
20. Marc H. Ross and Arthur H. Purcell, "Decline of Materials Intensiveness: the U.S. Pulp and Paper Industry," *Resources Policy*, pp. 235-249, December 1981.

Section 2.5.1

21. W.H. Carnahan *et al., Effect Use of Energy,* a report of the 1974 American Physical Society's Summer Study on Technical Aspects of Efficient Energy Utilization, Vol. 25 of the American Institute of Physics Conference Proceedings, 1975.
22. Marc H. Ross and Robert H. Williams, *Our Energy: Regaining Control,* McGraw-Hill, 354 pp., New York, 1981.

Section 2.5.2

23. Electric Power Research Institute, *Technical Assessment Guide,* Special Report No. PS-1201-SR, prepared by the Technical Assessment Group of the EPRI Planning Staff, July 1979.
24. U.S. Council of Economic Advisors, *Economic Report to the President,* Washington, D.C., 1981.
25. J. Hausman, "Individual Discount Rates and the Purchase and Utilization of energy-Using Durables," *The Bell Journal of Economics,* Vol. 10, pp. 33-54, 1976.
26. Alan K. Meier and Jack Whittier, "Consumer Discount Rates Implied by Consumer Purchases of Energy-Efficient Refrigerators," *Energy, The International Journal,* Vol. 8, No. 12, pp. 957-962, 1983.
27. J.E. McMahon and M.D. Levine, "Cost/Efficiency Tradeoffs in the Residential Appliance Marketplace," *Proceedings of the American Council for an Energy Efficient Economy's 1982 Summer Study of Energy Efficient Buildings, Santa Cruz, California, August 1982,* Energy Information Center, New York, 1983.

Section 2.5.3.1

28. Lee Schipper, Stephen Meyers, and Henry Kelly, *Coming In From the Cold: Energy-Wise Housing in Sweden,* 104 pp., Seven Locks Press, Cabin John, Maryland, 1985.
29. D.T. Harrje, G.S. Dutt, and J. Beyea, "Locating and Eliminating Obscure but Major Energy Losses in Residential Housing," *ASHRAE Transactions.* Vol. 85(II), pp. 521-534, 1979.

30. G.S. Dutt, "House Doctor Visits—Optimizing Energy Conservation Without Side Effects," *Proceedings of the International Energy Agency's Conference on New Energy Conservation Technologies and their Commercialization,* Springer, West Berlin, 1981.
31. G.S. Dutt, M. Lavine, B. Levi, R. Socolow, "The Modular Retrofit Experiment: Exploring the House Doctor Concept," Center for Energy and Environmental Studies Report No. 130, Princeton University, 1982.

Section 2.5.3.2

32. W.S. Johnson and F.E. Pierce, "Energy and Cost Analysis of Commercial Building Shell Characteristics and Operating Schedules," Oak Ridge National Laboratory Report No. ORNL/CON-39, April 1980.
33. Frank von Hippel, "The Energy Demand of U.S. Buildings," Center for Energy and Environmental Studies Report No. 119, Princeton University, Princeton, NJ, June 1981.
34. R.W. BARNES *et al.*, "A User's Guide to the ORNL Commercial Energy Use Mode," ORNL/CON-44, May 1980.
35. Solar Energy Research Institute, *A New Prosperity: Building a Sustainable Energy Future,* the SERI Solar/Conservation Study, Brickhouse, Andover, Mass (1981).
36. H. Ross and S. Whalen, "Building Energy Use Compilation and Analysis (BECA) Part C: Conservation Progress in Retrofitted Commercial Buildings," in *What Works: Documenting Energy Conservation in Buildings,* J. Harris and C. Blumstein, eds., Proceedings of the Second Summer Study of Energy Efficient Buildings, Santa Cruz, California, August 1982, American Council for an Energy-Efficient Economy, Washington, D.C., 1983.

Section 2.5.3.3

37. International Energy Agency, *Energy Statistics, 1975-1979,* Paris, 1981.
38. Charles L. Gray, Jr., and Frank von Hippel, "The Fuel Economy of Light Vehicles", *Scientific American,* May 1981, pp. 36-47.
39. Frank von Hippel and Barbara Levi, "Automotive Fuel Efficiency: the Opportunity and the Weakness of Existing Market Incentives," *Resources and Conservation,* Vol. 10, pp. 103-124, 1983.
40. C.N. Cochran and R.H.C. McClure, "Automotive Materials Decision: Energy, Economics and Other Issues," SAE Paper #820149, presented at the SAE International Congress and Exposition, Detroit, Michigan, February 22-26, 1982.
41. Philip Burgert, "Japan Aims for Non-Metallic Auto in '90s", *American Metal Market/Metalworking News,* April 25, 1983.
42. S.L. Fawcett and J.C. Swain, "Prospectus for a Consumer Demonstration of a 100 MPG Car," Battelle Memorial Institute Paper, March 1983.
43. Deborah Bleviss, "Prospects for Future Fuel Economy Innovation," report of the Federation of American Scientists, Washington, DC, 1985.
44. R.R. Sekar and R. Kamo [Cummins Engine Company, Columbus, Indiana] and J.C. Wood [NASA Lewis Research Center, Cleveland, Ohio], "Advanced Adiabatic Diesel Engines for Passenger Cars," Paper # 840434, presented at the International Congress & Exposition of the Society of Automotive Engineers, Detroit, Michigan, February 27 - March 2, 1984, and reprinted from SP-571, *Adiabatic Engines: Worldwide Review.*
45. _____, "Goodbye to Heavy Metal," *The New York Times,* November 17, 1985.
46. J.B. Smith, "Trends in Energy Use and Fuel Efficiencies in the U.S. Commercial Airline Industry", U.S. Department of Energy, 1981.
47. A.B. Rose, "Energy Intensity and Related Parameters of Selected Transportation Modes: Freight Movements," Oak Ridge National Laboratory, ORNL/TM-6700, 1979.

Section 2.5.3.4

48. International Energy Agency, *Energy Balances of OECD Countries, 1971/1981*, OECD, Paris, 1983.
49. Marc H. Ross, "Energy Consumption by Industry", *Annual Review of Energy,* Vol. 6, pp. 379-416, 1981.
50. G.J. Hane *et al.*, "A Preliminary Overview of Innovative Industrial Materials Processes," a report prepared for the U.S. Department of Energy, by the Pacific Northwest Laboratory, PNL-4505, UC-95f, September 1983.
51. Charles K. Hyde, *Technological Change and the British Iron Industry, 1700-1870,* Princeton University Press, Princeton, 1977.

52. Elisabeth K. Rabitsch, "Blast Furnaces and Steel Mills—SIC 3312" Chapter 24 in John G. Meyers *et al.*, *Energy Consumption in Manufacturing,* a report of the Conference Board to the Energy Policy Project of the Ford Foundation, Ballinger, Cambridge, 1974.

53. Elias P. Gyftopouios, Lazaridis, and Thomas F. Widmer, *Potential Fuel Effectiveness in Industry,* a report to the Energy Policy Project of the Ford Foundation, Ballinger, Cambridge, 1974.

54. United Nations, *Annual Bulletin of Steel Statistics for Europe, 1979,* Vol. VII, New York, 1980.

55. Lawrence C. Long (Director—Process Research, Corporate Research and Technology, Armco, Inc., Middletown, Ohio), "Steel-making in the Future," *Iron and Steel Engineer,* pp. 48-53, December 1981.

56. Robert V. Ayres, "Final Report on Future Energy Consumption by the Industrial Chemicals Industry (SIC 28), Appendix to *The Chemicals Industry,* Vol. 5 of 9, in *Industrial Energy Productivity Project Final Report,* prepared by Energy and Environmental Analysis, Inc., Arlington, Va., for the Assistant Secretary for Conservation and Renewable Energy Resources, U.S. Department of Energy, DOE/CS/40151-1, February 1983.

57. J.E. Burch, J.L. Otis, and R.W. Hale, "Pilot Study to Select Candidates for Energy-Conservation Research for the Chemical Industry," report prepared for the U.S. Department of Energy by the Battelle Columbus Laboratories, DOE/TIC-11118, Columbus, Ohio, 1979.

58. L. Riechert, "The Efficiency of Energy Utilization in Chemical Processes," *CES,* vol 29, p. 1613, 1974.

59. Avigdor M. Ronn, "Laser Chemistry", *Scientific American,* pp. 114-129, May 1979.

60. H.K. Lonsdale, *Membrane Separation in the 1980s,* Bend Research, Inc., Bend, Oregon, 1982.

61. C.T. Hill and C.M. Overby, "Improving Energy Productivity Through Recovery and Reuse of Wastes," in J.C. Sawhill, ed., *Energy Conservation and Public Policy,* Prentice-Hall, Englewood Cliffs, NJ, 1979.

62. J.D. Birchall and Anthony Kelly, "New Inorganic Materials", *Scientific American,* pp. 104-115, May 1983.

63. A.D. Little, Inc., *Energy Efficiency and Electric Motors,* prepared for the Office of Industrial Programs, U.S. Federal Energy Administration, August, 1976.

64. N. Ladomatos, N.J.D. Lucas, W. Murgatroyd, "Industrial Energy Use—I: Power Losses in Electricity-Driven Machinery," *International Journal of Energy,* Vol. 2, pp. 179-196, 1978.

65. N. Ladomatos, N.J.D. Lucas, W. Murgatroyd, "Industrial Energy Use—III: The Prospects for Providing Motive Power in a Machine Tool Shop from a Centralized Hydraulic System," *International Journal of Energy,* Vol 3, pp. 19-28, 1979.

66. N. Mohan, "Techniques for Energy Conservation in AC Motor-Driven Systems", report prepared for the Electric Power Research Institute, EPRI EM-2037, Palo Alto, California, U.S.A., September 1981.

67. D.J. Ben-Daniel and E.E. David, Jr., "Semiconductor Alternating-Current Motor Drives and Energy Conservation" *Science,* Vol. 206, pp. 773-776, 1979.

68. Dan Zegart, "AC Drive Costs Plummet, Spur HVAC Applications," *Energy User News,* Vol. 7, No. 10, March 26, 1981.

69. R.H. Williams, "Industrial Cogeneration," *Annual Review of Energy,* Vol. 3, pp. 313-356, 1978.

70. Eric D. Larson and Robert H. Williams, "Steam-Injected Gas Turbines," a Pamphlet Paper of the American Society of Mechanical Engineers, presented at the Gas Turbine Conference, Dusseldorf, West Germany, June 1986.

71. Robert H. Williams, "Potential Roles for Bioenergy in an Energy-Efficient World," *Ambio,* Vol. XIV, No. 4-5, pp. 201-209, 1985.

72. Alvin L. Alm and Kathryn L. Mowry, "PURPA: Purpose and Prospects," Discussion Paper No. E-83-03, John F. Kennedy School of Government, Harvard University, 1983.

73. *Cogeneration and Small Power Monthly,* Farragut Station, Washington, D.C., 1986.

74. K. Grubs and W.M. Alley, "Benefits of Electric Utility Participation in Industrial Cogeneration," *Proceedings of the Forest Products Research Society Energy Conference,* Atlanta, Georgia, 1980.

Section 2.6.1

75 P. Steen, T.B. Johansson, R. Fredricksson, and E. Bogren, *Energy—For What and How Much?,* Liber Forlag, Stockholm, 292 pp., 1981 (in Swedish); summarized in T.B. Johansson, P. Steen, E. Bogren, and R. Fredricksson, "Sweden Beyond Oil—the Efficient Use of Energy," *Science,* Vol. 219, pp. 355-361, 1983.

76 J. Norgard, *Husholdninger of energi,* Polyteknisk Forlag, Copenhagen, 1979 (In Danish).

77. T.B. Johansson and P. Steen, *Ambio* Vol. 7, pp. 70-74, 1978; *Bulletin of the Atomic Scientists,* pp. 19-22, October 1979. See also M. Lonnroth, T.B. Johansson, and P. Steen, "Sweden Beyond Oil: Nuclear Commitments and Solar Options," *Science,* Vol. 208, pp. 557-563, 1980. The full report *Solar Sweden* is available from the Secretariat

for Future Studies, Box 6710, S-11385 Stockholm. Reprinted by the U.S. House of Representatives, 96th Congress first session, June 14, 21, 1979 (No. 68), (Committee on Interstate and Foreign Commerce) Volume VI, Washington D.C., 1979.

78. Joseph T. Hamrick, "Development of Wood-Burning Gas Turbine Systems," *Modern Power Systems*, Vol. 4, No. 6, June 1984; Richard Layne, "This Gas Turbine Burns Sawdust," *Popular Science*, November 1984; J.T. Hamrick, "Installation of a Three Megawatt Wood Burning Gas Turbine at Red Boiling Springs, Tennessee," paper presented at the Institute of Gas Technology Symposium at Orlando, Florida, February 1,1984.

Section 2.6.2

79. Office of Policy, Planning, and Analysis, U.S. Department of Energy, "Energy Projections to the Year 2010: a Technical Report in Support of the National Energy Policy Plan," Report DOE/PE-0029/2, Washington D.C., October 1983.
80. Energy Information Administration, U.S. Department of Energy, "Residential Energy Consumption Survey: Consumption and Expenditures, April 1982 through March 1983, Part 1: National Data," DOE/EIA-0341(83), November 1984.
81. F.W. Sindèn, "A Two-Thirds Reduction in the Space Heat Requirements of a Twin Rivers Townhouse," *Energy and Buildings*, Vol. 1, pp. 243-260, 1978.
82. L. Schlussler, "The Design and Construction of an Energy Efficient Refrigerator," The Quantum Institute, University of California at Santa Barbara, June 1978.
83. Energy Information Administration, U.S. Department of Energy, "Non-residential Buildings Energy Consumption Survey: 1979 Consumption and Expenditures; Part 2: Steam, Fuel Oil, LPG, and All Fuels," U.S. GPO, Washington D.C., December 1983.
84. American Council for an Energy Efficient Economy, "Proceedings of the Panel on Financing Conservation: Energy Management Services, Shared Savings, and Other Mechanisms" (*Doing Better; Setting an Agenda for the Second Decade*, ACEEE 1984 Summer Study on Energy Efficiency in Buildings, Santa Cruz, California, August 1984).
85. L.K. Norford, A. Rabl, R.H. Socolow, and J.V. Spadaro, "Progress Report: Monitoring the Energy Performance of the Enerplex Office Buildings: Results for the First Year of Occupancy," PU/CEES Report No. 203, Center for Energy and Environmental Studies, Princeton University, Princeton, New Jersey, 1985.

Figure Captions, Tables, and Notes

Section 2.1

American Gas Association, *Monthly Gas Utility Statistical Report*, August 1983.

The British Petroleum Company, *BP Statistical Review of World Energy, 1984*, London, 1985 (and back issues).

Edison Electric Institute, *Statistical Yearbook of the Electric Utility Industry/1982*, Washington D.C., October 1982.

Energy Information Administration, U.S. Department of Energy, "Energy Price and Expenditure Data Report, 1970-1982 (State and U.S. Total)," DOE/EIA-0376(82), April 1985.

International Energy Agency, *World Energy Outlook*, Paris, 1982.

International Energy Agency, "Energy Balances of OECD Countries, 1971-1981," Paris, 1983.

International Energy Agency, "Energy Prices and Taxes: Second Quarter 1985," No. 4, 1985.

Arshad M. Khan and Alois Holzl, "Evolution of Future Energy Demands till 2030 in Different World Regions: an Assessment of Two IIASA Scenarios," Report No. RR-82-14 of the International Institute of Applied Systems Analysis, Laxenburg, Austria, April 1982.

Irving B. Kravis, Alan W. Heston, and Robert Summers, "Real GDP per Capita for More Than One Hundred Countries" *The Economic Journal*, Vol. 88, pp. 215-242, June 1978.

Organization for Economic Co-operation and Development, *Statistics of Foreign Trade, Series A*, 1982a (and back issues).

Organization for Economic Co-operation and Development, *Statistics of Foreign Trade, Series C*, 1982b (and back issues).

Organization for Economic Co-operation and Development, *Foreign Trade by Commodities, 1983*, Vol. I: Imports, 1984a (and back issues).

Organization for Economic Co-operation and Development, *National Accounts, 1960-1983*, Vol. I; Main Aggregates, Paris, 1984b.

Organization for Economic Co-operation and Development, *National Accounts, 1971-1983,* Vol. II: Detailed Tables, Paris, 1984c.

Organization for Economic Co-operation and Development, *Main Economic Indicators,* Paris, August 1985 (and back issues).

United Nations, *World Population Prospects as Assessed in 1980,* New York, 1981.

United Nations, *1981 Yearbook of World Energy Statistics,* New York, 1983.

Section 2.2

W.G. Dupree and J.S. Corsentino, "United States Energy Through the Year 2000 (Revised)," Bureau of Mines, U.S. Department of the Interior, December 1975.

Electric Power Research Institute, *Technical Assessment Guide,* Special Report No. PS-1201-SR, prepared by the Technical Assessment Group of the EPRI Planning Staff, July 1979.

Energy.Information Administration, *Annual Report to Congress, 1978, Volume Three,* DOE/EIA-0173/3 1979.

Energy Information Administration, U.S. Department of Energy, *Projected Costs of Electricity from Nuclear and Coal-Fired Power Plants, Vol. 1* DOE/EIA-0356/1, August 1982.

International Energy Agency, *World Energy Outlook,* Paris, 1982.

T.B. Johansson and P. Steen, Various Sources as given in: *Perspectives on Energy,* (in Swedish), Liber, Stockholm 1985.

Office of Policy, Planning, and Analysis, U.S. Department of Energy, "Energy Projections to the Year 2000," DOE/PE-0029, July 1981.

Swedish National Energy Administration, *Perspectives on Energy 1970-1995, Problems, Prognoses, Policies,* (in Swedish), Liber, Stockholm 1984.

United States Atomic Energy Commission, "Power Plant Capital Costs: Current Trends and Sensitivity to Economic Parameters", WASH-1345, October 1974.

U.S. Department of Energy, "The National Energy Policy Plan," DOE/E/S-0014/1, October 1983.

Matthew L. Wald, "9 States See Higher Rates Because of Nuclear Plants," *The New York Times,* February 26, 1984.

R.H. Williams, "Industrial Cogeneration," *Annual Review of Energy,* Vol. 3, pp. 313-356, 1978.

Section 2.3

John M. Anderson and John S. Saby, "The Electric Lamp: 100 Years of Applied Physics," *Physics Today,* pp. 32-40, October 1979.

Frank von Hippel and Barbara Levi, "Automotive Fuel Efficiency: the Opportunity and the Weakness of Existing Market Incentives," *Resources and Conservation,* Vol. 10, pp. 103-124, 1983.

Sction 2.4

Energy Information Administration, U.S. Department of Energy, *Housing Characteristics, 1980,* Residential Energy Consumption Survey Report No. DOE/EIA-0314, June 1982.

Energy Information Administration, "Monthly Energy Review," September, 1982.

Robert Herendeen and Anthony Sebald, "Energy, Employment, and Dollar Impacts of Certain Consumer Options," Chapter 4, pp. 131-163, in *The Energy Conservation Papers,* Robert H. Williams, ed., a publication of the Energy Policy Project of the Ford Foundation, Ballinger, Cambridge, 1975.

T.B. Johansson, P. Steen, E. Bogren, and R. Fredricksson, "Sweden Beyond Oil—the Efficient Use of Energy," *Science,* Vol. 219, pp. 355-361 1983.

T.B. Johansson, Peter Steen, *et al., Perspektiv pa Energi,* Liber Forlag, 381 pp., Stockholm, Sweden, 1985 (in Swedish).

Florentin Krause, "The Federal Republic of Germany in the Global Energy Context," paper presented at the Workshop on an End-Use Focussed Global Energy Strategy," Princeton University, April 21-28, 1982.

Eric D. Larson, Robert H. Williams, and Andrew Bienkowski, "Matrials Consumption Patterns and Industrial Energy Demand in Industrialized Countries," PU/CEES Report No. 174, Center for Energy and Environmental Studies, Princeton University, Princeton, New Jersey, December 1984.

David Olivier, Hugh Miall, Francois Nectoux, and Mark Opperman, *Energy-Efficient Futures: Opening the Solar Option,* Earth Resources Research, Ltd., Blackrose Press, London, 328 pp., 1983.

Marc H. Ross, Eric D. Larson, and Robert H. Williams, "Energy Demand and Materials Flows in the Economy," PU/CEES Report No. 193, Center for Energy and Environmental Studies, Princeton University, Princeton, New Jersey, July 1985.

Lee Schipper, Andrea Ketoff, and Stephen Meyers, "International Comparison of Residential Energy Use: Indicators of Residential Energy Use and Efficiency; Part One: the Data Base," Lawrence Berkeley Laboratory Report No. LBL-11703, UC-98, May 1981.

Lee Schipper and Andrea Ketoff, "Energy Efficiency in Homes: Progress in the Industrialized World and Encouraging Signs for the Less Developed Countries," paper presented at the Workshop on an End-Use Focussed Global Energy Strategy," Princeton University, April 21-28, 1982.

Solar Energy Research Institute, *A New Prosperity: Building a Sustainable Future, the SERI Solar/Conservation Study,* Brick House Publishing, Andover, Mass., 454 pp., 1981.

United Nations, *Statistical Yearbook 1979/80,* New York, 1981.

Section 2.5.1

March H. Ross and Robert H. Williams, *Our Energy: Regaining Control,* McGraw-Hill, 354 pp., New York, 1981.

Section 2.5.3.1

G.S. Dutt, J. Beyea, and F.W. Sinden, "Attic Heat Loss and Conservation Policy," paper no. 78-TS-5 presented at the ASME Energy Technology Conference, Houston, Texas, 1978.

G.S. Dutt, M. Lavine, B. Levi, and R.H. Socolow, "The Modular Retrofit Experiment: Exploring the House Doctor Concept," PU/CEES Report No. 130, Center for Energy and Environmental Studies, Princeton University, Princeton, New Jersey, 1982.

Margaret F. Fels and Miriam L. Goldberg, "Using Billing and Weather Data to Separate Thermostat from Insulation Effects," *Energy,* Vol. 9, No. 5, pp. 439-445, 1984.

Howard Geller, *Energy Efficient Appliances,* a report of the American Council for an Energy Efficient Economy, Washington, D.C., 1983.

Robert J. Hemphill, "Prototype of the Pulse Combustion Space Heater Being Readied for Field Testing", *Gas Research Institute Digest,* Vol. 6, no. 3, May/June 1983.

T.B. Johansson, Peter Steen, *et al., Perspektiv pa Energi,* Liber Forlag, 381 pp., Stockholm, Sweden, 1985 (in Swedish).

Jorgen S. Norgard, John Heeboll, and Jesper Holck, "Development of Energy Efficient Electrical Household Appliances; Progress Report No. 4 for the Period Sept. 15, 1982 to March 15, 1983," Technical University of Denmark, Lyngby, Denmark.

Jesse C. Ribot, Arthur H. Rosenfeld, F. Flouquet, and W. Luhrsen, "Monitored Low-Energy Houses in North America and Europe: A Compilation and Economic Analysis," pp. 242-256, in *What Works: Documenting Energy Conservation in Buildings,* J. Harris and C. Blumstein, eds., Proceedings of the Second Summer Study of Energy Efficient Buildings, Santa Cruz, California, August 1982. American Council for an Energy Efficient Economy Washington, D.C., 1983.

Lee Schipper, "Residential Energy Use and Conservation in Sweden," Lawrence Berkeley Laboratory, Berkeley, California, August 1982.

L. Schlussler, "The Design and Construction of an Energy Efficient Refrigerator," The Quantum Institute, University of California at Santa Barbara, June 1978.

Frank W. Sinden, "A Two Thirds Reduction in the Space Heat Requirements of a Twin Rivers Townhouse," *Energy and Buildings,* Vol. 1, no. 3, pp. 243-260, April 1978.

Peter Steen, Thomas B. Johansson, Roger Fredericksson, Erik Bogren, *Energy—What For and How Much?*, Liber Forlag, 292 pp., Stockholm, 1981 (in Swedish).

R.R. Verderber and F.R. Rubinstein, "Comparison of Technologies for New Energy-Efficient Lamps," paper present at the IEEE-IAS Annual Meeting, Mexico City, October 3-7, 1983.

R.H. Williams, G.S. Dutt, and H.S. Geller, "Future Energy Savings in U.S. Housing," *Annual Review of Energy,* pp. 269-332, 1983.

Section 2.5.3.2

R.W. Barnes *et al.,* "A User's Guide to the ORNL Commercial Energy Use Model," ORNL/CON-44, May 1980.

Jane E. Brody, "From Fertility to Mood, Sunlight Found to Affect Human Biology," *The New York Times,* p. Cl, June 23, 1981.

Lars-Goran Carlsson (Swedish National Energy Administration), "Energy Use in Homes and Buildings, 1970-1982" (in Swedish), draft, 1984.

Energy Information Administration, U.S. Department of Energy, "Residential Energy Consumption Survey; Housing Characteristics, 1980," DOE/EIA-0314, June 1982.

Energy Information Administration, U.S. Department of Energy, "Non-Residential Buildings Energy Consumption Survey: 1979 Consumption and Expenditures; Part 2: Steam, Fuel Oil, LPG, and All Fuels," U.S. GPO, Washington D.C., December 1983.

W.S. Johnson and F.E. Pierce, "Energy and Cost Analysis of Commercial Building Shell Characteristics and Operating Schedules," Oak Ridge National Laboratory Report No. ORNL/CON-39, April 1980.

K-Konsultant (a Swedish architect-engineer consulting firm), personal communication to Thomas Johansson, February 1984.

Leslie K. Norford, *An Analysis of Energy Use in Office Buildings: the Case of ENERPLEX,* PhD. thesis, Department of Aerospace and Mechanical Engineering, Princeton University, June 1984.

Howard Ross and Sue Whalen, "Building Energy Use Compilation and Analysis (BECA) Part C: Conservation Progress in Retrofitted Commercial Buildings," *Proceedings of the August 1982 Summer Study of Energy Efficient Buildings,* Energy Information Centre, New York, 1983.

Solar Energy Research Institute, *A New Prosperity: Building a Sustainable Energy Future,* the SERI Solar/Conservation Study, Brickhouse, Andover, Mass (1981).

K. Welmer, "A Method to Make Use of a Building's Heat Storage Capacity in a Controlled Manner to Save Energy" (in Swedish), Swedish Building Energy Council, BFR Report R104:1981, Stockholm, 1981.

Section 2.5.3.3

_____, "NHTSA Scraps Mandate for Air Bags or Automatic Belts: Peck Estimates Savings of $75-100 per Car", *Automotive News,* p. 1, November 2, 1981.

_____, "Save Passive Restraints, Suit Asks: Insurance Industry Seeks to Block Rescinding of Rule", *Automotive News.,* p. 2, November 30, 1981.

_____, "Cat's 3306B Makes It Big in the Real World," *Automotive News,* p. 16, November 7, 1983.

Edward A. Barth and James M. Kranig, (U.S. Environmental Protection Agency), "Evaluation of Two Turbo-charged Diesel Volkswagen Rabbits," October, 1979.

P. Baudoin, "Continuously Variable Transmissions for Cars with High Ratio Coverage," SAE Paper #790041, 1979.

Philip Burgert, "Japan Aims for Non-Metallic Auto in '90s," *American Metal Market/Metalworking News,* April 25, 1983.

U.G. Carstens and I. Isik (Sauer), and G. Biaggini and G. Cornetti (Fiat), "Sofim Small High-Speed Diesel Engines—D.I. vs. I.D.I., " SAE Paper #810481, 1981.

Hans Drewitz, (MAN), "Reducing Fuel Consumption of Heavy Truck Trains (Tractor Vehicle and Trailer) by Lowering Aerodynamic Drag", *Proceedings of the First International Automotive Fuel Economy Research Conference,* p. 444, U.S. GPO, 1979.

Energy and Environmental Analysis, "The Effect of Fuel Economy on the Used Car Pricing Mechanism," report prepared for the Office of Policy and Evaluation, U.S. Department of Energy, December 23, 1980.

Energy and Environmental Analysis, "The Highway Fuel Consumption Model," Sixth Quarterly Report, January 1982.

S.L. Fawcett and J.C. Swain, "Prospectus for a Consumer Demonstration of a 100 MPG Car," Battelle Memorial Institute Paper, March 1983.

Richard Feast, "Cars of Tomorrow Add Spice to Frankfurt Show: Innovation and Slippery Shapes Predominate," *Automotive News,* p. 1, September 28, 1981.

Federal Highway Administration, *Highway Statistics,* 1978.

W.L. Giles, "Expanded Applications, Wide Base Radial Truck Tires," SAE Paper 791044, 1979.

Hans-Wilhelm Grove and Christian Voy, "Volkswagen Lightweight Component Project Vehicle Auto 2000," SAE Technical Paper 850104, presented at the SAE International Congress and Exposition, February 25-March 1, 1985.

J.C. Houbalt, "Why Twin Fuselage Aircraft," *Astronautics and Aeronautics,* April 1982, p. 27.

International Energy Agency, *Energy Statistics, 1975-1979,* Paris, 1981.

Gail Klemer (Standards Development and Support Branch, Emission Control Technology Division, Office of Mobile Source Air Pollution Control, U.S. EPA), "Rolling Resistance Measurements for 106 Passenger Car Tires," August 1981.

R.E. Knight, "Tire Parameter Effects on Truck Fuel Economy", SAE Paper 791043, 1979.

R. Kamo and W. Bryzik, "Adiabatic Turbo-compound Engine Performance Prediction," SAE Paper 780068, 1978.

G. Kulp et al. *Transportation Energy Conservation Data Bank,* Edition 5, Oak Ridge National Laboratory Report No. ORNL-5765, 1981.

D.J. Maglieri and S.M. Dollyhigh, "We Have Just Begun to Create Efficient Transport Aircraft," *Astronautics and Aeronautics,* Feb. 1982, P. 26.

Rollfe Mellde, "Volvo LCP 2000 Light Component Project," SAE Technical Paper 850570, presented at the SAE

International Congress and Exposition, February 25-March 1, 1985a.

Rollfe Mellde (Volvo Car Corporation), personal communication to Frank von Hippel (Princeton University), at Goeteborg, Sweden, February, 1985a.

Motor Vehicle Manufacturers Association of the United States, *World Motor Vehicle Data, 1981*, Detroit, 1981.

National Highway Traffic Safety Administration, U.S. Department of Transportation, *Automobile Occupant Crash Protection, Progress Report 3*, p. 20, July 1980.

R.R. Ropelewski, "757 Key to Route Flexibility," *Aviation Week and Space Technology*, August 30, 1982, p. 36.

A.B. Rose, "Energy Intensity and Related Parameters of Selected Transportation Modes: Freight Movements," Oak Ridge National Laboratory, ORNL/TM-6700, 1979.

Ulrich Seiffert, Peter Walzer, and Herman Oetting (VW), "Improvements in Automotive Fuel Economy," *Proceedings of the First International Automotive Fuel Economy Research Conference*, (U.S. DOT, 1980), p. 95.

R.R. Sekar and R. Kamo [Cummins Engine Company, Columbus, Indiana] and J.C. Wood [NASA Lewis Research Center, Cleveland, Ohio], "Advanced Adiabatic Diesel Engines for Passenger Cars," Paper #840434, presented at the International Congress & Exposition of the Society of Automotive Engineers, Detroit, Michigan, February 27—March 2, 1984, and reprinted from SP-571, *Adiabatic Engines: Worldwide Review*.

Richard H. Shackson and James H. Leach, *Maintaining Automotive Mobility: Using Fuel Economy and Synthetic Fuels to Compete with OPEC Oil*, Arlington, Va: Mellon Institute, 1980.

J.B. Smith, "Trends in Energy Use and Fuel Efficiencies in the U.S. Commercial Airline Industry", U.S. Department of Energy, 1981.

G. Sovran and M.S. Bohn (General Motors), in "Formulae for the Tractive Energy Requirements of Vehicles Driving the EPA Schedules," Society of Automotive Engineers Paper 81084, 1981.

Thermoelectron, "Status Report on Diesel Organic Rankine Compound Engine for Long-Haul Trucks," Nov. 1980.

TRW Energy Systems Planning Division, *Data Base on Automobile Energy Conservation Technology*, Draft, 1979.

United Nations, *Statistical Yearbook, 1979/1980*, New York, 1981.

United Nations, *Yearbook of World Energy Statistics, 1979*, New York, 1980.

U.S. Department of Commerce, "United States Automobile Industry Status Rreport," submitted to the Subcommittee on International Trade of the U.S. Senate Committee on Finance, December 1981.

U.S. Department of Commerce, *Merchant Fleets of the World*, 1977.

U.S. Department of Commerce, *Statistical Abstract of the United States, 1980*.

U.S. Department of Transportation, *National Transportation Statistics*, 1980.

Volvo Car Corporation, "Volvo LCP 2000—Light Component Project," Goeteborg, Sweden, 1984.

Frank von Hippel and Barbara Levi, "Automotive Fuel Efficiency: the Opportunity and the Weakness of Existing Market Incentives," *Resources and Conservation*, Vol. 10, pp. 103-124, 1983.

B. Widemann and P. Hofbauer (VW), "Data Base for Light Weight Automotive Diesel Power Plants," Society of Automotive Engineers (SAE) Paper #780634, 1978.

L.J. Williams and P.J. Galloway, "Design for Supercommuters," *Astronautics and Aeronautics*, Feb. 1981, p. 20.

Section 2.5.3.4

Mattio Aario and Hannu Salakari, "Experiences from the Operation of Pressure Ground-wood Mills and Their Influence on the Economy of Paper Production," *1983 International Mechanical Pulping Conference*, (Proceedings of the Technical Association of the Pulp and Paper Industry, U.S.A., 1983).

Robert V. Ayres, "Final Report on Future Energy Consumption by the Industrial Chemicals Industry (SIC 28), Appendix to *The Chemicals Industry*, Vol. 5 of 9, in *Industrial Energy Productivity Project Final Report*, prepared by Energy and Environmental Analysis, Inc., Arlington, Va., for the Assistant Secretary for Conservation and Renewable Energy Resources, U.S. Department of Energy, DOE/CS/40151 1, February 1983.

J.D. Birchall and Anthony Kelly, "New Inorganic Materials," *Scientific American*, pp. 104-115, May 1983.

Bureau of the Census, U.S. Department of Commerce, "Fuels and Electricity Consumed," *Annual Survey of Manufacturers* and/or *Census of Manufacturers*, various years.

Bureau of Mines, U.S. Department of the Interior, *Mineral Facts and Problems*, 1980 Edition, Bureau of Mines Bulletin 671, 1980.

M.H. Chiogioji, *Industrial Energy Conservation*, Mercel Dekker, Inc., New York, 1979.

J.T. DiKeou, "Cement", in the U.S. Bureau of Mines, *1978-79 Minerals Yearbook, Volume I: Metals and Minerals*, 1980.

S. Eketorp *et al.*, "The Future Steel Plant," report of the National Swedish Board for Technical Development, Stockholm, Sweden, 1980.

Energy and Environmental Analysis, "The Iron and Steel Industry," Vol. 4 of 9 in *Industrial Energy Productivity*

Project—Final Report, Report DOE/CS/40151-1, prepared for the Assistant Secretary for Conservation and Renewable Energy, U.S. Department of Energy, February, 1983a.

Energy and Environmental Analysis, "The Pulp and Paper Industry," Vol. 2 of 9 in *Industrial Energy Productivity Project—Final Report*, Report DOE/CS/40151-1, prepared for the Assistant Secretary for Conservation and Renewable Energy, U.S. Department of Energy, February, 1983b.

Energy Information Administration, U.S. Department of Energy, "Projected Costs of Electricity from Nuclear and Coal-Fired Power Plants," Vol. 1, DOE/EIA-0356/1, August 1982.

Energy Information Administration, U.S. Department of Energy, "Monthly Energy Review," March 1984.

Energy Information Administration, U.S. Department of Energy, *Annual Energy Review 1983*, April 1984.

Food and Agriculture Organization of the United Nations, *Yearbook of Forest Products, 1967-1978*, Rome, 1980.

R.W. Foster-Pegg, Westinghouse Electric Corporation, and J.S. Davis, Struthers Wells Corporation, "A Coal Fired Air Turbine Cogeneration System," paper presented at the ASME Gas Turbine Conference, March 1983.

H.E. Gerlaugh *et al.*, "Cogeneration Technology Alternatives Study (CTAS)," report prepared by the General Electric Company for the National Aeronautics and Space Administration under contract for the U.S. Department of Energy; Vol. IV DOE/NASA/0031-80, NASA CR-159 765, GE 79 ET0102, January 1980.

Elias P. Gyftopoulos, Lazaridis, and Thomas F. Widmer, *Potential Fuel Effectiveness in Industry*, a report to the Energy Policy Project of the Ford Foundation, Ballinger, Cambridge, 1974.

Joseph P. Hamricks, "Development of Wood-Burning Gas Turbine Systems," *Modern Power Systems*, Vol. 4, No. 6, June 1984; Richard Layne, "This Gas Turbine Burns Sawdust," *Popular Science*, November 1984; J.T. Hamrick, "Installation of a Three Magawatt Wood Burning Gas Turbine at Red Bolling Springs, Tennessee," paper presented at the Institute of Gas Technology Symposium at Orlando, Florida, February 1, 1984.

G.J. Hane *et al.*, "A Preliminary Overview of Innovative Industrial Materials Processes," a report prepared for the U.S. Department of Energy, by the Pacific Northwest Laboratory, PNL-4505, UC-95f, September 1983.

International Energy Agency, *Energy Balances of OECD Countries, 1971/1981*. OECD, Paris, 1983.

Thomas B. Johansson, Peter Steen, Erik Bogren, and Roger Fredericksson, "Sweden Beyond Oil," *Science*, pp. 355-361, January 28, 1983.

John M. Kovacik (Industrial Sales Division, General Electric Company, Schenectady, New York), "Industrial Plant Objectives and Cogeneration System Development," paper prepared for the Texas Industrial Commission, Industrial Energy Conservation Technology Conference & Exhibition, Houston, Texas, April 17-20, 1983.

Florentin Krause, "Potential Energy Efficiency in the FRG—a Model Case", March 1981.

Arne Lindahl (Mo & Domsjo AB) and Pekka Haikala (OY Tampella AB), "Can Pressurized Ground-wood Save Energy, Improve Quality?" *Canadian Pulp and Paper Journal*, pp. 25-28, July 1978.

Arne Lindahl and Harald Wikström, "Pressurized Grinding—Promising Potential," *PPI* pp. 56-58, December 1978.

A.D. Little, Inc., *Energy Efficiency and Electric Motors*, prepared for the Office of Industrial Programs, U.S. Federal Energy Administration, August 1976.

A.D. Little, Inc., *Classification and Evaluation of Electric Motors and Pumps*, prepared for the U.S. Department of Energy, DOE/CS-0147, Washington D.C., February 1980.

Lawrence C. Long [Director—Process Research, Corporate Research and Technology, Armco, Inc., Middletown, Ohio], "Steel-making in the Future," *Iron and Steel Engineer*, pp. 48-53, December 1981.

National Swedish Industrial Board (SIND), "Forest Industry—Forest Energy". (in Swedish), SIND 1983:2, Liber, Stockholm 1983.

Office of Technology Assessment of the U.S. Congress, *Energy from Biological Processes*, 1980.

Hannu Paulapuro, "TMP-PGW Developments," *Know-How Wire*, SPCI Special, May 1981.

March H. Ross, "Energy Consumption by Industry," *Annual Review of Energy*, Vol. 6, pp. 379-416, 1981.

March H. Ross, "Industrial Energy Conservation," *Natural Resources Journal*, Vol. 24, pp. 369-404, April 1984.

March H. Ross, Eric D. Larson, and Robert H. Williams, "Energy Demand and Materials Flows in the Economy," PU/CEES Report No. 193, Center for Energy and Environmental Studies, Princeton University, Princeton, New Jersey, July 1985.

"33rd Annual Electrical Industry Forecast," *Electrical World*, September, 1982.

Haruki Tsuchiya, "Country Study: Japanese Case", paper presented at the Workshop on an End-Use Focussed Global Energy Strategy at Princeton University, April 1982.

U.S. Atomic Energy Commission, "Power Plant Capital Cost: Current Trends and Sensitivity to Economic Parameters," WASH-1345, October, 1974.

United Nations, *1979/1980 Statistical Yearbook*, New York, 1981.

United Nations, *Yearbook of Industrial Statistics, 1980 Edition, Volume II: Commodity Production Data, 1971-1980*, New York, 1982.

United Nations, *Annual Bulletin of Steel Statistics for Europe, 1979*, Vol. VII, New York, 1980.
R.H. Williams, "Industrial Cogeneration," *Annual Review of Energy,* Vol. 3, pp. 313-356, 1978.

Section 2.6.2

American Gas Association, "Gas Consumption by Residential Appliances," Report 1984-4, March 2, 1984.
R.W. Barnes *et al.*, "A User's Guide to the ORNL Commercial Energy Use Model," ORNL/CON-44, May 1980.
Bureau of the Census, U.S. Department of Commerce "Projections of the Population of the United States, 1982 to 2050," *Population Estimates and Projections*, Series P-25, No. 922, October, 1982.
Bureau of the Census, U.S. Department of Commerce, *Statistical Abstract of the United States, 1981*, December 1981.
Energy and Environmental Analysis, "The Highway Fuel Consumption Model, Eighth Quarterly Report," prepared for the Division of Conservation and Renewable Energy, Office of Policy, Planning and Analysis, U.S. Department of Energy, July 1, 1982.
Energy Information Administration, U.S. Department of Energy, "Residential Energy Consumption Survey; Housing Characteristics, 1980," DOE/EIA-0314, June 1982.
Energy Information Administration, U.S. Department of Energy, "Estimates of U.S. Wood Energy Consumption from 1949 to 1981," DOE/EIA-0341; August 1982.
Energy Information Administration, U.S. Department of Energy, "Nonresidential Buildings Energy Consumption Survey, 1979 Consumption and Expenditures: Part 1, Natural Gas and Electricity," DOE/EIA-0318/1, March 1983.
Energy Information Administration, U.S. Department of Energy, "Residential Energy Consumption Survey; Regression Analysis of Energy Consumption by End Use," DOE/EIA-0431, October 1983.
Howard S. Geller, *Energy Efficient Appliances,* American Council for an Energy Efficient Economy, Washington, D.C., 1983.
J.T. Hamrick, "The Design and Development of a Wood Burning Gas Turbine for Industrial Energy Production," *High Temperature Technology*, Vol. 2, No. 1, Butterworth & Company, England, February 1984.
G. Kulp and M.C. Holcomb, "Transportation Energy Data Book," prepared for the Office of Engine Research and Development, U.S. Department of Energy, Sixth Edition, 1982.
E.D. Larson, R.H. Williams, and A. Bienkowski, *Material Consumption Patterns and Industrial Energy Demand in Industrialized Countries,* Center for Energy and Environmental Studies Report PU/CEES 174, Princeton University, December 1984.
Marine and Industrial Engine Projects Department, General Electric Company, *Scoping Study: LM5000 Steam-Injected Gas Turbine,* report prepared for the Pacific Gas and Electric Company, July 1984.
Office of Policy, Planning, and Analysis, U.S. Department of Energy, *Energy Projections to the Year 2010: a Technical Report in Support of the National Energy Policy Plan,* Report DOE/PE-0029/2, Washington D.C., October 1983.
Robert S. Pindyck, *The Structure of World Energy Demand*, The MIT Press. Cambridge, 1979a.
Robert S. Pindyck, "The Characteristics of the Demand for Energy," in *Energy Conservation and Public Policy*, John C. Sawhill, ed., Prentice-Hall, Inc., Englewood Cliffs, New Jersey, pp. 22-45, 1979b.
Marc H. Ross, "Energy Consumption by Industry," *Annual Review of Energy*, 1981.
M.H. Ross and R.H. Williams, *Our Energy: Regaining Control*, McGraw-Hill, New York, 1981.
Marc H. Ross, Eric D. Larson, and Robert H. Williams, "Energy Demand and Materials Flows in the Economy," PU/CEES Report No. 193, Center for Energy and Environmental Studies, Princeton University, Princeton, New Jersey, July 1985.
The SERI Solar/Conservation Study Group, *A New Prosperity: Building a Sustainable Energy Future,* Brickhouse, Andover, Massachusetts, 1981.
Statistics Sweden, *Statistical Abstracts of Sweden*, Stockholm 1986, 1984, 1979, 1977, 1971, 1966, 1961, and 1955.
Robert H. Williams, "Industrial Cogeneration," *Annual Review of Energy*, Vol. 3, pp. 313-356, 1978.
Robert H. Williams, Gautam S. Dutt, an Howard S. Geller, "Future Energy Savings in U.S. Housing," *Annual Review of Energy*, Vol. 8, pp. 269-332, 1983.

3. Energy Strategies for Developing Countries

3.1 Introduction

The first question that arises in considering energy strategies for developing countries is whether these countries should be treated as a single group or considered in separate categories, such as "newly industrializing countries" and the "least developed countries", or as "low-income", "middle-income", "high-income" developing countries, or as "oil-exporting" and "oil-importing" developing countries.[1-2]

Different sub-groups of developing countries do have different interests in their relations with the industrialized countries. For instance, the major need of the poorest among the developing countries, mainly in South Asia and sub-Saharan Africa (sometimes referred to as the Fourth World), is for assistance in the form of additional finance and technical help. In contrast, the more industrialized among the developing countries need better access to capital, to markets for their manufactures, and to modern technology.

Despite their diversity of interests, there are strong common interests among these countries as well. They are poorer than developed countries, many have been colonies, and the international economic context in which they must operate is dominated by industrialized countries' interests.[3] And importantly, in terms of this analysis, most developing countries are highly stratified into "dual societies" with rich elites and poor masses. This polarization creates similar patterns of energy consumption among developing countries (Section 3.3.1).

An obvious difference between different developing countries that must be considered is the distinction between oil-exporting and oil-importing developing countries. The former usually enjoy balance-of-payments surpluses while the latter have serious debt problems, but this is not always the case. Mexico and Nigeria are examples of debt-ridden oil-exporters. Though oil-exporting and oil-importing developing countries have quite different oil-resource endowments, their patterns of energy consumption are largely similar.*

It is the similarity in the patterns of energy consumption which matters here and allows us to consider developing countries as a single group, without further disaggregation.

3.2 Low Levels of Energy Services in Developing Countries

One of the most striking features of the present global distribution of energy use is the enormous disparity between the developing and industrialized countries (Table 3.1). The 70 percent of humanity that lives in developing countries accounts for only 30 percent of global energy consumption. In fact, if only oil, natural gas, coal and electricity—the so-called "commercial" energy sources**—are considered, the developing countries' share of global energy shrinks to less than 20 percent.

The disparity is even more striking on a per capita basis. On the average, a person from a developing country consumes less than one-sixth the energy consumed by a person from an industrialized country, and less than one-tenth as much as the average U.S. citizen (Table 3.1). For some developing countries, the situation is even worse. Thus, the average Bangladesh citizen consumes less than 1.5 percent of the energy used by a U.S. citizen—in fact, the energy content of the food *eaten* by an average North American is more than the total energy used by a Bangladeshi for cooking, lighting, transportation, industry, agriculture, etc. And the per capita oil use in the United States for snowmobiles and recreational boats and vehicles (1.6 GJ per year) is almost as much as per capita oil use in India (1.8 GJ per year) for all purposes.

Since the efficiencies of energy use are also extremely poor in developing countries, the low levels of energy use correspond to abysmal levels of services provided by energy.

---

* The few low-population oil-exporting countries like Kuwait with economies which are almost exclusively based on oil are exceptions which prove the 'rule'.

** The term "commercial" is used to distinguish these marketed sources from fuelwood, animal wastes and agricultural residues which are referred to as "non-commercial" sources because they are usually gathered/collected outside the market economy.

Table 3.1. Global Distribution of Energy Use, 1980(a)

	Popu-lation(a) (million)	Per Capital Energy Use											Total Watts	Total Energy Use (TW)
		Non-Commercial Energy Watts	%	Oil Watts	%	Natural Gas Watts	%	Coal Watts	%	Hydro- and Nuclear Watts	%		
World	4371.5	308(b)	15	778	37	397	19	567	27	64	3	2114	9.24(g)
Industrialized Market Economies	795.1	NA(c)	—	2658	48	1323	24	1356	24	243	4	5580	4.44(h)
U.S.	227.6	220(d)	2	4093	41	3038	31	2249	23	281	3	9881	2.25
Western Europe	372.6	NA(c)	—	2035	52	689	18	979	25	199	5	3902	1.45(h)
Japan	116.8	NA(c)	—	2162	67	268	8	640	20	172	5	3242	0.38(h)
Centrally Planned Europe	377.8	NA(c)	—	1559	30	1371	27	2155	42	86	2	5171	1.95(h)
Developing Market Economies	2186.2	456(e)	52	272	31	68	8	65	7	20	2	881	1.93
Brazil	123.0	371(e)	34	513	48	12	1	63	6	118	11	1077	0.13
India	662.0	190(e)	52	51	14	2	1	115	31	9	2	367	0.24
Bangladesh	88.2	95(e)	69	22	16	18	13	2	1	1	1	138	0.01
China	939.3	317(e)	36	106	12	19	2	427	49	7	1	876	0.82

Notes to Table 3.1:

(a) Unless otherwise indicated, data are from United Nations, 1981.

(b) This is the world average value, assuming zero non-commercial energy use in industrialized countries other than the U.S.

(c) Not available.

(d) This is estimated wood consumption as fuel by the forest products industry plus firewood consumption. Source: OTA, 1980.

(e) Source: D.O. Hall *et al.*, 1982.

(f) Hydroelectric and nuclear energy are counted as 3.6 MJ per kWh (electrical).

(g) Excludes non-commercial energy in industrialized countries other than the U.S.

(h) Commercial energy only.

3.3 Energy in Dual Societies

3.3.1 *Dual Societies and Elites*

A wide chasm of incomes, consumption patterns, attitudes, aspirations, and lifestyles separates the elites of developing countries from the masses. The elite—industrialists, commercial traders, landlords, government officials and bureaucrats, upper echelons of the armed forces, professionals, and some skilled craftsmen—constitute only about 10 percent of the population of these societies, but to them goes the lion's share of these societies' economic rewards. And despite their small numbers, the elites are politically powerful, controlling virtually all aspects of the decision-making process in these countries. Their conspicuous affluence contrasts markedly with the abject poverty of the politically weak masses who are dispersed in the villages of the countryside and concentrated in the slums of the metropolises (Box 3.1). In short, developing countries consist of tiny islands of affluent splendour amidst vast oceans of miserable poverty.

The consumption patterns and life-styles of the elites of developing countries are strongly influenced by those in the industrialized countries. It is as if the elites of developing countries practise a philosophy best described thus: all that is rural is bad, all that is urban is better, and all that is current in the industrialized countries is best.

The stratification of these societies into elites and poor masses is the fundamental reality which has shaped the current energy systems of these countries. To ignore these realities, as all global energy strategies have done, and as international organizations and agencies unfortunately are obliged to do (on the grounds of non-interference in internal affairs), is to miss the essence of the situation. It is crucial, therefore, to explore further the impact of dual societies and their elites on the energy systems of developing countries.

3.3.2 *Elites, the Modern Sector and Commercial Energy*

The elites of developing countries in their desire for status-symbols and modern conveniences seek goods and services similar, if not identical, to those available in the industrialized countries. Initially, this desire can be satisfied by imports from the industrialized world. But such imports cannot be sustained for long because they result in balance-of-payments problems—a local productive base is required. A so-called "modern" sector involving import-substitution and modernization of the industrial, agricultural, transport, commercial, and residential sectors is necessary.

Of course, there are marked differences in the extent to which modernization of these sectors has taken place in different developing countries. In some of the countries, e.g., Brazil and India, the modernization started several decades ago; in others, e.g., Tanzania and Indonesia, it is a recent phenomenon. Generally, though, this modernization has tended to be along the lines found in industrialized countries.

In industry, the effort is to establish and/or increase the production of those basic materials and products for which the developing country has resources, e.g., iron and steel, cement, textiles, sugar, aluminium, etc., and other comparative advantages. These objectives are usually achieved with imports of technology from the industrialized countries—hence, the industrial sectors in developing countries are embryonic versions of the corresponding sectors in the industrialized countries, even retracing their course of development and repeating their mistakes—for example, their dependence on petroleum.

In agriculture, the attempt is to overcome the limitations of traditional agriculture and its low productivity (arising from its dependence on human labour, draft-animal power, and rainfed irrigation) by adopting the modern practices of mechanized ploughing, hybrid seeds, irriga-

---

Box 3.1: Poverty in a Developing Country—India[74, 75]

The population of India according to the 1981 census was 683.81 millions which corresponds to 5.25 persons per household or 124.5 million households, of which 76 percent are in rural areas.

But the existence of households does not necessarily mean the existence of houses as "permanent" structures. The 1971 census showed that, out of 104 million households in 1971, there were only 82.5 million units of usable housing stock, i.e., 21 percent were house-less.

Seventy-nine percent of the households lived in houses, but what is meant by houses needs some explanation. Only 38.4 percent of the houses were built of reasonably good material such as burnt bricks (24.2 percent), stone (13.8 percent) and cement concrete (0.3 percent), etc. The remaining 61.6 percent of the houses were of poor material such as grass, leaves, reeds or bamboo (10.3 percent), mud (42.0 percent), unburnt bricks (7.3 percent), wood (1.5 percent), and galvanized iron and other metal sheets (0.5 percent).

Some idea of the sizes of houses can be got from the fact that 40 percent of India's population lives in one-room tenements, 29 percent in two-room houses, and 31 percent in houses of three rooms or more.

An understanding of the incomes of these households is also important. Fifty percent of 124.5 million households had an annual household income of less than $375—these are the ones which are considered to be below the poverty line. Another 30 percent had incomes between $375-$625.

Foodgrains account for 66 percent of the total consumption expenditure of an average Indian. This corresponded in 1977 to a per capita availability of 403 grams per day of foodgrains, 42 kg per year of milk, 32 kg of vegetables and 18 kg of fruits, all resulting in an average daily intake of 2,100 calories.

The 1981 census figures for the birth and death rates were 36.0 per 1,000 people and 14.8 per 1,000 people. Life expectancy at birth was 52.5 years.

An analysis of the major causes of mortality reveals that about 48 percent of the deaths are due to diseases that are environmentally caused or promoted by, for instance, unsafe drinking water (10 percent); are the result of poor nutrition leading, for example, to anaemia (6 percent); preventible by immunization against diphtheria, for instance (20 percent); or associated with pregnancy, childbirth, and early infancy (12 percent). In other words, roughly half the deaths are preventible.

The situation is aggravated by the fact that, out of 576,000 villages in India, about 9 percent were without drinking water supply and another 23 percent have inadequate drinking water supply. Further, the 1973 figures indicate that per million of population there were 6.7 hospitals, 670 beds, 19 dispensaries, and 300 registered medical practitioners all involving a per capita annual expenditure on health of $1.6 (1973 $).

---

tion upon demand, fertilizer applications, crop care, harvesting and post-harvest crop-processing.

In the transport sector, the main thrust is towards transportation modes fuelled by petroleum products, i.e., towards trucks for freight, and buses and automobiles and airplanes for passenger transport.

In the commercial sector, the growth of modern business firms, particularly the local branches of transnational corporations, is leading to office buildings and practices with norms for illuminations, space cooling, etc., similar to those found in the industrialized countries.

In the residential sector, the trend is towards modern energy carriers, such as electricity and liquified petroleum gas (LPG) for cooking, electricity for lighting, and the use of electrical appliances such as refrigerators, electric fans, air conditioners, etc.

This process of modernization has led to a pattern of energy use in the modern sector of developing countries which is quite similar to the energy use pattern in the industrialized countries. On a sector-wise basis, the distribution of consumption of *commercial* energy in developing countries is not too different from that in the industrialized countries, with the main difference being the much lower share of energy use for commercial buildings in developing countries (Table 3.2). And it has been more convenient to use the same modern commercial energy carriers—electricity, oil, gas, and coal—as are used in the industrialized world, rather than traditional biomass fuels.

In sum, the demand of elites in developing countries for a modern sector to satisfy their desire for the goods and services of the industrialized countries has generated a similar pattern of consumption of commercial fuels. In contrast, the poor in developing countries must rely on traditional biomass energy sources and their pattern of energy use is totally different.

Table 3.2. Comparison of the Sectoral Distributions of Commercial Energy Use in the United States and India

Sector	Commercial Energy Consumption	
	U.S.[a]	India[b]
Industry	37.9%	54.5%
Transport	25.4%	19.8%
Domestic	21.1%	10.4%
Agriculture	2.1%	9.1%
Miscellaneous[c]	13.5%	6.2%

(a) *Source:* SERI, 1981.
(b) *Source:* A.K.N. Reddy, 1985a.
(c) The Miscellaneous sector in this table includes the commercial sector, which is very small in India but significant in the USA.

3.3.3 *The Poor and Non-Commercial Energy*

The poor in developing countries may be classified into two groups:

- The urban poor who inhabit the slums, shanty-towns, barios, and favelas of developing countries' metropolises, and
- The rural poor who live in the countryside.

In a country like India with 76 percent of the population in rural areas, the rural and urban poor can be considered to constitute about 48 percent and 5 percent respectively of the total population, i.e., the rural poor are roughly ten times more numerous than the urban poor.*

The urban poor in the developing world benefit far more from commercial energy and modernization than the rural poor. These benefits involve the use of modern fuels for cooking, electricity for lighting, and piped water either to the homes or to nearby locations. Of course, the extent of these benefits depends upon the overall standard of life. Hence, in a country like Brazil, about 84 percent of urban dwellers use LPG (liquefied petroleum gas) for cooking, in contrast to India where about 13 percent of the total firewood consumption takes place in towns and

---

* These figures have been arrived at on the following basis: about 60 percent of the rural population in India is below the poverty line and about 25 percent of the urban population lives in slums.

cities (Box 3.2), and where, in the city of Bangalore with a population of 3 million, 45 percent of the households use fuelwood either for cooking and water-heating or for water heating only.

Box 3.2: Firewood Consumption in Bangalore[76]

A study of the role of firewood in the city of Bangalore has been carried out by investigating its supply, transportation, distribution, and consumption. The average magnitude of firewood involved in these four phases of the fuel cycle is 1197 ± 51 tonnes/day, i.e., about 0.44 million tonnes/year. About 35 percent of this total firewood comes from sources located within a radius of 30-40 km, 50 percent from forests 120-150 km away, 6-7 percent from 300-400 km away, and 8-9 percent from 650-700 km away.

Private contractors account for 95 percent of the total supply of firewood, and the remainder comes from the Forest Department.

About 85 percent of the firewood is transported into the city by road, involving an average of 114 trucks/day; 10 percent by rail in an average of 10 wagons/day; and the rest by bullock-carts and head-loads.

The distribution of firewood is carried out by commission agents, 23 cooperative societies, and 1,400 (registered and un-registered) retail depots—accounting, respectively, for 10 percent, 5 percent, and 85 percent of the firewood sales.

Households alone account for about 78 percent of the total consumption, and households along with dyeing factories, bakeries, hotels, and industries, for 95 percent of the firewood used in Bangalore. Of this total firewood, about 53 percent is used for water-heating, 42 percent for cooking, and 5 percent for industrial process heat.

A 1,000-household survey showed that 65.5 percent of the households did not use firewood at all, but 16.1 percent depended on firewood for cooking and water-heating, and another 18.4 percent for water-heating only. However, the type of fuel used by households depends strongly on their incomes. By extrapolation from the sample to the whole of Bangalore, it is estimated that 82 percent of Bangalore's domestic firewood consumption of 970 tonnes/day is accounted for by households with a per capita income of less than Rs. 200 per month ($ 22), even though this income category constitutes only 32 percent of the households. Further, in the case of low per-capita-income households, the estimated expenditures on firewood is as high as 17 percent of their income.

The present supply, transportation, distribution, and consumption of about 1200 tonnes/day of firewood in Bangalore has impacts on: (1) forests—about 10 hectares have to be cleared every day to maintain the firewood supply, (2) the transport system—about 18 percent of the truck traffic into Bangalore, and 8 percent of the railway wagon traffic, are tied up with firewood transport, (3) diesel consumption—about 2.2 million litres/year of diesel are required to run the trucks and railway engines hauling firewood into the city, and (4) foreign exchange expenditures—about Rs. 3.55 millions/year ($0.4 million/year) in foreign exchange is required for the import of the diesel necessary to transport the firewood. If Bangalore's population grows to 4.4 million in 1991, as is likely, the firewood demand would rise to at least 1,760 tonnes/day or 0.64 million tonnes/year with consequent increased impacts on forests, transport, diesel, and foreign exchange. Therefore, the present pattern of firewood supply, transport, and consumption is not sustainable unless serious steps are taken.

Box 3.2 *(Contd.)*

Except for the growth of firewood for which the responsibility rests largely with the Forest Department, the remainder of the firewood-fuel cycle involving the extraction, transport and distribution is mainly in private hands—hence, the government/public sector can have little influence over these aspects of the problem which are controlled by the contractor-trucker-retailer nexus. Solutions must therefore be based on the consumption end of the fuel cycle. Since 78 percent of Bangalore's firewood is used in households—54 percent for cooking and water heating, and 24 percent for water-heating only, the main solution to this problem is to provide inexpensive and efficient cooking fuels and/or devices to the poorer households.

Immediate measures could include (1) improved firewood stoves (a) for cooking, (b) for water-heating, and (c) for achieving both cooking and water-heating in a single device, as well as (2) extremely cheap solar water-heaters consisting of black-plastic "water-pillows" which act both as flat-plate collectors and water tanks. Longer-term measures may include a forest belt around the city—a 1.0 km wide forest belt around Bangalore's proposed metropolitan area of 321 km would provide its firewood requirements in a sustainable manner, either directly or after gasification into producer gas which could be piped into homes along with methane-rich biogas obtained from the treatment of the city's sewage.

The study of firewood in Bangalore illustrates some general principles. In stratified societies, the poor are forced to degrade the environment because they have no alternative to surviving at its expense. Further, the total energy consumption of a stratified society is needlessly high because the poor are compelled to use inefficient fuels/devices. The solution to the environmental and energy problems of stratified societies is to satisfy basic needs and remove inequalities.

In contrast, the rural poor in most developing countries benefit only marginally from modernization and the energy carriers associated with it. The rural electrification that is widely emphasized in developing countries does not necessarily lead to the electrification of all homes (Box 3.3). With only a small percentage, say 10-15 percent, of homes in electrified villages being able to afford domestic electric connections, the poor have no choice other than depending on oil lamps for lighting. Vegetable oils used to be the main source of illumination, but they have been largely replaced by kerosene over the past few decades.

Even this small improvement has not affected cooking. The vast majority of the rural population in developing countries still depends mainly on biomass sources—chiefly firewood, but also agricultural wastes and cattle-dung cakes. These sources are known as "non-commercial" sources because, instead of being purchased, they are usually gathered in the form of branches, twigs, etc., often with considerable effort by women and children.[4] (There is, however, a noticeable trend for even these so-called "non-commercial" sources of cooking fuel to become marketed commodities as they grow scarcer.[5]) Drinking water is another necessity which still has to be transported in pots by the large majority of the rural poor; and, if only available below ground, it has to be lifted manually.

Thus, commercial energy, in the form of electricity, LPG, petroleum, etc., plays only a minor role in the life of the rural poor. Since a large percentage of the population of most developing countries is rural, this means that non-commercial energy plays a very significant role in the energy systems of these countries. On the average, non-commercial energy constitutes about half the total energy used in a developing country (Table 3.3). But it was only during the last decade that the importance of non-commercial energy use has been recognized and studied.[4,6,7]

Box 3.3: Rural Electrification in India

(a) All the 2,700 towns (with a population above 10,000), but only 51 percent of the 576,000 villages had been electrified by March 1982.

(b) Only about 16 percent of India's electricity goes to rural areas, where 76 percent of the people live.

(c) Rural electricity is used primarily (87 percent) for agriculture, i.e., for irrigation pumpsets, with only 13 percent being used by households.

(d) There is a strong correlation between farm-size and pumpset ownership. For example, in the states of Punjab, Haryana and Uttar Pradesh, the number (P) of pumpsets is correlated to the number (N) of holdings above 5 hectares (ha) as follows:

$$P = 1.1041 N + 0.3658 (r = 0.9768)$$

Hence, rural electrification has mainly served the pumpsets of the more affluent farmers.

(e) The percentage of households electrified in electrified villages is only about 14 percent.

(f) While the number of new electricity connections increases at the rate of about 1 million households per year, the number of new households increases at the rate of about 2.2 million households per year, i.e., the number of unelectrified homes is continuously increasing despite the decreasing number of unelectrified villages.

(g) The slow pace of progress with regard to electrification of villages is mainly because it is based on linking the vast number of small, remote and scattered settlements to an electricity grid. Not only does this approach lead to major transmission losses (for which the national average is about 20 percent) but it requires supply to load centres with low load factors of about 8 percent. Besides, with increasing energy costs, the costs of aluminium conductors are escalating and therefore the cost of transmission lines. All these factors work together to make State Electricity Boards implement and maintain rural electrification with great reluctance.

(h) The situation with regard to home electrification is unsatisfactory mainly because "the householders are not able to pay for the initial investment required to set up the switchboard and for the wiring arrangements to receive electricity".[77] This initial investment constitutes an exorbitant sum for households below the poverty line because it represents about 3 months of their total household income. In contrast, for their single-room tenements, the recurrent electricity charges at the domestic rate of $0.05/kWh (Rs. 9.30/$) for running one 25 W bulb for 3 hours/day would be much less than what they are now spending on kerosene lamps (about 4 litres/household/month at about $0.22/litre). The major social benefits that would accrue from electrification of homes is discussed further in Section 3.3.8.

3.3.4 *Energy Consumption Patterns in Developing Countries*

The inclusion of non-commercial energy, and therefore the rural poor, into the analysis of energy consumption patterns of developing countries (Table 3.4) alters the conclusions substantially.

Firstly, the percentage share of the *domestic* sector (which accounts for only about 10 percent of the total energy when the analysis is restricted to commercial energy) shows a marked increase after including non-commercial energy. The extent of this increase depends upon the frac-

Table 3.3. Summary of Selected National Energy Consumption Surveys[a]

Country	Commercial Energy	Non-Commercial Energy (NCE) kW/capita	Total Energy	Percentage of Energy from NCE Sources
Bangladesh	0.038	0.095	0.133	71
Niger	0.035	0.254	0.289	88
Gambia	0.098	0.222	0.320	69
Morocco	0.267	0.073	0.340	21
India	0.165	0.190	0.355	54
Ethiopia	0.019	0.371	0.390	95
Nepal	0.009	0.429	0.438	98
Somalia	0.092	0.476	0.568	84
Bolivia	0.340	0.263	0.603	44
Sudan	0.159	0.635	0.794	80
Thailand	0.305	0.524	0.829	63
Tanzania	0.060	0.810	0.870	93
China	0.778	0.317	1.10	29
Brazil	0.737	0.371	1.11	34
Mexico	1.29	0.127	1.43	9
Libya	1.76	0.095	1.86	5
Developing Countries (Average)	0.550	0.416	0.966	43

(a) This is a revised form of Table 2.3 in D.O. Hall *et al.*, 1982.

Table 3.4. Sectoral Distribution of Energy in Developing Countries when Non-Commercial Energy is Excluded and Included

Sector	India (a) Excl. NCE	India (a) Incl. NCE	Brazil (b) Excl. NCE	Brazil (b) Incl. NCE
Industry	54.5	34.1	41.8	39.2
Transport	19.8	9.1	23.5	20.5
Domestic	10.4	49.9	9.4	18.8
Agriculture	9.1	4.2	2.7	5.0
Miscellaneous	6.2	2.8	22.6	16.5

(a) *Source:* A.K.N. Reddy, 1985a.
(b) *Source:* J. Goldemberg, 1982.

tion of the total population that lives in rural areas. When the extent of urbanization is small, say between 10-20 percent as is the case in India, the domestic sector can consume over 50 percent of the total energy to become the largest consumer of energy. Even when urbanization is preponderant, as is the case in Brazil, for example, where the urban population makes up over 60 percent of the total population, the domestic sector's share can double.

Secondly, the industrial sector's share, which is the largest consumer of commercial energy, is reduced when non-commercial energy is also considered. And, in countries which are predominantly rural, the industrial sector may even assume second place to the domestic sector. Finally, the inclusion of non-commercial energy into the energy analysis of a developing country leads to upward revision of the per capita energy consumption.

Though there has been little work on energy end-use analysis in developing countries, it seems clear that, in countries where non-commercial energy plays a significant role, medium-temperature heating (about 100-250°C) for cooking is a very important end-use, in addition to industrial process heat and stationary power.

3.3.5 *The Unviability of Present Energy Consumption Patterns*

The patterns of energy consumption referred to above have led to three energy crises in developing countries:
(1)   an oil crisis (in oil-importing developing countries),
(2)   a fuelwood crisis, and
(3)   in some countries, an electricity crisis.

The oil crisis, which is the most well-known, is manifested in a rapidly worsening balance-of-payments situation with large fractions of the export earnings of developing countries going to the payments of oil imports (Figure 3.1 and Table 3.5). As a result, the economies of most oil-importing developing countries are hurting. Moreover, the unpredictability of future oil prices, and therefore of the impacts of these prices on internal economies, has made economic planning extremely difficult.

Table 3.5. Net Oil Imports and Their Relationship to Export Earnings for Eight Developing Countries, 1973-1984

	Net Oil Imports Million U.S. $, current prices						
	1973	1974	1977	1979	1981	1983	1984
Kenya	1	27	57	113	316	208	219
Zambia	11	30	53	72	63	274	454
Thailand	173	510	806	1,150	2,170	1,740	1,480
Korea	276	967	1,930	3,100	6,380	5,580	5,770
Philippines	166	570	859	1,120	2,080	1,740	1,470
Brazil	986	3,230	4,200	6,920	11,720	8,890	7,470
Argentina	83	328	338	351	302	—	—
Jamaica	71	193	242	309	490	—	—
India	308	1,170	1,750	3,067	—	—	—
Bangladesh	—	92	172	247	509	286	314
Tanzania	47	153	102	174	306	175	156

	Imports in Relation to Export Earnings %						
	1973	1974	1977	1979	1981	1983	1984
Kenya	0.1	4.1	4.8	10.2	26.9	21.2	20.3
Zambia	2.2	5.1	9.5	8.2	7.8	20.8	21.4
Thailand	11.1	20.9	23.1	21.6	30.9	27.3	20.0
Korea	8.6	21.7	19.2	20.6	30.0	22.8	19.7
Philippines	8.8	20.9	27.5	24.4	36.8	35.4	27.8
Brazil	15.9	40.7	34.7	45.4	50.4	40.6	27.7
Argentina	2.5	8.3	6.0	4.5	3.3	—	—
Jamaica	18.1	27.3	32.4	37.7	50.3	—	—
India	10.6	29.7	27.5	39.3	—	—	—
Bangladesh	—	26.5	36.1	37.4	64.6	39.4	33.6
Tanzania	12.8	38.0	20.2	34.8	52.7	47.0	42.3

Source: International Monetary Fund 1985

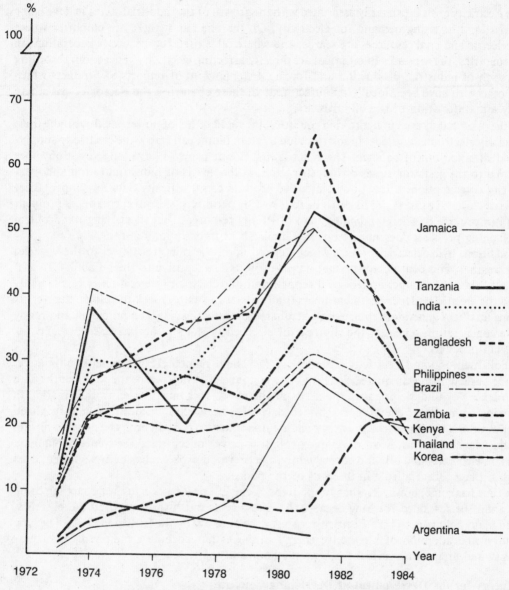

Figure 3.1. Oil import bill as a percent of export earnings.
Source: CNP/PETROBAS
Suma Estatistica Fev/85.

The crisis of non-commercial energy, often referred to as "the other energy crisis", not only involves severe scarcities of cooking fuel, but also the associated problems of tree-felling, deforestation, soil erosion and desertification. As already stated in Section 1.1.7, the Cassandra-like cry regarding this crisis goes thus: "Even if one can grow the food required by the growing population, there will not be enough fuel to cook it with". And, the conditions in the Sahelian and Sudanese zones of Africa (Figure 1.16) lend some credence to this concern.

The electricity crisis is primarily associated with the growth of the industrial sector in developing countries, and its rising demand for electricity. On the one hand, there are countries where hydroelectric and coal resources are scarce, and where oil is used for electricity generation. In these countries, the increases in oil prices are directly affecting electricity generation, throttling the growth of industrial production and forcing a stepping-up of imports of products which could otherwise have been locally manufactured. In these countries, the electricity crisis here is only a manifestation of the oil crisis.

On the other hand, even in hydro-rich countries, the rapid influx of power-intensive industries, particularly electrometallurgical industries such as aluminium, can throw the electricity supply-demand situation out of balance. The point is that power projects have long gestation times compared to the gestation times of industries and, at low levels of industrialization and total electricity consumption, a few power-intensive projects can drastically outstrip supply. Brief periods of electricity excess are likely to be followed by periods of shortage, creating a problem which is more one of energy *planning* than of energy resources. But identifying the problem as a planning problem does not in any way diminish its intensity and impact.

In addition, hydroelectric projects which depend upon the rain are subject to the vagaries of the weather, and coal-based thermal power projects are sensitive to the reliability of coal supplies. When the weather and/or coal supplies fail, the electricity generation system is badly hit and the electricity-dependent consumers (industry, for example) seek to resolve their crisis by using auxiliary generators. Since these auxiliary generators usually run on diesel, and rarely involve cogeneration schemes, the shortage of electricity results in an increased demand for petroleum products.

The oil-fuelwood and electricity crises are, in fact, inter-related. They originate, ultimately, from the basic tendency in dual societies to bias economic growth towards the wants of the elite and to ignore or under-emphasize the needs of the poor. This bias has led to the creation in developing countries of energy-norms which may have been economical in the industrialized countries during the era of cheap energy but are uneconomic today in both the developing and the industrialized worlds. Most developing countries have concentrated their energy planning efforts on electricity and oil, and consequently, the firewood problem has been neglected, even though it plays a crucial role in the lives of the poor.

The time has come, however, when the patterns of energy consumption in the developing countries cannot be sustained for long because the elites are as much threatened by the oil crisis, and (in many countries by) the electricity crisis, as the poor are by the fuelwood crisis. The consideration and adoption of alternative energy strategies has become a matter of immediate necessity and long-term survival for both the poor and the rich.

3.4 Energy for the Development of Developing Countries

3.4.1 *Energy for Basic Needs and for Much More*

It has been stressed in Chapter 1 that energy in developing countries must become an instrument for development. This means that energy must become an instrument for satisfying basic human needs. Hence it is important to estimate how much energy is required per capita for this purpose in order to get an idea whether the provision of this energy is feasible.

A rigorous estimation of the per capita energy requirement for the satisfaction of basic human needs is extremely difficult. Nonetheless, an estimate is essential because it permits a preliminary assessment of the global requirements of energy if developing countries set out to satisfy basic needs as a principal development objective.

A systematic approach to the quantification of direct energy needs may involve the following steps:

(1) listing the various energy needs such as cooking, food preservation, lighting, space heating, space cooling, water pumping, water heating, recreation and social communication;

(2) defining minimum physical targets, such as temperature, illumination, etc., to satisfy each need;

(3) estimating the useful energy required for the attainment of these targets taking into account climatic conditions; and

(4) calculating the direct primary energy needs from the useful energy requirements, assuming efficiency values for the technologies of energy conversion, distribution and end-use, as well as conventional energy carriers.

The results of such computations using average present-day technologies for hot (25°C), temperate (18°C) and cold (10°C) climates lead to per capita *direct* energy needs of 0.3-0.8 kW per capita (Table 3.6).[8]

But the indirect energy embodied in all of the goods and services needed for the satisfaction of basic needs (but not in conspicuous consumption) needs to be taken into account. This requires that the per capita energy consumption and the quality of life be correlated. Then the question arises as to how the quality of life can be quantified. In this context, an indicator known as the Physical Quality of Life Index (PQLI) has been used.[9] The PQLI is a composite index which gives equal weights to indices for life expectancy, the infant mortality rate, and literacy. These indices are assigned values on a scale of 1 to 100, with the lowest and the highest values corresponding to the worst and best world performances, respectively. When the PQLI indices of various countries are plotted against their per capita energy consumption (Figure 3.2), it is observed that above the range of 1.1 to 1.3 kW per capita, the improvements in PQLI with increasing energy consumption are only marginal.[10] Hence, it can be concluded that, if there is continued use of the energy technologies currently in vogue in these countries, an energy consumption rate of 1.1 to 1.3 kW per capita is required to satisfy basic human needs.

A more sophisticated approach to the question of basic needs was pursued in the Latin American World Model proposed by the Bariloche Foundation.[11] This model sought to determine whether current economic resources are likely to impose physical limits on the achievement of a society directed towards satisfying basic needs.

Table 3.6. Calculation of Direct Energy Needs for Various Climatic Regions[a]

Basic Need	Per Capita Requirement (kW)		
	Hot Climate	Temperate Climate	Cold Climate
Lighting	0.06	0.07	0.11
Space Heating	—	0.09	0.34
Space Cooling	0.02	0.01	—
Food Preservation	0.05	0.04	—
Cooking	0.07	0.09	0.11
Hot Water	0.04	0.09	0.13
Leisure, etc.	0.05	0.05	0.07
Total	0.29	0.44	0.76

(a) *Source:* V. Bravo et al., 1979.

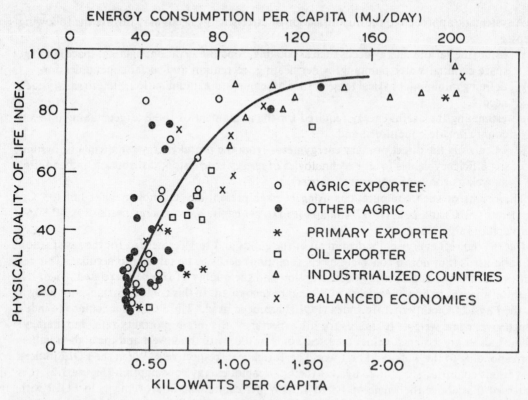

Figure 3.2. The Physical Quality of Life Index vs. total (commercial plus non-commercial) per capita energy use. *Source:* Palmedo *et al.,* 1978.

This objective was accomplished by considering a five-sector economic system: (1) nutrition; (2) housing, (3) education, (4) other services and consumer goods, and (5) capital goods. The first three sectors are specified respectively by the production of calories and protein, the number of dwellings, and the places available for the first twelve years of formal education. The sector designated as "other services and consumer goods" includes clothing, furniture, household utensils, health care, transportation, leisure activities, public and administrative services, and all educational activities not considered in the education sector. The capital goods sector includes housing construction, the infrastructure of cities, public buildings, the infrastructure of transportation, of communication and other basic services, machinery and vehicles, etc.

The model serves to describe the time evolution of population as well as other demographic variables and of calorie and protein intake per capita per day, housing per family, school enrolment, etc. Starting from 1960, the model is used to compute the values of the various indicators year by year for any given period. The computation involves calculations of (1) the availability of capital, (2) the magnitude of the labour force, and (3) the distribution of the total capital and the labour force between five sectors.

The capital available is given by the previous year's capital minus its depreciation plus the output of the capital goods sector. The size of the labour force is calculated from a population sub-model. This sub-model yields year by year the population and its structures by age and sex as a function of seven socioecomic variables: population employed in the secondary sector, school enrolment, houses per family, calorie and protein intake, population employed in agriculture, and urbanization.

The maximization of the output of basic needs proved inconvenient in this model because of the geographical differences in the nature of these needs and in the requirements for satisfying them. Life expectancy at birth was therefore used as a proxy for basic needs, so that maximization of life expectancy at birth became the optimization criterion for allocating capital and labour between the sectors. The approach also involved sectoral targets, viz., a daily intake of 3,000 kcal and 100 grams protein per person, 7 square metres of housing per person, and 12 years of basic education for all persons between 6 and 17 years. Indeed, these sectoral targets in food, shelter, and education may be regarded as the basic core of human needs for productive survival.

The Latin-American World Model considers three regions of the developing world—Latin America, Africa and Asia—and estimates the period (in years) required, after the base year of 1970, for basic needs to be satisfied. Estimates are also obtained of the GNP per capita and populations at the end of the period (Table 3.7).

The increase in per capita GNP associated with the process of satisfying basic needs in the various regions has been used to estimate an upper limit to the increases in per capita energy consumption required for the process.[11] This calculation can be most simply achieved by using correlations between per capita energy consumption and per capita GNP derived from past experiences, noting that such correlations are valid only when the composition of the economy's output and the energy-using technologies remain the same. If these assumptions are made, the result is that between 1.0 and 1.4 kW per capita of commercial energy is required for the satisfaction of basic needs.

This result is incomplete, because, in addition to its other limitations, the energy-GNP correlation is restricted to commercial energy. Hence, the per capita energy consumption required for the satisfaction of basic needs must also include the non-commercial energy consumption. As a first approximation, this can be done by adding the estimated current per capita consumption of non-commercial energy in developing countries,* which can be taken as 0.4 kW per capita. Thus, the total per capita energy consumption required for the satisfaction of basic needs in Africa, Asia, and Latin America may be taken as 1.4, 1.8, and 1.7 kW per capita, respectively (Table 3.7). Taking into account the present populations in these three regions, the weighted average of the per capita energy consumption for satisfying basic needs works out to be 1.75 kW per capita.

It is important to keep in mind, though, that this estimate of the energy required to meet basic needs only represents an upper limit because it involves the use of an energy-GNP correlation based on the past and present performance of the energy sector. That is, it is based on the technologies which have been, and are being, used, and it assumes that their low efficiencies of energy use will persist into the future, even though it is known that major improvements in energy-efficiency are possible. In fact, the per capita energy required for satisfying basic needs can be reduced drastically through the implementation of improved technologies. How much it can be reduced depends upon the particular energy-using technologies that are deployed and the extent to which energy-efficiency improvements are implemented. However, a good idea of the possibilities can be revealed through the following thought-experiment.

---

* The drawback of this assumption is that the non-commercial energy is being used mostly in fuelwood stoves with low efficiencies of the order of 10 percent, and therefore the use of current non-commercial energy consumption norms is tantamount to acceptance of the low stove-efficiencies as the "permanent" state of affairs.

Table 3.7. Per Capita Energy Requirements Associated with the Satisfaction of Basic Human Needs, Based on Historical E/GNP Correlations and the Use of Latin American World Model for Future Economic Growth[a]

Region	1970 Per Capita GNP (1960 $)	1970 Per Capita Energy Use (b) (kW)	Commercial Energy Intensity in 1970 (Watts per 1960 $)	Required Increment in Per Capita GNP (c) (1960 $)	Energy Use (d) (kW)	Date by which BHN could be Satisfied (e)	Per Capita Energy Use Required to Satisfy BHN (kW)
Latin America	440	1.1	1.67	369	0.6	1992	1.7
Africa	154	0.7	1.83	405	0.7	2008	1.4
Asia	112	0.7	2.89	394(f)	1.1	2020	1.8

(a) *Source:* A.O. Herrera *et al.*, 1976.
(b) Includes an estimated 0.4 kW per capita of non-commercial energy use.
(c) This is the increment in per capita GNP, above the 1970 level, required for the satisfaction of basic human needs, as estimated with the Latin American World Model.
(d) Based on the commercial energy intensity in 1970.
(e) As estimated with the Latin American World Model, assuming that implementation of the BHN policy begins in 1980.
(f) For the case in which the maximum annual yield of edible products is assumed to be increased from 4 to 6 tonnes per hectare, to avoid collapse of the society.

The One Kilowatt Per Capita Thought-Experiment

Suppose that a developing country has a set of activity levels which corresponds to those in the WE/JANZ region (Western Europe, Japan, Australia, and New Zealand) in the mid- to late 1970's (Table 3.8). In other words, all families, on average, live in reasonably solid houses with about 25 m² per capita and water supplies and sanitation. Further, all homes have a clean, easy-to-use cooking fuel (for example, gas), are illuminated with electric lights, and they come

Table 3.8. Activity Levels for a Developing Country in a Warm Climate, with a Level of Amenities (except for Space Heating) Comparable to that in the WE/JANZ[a] Region in the 1970s

Activity	Activity Level
Residential[b]	4 persons/household (HH)
Cookng	Typical cooking level w/LPG stoves[c]
Hot Water	50 l of hot water/capita/day[d]
Refrigeration	One 315 l refrigerator-freezer/HH
Lights	New Jersey (US) level of lighting[e]
TV	1 colour TV/HH, 4 hours/day
Clothes Washer	1/HH, 1 cycle/day
Commercial	5.4m² floor space/capita (WE/JANZ ave, '75)
Transportation	
Automobiles	0.19 autos/capita, 15,000 km/car/yr (WE/JANZ ave, '75)
Intercity bus	1850 p-km/capita (WE/JANZ ave, '75)
Passenger train	3175 p-km/capita (WE/JANZ ave, '75)[f]
Urban Mass Transit	520 p-km/capita (WE/JANZ ave, '75)[g]
Air Travel	345 p-km/capita (WE/JANZ ave, '75)
Truck Freight	1495 t-km/capita (WE/JANZ ave, '75)
Rail Freight	814 t-km/capita (WE/JANZ ave, '75)
Water Freight (incl. bunkers)	1/2 OECD Europe ave, '78[h]
Manufacturing	
Raw Steel	320 kg/capita (OECD Eur ave, '78)
Cement	479 kg/capita (OECD Eur ave, '80)
Primary Aluminium	9.7 kg/capita (OECD Eur ave, '80)
Paper and Paperboard	106 kg/capita (OECD Eur ave, '79)
Nitrogenous Fertilizer	26 kg N/capita (OECD Eur ave, '79/80)
Agriculture	WE/JANZ ave, '75
Mining, Construction	WE/JANZ ave, '75

(a) Here WE/JANZ stands for Western Europe, Japan, Australia, New Zealand, and South Africa. The WE/JANZ 1975 average values are from (Khan and Holzl, 1982).

(b) Activity levels for the residential sector are estimates, owing to lack of data for the WE/JANZ region.

(c) In Brazil cooking with LPG averages one 13 kg cannister per month for a family of 5, corresponding to per capita fuel consumption rate of 49 Watts, for an ordinary gas stove with a burner efficiency of about 50 percent.

(d) For water heated from 20 to 50°C. In the U.S. the average is about 100 litres per capita per day.

(e) It is assumed that 5 compact fluorescent light bulbs are used on average 4 hours a day.

(f) In 1975 the diesel/electric mix was in the ratio 70/30.

(g) In 1975 the diesel/electric mix was in the ratio 60/40.

(h) The 119 kg of oil use per capita for water freight in 1978 in OECD Europe is assumed to be reduced in half because of reduced oil use (58 percent of Western European import tonnage and 29 percent of that of exports were oil in 1977) and emphasis on self-reliance.

Source: J. Goldemberg *et al.,* 1985.

equipped with all the basic electric appliances—a refrigerator/freezer, a water heater, a clothes washer, and a television set. Also, there is one automobile for every 1.2 households on average; and the average person travels by air to the extent of 350 km per year. All this cannot be sustained without well-developed industries for the processing of basic materials and a large service sector— hence, it is visualized that this infrastructure has been established and is in operation.

It is clear that these activity levels are more than sufficient to meet the basic needs of the population; in fact, they go very much farther to provide for major improvements in the quality of life.

Most of the energy-utilizing technologies that are envisaged to achieve the above activities are examples of the "best available" technologies in terms of their energy performance—for example, the most energy-efficient stoves, water-heaters, refrigerators/freezers, light bulbs, commercial buildings, cement plants, paper mills, nitrogen fertilizer plants (Table 3.9). Because these technologies are available on the market today they can be considered to be economically viable at present energy prices. A few of the indicated technologies are "advanced technologies" that could be commercialized over the next decade—hence, they are not contingent on the achievement of technological breakthroughs. Indications are that these technologies would be cost-effective at present energy prices.

The final steps in the thought-experiment consist of multiplying each activity level by the corresponding *specific energy demand,* that is, the energy demand for unit level of the activity, and then of summing up all the activities.

It turns out that, roughly speaking, *the total final energy demand for the WE/JANZ set of activity levels and the menu of energy-efficient technologies is only about 1 kW per capita* (Table 3.10). This is both a surprising and remarkable result, because this level of final per capita energy use is only about 20 percent more than the actual per capita energy use rate in developing countries in 1980. The interesting implication of this result is that with 1 kW per capita of energy, developing countries can provide any standard of life ranging from the present low level, in which even basic human needs are not satisfied for the majority of the population, to a level as high as in the WE/JANZ region in the mid- and late 1970's.

Table 3.9. Technological Opportunities for a Developing Country in a Warm Climate to Use Currently Best Available or Advanced Energy Utilization Technologies

Activity	Technology, Performance
Residential	
Cooking	70% efficient gas stove[a]
Hot Water	heat pump, WH, COP = 2.5[b]
Refrigeration	Electrolux, 475 kWh/year[c]
Lights	Compact Fluorescent Bulbs[d]
TV	75 Watt unit
Clothes Washer	0.2 kWh/cycle[e]
Commercial	Performance of Harnosand Building (all uses, ex. space heating)[f]
Transportation	
Automobiles	Cummins/NASA Lewis Car @ 3.0 1/100 km[g]
Intercity bus	3/4 energy intensity in '75[h]
Passenger train	3/4 energy intensity in '75[i]
Urban Mass Transit	3/4 energy intensity in '75[j]
Air Travel	1/2 U.S. energy intensity in '80[k]
Truck Freight	0.67 MJ/tonne-km[l]
Rail Freight	Electric rail @ 0.18 MJ/tonne-km[m]
Water Freight (incl. bunkers)	60% of OECD energy intensity[n]

Manufacturing

Raw Steel	ave, Plasmasmelt & Elred Processes[o]
Cement	Swedish ave in 1983[p]
Primary Aluminium	Alcoa process[q]
Paper and Paperboard	Ave of 1977 Swedish designs[r]
Nitrogenous Fertilizer	Ammonia derived from methane[s]

Agriculture	3/4 of WE/JANZ energy intensity[t]
Mining, Construction	3/4 of WE/JANZ energy intensity[t]

(a) Compared to an assumed 50 percent efficiency for existing gas stoves. 70 percent efficient stoves having low NOx emissions have been developed by Thermoelectron Corporation for the Gas Research Institute in the United States (Shukla and Hurley, 1983).

(b) The assumed heat pump performance is comparable to that of the most efficient heat pump water heaters available in the U.S. (Williams *et al.*, 1983). These heat pump water heaters would provide a modest degree of air-conditioning comfort as well, extracting some 5.5. GJ of heat from the living space each year.

(c) This Electrolux model was the most energy-efficient 2-door, 315 litre refrigerator/freezer available in Europe in 1982.

(d) These bulbs, which can be screwed into ordinary incandescent sockets, draw 18 W but put out as much light as 75 W incandescent bulbs.

(e) Typical value for U.S. washing machines.

(f) The Harnosand Building was the most energy-efficient commercial building in Sweden in 1981, at the time it was built. It used 0.13 GJ of electricity per m^2 of floor area for all purposes other than space heating. For details, see Chapter 2.

(g) The Cummins/NASA Lewis car is a design for a 1,360 kg, 4-5-passenger car in a 1984 Ford Tempo body, with a 4-cylinder, direct-injection, spark-assisted, multi-fuel capable, adiabatic diesel engine with turbo-compounding. For details, see Chapter 2 (von Hippel, 1981).

(h) In 1975 the average energy intensity of intercity buses was 0.60 MJ/p-km. A 25 percent reduction is assumed from the introduction of adiabatic diesel engines with turbo-compounding.

(i) In 1975 the average energy intensity of passenger trains was 0.60 MJ/p-km for diesel units and 0.20 MJ/p-km for electric units. A 25 percent reduction in energy intensity is assumed, arising from a switch to adiabatic diesels with turbo-compounding and the use of electric motor control technology.

(j) In 1975 the average energy intensity of urban mass transit was 1.13 MJ/p-km for diesel buses and 0.41 MJ/p-km for electric mass transit. A 25 percent reduction in energy intensity is assumed, arising from a switch to adiabatic diesels with turbo-compounding and the use of electric motor control technology.

(k) In 1980 the U.S. average energy intensity for air travel was 3.8 MJ/p-km. With the various improvements described in Chapter 2 this could be reduced by half (von Hippel, 1981).

(l) The assumed energy intensity is 1/3 less than the simple average today in Sweden for single-unit trucks (1.26 MJ per tonne-km) and combination trucks (0.76 MJ per tonne-km), to take into account improvements via use of adiabatic diesels with turbo-compounding.

(m) The average energy intensity for electric rail in Sweden, with an average load of 300 tonnes and an average load factor of about 40 percent.

(n) A 40 percent reduction in fuel intensity is assumed, reflecting innovations such as the adiabatic diesel and turbocompounding.

(o) Assuming a 50/50 mix of the Elred process [requiring 10.7 GJ of fuel and 1.3 GJ of electricity per tonne] and the Plasmasmelt process [requiring 4.6 GJ of fuel and 4.2 GJ of electricity per tonne] for steelmaking. See Chapter 2.

(p) Assuming an energy intensity of 3.56 GJ of fuel and 0.40 GJ of electricity per tonne, the average for Sweden in 1983.

(q) Assuming an energy intensity of 84 GJ per tonne of fuel (the U.S. average in 1978) and 36 GJ of electricity [the requirements for the Alcoa process now being developed (Beck, 1977)].

(r) Assuming an energy intensity of 7.3 GJ of fuel and 3.2 GJ of electricity per tonne (the average for 1977 Swedish designs—see Chapter 2).

(s) Assuming an energy intensity of 44 GJ of fuel per tonne of nitrogen in ammonia, the value with steam reforming of natural gas in a new fertilizer plant (Waitzman *et al.*, 1978).

(t) Assuming a 25 percent reduction in energy intensity, owing to innovations such as the use of advanced diesel engines.

Source: J. Goldemberg *et al.*, 1985.

It is possible to achieve the large improvements in living standards characterizing this scenario without increasing energy use in part because enormous increases in energy efficiency arise simply by shifting from traditional, inefficiently used, non-commercial fuels (which at present account for nearly half of all energy use in developing countries) to modern energy carriers (electricity, liquid and gaseous fuels, processed solid fuels, etc.).

Table 3.10. A Final Energy Use Scenario for a Developing Country in a Warm Climate, with a Level of Amenities (except for Space Heating) comparable to that in the WE/JANZ[a] Region in the 1970's, but with Currently Best Available or Advanced Energy Utilization Technologies

Activity	Average Rate of Energy Use (Watts Per Capita)		
	Electricity	Fuel	Total
Residential			
Cooking		34	
Hot Water	29.0		
Refrigeration	13.5		
Lights	3.8		
TV	3.1		
Clothes Washer	2.1		
Subtotal	51	34	85
Commercial	22	—	22
Transportation			
Automobiles		107	
Intercity bus		26	
Passenger train	4.5	32	
Urban Mass Transit	2.0	8	
Air Travel		21	
Truck Freight		32	
Rail Freight	5		
Water Freight (incl. bunkers)		50	
Subtotal	12	276	288
Manufacturing			
Raw Steel	28	77	
Cement	6	54	
Primary Aluminum	11	26	
Paper and Paperboard	11	24	
Nitrogenous Fertilizer	—	36	
Other[b]	65	212	
Subtotal[c]	121	429	550
Agriculture	4	41	45
Mining, Construction	—	59	59
Totals	210	839	1049

(a) Here WE/JANZ stands for Western Europe, Japan, Australia, New Zealand, and South Africa. For the WE/JANZ 1975 average values for activity levels are indicated in Table 3.8 and the energy intensities given in Table 3.9.

(b) This is the residual, the difference between the manufacturing total and the sum of the items calculated explicitly. Energy usage associated with "other" for the non-manufacturing sectors is negligible and thus is not shown explicitly in this table.

(c) It has been estimated that at Sweden's 1975 level of GDP, final energy demand in manufacturing would have been 1.0 kW (half the actual value), had advanced technology been used (Steen *et al.,* 1981). The value assumed here is 45 percent less, since the average per capita GDP was 45 percent less for W. Europe than for Sweden in 1975. Also, 22 percent of final manufacturing energy use is assumed to be electricity, the Swedish value for 1975. *Source:* J. Goldemberg *et al.,* 1985.

The importance of modern carriers is evident from the fact that for Western Europe, where non-commercial fuel use is very small, per capita final energy use for purposes other than space heating in 1975 was only 2.3 kW*, about 2 1/2 times that in developing countries, even though per capita GDP was 10 times as large.

The importance of modern energy carriers can be illustrated via an end-use analysis of cooking, which accounts for most non-commercial energy use in developing countries. The per capita energy use rate for fuelwood stoves, some 0.25 to 0.6 kW (0.4 to 1.0 tonnes of wood per year) per capita, is far in excess of the 0.05 kW per capita rate which is typical for cooking with LPG or natural gas in developing and industrialized countries alike (Figure 3.3). The much lower energy-use rates for cooking with modern energy carriers reflects both the better efficiency (40 to 50 percent versus 12 to 18 percent for traditional fuelwood stoves) and the better controllability of stoves fuelled with modern energy carriers.

In addition to the energy savings associated with the shift to modern energy carriers, considerable further savings can be gained by adopting more energy-efficient technologies that have recently become available, as described in Chapter 2.

This thought-experiment is not intended to establish activity-level targets for developing countries, to be achieved by some particular future date. The appropriate mix and levels of activities for the future in developing countries may well have to be different to be consistent with overall development goals. Rather its purpose is to show that it is possible not only to meet

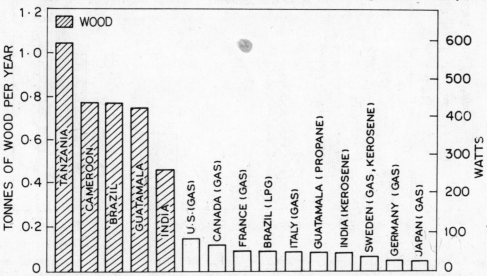

Figure 3.3. Per capita energy-use rates for cooking. For both wood stoves and stoves involving high-quality energy carriers, the per capita energy-use rate for cooking is expressed in Watts. The wood consumption rate is also given in tonnes of dry wood per year.
Source: R.H. Williams, 1985b.

---

* In addition, about 1 kW per capita was used for space heating in the WE/JANZ region for space heating—essential in cold climates, but not needed in most developing countries.

basic human needs but also to provide improvements in living standards that go far beyond the satisfaction of basic needs, without significant increases in per capita energy use.

The value of this "thought-experiment" in making this point would be less if it turned out there are energy-intensive activities which are likely to be important for developing countries in the coming decades and which are characterized by activity levels in excess of those assumed for this scenario.

A critical question is whether at intermediate times, during the intensive "infrastructure-building" phase of development, there would be need for more energy than what is now required to support the economies of modern Europe.

The prominent role of the industrial sector in the 1 kW scenario—accounting for 58 percent of total final energy use, compared to 28 percent, 4 percent, and 37 percent for India, Tanzania, and Brazil, respectively, or, alternatively 26 times, 6 times, and 1.75 times as much energy per capita as in these same countries today (Figure 3.4)—indicates that this scenario is consistent with considerable infrastructure-building. Nevertheless, this question is worthy of close attention,

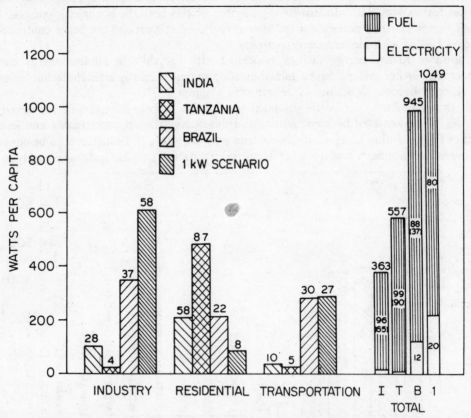

Figure 3.4. Final energy use per capita by sector and energy carrier, for India in 1978 (A.K.N. Reddy, 1985), Tanzania in 1981 (M.J. Mwandosya and M.L.P. Luhanga, 1985), Brazil in 1982 (Republica Federativa do Brasil Ministerio das Minas e Energia, 1983), and the 1 kW scenario presented in Table 3.10. For the three sets of columns on the left, the numbers at the top are the sectoral shares (in percent) of total final energy use. For the four columns on the right, which give total final energy use per capita by carrier for India (I), Tanzania (T), brazil (B), and the 1 kW scenario (1), the numbers at the top are the total final energy use per capita, while the numbers in the columns are the carrier shares (in percent). The numbers in parentheses in these columns give the non-commercial energy share (in percent) of total final energy use.
Source: J. Goldemberg *et al.,* 1985.

as it is certainly true that Western Europe has largely completed the infrastructure-building period of its development.

The infrastructure-building period of development is characterized by rapid growth in the production and consumption of basic materials such as steel and cement, which provide the building-blocks for construction of factories, commercial buildings, roads, railroads, bridges, etc. During this period the production and use of these materials will tend to grow much faster than the economy as a whole (cf. Section 2.4). Because the basic materials processing industries are so energy-intensive (accounting for most industrial energy use even in highly industrialized countries—cf. Figures 2.13 and 2.19), industrial energy use would typically grow much more rapidly than GNP during the infrastructure-building period.

However, the absolute levels of basic materials production and use are not likely to be higher during the infrastructure-building period than in the beginnings of the post-industrial phase (e.g., the mid-1970's for Western Europe), because although basic materials play a lesser and lesser *relative* role in economic activity as the economy matures, these materials continue to play an increasing *absolute* role for a long time thereafter, as wider uses are found for these materials.

The history of the per capita use of seven important basic materials (both traditional and modern) for some Western European countries is shown in Figure 2.17. The values chosen for the per capita consumption rates of basic materials in the 1 kW scenario are from the period near the peaks of these curves.

This analysis thus suggests the feasibility of achieving a standard of living in developing countries anywhere along a continuum from the present one up to a level of amenities typical of Western Europe today, without departing significantly from the average energy use per capita for developing countries today, depending on the level of energy efficiency that is emphasized. While the standard of life actually achieved depends on many considerations, this thought-experiment implies that energy supply availability *per se* need not be a fundamental constraint on development.

3.4.2 *Rural and Urban Energy Consumption*

The discussion of energy for basic needs presented above must now be elaborated to take into account differences between rural and urban energy consumption patterns. The basic feature of rural areas in developing countries is that the populations living in the countryside are quite removed from the amenities of modern cities and towns. They are forced to lead simple lives, extracting virtually all their energy requirements from their environment. The price of this energy self-sufficiency, particularly for the rural poor, can be quite enormous in terms of human effort for such apparently trivial, but crucial, tasks as obtaining cooking fuel, processing drinking water, and carrying out agriculture.[12, 13] This tedious and arduous labour (Box 3.4) is aggravated by the low efficiencies of end-use devices, the most notable example being the woodstove.[14] As a result:

- biomass, especially in the form of non-commercial fuelwood, is the principal fuel,
- animate energy from human beings and draught animals is the chief source of mechanical energy.
- the domestic sector is the main consumer of energy, and
- cooking is the dominant end-use.

Many of the tasks which are so crucial to rural energy consumption, for example, gathering cooking fuel, are unnecessary in cities. In addition, the efficiencies of energy carriers and end-use devices commonly available in urban settlements are also much higher. All this is part of the driving force for migration to the cities—hence, the marked trend towards urbanization in developing countries, which is particularly significant in Latin America.[15]

Box 3.4: Energy Consumption in a South Indian Village

Pura (latitude; 12 49′000″ N, longitude: 76 57′49″ E, height above sea level: 670.6 m, average annual rainfall: 50 in/year, population (in September 1977): 357, households: 56) is one of the villages in Kunigal Taluk, Tumkur District, Karnataka State, South India.

The energy-using activities in Pura are*: (a) agricultural operations (with ragi and rice as the main crops), (b) domestic activities, via grazing of livestock, cooking, gathering firewood and fetching water for domestic use, including drinking, (c) lighting, and (iv) industry—pottery, flour mill, and coffee shop. These activities are achieved with the following *direct* sources of energy: human beings, bullocks, firewood, kerosene, and electricity.

An aggregated matrix showing how the various energy sources are distributed over the various energy-using activities is presented in Table B3.4.1.

Table B3.4.1. Pura energy source-activity matrix (million kcals/year)

	Agriculture	Domestic	Lighting	Industry	Total
Human	7.97	50.78	—	4.97	63.72
(Man)	(4.98)	(20.59)	—	(4.12)	(29.69)
(Woman)	(2.99)	(22.79)	—	(0.85)	(26.63)
(Child)	—	(7.40)	—	—	(7.40)
Bullock	12.40	—	—	—	12.40
Firewood	—	789.66	—	33.93	823.59
Kerosene	—	—	17.40	1.40	18.80
Electricity	6.25	—	2.65	0.71	9.61
Total	26.62	840.44	20.05	41.01	928.12

Total energy = 928 million kcal/year = 2955 kWht/day = 8.25 kWht/day/capita

The matrix yields the following ranking of sources (in order of percentage of annual requirement): (a) firewood 89 percent, (b) human energy 7 percent, (c) kerosene 2 percent, (d) bullock energy 1 percent, (e) electricity 1 percent. The ranking of these activities is as follows: (a) domestic activities 91 percent, (b) industry 4 percent, (c) agriculture 3 percent, (d) lighting 2 percent.

Human energy is distributed thus: domestic activities 80 percent (tending livestock 37 percent, cooking 19 percent, gathering firewood 14 percent, fetching water 10 percent), agriculture 12 percent, and industry 8 percent. Bullock energy is used wholly for agriculture, including transport. Firewood is used to the extent of 96 percent (cooking 82 percent and heating bath water 14 percent) in the domestic sector, and 4 percent in industry. Kerosene is used predominantly for lighting (93 percent), and to a small extent in industry (7 percent). Electricity flows to agriculture (65 percent), lighting (28 percent), and industry (7 percent).

There are several features of the patterns of energy consumption in Pura which must be highlighted:

(contd.)

* Transport has been included in agriculture because the only vehicles in Pura are bullock carts and these are used almost solely for agriculture-related activities such as carrying manure from backyard compost pits to the farms and produce from farms to households.

(i) What is conventionally referred to as *commercial* energy, i.e., kerosene and electricity in the case of Pura, accounts for a mere 3 percent of the inanimate energy used in the village, the remaining 97 percent coming from firewood.* Further, firewood must be viewed as a *non-commercial* source since only about 4 percent of the total firewood requirement of Pura is purchased as a commodity, the remainder being gathered at zero private cost.

(ii) *Animate* sources, viz., human beings and bullocks, only account for about 8 percent of the total energy, but the real significance of this contribution is revealed by the fact that these animate sources represent 77 percent of the energy used in Pura's agriculture. In fact, this percentage would have been much higher were it not for the operation of *four* electrical pumpsets in Pura which account for 23 percent of the total agricultural energy.

(iii) Virtually all of Pura's energy consumption comes from traditional renewable sources—thus agriculture is largely based on human beings and bullocks, and domestic cooking (which uses about 80 percent of the total inanimate energy) is based entirely on firewood.**

(iv) However, this pattern of dependence on renewable resources is achieved at an exorbitant price: levels of agricultural productivity are very low, and large amounts of human energy are spent on firewood gathering—on the average of about 2-6 hours and 4-8 km per day per family to collect about 10 kg of firewood.

(v) Fetching water for domestic consumption also uses a great deal of human energy— an average of 1-5 hours and 1-6 km per day per household—to achieve an extremely low *per capita* water consumption of 17 litres per day.

(vi) 46 percent of the human energy is spent on tending livestock—5-8 hr/day/household—which is a crucial source of supplementary household income.

(vii) Children contribute a crucial 30 percent, 20 percent, and 34 percent of the labour for gathering firewood, fetching water, and tending livestock respectively. Their labour contributions are vital to the survival of families, a point often ignored by population and education planners.

(viii) Only 25 percent of the houses in the "electrified" village of Pura have acquired domestic connections for electric lighting; the remaining 75 percent of the houses depend on kerosene lamps, of which 78 percent are of the open-wick type.

(ix) A very small amount of electricity, viz., 30 kWh/day, flows into Pura, and even this is distributed in a highly unequal way—65 percent of this electricity goes to the 4 irrigation pumpsets of 3 landowners, 28 percent to illuminate 14 out of 56 houses, and the remaining 7 percent for one flour-mill owner.

An end-use analysis for Pura is shown in Table B3.4.2 which also contains the output energies, taking into account the efficiencies of energy use.

In the first part of the table, the end-uses of inanimate and animal energy (along with indicative temperatures corresponding to these tasks) are ranked in order of decreasing magnitude of energy used.

---

* Pura uses about 217 tonnes of firewood per year, i.e., about 0.6 tonnes/day for the village, or 0.6 tonnes/year/capita.

** Unlike some rural areas of India, dung cakes are not used as cooking fuel in the Pura region. In situations where agro-wastes, e.g., coconut husk, are not abundant, it appears that, if firewood is available within some convenient range (determined by the capacity of head-load transportation), dung-cakes are never burnt as fuel; instead, dung is used as manure.

Table B3.4.2. End-Uses of Inanimate and Animal Energy in Pura

End-use	Input Energy (million kcal/year)	Estimated Efficiency	Output Energy (million kcal/year)
1. Heating (95-250 C)	688.9	5	34.4
2. Heating (~55 C)	112.4	5	5.6
3. Heating (~800 C)	23.8	5	1.2
4. Lighting	20.1	2-5	0.5
4.1. Lighting (electrical)	(2.7)	10	(0.3)
4.2. Lighting (kerosene)	(17.4)	1	(0.2)
5. Mobile power	12.4	20	2.5
6. Stationary power	7.0	80	5.6
6.1. Water lifting	(6.3)	80	(5.0)
6.2. Flour milling	(0.7)	80	(0.6)
Total	864.6		49.8

Table B3.4.3 shows the human time and energy expenditures by activity in Pura. It is obvious that the inhabitants of Pura have to endure burdens which have been largely eliminated in urban settings by the deployment of inanimate energy. For example, gathering firewood and fetching water can be eliminated by the supply of cooking fuel and water respectively.

Table B3.4.3. Expenditures of Human Energy in Pura

Human Activity	Hours/Year	Hours/Day/HH	Million kcal/Year
1. Domestic	255,506	12.5	50.8
1.1. Tending livestock	(117,534)	(5.7)	(23.4)
1.2. Cooking	(58,766)	(2.9)	(11.7)
1.3. Firewood gathering	(45,991)	(2.3)	(9.1)
1.4. Fetching water	(33,215)	(1.6)	6.6
2. Agriculture	34,848	1.7	8.0
3. Industry	20,730	1.0	5.0
Total	311,084	15.2	63.8

The poorer migrants from the countryside leave behind a tough rural life usually without access to electricity, running water, schools, and medical care, a situation aggravated by the vagaries of traditional agriculture. It does not seem to matter to them that they are forced to live in urban slums, shantytowns, barrios, and favelas, the poverty of which is striking because it "coexists" with the affluent sections of the cities. The living conditions of these slum-dwellers may appear unbearable to middle- and upper-class urbanities, but these conditions nevertheless seem to represent significant improvements to the migrants. Even in slums, many of them can enjoy the benefits of electric lights, piped water, schools, modern medicine, entertainment through movies, radio and (in the more affluent developing countries) television, and the hope at least of regular, rather than seasonal, employment. Also, slum-dwellers are constantly subject, through direct observation and indirect exposure to the media, to the "demonstration effect" of the affluence of the urban rich, with their luxurious food, homes, clothes, domestic appliances, automobiles, etc., and aspire to share in this conspicuous consumption.

These disparities in the life-styles of the rural and urban populations are manifest in differences in energy consumption patterns.[15,16] In general, the per capita energy consumption tends to rise with per capita income, at first "linearly", but there is generally a flattening out in the energy-income curves (Figure 3.5). But the curve for rural areas is usually above the corresponding curve for urban areas.[16] This means that, *for any given income/expenditure,* the per capita consumption of direct energy is *higher in rural areas than in cities.*

The reason for this result is simple: cooking is a major end-use of domestic energy in developing countries; the use of biomass, particularly fuelwood, as a cooking fuel is far more common in rural areas; and this non-commercial energy is used at low efficiencies in fuelwood stoves. The tendency in cities is to shift to more efficient cooking fuels, often in this sequence: fuelwood to charcoal to kerosence to LPG. And the fuel efficiencies, with current technologies, are in the same sequence. Basically, the same type of effect takes place in the case of lighting too, because the percentage of kerosene-illuminated houses is higher in rural areas, and the tendency in cities is to shift to more efficient electric illumination. Thus, the lower urban energy-consumption for a given income level corresponds to greater efficiencies, and a better quality of life for urban households.

If the rural-urban comparison is made at the prevailing average per capita incomes in rural and urban areas, then the urban per capita energy consumption can turn out to be almost the same as the corresponding rural figure, or at least not substantially lower as is the case at a given income. This is due to the fact that it is the low-income groups which tend to migrate, and though the migration from village to city is generally associated with an increase in average per capita income, the associated enhanced energy demand is offset by a demand reduction arising from the shift to more efficient, but costlier, fuels.

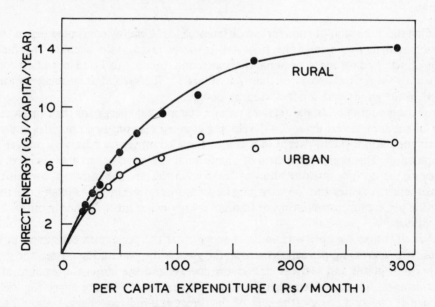

Figure 3.5. Comparison of rural and urban per capita energy use in India versus per capita income. *Source:* G. Sumithra, 1981.

Instead of analyzing income/expenditure effects, estimates can be made of the direct energy required in the context of technologies prevailing today to satisfy basic needs for urban and rural populations. Such calculations confirm (Table 3.12) that per capita energy use is higher in rural areas.[8] The end-uses which depend upon the combustion of fuel require a greater consumption of primary energy in rural areas, mainly because the corresponding end-use technologies are less efficient.

Table 3.12. Calculation of Differences in Direct Energy assumed to be needed in the "Bariloche model" to Satisfy Basic Needs in Rural and Urban Areas[a]

Basic Need	Per Capita Energy Requirement (MJ/day)					
	Hot Climate[b]		Temperate Climate[b]		Cold Climate[b]	
	Rural	Urban	Rural	Urban	Rural	Urban
Lighting	1.9	4.9	2.3	6.2	2.7	9.9
Space Heating	—	—	25.1	7.4	94.1	29.7
Space Cooling	0.8	1.3	0.3	0.6	—	—
Food Preservation	3.8	4.4	2.9	3.1	—	—
Cooking	21.8	6.2	26.2	7.4	32.7	9.3
Water Pumping	0.2	—	0.2	—	0.1	—
Hot Water	7.1	3.7	13.5	7.4	21.2	11.1
Recreation, etc.	3.0	3.1	3.0	3.1	3.8	4.9
Total	38.6	23.6	73.7	35.2	154.6	64.9

(a) *Source:* V. Bravo *et al.,* 1979.

(b) Hot, temperature and cold climates are characterized by mean annual temperatures of 25°C, 18°C, and 10°C respectively.

Extending the discussion of rural/urban differences to the energy consumed beyond the home can be achieved in two steps. The first step involves taking into account the fuel energy (gasoline/diesel) used for mass/personal transportation, because this transportation component is significantly larger for urbanites than for villagers. The second step concerns the energy embodied in the goods and services used by people.

Most of the modern goods and services require commercial energy for their production, and the bulk of the commercial energy in developing countries is consumed in cities and towns by the urban population which, except in a few of these countries, is relatively smaller than the rural population. This preferential flow of commercial energy to urban areas is inevitable, not only because the decision-making elites of these countries are urban-based, but also because the infrastructures of cities and towns require larger inputs of energy than villages. All this would result in the per capita consumption of indirect energy being much higher in cities and towns than in villages.

However, it is also the case with indirect energy that the per capita consumption is bound to increase with increasing per capita income: the greater the purchasing power, the greater the consumption of goods and services embodying energy, and the larger the amount of indirect energy consumed. Thus, indirect-energy-income curves, as well as the curve for the income-dependence of the total energy (the sum of the direct and indirect energy), should be similar to those seen in the case of direct energy. And, indeed this is the case for the state of Sao Paulo in Brazil (Figure 3.6).

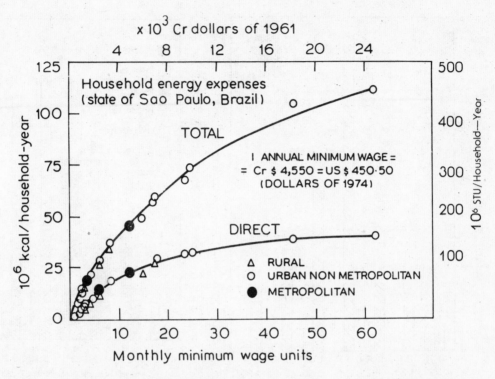

Figure 3.6. Household energy use in the State of Sao Paulo, Brazil, versus household income. *Source:* J. Goldemberg *et al.*, 1984.

Though considerable work needs to be done to expand understanding of rural-urban differences in energy consumption, these conclusions are probably adequate to suggest guidelines for energy strategies for developing countries.

3.4.3 *The Conventional Wisdom of Dual Societies on Energy Strategies for Developing Countries*

Developing-country elites drive modern automobiles, live in spacious air-conditioned houses, and benefit from the modern technologies imported from the industrialized countries, irrespective of whether the luxurious areas in which they reside are located in Bombay, Jakarta, Nairobi, or Sao Paulo. Consequently, they tend to have the same basic energy consumption patterns as people in the industrialized countries, and depend largely on oil and electricity. The poor in developing countries may crave for the consumption patterns of the elite, but they have to rely mostly on traditional technologies that are invariably based on local biomass fuels—fuelwood, charcoal, animal wastes, and agricultural residues.

These differences in energy consumption are well illustrated by the type and amount of energy carriers consumed by different income groups. For example, Figure 3.7 reveals how the direct consumption of oil derivatives, electricity and firewood in Brazil in 1979 depended upon the income bracket—the elite uses more oil-derivatives and electricity than the poor who in turn use more fuelwood than the rich.

A preoccupation with the energy problems of the elite leads naturally to the conventional wisdom on formulating energy strategies for developing countries, which equates growth with energy consumption and energy consumption with electricity or oil:

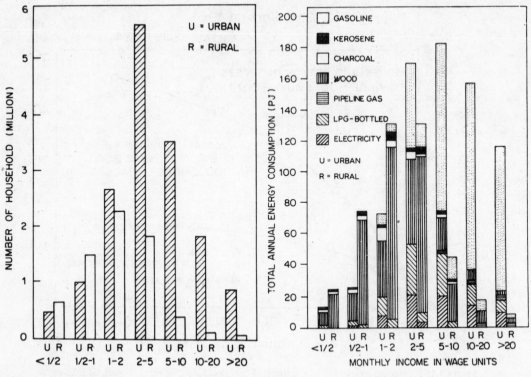

Figure 3.7. Distribution of the number of households (left) and total annual energy consumption (right) by income bracket, for Brazil in 1979.
Source: J. Goldemberg *et al.,* 1984.

$$\underset{1}{\text{Development}} = \underset{2}{\text{Growth}} = \underset{3a}{\text{Energy}} = \underset{4a}{\text{Electricity}} = \text{Centralized Electricity} + \text{Grid Transmission,}$$

with a modern addendum:

$$\underset{1}{\text{Development}} = \underset{2}{\text{Growth}} = \underset{3b}{\text{Energy}} = \underset{4b}{\text{Oil}} = \text{Engines/Furnaces/Heating Devices fuelled with}$$

petroleum derivatives

This results in a neglect of crucial aspects of the lives of the rural and urban poor—their basic human needs, their settlements (slums and villages), their fuels (particularly fuelwood), and the end-uses (cooking and lighting), and their energy-using devices (cooking stoves). And, instead, the energy carriers of the elites—oil and electricity—are overemphasized.

Quite apart from this distortion in energy planning, some steps in the arguments underlying the conventional wisdom are highly questionable, if not patently false, and others have been presented as if they are the only feasible options when in fact there are alternatives available.

(1) *As argued earlier, if the aim of development is to lead to the eradication of poverty, then the structure and content of growth, as well as the distribution of its benefits among the various sections of society, is as important as the magnitude of growth. In other words, there can be economic growth without eradication of poverty when the main beneficiaries of growth turn

---

* These numbers correspond to the steps in the conventional wisdom shown above.

out to be—as is often the case—the affluent rather than the needy. Thus, the emphasis of development must be on the immediate and direct satisfaction of basic human needs, starting from the needs of the neediest.

(2) The so-called "correlation" between energy consumption and GDP ignores the fact that it is possible for them to be decoupled by (Note 3.4.A):

- changes in the product-mix of the economy (for example, a trend away from basic materials to finished products), and
- Improvements in end-use efficiencies.

Energy strategies can be designed which deviate from the energy-GDP correlation by both of these paths, as the experience of the industrialized market economies has already shown (Chapter 2).

(3a) The elites in developing countries have been so preoccupied with electricity that, until recently, electricity planning was considered to be synonymous with energy planning and vice versa. The outlook began to be abandoned with the growth of the petroleum-consuming transport sector, and was completed by the oil-price hikes of the 1970's. In addition, energy planners began to recognize that from the standpoint of technical efficiencies, energy sources should be matched to energy-using tasks, and since these tasks are varied in nature, a mix of energy sources should be exploited.

(4a) Even where electricity is the appropriate source, centralized electricity generation followed by grid transmission is *not* the only option. The costs of transmission increase with distance, and beyond a break-even distance, decentralized generation from local sources may turn out to be more economical than centralized generation.

(3b) The most important energy development in developing countries has been the rising share of oil in the transport, domestic, industrial and agricultural sectors—especially for vehicles, for cooking and space heating, and for furnaces and turbines. Despite this, the use of petroleum derivatives is neither inevitable nor unavoidable—alternative fuels and/or devices are already attractive.

(4b) In particular, vehicles can be fuelled with ethyl or methyl alcohol; cooking can be done in highly fuel-efficient woodstoves and/or with producer gas or biogas; and furnaces and turbines can be run on producer gas generated from the gasification of wood.

These shortcomings in the conventional wisdom underscore the need to pursue an alternative approach which can be stated in the form of guidelines for energy strategies for developing countries. Below we discuss guidelines for adapting to the special conditions of developing countries the features of energy strategies needed to achieve a sustainable world (Section 1.5).

3.4.4 *Guidelines for Energy Strategies in Developing Countries*

(i) *Contribution to the Solution of Other Global Problems*

Among the solutions to the major global problems, there are some that the developing countries can do very little about, and others to which they can make substantial contributions. For example, the developing countries do not have a significant role to play in the solving of the problems of global climatic change and the danger of nuclear war associated with the vertical proliferation of nuclear weapons. The problems developing countries can make major contributions to solving are the problems of their poverty, their food supplies and nutritional levels, their population growth, and their environment. But this will require them to make commitment to genuine development, that is, to *a need-oriented, self-reliant, environmentally sound and peaceful socio-*

economic process. The implication of such a commitment as far as energy strategies are concerned is that *energy must be an instrument of development.*

(ii) *Energy for the Satisfaction of Basic Human Needs, Starting from the Needs of the Rural and Urban Poor*

In the case of the rural poor, the principal *direct* energy needs follow from the requirements of the main end-uses—cooking, lighting, and provision of water for human and domestic consumption (Box 3.4), and the *indirect* energy needs involve the requirements of food production and distribution, house construction and maintenance, fibre production and cloth manufacture, sanitation, education and transport, and protection of water supplies from health hazards.

The urban poor have all these direct and indirect energy needs, but they also have extra energy needs arising from the fact that the slums in which they live are inseparable parts of cities. The satisfaction of these other energy needs has to be an integral part of the solution of the energy problem of cities. To a large extent, this solution requires an improvement of the infrastructure of the city—its food distribution, its supply of cooking fuel, its housing, lighting, water supply and sewers, health and educational facilities, transport, etc.—and an extension of these improvements to those parts of the city where the urban poor live.

In following the guideline that energy must become an instrument for satisfying the basic needs of the rural and urban poor, a cautionary note is in order. There is a risk that an *exclusive* concern for the immediate energy needs of specific target groups, for example, the rural and urban poor, may result in a narrow and short-sighted view in which remote and long-term linkages to the basic needs of even these target groups are ignored. One pitfall, for example, occurs when subsidies are given to the poor, not as a transitional measure leading to their self-reliance, but as a quasi-permanent substitute for genuine developmental changes. Or, short-range planning may ignore or give insufficient attention to basic goods and infrastructural services, such as steel and transport, so that the satisfaction of basic needs cannot be sustained over the long run.

To achieve a balanced concern over long- and short-term energy objectives, it is vital to address the energy requirements of agriculture, industry, and transport, in addition to the energy demands associated with the basic needs of target groups.

(iii) *Energy and Employment Generation*

In insisting that the emphasis on energy for immediate basic needs must not preclude the provision of energy for industry (and other sectors), we are not arguing for an uncritical transfer of the labour-saving, high-productivity technologies now prevalent in the industrialized countries. Such a blind imitation of the path of industrialization followed by the developed countries is inconsistent with the severe unemployment problems confronting most developing countries. And employment is one of the most important basic needs because it provides purchasing power which enables the employed to have access to the other basic needs. Hence the generation of employment must be a major objective of the development process in general and industrialization in particular.

In this context, it must be noted that various industries differ radically in terms of their employment potential.[17] That is, the employment generated per unit of power (kW) installed or energy (kWh) supplied depends in a significant way upon the particular industry. There are highly energy-intensive industries where energy is virtually a "raw material" (aluminum smelting for instance) and non-energy intensive industries where energy is mainly an amplifier of labour or controller of information (Table 3.13).

It follows that indiscriminate industrialization can impede the achievment of development goals. If the bulk of energy is diverted to energy-intensive industries, then the production and distribution

Table 3.13. Capital Employment-Energy Demand Ratios for selected Industries in Karnataka State, India

NIC Code[a]	Item(s) Manufactured	GFC[b] per Employee ($)[c]	Employees per 100 kW of Energy Demand	GFC per kW of Energy Demand ($)
20-21	Food Products	1,960	86	1,680
23	Cotton Textiles	822	80	656
27	Wood, Wood Products, Furniture & Fixtures	640	104	667
28	Paper, Paper Products & Printing, Publishing & Allied Industries	3,370	29	984
30	Rubber, Plastic, Petroleum & Coal Products	2,670	29	765
31	Chemicals & Chemical Products (exc. products of petroleum and coal)	15,780	7	1,140
32	Non-metallic Mineral Products	4,260	13	551
33	Basic Metals & Alloys	5,770	5	295
34	Metal Products & Parts (exc. Machinery & Transport Equipment)	1,910	46	89
35	Machinery, Machine Tools and Parts (exc. Electrical Machinery)	3,720	83	3,100
36	Electrical Machinery, Appliances, Supplies and Parts	1,970	128	2,510
37	Transport Equipment and Parts	2,060	98	2,020
	ALL INDUSTRIAL ACTIVITIES	2,980	28	821

(a) NIC = National Industrial Classification
(b) GFC = Gross Fixed Capital
(c) 1 $ = Rs. 10
Source: Government of Karnataka, India

of energy may make little contribution to direct employment generation, and therefore to development.*

A recent report pertaining to the State of Karnataka in India illustrates the energy-intensity/employment issue.[18] It showed that not only did the industrial sector use 74 percent of the State's electricity output of 6,181 million kWh but just 18 major electro-metallurgical industries consumed two-thirds of the industrial electricity and gave direct employment to 4,000 people; in contrast, 1,200 other electricity-using industries used the remaining one-third of the electricity but provided employment for 250,000 people (Table 3.14).

Thus, another important guideline in countries beset with unemployment problems is that energy must be used as a mechanism for employment generation.

A corollary guideline is that the spatial allocation of energy, particularly electricity, can be used to influence the rate of growth of human settlements. For instance, by ensuring the supply of energy to small settlements which can provide local employment opportunities and be developed as alternative growth centres, the migration to already overburdened metropolitan centres can perhaps be slowed.

---

* In analyzing industrial development, the ''down-stream'' employment of an energy-intensive industry has to be taken into account. For example, associated with aluminium making in some cases is the employment provided by those who manufacture aluminium cables, apart from those involved in smelting. Often, though, the aluminium or other basic material is exported and then there is no ''down-stream'' employment.

Table 3.14. Industrial High-tension Electricity Consumption (1978-79) in Karnataka State, India

Category	Number	Electricity Consumption (GWH)	Employment
Power-intensive units*	20* (3.1%)	2,307 (68.7%)	18,270 (7.8%)
Non-power-intensive units	615 (96.9%)	1,049 (31.3%)	217,027 (92.2%)
	635	3,356	235,297

*Out of these 20, one iron and steel unit used 47 GWH and employed 12,102 persons while an aluminium plant used 76 GWH and employed 1,013 persons.
Source: Government of Karnataka, India.

(iv) *Animate Energy*

Animate energy still plays a significant role in developing countries—roughly 10 percent of the total energy comes, on the average, from human and animal work (Table 3.15). Actually, this figure is misleading, because if the agricultural sector alone is considered, the contribution of animate energy may be as high as 90-100 percent. In fact, traditional agriculture is carried out without any *direct* inputs of inanimate energy, and the price paid for this is low productivity. A superficial approach, therefore, would consist of increasing agricultural productivity by replacing animate energy with inanimate energy sources. But this approach can give rise to serious problems.

Table 3.15. Energy Inputs in Rural Areas from Human and Animal Work[a]

Country	Village	Human Work	Animal Work kW/Capita	Total Energy Consumption	Percentage of Total Energy supplied by human & animal work
China (c)	Peipan	0.0317	0.0444	0.847	9
Tanzania (c)	Kolombero	0.0285	—	0.755	4
Nigeria (c)	Batagawara	0.0285	0.0063	0.533	7
Mexico (c)	Arango	0.0349	0.0634	1.776	6
Bolivia (c)	Quebrada	0.0349	0.0888	1.224	10
Bangladesh (b)	—	0.0317	0.0476	0.250	32
India (b)	—	0.0317	0.0476	0.343	23
India (d)	Pura	0.0222	0.00317	0.343	7
India (e)	Ungra	0.0285	0.0127	0.336	12

(a) Figures for human and animal work are an approximation calculated using the method of Revelle, in which only the energy expended in performing work is counted. This is assumed to be 33% and 40% of the energy content of the food eaten by humans and draught animals, respectively. If the entire food energy content were included, the figures would be substantially higher.
(b) *Source:* R. Revelle, 1976.
(c) Adapted from A. Makhijani and A. Poole, 1975, by Revelle using the assumptions in (a).
(d) *Source:* A.K.N. Reddy and D.K. Subramanian, 1979.
(e) *Source:* N.H. Ravindranath *et al.,* 1980.

In the first place, uncritical agricultural mechanization (for example, tractorization) can lead to the redundancy of human labour, which may benefit the proprietors of the machines, but will only increase unemployment.[19, 20] This is why agricultural mechanization has to be linked, in development planning and implementation, to the generation of on-farm and off-farm employment opportunities in rural areas and/or the absorption of labour by urban industry and services.

Similarly, the sudden displacement of draught animals (mainly bullocks) from the village scene may prove to be disadvantageous for several reasons. First, bullock power is a substitute for and saves inanimate energy sources. Second, cows and bulls play an intricate role in an agrarian economy by serving several economic purposes—(1) consumer goods (they can be eaten as beef), (2) machines for producing consumer goods such as milk, (3) equipment yielding intermediate goods (dung which yields biogas energy and fertilizer) and services (transportation and traction), and (4) they reproduce, creating more cattle. Cattle are a crucial part of agricultural ecosystems entering in many ways into the flows of matter and energy.[21]

It seems, then, that for a considerable transition period, human labour and draught animals will continue to play a significant role in rural areas along with selective mechanization. If so, major improvements have to be made in the productivity of human and animal power.[22]

Continued reliance on human labour would be turning back the clock of history only if the arduousness, drudgery and low productivity of labour are preserved in their traditional and original form. It is possible, however, to use modern science and engineering to lighten the burden and increase per capita productivity in the productive activities involving human effort. The *modus operandi* for attaining this target has to be based on exploiting the mechanical advantage of what are known in physics as *simple machines*—levers, pulleys, wheels, cranks, gears, etc.— which do not require inputs of harnessed energy.

One of the most important possibilities in this mission of finding better ways of using human energy is "pedal power" involving the principles and parts of bicycles and bicycle mechanisms, e.g., pedals, sprockets, chains, axles, wheels, etc.[23] There are two main categories of pedal power possibilities:

- stationary applications—e.g., water-pumps, winches, electrical generators, refrigerators, winnowers, corn grinders, mechanical power take-offs, hydraulic pumps, etc.—and
- transport applications—e.g., bicycle-drawn trailers, cycle rickshaws for passenger transport, pedal rovers for unprepared terrains, wheelbarrows, etc.

Above all, a widespread increase in pedal power will significantly augment the effectiveness of a vital source of energy in the rural developing world—human energy—without requiring large inputs of commercial or non-commercial energy.

Similarly, the productivity of bullock power can also be enhanced by creative engineering on the devices which transform this animate energy source into desired end-uses.

Thus, the guideline is that animate energy—the energy of human beings and draught animals— has to be given its appropriate place in the spectrum of energy sources.

(v) *Inanimate Energy for Agriculture*

Notwithstanding these considerations regarding a place for animate energy sources, the fact remains that, particularly in rain-fed agriculture, the inadequate power of draught animals and humans may be the single most important obstacle preventing multiple cropping.[24] Inputs of inanimate energy for land preparation, for example, in the form of fuel for the tractors used for ploughing, may make possible double cropping and thereby increase crop output per hectare and thus generate the additional employment associated with an extra cropping season. The use of inanimate energy for the water-lifting aspects of irrigation can have similar beneficial effects. In other words, the prospects of increased productivity (more food), income, and employment make the replacement of draught animals with machines and the introduction of mechanical water-lifting devices for irrigation sound development decisions in most situations.

Fortunately, the agricultural sector rarely accounts for much of the inanimate energy used in a country. The total amount of energy used in 1980 by the agricultural sector of all developing countries was 35 million tonnes of oil equivalent, which was 2.4 percent of the total commercial energy used. Therefore, enhanced inputs of inanimate energy in agriculture should not be denied on the grounds that they will lead to major increases in the country's total energy consumption, for the energy required will amount to only a small part of overall energy use. The agricultural sector in developing countries must be deemed to be a priority sector in the formulation of energy strategies.

The guideline is, therefore, that while improvements in agricultural productivity must be achieved cautiously taking care to use animate energy sources optimally, energy must be provided for the needs of the agricultural sector in order to enhance agricultural productivity and achieve more food self-sufficiency.

(vi) *Renewable Energy Sources*

It is clear that biomass resources, particularly fuelwood, have played and are playing a significant role in the energy supply mix of developing countries, accounting for about 40 percent of their total energy consumption (Table 3.16). As previously noted, the biomass energy resources consist of fuelwood as well as animal wastes and agricultural residues such as stalks from food

Table 3.16. Contribution of Biomass Energy in Developing Countries[a]

Country	Commercial Energy	Biomass Energy kW/Capita	Total Energy	Percentage of Energy from Biomass
Bangladesh	0.038	0.095	0.13	71
Niger	0.035	0.25	0.29	86
Gambia	0.098	0.22	0.32	69
Morocco	0.27	0.073	0.34	21
India	0.17	0.19	0.36	53
Ethiopia	0.019	0.37	0.39	95
Nepal	0.0095	0.43	0.44	98
Somalia	0.092	0.48	0.57	84
Bolivia	0.34	0.26	0.60	44
Sudan	0.16	0.63	0.79	80
Thailand	0.30	0.52	0.83	63
Tanzania	0.060	0.81	0.87	93
China	0.78	0.32	1.1	29
Brazil	0.74	0.37	1.1	34
Mexico	1.3	0.13	1.4	9
Libya	1.8	0.095	1.9	5
Developing Countries (Average)	0.549	0.416	0.956	43

(a) *Source:* D.O. Hall *et al.*, 1982.

and fibre crops, rice husks, sugar bagasse, etc. The question is whether they will continue to play a significant role in the future. The answer to this question depends on the future *availability* of biomass for energy use. Availability, in turn, is a matter of how much biomass is there but is going unused, and whether the biomass resources can be used in a renewable way.

Even though these biomass resources constitute a harvesting of solar energy, they are neither automatically available for energy purposes nor *ipso facto* renewable.

For instance, virtually all food and fibre crops give rise to residues with significant energy contents. Typical yields of residues from cereal crops in developing countries are 3.2 ± 1.3 tonnes/hectare/year (Table 3.17). From these yields, an estimate can be obtained of per capita

Table 3.17. Production of Residues from Cereal Crops in Developing Countries (per Hectare)[a]

Crop	Crop Yield[b] tonnes/ha/yr range	average	Residue:Crop Ratio[c]	Residue Production[d] tonnes/ha/yr range	average
Rice	0.7-5.7	(2.5)	2	1.4-11.4	(5.0)
Wheat	0.6-3.6	(1.5)	1.75	1.1-6.1	(2.6)
Maize	0.5-3.7	(1.7)	2.5	1.3-9.3	(4.3)
Sorghum	0.3-3.2	(1.0)	2.5	0.8-8	(2.5)
Barley	0.4-3.1	(2.0)	1.75	0.7-5.4	(3.5)
Millet	0.5-3.7	(0.6)	2	1.0-7.4	(1.2)

(a) *Source:* D.O. Hall *et al.,* 1982.
(b) 1979 FAO Production Yearbook. Figures refer to the range between countries of national crop yield averages. The numbers in parentheses represent average yield figures for the developing countries as a whole.
(c) World Bank, 1979. Figures refer to approximate residue: crop ratios at field moisture content.
(d) Calculated from columns 1 and 2. These figures are, at best, approximate since residue: crop ratios are likely to show regional as well as varietal variations. The highest values for residue production are likely to be somewhat exaggerated because high-yielding crop varieties tend to have lower residue: crop ratios.

production of cereal crop residues in selected developing countries—on the average, it is about 0.18 kW/capita (Table 3.18).

It does *not* follow, however, that all these residues are available for energy purposes; there are competing demands. In particular, crop residues are often used as fodder for livestock, and there are many food-processing wastes, particularly sugarcane bagasse, cotton husks and groundnut husks, which are already being used for energy purposes (Table 3.19). In addition, crops which are simply left in the land, for ploughing under later, protect the soil from erosion during fallow periods.

There is also competition for animal wastes which are available in considerable quantities—about 0.178 kW per capita on the average (Table 3.20)—but are also used as organic fertilizers. In Pura, India (Box 3.4), for example, dung-cakes are never burnt as fuel if firewood is available within some convenient range—dung is used instead as fertilizer.

These questions of current non-availability (for energy purposes) of otherwise available biomass energy sources will be discussed later in a treatment of food, fuel, fodder, fertilizer, fibre, and foreign-exchange interactions (Section 3.5.7). Suffice it to say here that biomass resources presently unavailable for energy can become available if alternative technologies are deployed.

Table 3.18. Per Capita Production of Cereal Crop Residues in Selected Developing Countries[a]

Country	Energy Content of Residues from Cereal Crops[b] (kW/capita)
Argentina	].793
Thailand	0.295
Malawi	0.273
Brazil	0.257
Nepal	0.225
China	0.216
India	0.174
Upper Volta	0.162
Bangladesh	0.136
Ethiopia	0.105
Peru	0.0920
Somalia	0.0666
Congo	0.349
Zaire	0.0285
Developing Countries (Average)	0.178

(a) *Source:* D.O. Hall *et al.,* 1982.
(b) Figures based on cereal crop production data from 1977 FAO Production Yearbook, and grain/residue ratios as shown in Table 3.15. Energy content of residues assumed to be 13 GJ per tonne. Average figure calculated from 1979 FAO Production Yearbook. Note that cereal residues are not the only farm agricultural residues. Significant amounts of residues from roots and tubers, vegetables, nuts, cotton, sugar cane, etc., are produced in some countries.

Table 3.19. Estimates of the Production of Selected Food Processing Wastes in Developing Countries (1975)[a]

By-Product	Estimated Production million tonnes/yr	Approximate Total Energy Content PJ per year	Present Level of use for Energy Purposes
Sugarcane Bagasse	110	1060	High
Rice Hulls	55	790	Low
Coconut Husks	13	185	Low
Cotton Husks	6	110	High
Groundnut Shells	6	100	High
Coffee Husks	2	35	Low
Oil-palm Husks	2	35	High
Oil-palm Fibres	3	20	High

Total = 2330 PJ
(Equivalent to 2.5% of total energy use in developing countries)

(a) *Source:* FAO, 1979.

The classic example of such a technology is one that avoids the competition between dung-cakes-for-fuel versus dung-for-fertilizer by processing the dung in a biogas plant to produce methane-rich biogas fuel as well as sludge for fertilizer—it contains all the nitrogen, phosphorus, and potassium of the animal waste and actually enhances the stability of the nitrogen.

Table 3.20. Per Capita Production of Animal Dung in Selected Developing Countries[a]

(kW/Capita)

Botswana	1.744
Argentina	1.268
Somalia	0.888
Sudan	0.698
Bolivia	0.602
Brazil	0.507
Nepal	0.412
Peru	0.241
Upper Volta	0.232
India	0.200
Bangladesh	0.158
Thailand	0.124
China	0.108
Malawi	0.0888
Zaire	0.0349
Congo	0.0317
Developing countries (average)	0.181

(a) *Source:* Table 1-4 in (b). Figures based on livestock data from 1977 FAO Production Yearbook using the residue production factors indicated in (c) and assuming an energy content of 15 GJ per tonne for dry dung. Average figure calculated from 1979 FAO Production Yearbook.

(b) World Bank, 1980.

(c) The following are manure production rates (in tonnes per head per year) for domesticated animals from (b) used to produce the above table.:

Cattle, buffaloes, camels	1.0
Horses, donkeys, asses	0.75
Pigs	0.3
Sheep, goats	0.15
Chickens, poultry	0.005

On the question of renewability, fuelwood is a renewable resource only if it is consumed at rates which are not more than the rate at which it is regenerated. There are a number of methodological problems in determining whether fuelwood in a particular developing country is being used in the renewable mode or not. Nevertheless, a rough *prima facie* estimate can be obtained from the actual annual per capita fuelwood yield from forests. The data from a large number of developing countries can be represented in a scatter plot of fuelwood growth versus fuelwood use. The resulting diagram (Figure 3.8) shows that there are three zones—a zone of renewable usage, a second zone where fuelwood consumption is presently renewable but can become non-renewable if preventive measures are not taken, and a third zone where fuelwood is being consumed in the nonrenewable mode. However, the situation may be better or worse than is actually the case.

For example, it is usually assumed that fuelwood trees are found only in forests, that fuelwood is not used in the form of twigs, and that the present low efficiencies of fuelwood use as well as the low productivities of forests will persist into the future. If alternative conditions are established—for example, if improvements are made both in the efficiencies of fuelwood use and forest productivities—then fuelwood renewability might be ensured.

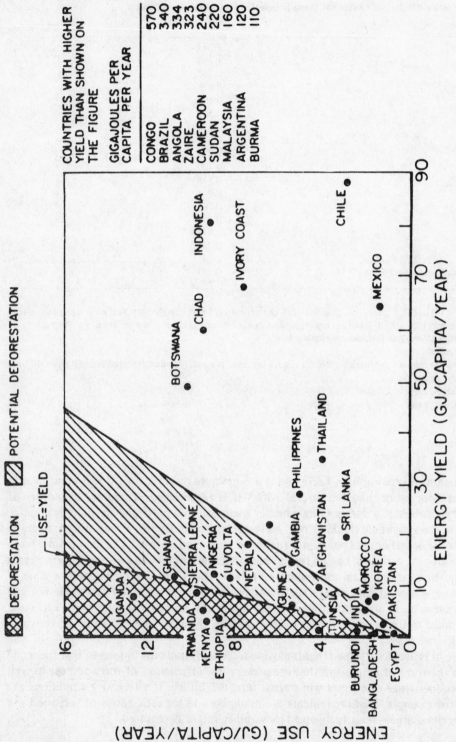

Figure 3.8 Fuelwood growth versus fuelwood use in some developing countries.
Source: Shell Briefing Service, 1980.

Energy strategies for developing countries, therefore, should fully exploit biomass energy resources but only in a renewable manner. In particular, the present pattern of inefficient usage of energy biomass cannot be perpetuated. These sources are being used predominantly in the domestic sector as fuel for cooking and water-heating and this usage is achieved at extremely low efficiencies—under 10 percent. Hence, a new role has to be defined in which there are dramatic improvements in the efficiencies of biomass use and biomass sources are diverted to other sectors.

(vii) *The Importance of Efficiency Improvements*

There is a widespread belief, particularly common in developing countries, that improvements in the efficiency of energy use are unimportant, or even irrelevant, to these countries. This attitude probably stems from their reluctance to undertake reductions in resource use in a situation where the bulk of the world's non-renewable resources is consumed by the industrialized countries. But our analysis leads to a fundamentally different conclusion.

For the same income/expenditure, the per capita energy consumption of the urban poor is less than that of the rural poor, but the contribution of various energy carriers is basically different. In general, the urban poor use less fuelwood, more electricity, and more petroleum products (LPG and kerosene) than the rural poor. These fuel shifts result in greater efficiencies of energy use for the same tasks. Thus, the urban poor have access to larger amounts of useful energy (for cooking, lighting, etc.) than the rural poor for the same amount of, or even less, primary energy.

Since, for a given task, it is the useful energy which determines the extent of satisfaction of energy needs, and of improvement of the physical quality of life, a crucial conclusion follows: *improvements in the efficiencies of energy use permit significant improvements in the physical quality of life without increases, or even with decreases, in the amount of primary energy which has to be supplied.* It is noteworthy that this conclusion has been completely missed when the sole preoccupation was the supply of primary energy and when it was assumed that the quality of life can be improved only by increases in energy supply.

In this context, another complementary approach to fuel savings is to change the definition of the task which must be performed to satisfy a human heed. For instance, instead of cooking at one atmosphere pressure in conventional vessels, pressure cookers can be used, and instead of cooking with raw cereals and beans, they can be pre-fermented and sprouted.

Improvements in end-use efficiencies and changes in task definition need not be restricted to the domestic sector. Vast possibilities exist in transportation, agriculture, and industry. And, just as the urban areas hold out lessons for efficiency improvements in the rural areas, the work in the industrialized countries on energy efficiency (Chapter 2) can be used very effectively in developing countries.

Another conclusion implicit in the above discussion is that when a particular energy carrier is being used very *in*efficiently, its much more efficient use, or a shift to a more efficient energy carrier results in the *release* the inefficiently used source for alternative purposes. This otherwise obvious conclusion is particularly important in the case of fuelwood in developing countries. By adopting high efficiency cooking fuels and devices, enormous amounts of fuelwood become available for alternative end-uses. It will be shown further on (Section 3.5.5) that this possibility can become the basis for a powerful energy strategy in developing countries.

(viii) *Fair Competition between all Energy Sources and between Energy Supply-increasing and Demand-reducing Technologies providing the Same, or even better Energy Services*

The argument for renewable energy sources should not be construed as a case for ignoring economic efficiency in the choice of energy supply technologies. In fact, one of the generic energy strategies for a sustainable world that has been stressed in the introduction (Chapter 1) is intended to remove various advantages and handicaps that bias technology choices for or against specific technologies, and thereby subsidize inefficiencies at great social cost. Such biases are found in both the developed and developing countries, but the concern at this point is for special strategies to ensure that distorting tendencies are minimized, if not removed altogether. Hence, an important guideline for energy strategies in developing countries is fair economic competition between all energy sources and between energy supply-increasing and demand-reducing technologies.

In particular, it is important to identify, on the energy supply side, subsidies and preferences that are given to various energy carriers—for example, electricity, diesel, and kerosene—and on the energy demand side, the biasing of choices towards specific end-use technologies—for example, freight transport with diesel-fuelled trucks and cooking with LPG. And just as important is the identification of biases towards investments on energy supplies in situations where the choice is between augmenting energy supply or reducing energy demand through efficiency increases.

(ix) *Synergisms*

Conventional energy strategies invariably tend to treat the various energy-using sectors as independent of each other. Thus, energy planning for each sector is usually done in isolation. As a consequence, conventional energy strategies rarely exploit the interdependence of energy-consuming sectors. In fact, though, this interdependence can become the basis of what have been termed *synergisms* in which the combined and cooperative effect of measures in two or more sectors is totally different from the sum of the separate effects if uncoordinated measures are taken in each of the sectors.

One class of synergisms is between energy sectors. Several illustrative examples can be cited. For example, the agricultural sector can be used to produce feedstocks for the production of liquid fuels for the transport sector—this synergism is the basis of the Brazilian Ethanol Program.[25] And, the introduction of high-efficiency devices for cooking in the domestic sector can release vast quantities of fuelwood which can be converted into liquid fuels for the transport sector—this synergism is the basis of a proposal for India.[26] Further, uniform regional development of agriculture and industry can drastically reduce freight traffic and therefore energy consumption in the transport sector.

Also, the use of agricultural wastes to produce materials which substitute for present-day industrial products can result in a synergism between the agricultural and industrial sectors. There are a vast number of possibilities with this type of synergism. For example, agricultural products can be processed to give the same starting materials as are currently being used by the petrochemical industry. Or, to cite a very specific possibility, rice husks can be burnt to give an ash which yields, when ground with lime, a cement-like material that can substitute for portland cement, a highly energy-intensive industrial product.[27] Another synergism between the agricultural and industrial sectors involves the use of gas derived from biomass to fuel irrigation pumpsets and to divert the saved electricity to industry.[28]

There are also synergisms possible within sectors. For example, since railways consume almost the same amount of fuel as buses for passenger traffic, but almost 80 percent less fuel, compared to trucks, for *freight* traffic, a diversion of short-distance (under 300 km) passenger traffic from railways to buses and the use of the released rail-haulage capacity for freight transport would result in a saving of approximately 1 million tonnes diesel for every 50 million passenger kilometres.[29] Or, as noted above, agricultural residues can be fed into gasifiers and the producer gas which is generated can run pumpsets for irrigating agricultural crops.[30, 31]

Wider synergisms are also possible. For example, the synergism between water and energy can be achieved in a number of ways. Large irrigation projects involve canals running for tens of kilometres with very small gradients to ensure low flow velocities despite large gradients of the terrain. This design requires that the accumulating potential energy of the water be eliminated at regular intervals by making the canal bed undergo sharp drops to dissipate the energy of the water. Efforts are now in progress to incorporate lowhead turbines at these drops to generate electricity.[32] Or, advantage can be taken of the fact that the energy required for ploughing land decreases with increases in its soil-moisture content. Thus, small amounts of irrigation water lead to large reductions in the energy used for ploughing. A further example is that of using irrigation to increase the fuelwood yield of energy plantations.

Synergisms between housing and energy can arise through the architecture of houses and the materials of construction. Whereas the traditional architecture of tropical climates was evolved to exploit natural air-flows by ventilation, modern housing in these areas imitates the tightly-closed building of temperate climates, thereby requiring the consumption of energy for space cooling. Instead, a scientific understanding of traditional architecture can lead to designs which reduce or even eliminate energy for space cooling. Or, the use of waterproofed, manually compacted, unfired mud-blocks and of new roofing designs can eliminate the need for energy-intensive bricks and cement.[33]

It is clear, therefore, that synergisms must play a crucial role in energy strategies for developing countries. But to exploit fully the many possibilities for synergisms is a creative process which hinges on detailed disaggregated studies at the local, sub-national, country, and regional levels. Of necessity, global scenarios require aggregation of data, and the process of aggregation inevitably suppresses precisely those disaggregated details from which synergisms can be identified.

(x) *The Explicit Concern for the End-uses of Energy*

The guidelines listed above have the some very general implications concerning the approach to, and methodology of, energy analysis, planning, and management. The most important implication is that there must be as much concern for human beings as for technical systems. This reinforces two previously made points:

- the conventional obsession with the *supply* aspects of energy must be abandoned, and
- there must be a balanced emphasis on energy *demand* and a necessary focus on the *end-uses* of energy.

(xi) *Energy for Self-Reliance*

Insofar as another crucial development objective is self-reliance, it is vital that energy becomes an instrument for self-reliance. Since this is not the same objective as that of *energy self-reliance,* and since 'self-reliance' must not be equated with 'self-sufficiency', a brief discussion of these distinctions is necessary.

''Self-sufficiency'' denotes a situation in which a village, town, city, province, country, or region has enough resources within its domain to meet its requirements, and it is not a necessary condition for the achievement of ''self-reliance''. Self-reliance does not preclude one unit of society from interacting with another, i.e., having import-export dealings with them. Indeed, it may permit or even encourage this interdependence so long as overdependence does not result; the stipulation is that one unit of society must not become dependent on other units in such a way that the result is an external control over judgments, decisions and actions, and a subservience which erodes the ability to make independent decisions and carry them out. Obviously, the whole matter of interdependence versus overdependence is much more important the larger the social unit involved. A country, for example, does not want to become so reliant upon another for its energy sources that its ability to act independently is impaired. Within a country, the overdependence of one part upon another may put it at a permanent political disadvantage. Interdependence has its potential political pitfalls as well as its economic benefits.

It is clear that not every country can be energy self-sufficient and to attempt to become self-sufficient makes little sense economically if another country has an abundant supply of, say, oil and is willing to sell it at a cost which is below the cost of indigenous energy resources—at least to a point. The policy objective here is to keep the import level low enough that the buyer is not subject to political or economic blackmail by the supplier. That is self-reliance. Several considerations enter into this matter, including the diversity of suppliers, the availability of technologies to exploit indigenous resources (especially biomass), etc. Exact determination of what level of imports or degree of development of indigenous resources constitutes self-reliance is not possible. Even more difficult is the issue of self-reliance within a country. No part wants to be "exploited" by another. And yet, parts of a country poorly endowed with material resources may need a net inflow of energy "imports" in order to produce an outflow of "exports" from that region. Energy strategies should support, as much as possible, the balanced growth of all parts of a country.

Rural settlements pose a special challenge in regard to making energy strategy an instrument of balanced growth. They are small and scattered (Box 3.5) and also quite restricted in their range of economic activities. The result is that they present fairly small energy loads (of the order of 1 MW) and low load factors (of around 10 percent). Most often, therefore, in the early stages of development of these settlements, it is not worthwhile to link up to centralized electricity generation/production units because of the high transmission/distribution costs. In fact, as noted earlier, beyond a certain break-even distance from such centralized sources, decentralized energy generation/production units located in the villages may become an attractive proposition, if sufficient amounts of local energy sources are available. The decentralized energization of rural settlements is important from another perspective also—it provides a basis for rural industrialization which is a strong check against rural underemployment, migration, and excessive urbanization.

The situation is fundamentally different in the case of cities. With their large concentrations of urban populations and their broad array of diverse activites, metropolitan areas generally require large centralized energy generation/production systems to meet their energy needs. This means that unlike remote rural settlements, cities usually cannot become energy self-sufficient, nor do they need to. They can import the fuels for their electricity-generating plants and for cooking and transportation, etc.

In the case of industry, two categories must be considered. Small industry is usually dispersed within cities, and its energy requirements become part of metropolitan energy demand. But a large power-intensive industry–an integrated steel plant or an aluminum complex—not only requires energy of roughly the same order of magnitude as a city, but may often demand an entire settlement dedicated to the industry, for example, a steel town. Such large industries, like cities, are usually serviced by huge centralized energy generation/production systems.

However, even large power-intensive industries—such as aluminum or steel plants—can be

Box 3.5: Settlements in India

The 1971 Census showed a population of 548.2 million of which 77 percent lived in 576,000 villages and 23 percent in towns and cities. The distribution of this population over the various settlement sizes is shown in Table B3.5.1.

This distribution shows that the bulk of the country's population is in villages with a population in the range of 500-5,000. But, about one-quarter of the population is in towns and cities—half this urban population is in towns with a population of 10,000-100,000 and another significant fraction in cities with a population of 200,000 to 7,000,000.

Table B3.5.1. Distribution of Settlement in India

Population Range	Number of Settlements	Population Millions	%
I. VILLAGES			
200	150,100	15.2	2.9
200-499	168,600	56.6	10.3
500-999	133,000	94.4	17.2
1,000-1,999	82,000	113.2	20.6
2,000-4,999	36,200	105.1	19.2
5,000-9,999	6,100	37.9	6.9
Subtotal	576,000	422.4	77.1
II. TOWNS			
10,000-19,999	2,274	34.7	6.3
20,000-49,999	582	17.8	3.2
50,000-99,999	183	12.5	2.3
Subtotal	3,039	65.0	11.8
III. CITIES			
100,000-200,000	75	11.3	2.0
200,000-999,999	65	23.3	4.3
1,000,000	8	26.2	4.8
Subtotal	148	60.7	11.1
Total	579,187	548.2	

installed in "enclaves" that are self-sufficient in energy. For example, in southeast Brazil, approximately one half of the production of steel is based on the use of charcoal which comes from eucalyptus plantations surrounding the steel mills.

This discussion of the energy requirements of rural and urban settlements, and of large energy-intensive industry, leads to an important conclusion: *developing countries require a mix of centralized and decentralized energy systems*. Exclusion of either the centralized or the decentralized component inevitably results in the energy needs of either the urban or rural settlements respectively being underemphasized or even bypassed.

The energy demands of the transport sector require special attention, with the trend of decreasing contribution from the railways and the total dependence of road transport on oil-derived fuels. This sector has become increasingly petroleum-reliant. Further, since very few of the developing countries have petroleum resources, the vast majority of them must import oil to sustain their transport sectors.

This has had profound economic and political effects on oil-importing developing countries. For instance, it has forced them into such enormous foreign debts, and even servicing these debts has become a staggering task. As a result, numerous developing countries have been forced to permit external institutions to decide their internal policies, i.e., they have lost their self-reliance. Hence, energy strategies for developing countries must attempt to achieve the highest economically feasible degree of energy self-sufficiency in the transport sector.

(xii) *Summary of Guidelines*

It is now possible to summarize the guidelines for energy strategies for developiing countries.

In the first place, energy must be deployed as an instrument for development. Thus, energy must be directed principally, though not solely, towards the satisfaction of basic human needs, and in particular the basic needs of the rural and urban poor. This means assigning high priority to the provision of energy for end-uses such as cooking, lighting, etc. At the same time, however, there must be a balanced concern for the energy requirements of the industrial, agricultural, and transport sectors so that the satisfaction of basic needs can be sustained and enhanced over the long run.

In supplying energy to industry, it is important to ensure that energy is used as a mechanism for creating employment, and that adequate energy is provided to industries with high employment potential.

The stress on inanimate energy must not prevent attention being given to the role of animate energy—the energy of human beings and draught animals—in developing countries; otherwise, the serious problem of rural unemployment and underdevelopment will only be aggravated. Deliberate efforts must be made towards enhancing the productivity of animate energy. However, in view of the overwhelming importance of augmenting food supplies to improve the nutritional status of growing populations, there should be no hesitation to increasing inputs of inanimate energy to the agricultural sector when greater productivity and income and employment can be achieved.

In order to enlarge the renewable base of their energy systems, advantage must be taken of energy from biomass sources, which already play a major role in developing countries' energy supplies. But to achieve this objective, these resources cannot be used in traditional ways, particularly with extremely low efficiencies. New high-efficiency technologies must be deployed extensively. This should free up biomass sources for other purposes and help ensure renewable use.

Such strategies must involve therefore improvements in the efficiencies of energy use in all sectors as essential and integral components.

The emphasis on renewable energy sources and on energy-efficiency improvements should not exclude a concern for economic efficiency; in fact, fair competition should be encouraged between various possible energy carriers and also between the options of increasing energy supplies or reducing energy demand by providing the same or even better energy services through improved energy efficiencies.

Narrow sectoral thinking must be abandoned, and conscious attempts must be made to exploit synergisms between and within sectors.

None of the above guidelines can be implemented until the present supply bias in energy analysis, planning, and management is eliminated. This bias must be replaced with a balanced emphasis on energy demand and a focus on the end-uses of energy.

A crucial objective for energy strategies should be the achievement of self-reliance at the national level. Within a country, the quest for self-reliance requires a mix of decentralized energy sources, which are particularly suited for rural areas, and centralized sources, which are essential for urban concentrations and large-scale energy-intensive industries, as well as energy self-sufficiency in the transport sector.

3.5 Opportunities for Synergisms

3.5.1 *Synergisms involving the Transport and Agriculture Sectors*

Oil use in the transport sector of developing countries has grown until it now accounts for about a third of the energy consumed in these countries. The main factors responsible for this increase in oil consumption in the transport sector were the deemphasizing of rail transport, the development of roads, the increasing share of trucks in freight transport, and the rapid increase in the number of vehicles, particularly automobiles for personal transportation (Figure 3.9).

All this reflects, not only the tendency of the growing elites to follow the consumption patterns of the industrialized countries, but also the availability of cheap oil. Most developing countries, however, either do not produce any oil or meet ever decreasing fractions of their growing requirements of this commodity from indigenous production. Inevitably, therefore, their increasing petroleum demand economies had to be sustained through oil imports. Prior to 1973, these imports were not very important because they only constituted small fractions of the export earnings. After the oil shocks of 1973 and 1979, though, the oil import bills became increasingly and unmanageably large. In fact, these bills have drastically affected balances of payments, resulting in ever increasing debts which threaten the financial survival of debt-ridden nations.

The increasing costs of oil imports have affected individuals, both rich and poor. The elites see their affluence at stake with automobile fuel bills consuming larger and larger fractions of their disposable incomes. And the poor suffer, not only from increased costs of petroleum products, including kerosene for lighting, but also from escalations in the prices of all essential commodities due to the increased transport costs.

Under these conditions, it has become imperative for oil-importing developing countries to reduce oil consumption and to find substitutes for oil derivatives.

Of the several technical alternatives for petroleum substitutes in the transport sector, there are four options worth mentioning here:

- producer gas (primarily a mixture of carbon monoxide and hydrogen).
- biogas (a mixture of methane and carbon dioxide).
- ethanol (ethyl alcohol).
- methanol (methyl alcohol).

Producer gas generated *in situ* with charcoal/wood gasifiers was used widely during the Second World War in trucks and buses.[30, 31] For transportation applications the disadvantage of producer gas is that it requires the use of solid fuels (wood, charcoal, or densified biomass) which are not conducive to the development of widespread distribution networks. Hence, producer

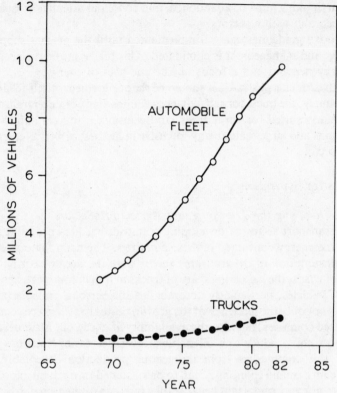

Figure 3.9 Evolution of the automobile and truck fleets in Brazil.
Source: J. Goldemberg *et al.*, 1984.

gas is really suitable only for vehicles operating up to distances that are accessible with a single load of solid fuel from the biomass production site.

Biogas is a gaseous fuel which can be stored in balloons or metal containers and used to run internal combustion engines.[34, 35] Because of its low energy per unit volume, however, biogas is suitable only for transportation over very short distances from the biogas plant. Also, it can be liquified only by going to cryogenic temperatures.

For these reasons, it is only the liquid fuels—ethanol and methanol—that are widely applicable alternative fuels for the transportation sector. They can be distributed through the nation-wide network already established for gasoline and diesel, so that the range of ethanol- or methanol-fuelled vehicles is essentially determined by the spatial coverage of the network. Ethanol and gasoline mixtures—so-called "gasohol"—were in fact used widely during the Second World War as gasoline extenders. And pure methanol, although never used extensively, is, like pure ethanol, an excellent fuel for internal combustion engines.

Whichever of these four fuels, or combinations of them, are chosen, they can all be obtained from biomass sources: producer gas and methanol from cellulosic materials such as fuelwood and agricultural wastes; biogas from the anaerobic fermentation of non-ligneous cellulosic materials; and ethanol from sugarcane or other feedstocks* such as corn, etc.

---

* Ethanol can also be produced from cellulosic feedstocks, although at a higher investment cost compared to methanol.

A strong possibility arises therefore of a synergistic coupling between the transport sector and the agricultural sector whereby the output of the latter can be converted into a fuel for the former. In principle, "fuel farms" can be established to supply the needs of transportation in the same way that cereals are produced by the rural areas for urban demands.

3.5.2 *The Brazilian Example* [25, 36-46]

(i) *Introduction*

Like most other developing countries, Brazil (Box 3.6) had shifted away from traditional biomass fuels and become a petroleum-based economy before the oil-price increase of 1973 (Figure 1.2). This evolution was possible because Brazil was a conspicuous beneficiary of low international petroleum prices and of the great increase in international trade leading to a rapid 'internationalization' of the Brazilian economy.

Since Brazil's petroleum resources are meagre, the domestic production of oil could not, and still cannot, meet the country's requirements. The 1983 oil consumption was approximately

---

Box 3.6: Brazil

Brazil is a large tropical country with a population of 125 million inhabitants (1983) and an area of 8.5 million square kilometres. In 1984, the average per capita energy consumption was 1.6 kW/capita. (For comparison, it may be noted that the corresponding figure for the United States is 10.1 kW/capita.)

This average per capita energy consumption hides the fact that there are extremes of wealth and poverty in Brazil. Many districts in the great industrial cities of the southeast (Sao Paulo, Rio de Janeiro, and Belo Horizonte) are modern and prosperous, resembling cities in the industrialized countries; but at the same time, there are large slums around these cities. And the rural population, mainly in the northeast, is very poor. One could almost say that Brazil consists of one country within another ("a Belgium inside an India") in which approximately 20 million people have a standard of life comparable to urban dwellers in Belgium and the remaining 100 million have one comparable to the rural poor in India. For this reason, one and the same developing country has a modern sector, which consumes commercial sources of energy (hydroelectricity, coal, and petroleum) and a traditional sector which depends on non-commercial sources of energy, mainly fuelwood.

In the past 30 years, economic growth has been explosive in the urban centres, whereas most rural areas have remained relatively stagnant. In 1983, 72 percent of the population lived in urban settlements. Thirty years ago, however, as much as 60 percent of the country's population lived in rural areas.

This growth was fuelled with petroleum. Indeed the modernization of the country which took place after 1950 corresponds to the adoption of a petroleum-based economy. An automobile industry was installed in Brazil through fiscal incentives from the Government, and a network of roads was built. The result was a type of development which brought about patterns of consumption similar to the ones in the United States for the 10-20 million people who constitute an elite who control all political power. Priorities are set by a highly centralized government channelling the income from taxes to policies suited to the preservation of the system.

133,000 tonnes per day, of which only 40 percent was produced internally; hence, the country had to go in for borrowing from abroad to pay for the oil imports necessary to compensate for the inadequate domestic production. Consequently, the foreign exchange to pay for imported oil became a fundamental problem, and a policy of further heavy borrowing abroad was adopted as the solution. Although exports of raw materials and semi-manufactured products increased as a deliberate policy of the government, Brazil was not able to repeat the successes of Japan and Germany in exporting sufficient manufactured goods to compensate for the petroleum imports. Thus, an increasingly larger percentage of the income from these exports has to be used to pay for the mounting petroleum bills. For example, 47 percent of the export earnings in 1982 were used up for oil imports (Table 3.21).

As heavy borrowing from abroad continued, the foreign debt climbed to $100 billion by 1984, and this, together with the high interest charges for debt servicing, threatened to lead the country into bankruptcy. That the external debt can be attributed in large part to the accumulated oil bill since 1973 can be seen in Figure 3.10.

The extreme gravity of the situation made the rulers of Brazil receptive to suggestions for alternative solutions. It also induced them to enact non-conventional energy policies. Since these solutions and policies are not only innovative but likely to have applicability in other countries or regions of the world, they merit elaboration. In addition, the flexibility of Brazil in facing the oil crisis constitutes an interesting case study of how developing countries can respond when confronted with unexpected and serious crises.

The essence of the Brazilian example is a two-pronged strategy:

(1) reducing the consumption of petroleum derivatives through pricing policies, and
(2) producing large quantities of an alternative fuel from biomass sources (ethanol from sugarcane), thereby replacing a fossil fuel with a renewable fuel.

The strategy was based on the use of the agricultural sector to solve the problems of the

Table 3.21. Oil Imports and Trade Balance for Brazil (millions of dollars)

Year	Exports	Imports	Debt External	Oil Imports	Oil Imports Accumulated	Oil Imports/ Exports (percent)
1971	2,900	3,200		280		9.6
1972	4,000	4,200	9,500	380	660	9.4
1973	6,200	6,200	12,600	720	1,400	11.7
1974	8,000	12,600	17,200	2,800	4,200	35.2
1975	8,700	12,200	21,200	2,750	7,000	31.6
1976	10,100	12,300	26,000	3,460	10,400	34.0
1977	12,000	12,000	32,800	3,660	14,100	30.1
1978	12,700	13,700	43,500	4,090	18,200	32.3
1979	15,200	18,000	49,900	6,190	24,300	40.5
1980	20,100	23,400	53,800	9,370	33,700	46.5
1981	23,300	22,100	61,400	10,600	44,300	45.5
1982	20,200	19,400	70,200	9,600	53,900	47.4
1983	21,900	15,400	91,600	7,800	61,700	35.7
1984	27,000	13,900	100,200	—	—	—
1985	27,000	14,500	97,900	—	—	—

Source: Suma Economica, FEV/85
        CNP/PETROBAS

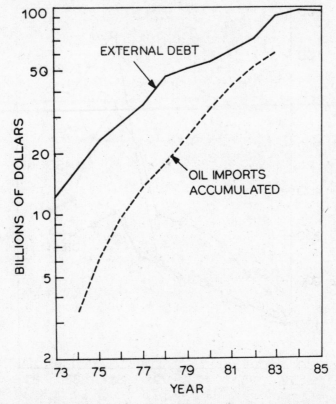

Figure 3.10 The evolution of Brazil's external debt and the influence of oil imports.
Source: J. Goldemberg, *et al.*

transport sector. Even if not applicable in detail to other countries, this strategy indicates the opportunities for synergisms between these two sectors.

(ii) *Pricing Policies*

Even before the production of ethanol became significant, pricing policies had a great impact on the consumption of gasoline in the period 1975-1979.

Starting in 1974, the Government increased the price of gasoline to a very high level, but kept the price of diesel and fuel oil very low and in fact subsidized them (Figure 3.11). The rationale for this policy was that automobile owners (who use their cars mainly for personal convenience and recreation) cannot pass on the increases in gasoline prices to other sections of society; in contrast, the consumers of diesel for freight and passenger transport and of fuel-oil in industry can transfer the increases in the costs of their services and products on to their consumers—thus contributing to inflation.

The sharp rise in the gasoline price had a dramatic dampening effect on gasoline consumption (Figure 3.12). In the beginning, gasoline consumption stopped growing and stabilized as early as 1974, in part because the large fleet of gasoline trucks shifted to diesel through engine replacement. In addition, the average consumption per vehicle was reduced by almost a factor of two between 1974 and 1979 because of a number of factors such as improvements in fuel economy, better roads, better traffic systems, changes in driving habits encouraged by the high

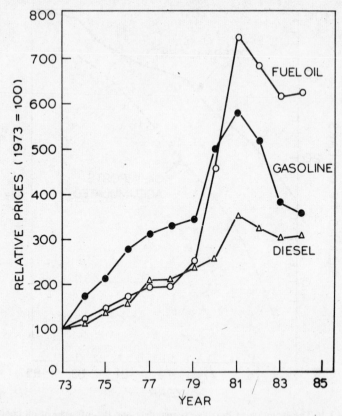

Figure 3.11 Prices for petroleum products in Brazil relative to 1973 prices (in current dollars)
Source: CNP, Brazil.

cost of gasoline, and the prohibition of its sale in the evenings and nights and on weekends, etc.

But the production of petroleum derivatives from the refineries of PETROBRAS, the government-owned oil-refining company, did not follow the changes in consumption of petroleum-derived fuels (Table 3.22). The refinery profile has changed little since 1975, except that the gasoline fraction has decreased from 32 percent to 18 percent while the use of naphtha for petrochemical purposes has increased from 3.5 to 21 percent. From a technical viewpoint,

Table 3.22. Refining Profiles in Brazil

Oil Derivatives	1975	1979	1982	1983	1984
Gasoline	32	23.1	20	17	18
Diesel Oil	26	30	32	33	32
Fuel Oil	32	32	24	22	21
LPG	6.5	6	7	8	8
Others (mainly naphtha)	3.5	9	17	20	21
Total	100.0	100.0	100.0	100.0	100.0

Source: CNP/PETROBRAS

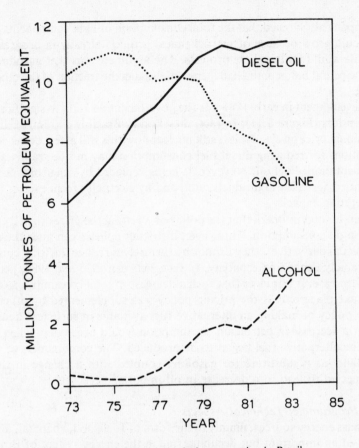

Figure 3.12 Transport fuel consumption in Brazil.
Source: J. Goldemberg

however, it is possible to have refineries which will produce either more diesel and less gasoline, as is the case in Western Europe, or more gasoline and less diesel, as is the case in the United States (Table 3.23).

Probably one of the weakest points of Brazil's pricing approach up until 1981 was its failure to bring about immediate changes in the output from the refinery operations. PETROBRAS was unwilling to make such changes on the grounds that they would require high investments and that the company has been able to market its excess gasoline abroad, purchasing heavy oils

Table 3.23. Percentage Refinery Yields on Crude Oil (1979)

Products	Percentage by Weight		
	North America	Western Europe	Japan
Gasolines (light distillates)	38	19	19
Diesel Oil (medium distillates)	28	35	28
Fuel Oil (heavy distillates)	13	31	41
Other Products (incl. Refinery Fuel and Loss)	21	15	12
Total	100	100	100

in return. The company also argued that the total consumption of light components of petroleum (e.g., naphtha) would grow because of petrochemical projects already in progress and that a new refinery profile would inhibit these projects. The argument was that gasoline production had decreased in the period but naphtha had increased, leaving the fraction of the other derivatives almost untouched.

An important development in early 1981 was the government decision to significantly increase diesel and fuel oil prices (Figure 3.11). In fact, diesel oil in Brazil is currently selling at prices above the international price levels, but it is still unclear how this will affect diesel consumption, since advanced options for reducing diesel fuel consumption may not be readily accessible for immediate implementation. Fuel oil, however, is being replaced by other biomass fuels, such as fuelwood, densified agricultural products, etc., and by electricity from excess "wet season" hydroelectric-generating capacity.

It was realized quite soon in Brazil that the policy of "getting the prices right" can only have a limited impact on oil consumption. Thus, even if pricing policies can alter the relative consumption of petroleum derivatives, they cannot by themselves reduce the total consumption of oil products in a rapidly expanding economy. In fact, between 1975 and 1980, petroleum consumption did not decrease. It was clear that such a decrease in total consumption (of petroleum products) would take place only if the pricing policy was supplemented by other measures.

Brazil chose the policy of making an alternative fuel available in sufficient quantities.* The reasoning was that a decrease in petroleum consumption would occur only when very significant amounts of the alternative fuel began to be produced. The combination of an increased availability of ethanol as a substituté for gasoline, coupled with a change in the output of refineries, was expected to lead to a decrease in oil imports.

(iii) *The Ethanol Programme: Technical Aspects*

Renewable biomass energy sources, mainly in the form of fuelwood, charcoal, and sugarcane bagasse, have played an important but declining role in the energy system of Brazil, as is the case in most of the developing countries. While biomass sources represented almost 80 percent of total energy consumption in 1940, this fraction had decreased to 30 percent in 1982 (Figure 1.2). While most of this biomass consumption occurred outside of the market as *non-commercial* fuels, these biomass sources have been very important to the country's energy supply.

Biomass, used as solids in the forms mentioned above, is inconvenient to transport due to its low bulk density, and therefore low energy content per unit volume of solid fuels. In addition, solid biomass is difficult to use in internal-combustion engines which have been designed for liquid fuels. Although producer gas can be produced from solid biomass and used as a fuel in automobiles and trucks, the gasifiers necessary for the gasification are often cumbersome and heavy retrofits. Besides, distribution networks for solid fuels have yet to be established, whereas extensive networks already exist for liquid fuels.

For these reasons, the energy policy followed in Brazil was not based on turning back to the past and expanding the use of the traditional and inefficient technologies of utilizing biomass as fuelwood, charcoal or bagasse. Instead, the policy was to produce ethanol, a high-quality liquid fuel, from one particular type of crop, namely, sugarcane.

Fortunately, even before 1975, extensive experimentation at the Aerospace Technological Center

---

* This was not the only option. Brazil could have shifted instead, or in addition, to highly fuel-efficient cars—and this is what is happening now.

of the Air Force, in Sao Jose dos Campos, had demonstrated the technical viability of using ethanol either in existing or in retro-fitted engines. In fact, engines specially designed for the use of ethanol show superior performance. Automobiles in Brazil use Otto engines with a compression ratio of approximately 6-7 which is suited to the low-octane gasoline in use (62 octane), but ethanol, which is 98 octane, allows the use of a higher compression ratio (approximately 10) and therefore produces higher engine efficiency. In general, the efficiency for Otto engines using ethanol is about 38 percent, while for gasoline it is only about 28 percent. (Incidentally, it may be noted that such gains from the use of pure ethanol cannot be realized in countries such as the United States where automobile engines already have high compression ratios of approximately 9.)

In practice, therefore, although the energy content of ethanol is only about 60 percent of the energy content of gasoline, the decrease in the mileage per litre of fuel is only 20 percent when Brazilian automobiles are converted to the use of hydrated ethanol (91-93 per cent ethanol plus water). Further, since ethanol is sold at a price per litre which is up to 59 percent of that of gasoline, a shift to ethanol brings in a net benefit to the consumer of about 30 percent.

(iv) *The Ethanol Programme: Implementation*

In 1975, the Brazilian production of ethanol from sugarcane was only 903 million litres per year. This ethanol was used mainly for industrial purposes, and was quite inadequate to meet the requirements of the transport sector. It was necessary, therefore, to launch a large financing programme for the production of new varieties of sugarcane and the establishment of increased capacity for manufacturing the ethanol necessary to meet transportation fuel requirements.

Ethanol can be produced either in *autonomous* distilleries dedicated exclusively to ethanol production or in *annex* distilleries which are adjuncts to sugar refineries. Annex distilleries have the advantage of requiring smaller investments and yielding results in a shorter time, but their disadvantages are the limited capacity of the industrial units already in place and available land. Autonomous distilleries offer more freedom, but the investments are larger and a longer gestation time is required for the projects to reach maturity.

Phase I (1975-79) of the implementation of the Brazilian Ethanol Programme was directed towards:

- expansion of the capacity of *annex* distilleries, that is, the distilleries built as adjuncts to existing sugar mills;
- continuation of the blending of ethanol with gasoline, with the goal of reaching 20 percent ethanol in blend by 1980.

The push for annex distilleries was favoured because, in 1974, the price of sugar in the international market was very low (approximately $200 per tonne). Therefore, the diversion of some sugar to ethanol production (one kg. of sugar is approximately equivalent to 0.7 litres of ethanol) was considered to be a method of increasing the value of this commodity and using the considerable idle capacity in the sugar refineries. (Incidentally, the price of sugar went up to $608 per tonne in 1975, but it went down again in 1979 to $120 per tonne—an example of the type of fluctuation in commodity prices referred to in Section 1.3.2 on North-South disparities.)

Benefitting from government subsidies, the programme picked up speed, and all gasoline used in the country was rapidly converted to "Brazilan gasohol", a mixture of up to 20 percent ethanol and 80 percent gasoline.

In 1979, the Iraq-Iran war threatened the stability of oil supplies from the Middle East, and the automobile manufacturers, encouraged by the Government, took the bold step of producing cars with new engines adapted for the use of pure ethanol. The Brazilian Ethanol Programme thus entered a Phase 2 from which there was no return. In Phase 2, the reliance shifted to *autonomous* distilleries.

The Brazilian production of ethanol in 1981 reached a total of 4.08 billion litres, out of which 1.88 billions were consumed as hydrated ethanol (91-93 percent ethanol plus water) in more than 300,000 automobiles and the remainder of the fleet used 2.2. billion litres of anhydrous ethanol mixed with gasoline in the proportion 10-20 percent. In 1984/85, the production of ethanol reached 9 billion litres and the number of cars running on hydrated alcohol reached 2 million. There are presently close to 11 million cars in Brazil.

The distribution of (hydrated) ethanol benefited from the infrastructure previously used for high octane gasoline. In March 1, 1983, there were 13,415 service stations in all the 22 states offering ethanol (in addition to gasohol) out of 19,588 service stations selling gasoline.

The Executive Committee of the National Alcohol Commission (CENAL), which oversees the implementation of the programme, approved a total of 387 ethanol production units up through October 1982, with a total capacity of 7.9 billion litres per year. Adding this to the previously existing capacity for ethanol production (0.9 billion litres in 1975), Brazil had already approved by October 1982 a capacity of 8.8 billion litres (about 132,600 barrels per day). By March 1984, Brazil had already approved the target capacity of 10.7 billion litres.

In Brazil, the trend is for new capacity for the production of ethanol to be diverted from annex to autonomous distilleries. It is estimated that the traditional sector could produce an additional 3.2 billion litres in annex distilleries in 4 to 5 years. The establishment of new autonomous distilleries could produce another 3.5 billion litres in about three years after approval by CENAL.

The initial goal (set in 1979) for the production of 10.7 billion litres of ethanol per year in 1985 will probably be reached in 1986. Out of this quantity, it is expected that 9.2 billion litres will be used as fuel in the transport sector and 1.5 billion litres in industry. At this production level, the ethanol would be sufficient to replace about 60 percent of the gasoline that would be required in the absence of the ethanol programme.

The Brazilian Ethanol Programme appears to have met its original objectives.

(v) *Other Biomass-derived Fuels*

The successes of the Ethanol Programme stimulated many other initiatives for using traditional biomass sources in the form of high quality fuels such as methanol, methane, densified agricultural wastes, vegetable oils, and many others—opening new perspectives for all of these fuels. Most of these initiatives are geared to finding substitutes for diesel oil, since fuel oil can be replaced by coal or woodfuel with existing technologies. The most important programmes in this area are the use of vegetable oils and methanol (from biomass).

The Oilseeds Programme was established in 1980 and its original goals were to substitute 16 percent of the diesel oil with vegetable oils by 1985 and 30 percent by 1990. For this purpose, an additional 1.5 million tonnes of vegetable oil for energy purposes was targeted for 1985, so that total production that year would be double the normal amount expected in that year for food purposes. The programme was to be based on soyabeans and peanuts starting in 1981, colza and sunflower starting in 1982, and palm oil starting in 1986.

It was hoped that this programme would emulate the ethanol programme and provide increasing amounts of oils to substitute for diesel. From a strategic point of view, this is attractive because of the flexibility in the vegetable oil plus diesel mixture. More or less vegetable oil may be added to diesel depending on the availability of vegetable oil, without the need for irreversible commitments to particular quantities of substitutes. No changes are contemplated in existing diesel engines and the new fuels would have to be adapted to them.

However, economic problems associated with the high cost of vegetable oil compared to diesel and the lack of funding have brought the Oilseeds Programme to a standstill. Technical problems such as carbon deposition inside the engine cylinder have been solved through chemical changes (trans-esterification) in the vegetable oil. But this additional step in the processing of the vegetable oil has added to the cost of the final product making it even less competitive with diesel.

Methanol production from natural gas or coal is a well-mastered technology, and this fuel is favoured in some North European countries as the best replacement for gasoline. Most probably this will not be the route followed in Brazil, which has already adopted ethanol as the substitute for gasoline. It is unlikely therefore that both ethanol and methanol will be used for automobiles in the near future. What is more probable is the use of ethanol for automobiles, and methanol for diesel engines. It is imperative, however, that broad-fuel tolerance engines are developed as soon as possible.

Burning of pure methanol in present-day diesel engines can be achieved with small modifications in the engine, mainly the introduction of a glow-type spark plug. Extensive testing of this is taking place with encouraging results. In addition, the use of methanol in engines for locomotives or industry offers great opportunities.

Since the Ethanol Program seems well entrenched in Brazil, a number of experiments have also been made to use ethanol in diesel engines by, for instance, adding an explosive such as 12 percent of hexile nitrate or 4-5 percent triethylene glycol to the ethanol. Very good results have been reported with these fuels, although questions of price remain to be solved since the consumption level is 60 percent higher than diesel in volume. The additives being seriously considered at present can be produced from sugarcane bagasse.

The possibility of using a mixture of diesel oil and pure ethanol in dual-injection engines is also being investigated.

In addition to ethanol being produced from sugarcane, its production from wood by means of acid hydrolysis is an attractive proposition because wood from marginal land could be used and then the lands reforested. However, the economics of this alternative alcohol-producing process has still to be proven. Capital investment is about double that for ethanol derived from sugarcane. A state-owned company named COALBRA was established by the Government in 1980 to build and operate the first prototype plants based on this technology. The first one, a 30,000 litre/day plant, is operating in Uberlandia, Minas Gerais.

(vi) *Criticisms of the Brazilian Ethanol Programme*
Four main criticisms have been levelled against the Brazilian Alcohol Programme:
(1) it is competing with food production;
(2) it is uneconomical;
(3) it yields less energy than goes into it; and
(4) it is leading to the concentration of land ownership.

(1) *Biomass for Fuel vs Biomass for Food*
Between 1976 and 1982, there was an increase of 5.4 million hectares of land under cultivation

in Brazil—a 12 percent increase (Table 3.24). Of this increase, sugarcane accounted for 24.1 percent, food crops for 35.2 percent, and export crops for 40.7 percent (soya—33.3 percent, and other export crops—7.4 percent). This pattern of allocation of the additional land brought under cultivation is partly explained by the differential *fall* in the prices of various crops—food crop prices dropped about 38 percent, export crops other than coffee about 17 percent, and sugarcane 9 percent.

The increase in area for food crops between 1976 and 1982 was only 6.9 percent compared to a population increase of 16 percent. Though the rate of growth of food crops has to be stepped up to keep pace at least with the population increase, it appears that food crops faced a much greater competition from export crops, particularly soya, than from sugarcane production for the Ethanol Programme.

(2) *The Cost of Ethanol Production*

The true economic cost of ethanol production is very difficult to compute. The cost estimates vary from $35 to $90 per barrel of gasoline replaced. The wide variation arises not only from conflicting primary data, but also from major assumptions that go into the cost accounting such as the exchange rate and the treatment of government subsidies. However, a number of recent studies as well as actual government payments during 1983 point to a cost in the range of $45

Table 3.24. Trends in the Land Area Used for Crops in Brazil[a]
Area Cultivated (million ha)

Product	1976	1979	1982	1982/1976
Basic Foods	27.5	26.8	29.4	1.069
Corn	11.1	11.3	12.6	1.135
Rice	6.7	5.4	6.0	0.90
Beans	4.1	4.2	5.9	1.44
Wheat	3.5	3.8	2.8	0.80
Manioc	2.1	2.1	2.1	1.00
Export Crops	13.4	15.9	15.6	1.16
Soybeans	6.4	8.3	8.2	1.28
Cotton	3.4	3.6	3.6	1.06
Coffee	2.2	2.4	1.9	0.86
Cocoa	0.4	0.4	0.5	1.25
Oranges	0.4	0.5	0.6	1.5
Castor Oil	0.3	0.4	0.5	1.67
Tobacco	0.3	0.3	0.3	1.1
Sugar Cane[b]	2.6	3.1	3.9	1.5
Other	1.6	2.0	1.6	1.0
Total	45.1	47.8	50.5	1.12

(a) *Source:* H. Geller, 1985.
(b) The total land area devoted to sugarcane is given here. This is approximately 25% greater than the sugar cane area harvested annually.

to $60 per barrel of gasoline replaced (Table 3.25). By comparison, gasoline derived from $29 per barrel of imported petroleum is estimated to cost about $41 per barrel ex-refinery in Brazil, after considering shipping and refining costs at $2 and $10 per barrel. On the surface, therefore, Brazil's ethanol produced from sugarcane does not appear to be competitive at present with gasoline from imported petroleum.

However, this direct cost comparison neglects a number of important factors.[39] Incremental oil imports are debt-financed and place a number of other strains on the Brazilian economy. Recognizing these problems, the Brazilian Government has implemented tight import restrictions and in 1984 applied a 20 percent surcharge to companies requiring hard currency to purchase foreign goods. Adding the 20 percent surcharge to imported oil brings the total cost of gasoline from imported oil to $47 per barrel. This still leaves the cost of ethanol, based on full gasoline replacement, costing 6 to 19 percent more than gasoline (see Table 3.25). But, if the world oil price had remained at the high 1981 price level, ethanol would be clearly cost-competitive as a full replacement for gasoline.

Table 3.25. Overall Cost Comparison for Ethanol Production and Gasoline Based on Imported Petroleum[a, b]

Sugarcane Cost	$10-$12 per tonne
Ethanol Yields	65 litres per tonne
Distillation Cost	$0.09-$0.11 per litre
Total Ethanol Cost	$0.264-$0.295 per litre
Replacement Ratio	
Full Gasoline Replacement	1.2 litres of ethanol per litre of gasoline
20% Ethanol, 80% Gasoline Blend	1.0 litre of ethanol per litre of gasoline
Total Ethanol Cost	
Full Gasoline Replacement	$50-$56 per barrel of gasoline replaced
20% Ethanol, 80% Gasoline Blend	$42-$47 per barrel of gasoline replaced
Imported Petroleum	$29 per barrel
Transport	$2 per barrel
Import Surcharge	$6 per barrel
Refining Cost	$10 per barrel
Total Gasoline Cost	$47 per barrel

(a) Costs are in 1983 U.S. Dollars.
(b) *Source:* H. Geller, 1985.

To justify full gasoline replacement at the 1984 world oil price, policy-makers can look to several other social benefits of ethanol production: employment generation, rural development, increased self-reliance, and reduced vulnerability to future crises in the world oil market.

(3) *Energy Balance for Ethanol Production*

In Brazil, the energy balance for ethanol production, i.e., the amount of energy required to produce a unit of alcohol relative to the energy content of that same unit, appears to be quite favourable. The energy balance for ethanol production in Brazil is approximately the same as that for the extraction and refining of gasoline in the United States. A comprehensive assessment of the direct and indirect energy consumption in the agricultural and industrial phases of fuel production shows that the input energy is only about 25 percent of the energy content of the final product. Also, about 70 percent of the energy inputs go into the agricultural phase, and about 40 percent of these inputs are in the form of fuel. The energy balance of ethanol production is much more favourable in Brazil than in the United States, for example, because Brazilian farming is very much less energy-intensive.

(4) *Concentration of Land Ownership*

As a result of the Ethanol Programme, large sugarcane plantations are being established in regions where previously many small farms existed. This government policy is favoured by the fact that sugarcane production is well suited to mechanized techniques. As a result, the subsistence crops of small farms—corn, vegetables, blackbeans, etc.—are being eliminated, leading to the import of food from distant regions. This has had the negative social consequence of forcing an exodus of small farmers and field labourers to cities where it is difficult for them to get jobs, or of making them seasonal labourers for the large plantations where sugarcane cultivation occupies only six to seven months in a year.

The encouragement of large farms for technical reasons, and the availability of government-subsidized credits for ethanol production, have also led to the growth of a few large companies that hold most of the land in many regions of Brazil. This has had a negative effect on income distribution by concentrating resources in the hands of a few entrepreneurs.

An alternative would have been the establishment of a system of cooperatives in which individual farms could grow sugarcane and process it in a cooperatively-owned distillery.

This alternative has been tried with poor results thus far. In Brazil, co-operatives have been a good instrument for promoting the participation of small and medium farmers in the production of the raw material (sugarcane), but new approaches to the structure and functioning of cooperatives involving the industrial phase of the project are needed. This is because an industrial enterprise such as a large distillery (typically producing 120,000 litres per day) requires large capital and unified command. In this context, it is worth considering the experience of the Anand milk cooperatives in India where a combination of centralized management and operation of large dairies and decentralized ownership of cattle and production of milk has been a significant success.

Though the Ethanol Programme has further concentrated income (and political power) in the hands of the few hundred groups owning distilleries, the trend may be reversed by the development of microdistilleries and governmental measures to correct the bias towards large distilleries.

In contrast to a typical 120,000 litres per day distillery which requires sugarcane from an area of about 5500 hectares in Sao Paulo State, microdistilleries producing less than 5000 litres per day of ethanol have been developed. Since microdistilleries require only about 200 hectares of sugarcane plantation to supply them with the necessary feedstock, the large-scale deployment of microdistilleries could reverse the current trend of land concentration. In addition, a microdistillery requires only about one-third the investment and employs five times as many workers unit of output capacity as the typical large unit. Thus, the capital investment required for the Ethanol Programme could be greatly reduced by shifting to microdistilleries since most of the capital costs are associated with the distillery phase. Finally, overall microdistillery production costs are estimated to be competitive with large-scale distilleries provided that their output is at least 400,000 litres per year, yields are at least 52 litres per tonne, three shifts are used and the interest rates are under 14 percent.

Several government measures are aiding the growth of distilleries smaller than the typical large-scale units with a capacity of 120,000 litres per day. Government financing has been made available for mini-distilleries with capacities of 10,000 to 50,000 litres per day. And since 1981 the construction of microdistilleries can be taken up without governmental ''red-tape''

(vii) *Benefits from the Brazilian Ethanol Programme*

The government projections for energy consumption do not contemplate structural changes in the sectors which generate energy demand. It might be concluded, therefore, that the energy policies adopted in Brazil are nothing but *interfuel* substitutions. This, however, is not the case because it is expected that these policies will have many positive impacts.

(1) *Reduction of Oil Imports and Saving of Foreign Exchange:* Table 3.26 shows that in 1984 oil imports were reduced to 65% of the 1979 level. Moreover, imports in 1985 are expected to be only 46% of the 1979 level. Such reductions of oil imports in just a few years is a remarkable achievement, considering that it is based entirely upon a new industry of a type not found in the industrialized countries.

Table 3.26. Recent Trends in Brazilian Energy Use

Primary energy	10^3 TEP/year		
	1979	1984	Increment
1) Fossil	64,691	67,079	+ 2,388
Imported petroleum	50,049	32,395	− 17,654
Local production	8,262	23,135	+ 14,873
Gas	983	2,607	+ 1,624
Coal	5,397	8,942	+ 3,545
2) Renewables	72,931	104,833	+ 31,902
Alcohol	2,037	6,154	+ 4,117
Bagasse	6,691	12,072	+ 5,381
Fuelwood	27,265	32,293	+ 5,028
Charcoal	3,320	5,753	+ 2,433
Hydro	33,382	47,970	+ 14,588
Other	236	591	+ 355
Total	137,622	171,912	+ 34,290

Source: BEN/1985

The quantity of ethanol produced in 1985 will reach 145,000 barrels per day of petroleum replacement, which at the price of U.S. $40 per barrel (i.e., $30 per barrel + $10 financing charges), corresponds to approximately $2.1 billion. This is the amount of savings expected in the 1985 petroleum foreign-exchange bill. The Ethanol Program, therefore, has made a major economic contribution, improving the balance-of-payments situation and freeing foreign exchange for development-oriented investments inside the country.

(2) *Strengthening of Self-Reliance:* To the extent that it reduces petroleum imports, the Brazilian Ethanol Programme reduces the burdens of foreign debt and debt servicing. Such an achievement corresponds to a strengthening of self-reliance because both strategic vulnerability and foreign indebtedness are reduced.

(3) *Development of Technological Capability:* Self-reliance is also strengthened because the establishment of a national ethanol industry with indigenous know-how has led to the growth of local technological capability.

(4) *Employment Generation and Development of Backward Regions:* Many new jobs will have to be created to produce the planned amount of 10 billion litres of ethanol. It is estimated that 475,000 direct jobs (700,000 at the peak of the harvesting season) in agricultural and industrial activities will be generated by 1985 and another 100,000 indirect jobs in commerce, services, and government.

This employment generation is primarily because ethanol production from sugarcane is highly labour-intensive in Brazil. The labour absorption occurs primarily in the agricultural phase of the production process, and the extent of labour absorption depends upon the degree of mechanization of agricultural operations. In the North-East part of Brazil ethanol production is over three times as labour-intensive as in the Central-South region, reflecting the fact that in the North-East, where there is much less agricultural mechanization, as much as 90 percent of the total labour requirements are for land-preparation and for sugarcane planting and harvesting—in contrast to 75 percent in the Central-South region.

Thus, it is likely that the Brazilian Ethanol Programme will generate more employment precisely in the backward North-East region where it is most needed. This is a significant step towards equity and the reduction of regional disparities. It should be stressed here that the employment potential of the Programme derives from the fact that, unlike petroleum production, the industrial cane-crushing and distillation process has to be preceded by the growth of sugarcane which is an agricultural activity.

When the employment potential of sugarcane is compared with other crops, it turns out that soyabean generates only 25 percent of the employment per hectare cultivated as does sugarcane. In fact, sugarcane is more labour-intensive than any of the other major crops dominating the Brazilian land-use pattern—corn (by about 50-58 percent), rice, and beans.

(5) *Effective Investment Strategy:* The investment required in ethanol production for the generation of one unit of employment can also be compared with other alternative investments. Using a parity exchange rate and including land costs, the investment in the distillery and farm equipment turns out to be $6000-7000 per person-year in the North-East and $23,000-28,000 per person-year in the Central-South region. By contrast, it requires, on average, an investment of $42,000 to generate one person-year's employment in the Brazilian industrial sector as a whole; in the basic materials-processing industry, it requires $70,000 per person-year; and at Camacari, a $5 billion Brazilian refining-petrochemical complex, it is $200,000 per person-year. In fact, this Camacari complex has resulted in only 5 percent of the jobs generated by the Ethanol Programme though its investment cost is roughly equal to the entire programme up to 1985. Thus, the Brazilian Ethanol Programme also appears to be an effective use of scarce capital resources for employment generation.

3.5.3. *The Replicability of the Agriculture-Transport Synergism*

The stage of development reached by Brazil is typical of a number of developing countries. The lack of adequate fossil-fuel reserves, the abundance of forests, a highly developed urban sector, a skewed income distribution, and a mounting external debt, are common characteristics of many Latin-American and some African and Asian countries.

Though the increased use of biomass to produce ethanol for use as an energy source is thus far a development unique to Brazil, the strategy can be seriously considered in a number of other countries. The Brazilian approach has hitherto
- restricted itself to ethanol production from sugarcane, and
- used about 40 percent of the sugarcane crop for ethanol production.

(Brazil produced 117,500 barrels of ethanol per day in 1983 using almost 2 million hectares of land, which is about 5.0 percent of the total cultivated land). If the Brazilian approach constitutes a reasonable strategy for India, Cuba, China, Mexico, and Pakistan to follow, world production of ethanol would go up to 40,000 barrels per day (Table 3.27).

Table 3.27. Sugarcane Production in Developing Countries[a]

	Area used (thousand ha)	Production[b] (thousand tonnes)	Total arable & Permanent cropland (thousand ha)	% of cropland occupied by sugarcane
India	3,170	177,000	168,400	1.9
Brazil	3,860	241,500	74,700	5.2
Cuba	1,400	75,000	3,220	43.5
China	745	46,200	100,900	0.74
Mexico	525	36,500	23,600	2.2
Pakistan	897	34,300	20,500	4.4
Colombia	280	24,000	5,700	4.9
Philippines	475	20,100	11,300	4.2
Thailand	577	24,900	19,400	3.0
Indonesia	278	23,700	20,300	1.4
Argentina	319	15,500	35,700	0.89
Dom. Rep.	188	11,750	1,460	12.9
Bangladesh	165	7,350	9,140	1.8
Total	12,900	737,800	494,320	2.6 (ave.)

(a) *Source:* FAO, 1984.

(b) The yield varies from country to country, in the range 50 to 90 tonnes per hectare per year; for this reason, the numbers in the second column are not strictly proportional to the ones in the first column.

There are also a number of developing countries in which forests occupy a large fraction of their area, say, more than 50 percent. In such countries, the forests can be used renewably for the production of charcoal, methanol, or ethanol—this would be a very promising approach in Indonesia, Zaire, Colombia, Peru, Angola, and Burma (Table 3.28).

Thus, in many countries, there is the possibility of supporting economic development to a considerable extent with locally-produced liquid fuels derived from renewable biomass resources, thus greatly reducing dependence on imported fossil fuels. Since the cost of these locally-produced liquid fuels might be high and they might drain resources needed for other purposes, a balance must be found between security of supply from local sources and vulnerable dependence on imports. In this context, the experience of Brazil with biomass might stimulate other countries to attempt their own large-scale use of biomass resources.

Thus far, the Brazilian Programme has used good agricultural land for producing a liquid fuel for the transport sector. In other words, the opportunity of growing more food has been sacrificed for the sake of producing fuel. The annual yield of ethanol in Brazil has been about 3500 litres of ethanol per hectare, and roughly this quantity of ethanol is required (according to Brazilian norms) to run an automobile for a year. But, a hectare produces on the average about 2500 kg of rice per year which at current Indian consumption levels of about 165 kg per

Table 3.28. Forested Areas in Selected Developing Countries[a]

	Forests and woodlands (million ha)	Land area (million ha)	Fraction of land area (%)
Brazil	568	846	67.1
Indonesia	122	181	67.2
China	131	933	14.0
Zaire	177	227	77.9
Sudan	48.0	238	20.2
Colombia	50.8	104	48.9
Peru	70.2	128	54.8
Angola	53.5	125	42.9
Mexico	46.9	192	24.4
India	67.4	297	22.7
Argentina	60.0	274	21.9
Bolivia	56.0	108	51.7
Venezuela	34.1	88	38.7
Burma	32.1	65.7	48.9
Total	1,516	3,806	39.8 (av.)

(a) *Source:* FAO, 1984.

capita per year can feed about 15 persons. *The conflict between running one automobile and feeding 15 persons can be avoided only if there is no competition between food and fuel for the same land.* To satisfy this condition, it is necessary to have a low population density. Thus, a strategy such as the one followed by Brazil to solve the liquid-fuel problem is applicable only to countries with abundant land resources. Otherwise, what could be a synergism between the agriculture and transport sectors can become a conflict—a point that is elaborated upon in Section 3.5.6.

It has been possible for Brazil to make ethanol a major energy resource by increasing the amount of land for sugarcane production, because Brazil has large areas of land yet unopened for agriculture. But so doing would be difficult (or even impossible) in many other countries which are not so well endowed with agricultural land resources.

Even in the case of Brazil, competition for good land between fuel and food is taking place in some regions. For this reason, a large research and development effort has been initiated there to produce liquid fuels from cellulosic materials, such as fuelwood, and agricultural and urban wastes. Ethanol can be produced from these materials through the saccharification of cellulose, and methanol can be produced via pyrolysis. These are two important technologies that could change radically the prospects for large-scale biomass use in many developing countries, thus avoiding the conflicts that arise from the use of high-quality land for fuel production. An idea of the potentialities of this approach can be obtained from countries with large forest areas (Table 3.28).

In conclusion, there are two crucial aspects of the Brazilian strategy which can be considered for emulation by most developing countries. The first is the use of renewable biomass resources for producing liquid fuels for the transport sector and thereby reducing oil imports. The second is that a developing country can, acting on its own, establish in a short period a major industry to form the basis of a modern renewable resource strategy involving a transition away from petroleum. There is, however, a negative aspect of the Brazilian strategy—this involves the use of good agricultural land for the production of transport fuel rather than food. It may turn out, therefore, that the Brazilian strategy can only be accepted as a transitional solution and even then only in countries with abundant land resources.

3.5.4 *Synergism involving the Domestic, Agriculture and Transport Sectors*

The agriculture-transport synergism, in which the agriculture sector is used to produce liquid fuels for the transport sector, addresses the oil crisis. But—as has been pointed out in Section 3.3.5—this is only one of the serious energy crises facing developing countries, the other being the fuelwood crisis which has a particularly adverse impact on the rural poor.

As long as the agricultural-transport synergism is restricted to cropland through crops such as sugarcane, it may conflict with food production in land-scarce countries, but it does not aggravate the fuelwood situation. When, however, attempts are made to avoid the fuel-food conflict by turning away from agricultural land to cellulosic resources, particularly fuelwood, to produce methanol and/or ethanol, then a conflict can develop between the transport and domestic sectors through a competition for fuelwood and other agricultural residues.

The essence of this conflict is: fuelwood to produce liquid fuel for the vehicles of the elite? Or fuelwood to be used as cooking fuel in the homes of the poor? Given the current distribution of political power in most developing countries, it is almost certain that the more successful the conversion of fuelwood into transportation fuels, the less access will the poor have to fuelwood as a domestic cooking fuel. In other words, the alcohol-from-fuelwood solution to the oil crisis can aggravate the domestic fuelwood crisis.

Cooking fuel for homes, however, is one of the basic energy needs, and the satisfaction of this need has to be an essential feature of a development-oriented energy strategy. Hence, *the solution to the oil crisis must be compatible with the solution to the fuelwood crisis.*

One way of achieving such a compatible solution is to extend in two steps the synergism between the agriculture and transport sector is to include the domestic sector also.

The first step involves establishing a synergism between the domestic and transport sector is based on the fact that in developing countries fuelwood is used for cooking at extremely low efficiencies—typically of the order of 10 percent. If alternative high-efficiency fuels are provided for cooking, or the efficiencies of fuelwood stoves are radically improved, then the resulting drastic reductions in fuelwood consumption can free a vast fuelwood resource base for the production of liquid fuels for the transport sector.

Gaseous cooking fuels are easy to light and extinguish. They also offer a tremendous advantage—the rate of gas flow, and therefore the rate of combustion and the rate of release of heat energy to the cooking vessel, can be very rapidly altered and easily controlled.[47, 48] This is extremely convenient because there are cooking operations such as boiling which require a high power output from the stove, and others such as simmering which need a low power output. It is this convenience that perhaps more than any other factor is responsible for the popularity of cooking with LPG. In addition to the control over the gas flow rate which enables easy and quick variations of the power output, a gaseous cooking fuel permits manipulation of the air to fuel ratio and ensures the completeness of the combustion process. The net result of these advantages is that stoves using gaseous cooking fuels can achieve efficiencies which can be *five* to *eight times* the efficiency of traditional firewood stoves.

One obvious gaseous cooking fuel is biogas obtained from the anaerobic fermentation of non-ligneous cellulosic material, e.g., animal wastes.

In this context, it must be noted that, after a long period of neglect, the problem of improving fuelwood stoves—a problem which is of concern to two-thirds of the world's population—

has recently attracted research and development efforts from technical personnel with the requisite expertise.[49-51] These efforts have opened up the possibility of achieving anywhere between 30-50 percent fuel efficiency in smokeless stoves.

Neither the biogas nor woodstove options are problem-free. The biogas approach is associated with several problems:

- family-size biogas plants lose the economies of scale which are very significant,
- the more economical community-scale plants can bring in their wake difficulties of organization and possibly issues of equity,
- the low body weight of cattle in developing countries, particularly in the drought-prone areas, can make the cattle waste resource inadequate to meet cooking energy needs even though the cattle-human population ratio may seem satisfactory.

The dissemination of fuel-efficient woodstoves has recently started becoming very widespread, but the fact is that fuel-efficient woodstoves are unlikely to be accepted by the urban rich who are quickly switching over to LPG cooking fuel. Hence, the dissemination of woodstoves is likely to lead to a dual-fuel cooking-energy system (LPG for the urban and perhaps rural rich, and fuelwood for the rural and urban poor)—yet another manifestation of the dual society. Nevertheless, the benefits accruing to the poor from fuel-efficient stoves—major savings in fuelwood consumption which would cut fuelwood-gathering effort or lower expenditures on cooking fuel, reduced cooking times, and smokeless cooking—are so large and so quickly achievable that the dissemination of such stoves may be invaluable as a *transitional* step in the improvement of the quality of life of the poor.

Either way, whether fuelwood is completely replaced for cooking with biogas or used at greatly enhanced efficiencies in improved woodstoves, large amounts of liquid fuels can be generated from the *saved fuelwood*. Taking the current sustainable yield of fuelwood as 2 tonnes per ha per year, there is scope for the annual production in developing countries of about 3 billion tonnes of methanol if all the fuelwood currently being used is released for methanol production, and for about 2.5 billion tonnes of methanol if four-fifths of the fuelwood becomes available due to a five-fold efficiency improvement in wood-stoves. This methanol is equivalent to 1.8 times the present oil consumption of developing countries.

The sustainable fuelwood yield used in the above estimate is based on the natural untended growth of forests. But, once the pressure on forests as a source of cooking fuel decreases, conditions become established for managing the growth of forests, and dramatically improving their fuelwood yields, say, by a factor of five. In other words, silvicultural practices—agriculture in the general sense—can be implemented to increase fuelwood availability. This is the second step in the extension of the synergism and consists of inlcuding agriculture in the domestic-transport synergism.

3.5.5. *A Proposed Energy Strategy for India*[26, 52-57]

A specific example of synergism between the domestic, transport and agriculture sectors is a two-pronged energy strategy which has been proposed for India.

The main thrust of the first prong is towards satisfying the most important energy need of the poor—*fuel for cooking*.

For villages, there are two options for providing cooking fuel:

(1) the wastes from India's 235 million cattle, most of which are concentrated in rural areas, can be anaerobically fermented to generate biogas for distribution as cooking fuel, and

(2) improved woodstoves, which consume only about one-half the fuelwood required by traditional woodstoves, can be installed in fuelwood-using homes.

Though both the biogas and high-efficiency fuelwood-stove options are associated with problems, either one or a mix of them (with biogas in some places, perhaps coupled with a dairy development programme to increase cattle-waste yields, and improved woodstoves in others) can be introduced in the country's 576,000 villages. Since about 110 million tonnes of fuelwood are currently being used in rural areas, there should be a saving of between 110 to 55 million tonnes of firewood, the two values corresponding to the two extreme cases of all-biogas and all-(improved)-woodstoves, respectively.

For cities and towns, the provision of cooking fuel is a more complicated matter. The use of kerosene for cooking has to be avoided for reasons which will be stated below in describing the second prong of the proposed energy strategy. This leaves three options for urban areas:

(1) provision of LPG even for the urban poor—this has been the approach adopted in Brazil,

(2) supply of electricity for cooking, and

(3) piped mixtures of sewage and coal/producer gases.

The LPG option has been extended with increasing rapidity in recent years, and in fact would have been even faster if the availability of cylinders had kept pace with the expansion of indigenous LPG production. (Though all the large metropolises had LPG distribution networks by June 1983, only 162 (22 percent) towns will be covered out of the 739 towns with a population between 20,000 and 50,000. And in 1981, LPG was available only to 12.7 percent or 3.8 million out of 29.8 million urban households). The expansion of LPG supplies will lead to an enormous increase in demand for a valuable resource that can perhaps be better used as a feedstock for fertilizer and petrochemicals production. In particular, compared to the 1982 consumption of 0.62 million tonnes of LPG, a 19-fold increase to about 12 million tonnes by the year 2000 is necessary to supply all the projected 52 million urban homes.

The electricity option does not merit much attention, not only because it is a precious energy carrier which can be used more usefully for other end-uses, but also because it will only aggravate the electricity crisis which is crippling industry.

The piped gas option can be based on sewage gas, i.e., the biogas generated from the treatment of urban wastes and currently being wasted through flaring. But on a per capita basis, sewage gas can only meet about 20 percent of the requirements of cooking; hence, it has to be supplemented with other fuels. Two possible supplements (after ensuring that safety standards are met) are producer gas generated from fuelwood grown in energy-forest "green" belts around cities and towns, and (for arid regions) coalgas obtained from soft coke or coal. However, whichever one of these urban options is adopted, the 23 million tonnes of fuelwood now being used by cities and towns would become redundant.

In all, therefore, the provision of high-efficiency cooking fuels and/or devices in rural and urban areas would make available 78 to 133 tonnes of fuelwood, provided that all the firewood being used today for cooking can still be collected. This saved fuelwood can be converted either into producer gas (which is an intermediate in the manufacture of methanol) or into methanol itself. Further, these biomass-derived fuels can be deployed:

(1) to fuel all the trucks and buses,

(2) to run the 2.7 million diesel-fuelled irrigation pumpsets presently in use, as well as the additional 1.7 million sets of this type projected to be installed by the year 2000, and

(3) to operate the 7.4 million additional pumpsets that will be required if the number of electrical pumpsets are frozen at their present level of 3.6 million.

These fuel substitutions would require (Table 3.29 and Figure 3.13) 118 million tonnes of fuelwood to be converted into producer gas and/or methanol—compared to the 67-133 million tonnes of fuelwood saved by providing high-efficiency cooking fuels/stoves. But, at the same

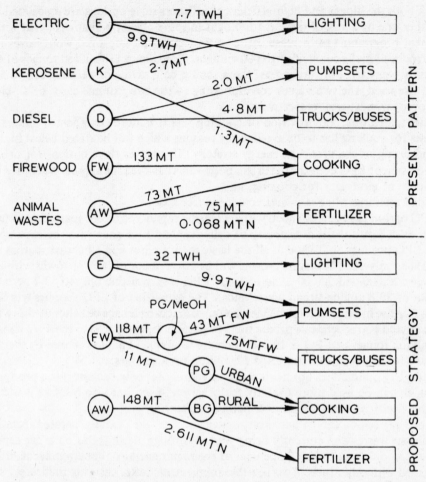

Figure 3.13 Comparison between India's present energy consumption and the proposed strategy.
Source: A.K.N. Reddy, 1985a.

time, the switch to biomass-derived fuels for transport vehicles and for agricultural pumpsets would save 6.8 million tonnes of diesel, 1.3 million tonnes of kerosene and 17 TWh of electricity.

The second prong of the proposed energy strategy focuses on another crucial energy need of the poor—*illumination of all homes.*

About 88 percent of rural households and 51 percent of urban households, together constituting four-fifths of the country's 116 million households, depend wholly on kerosene for lighting (Figure 3.14). The amount of kerosene used was about 4 million tonnes in 1978-79 in comparison with about 10 million tonnes of diesel. Further, about 40 percent of the country's kerosene was imported in 1980. The consumption for lighting is, on the average, about 2.2 litres/month/household. Kerosene lamps have an extremely low luminous efficiency output of light energy (in lumens) per unit of energy consumed in the lighting device—the efficiency of electric bulbs (incandescent lamps) is about 200 times the efficiency of kerosene lamps. Though electric lighting will result in a dramatic improvement in the quality of life, the number of *un*-electrified homes in the country is continuously increasing despite the decreasing number of unelectrified villages (Box 3.3).

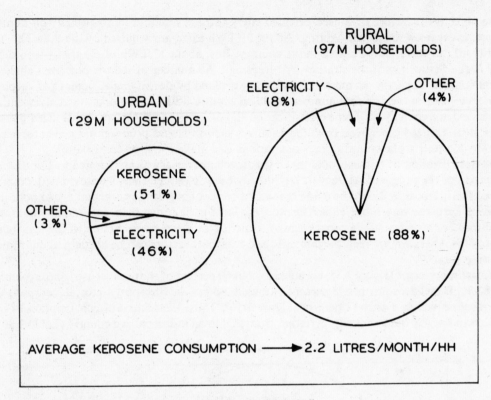

Figure 3.14. Distribution of energy sources for illumination in India.
The average consumption rate for houses using kerosene is 2.2 litres per month per household.
Source: A.K.N. Reddy, 1985a.

Table 3.29. Energy Sources and End-Uses in Proposed Strategy for India[a]

	Unit	Cooking	Lighting	Rail	Trucks/ Buses	IP[b] Sets	Total
Fuelwood	MT	0	0	0	75	43	118
		(133)	(0)	(0)	(0)	(0)	(133)
Animal Wastes	MT	148[c]	0	0	0	0	148
		(73)	(0)	(0)	(0)	(0)	(73)
Kerosene	MT	0	0	0	0	0	0
		(1.3)	(2.7)	(0)	(0)	(0)	(4.0)
Electricity	TWh	0	32	3.6	0	9.9	45.5
		(0)	(11)	(3)	(0)	(9.9) + 17.0[d]	(40.9)
Diesel	MT	0	0	1.5	0	0	1.5
		(0)	(0)	(1)	(4.8)	(2.0)	(7.8)

(a) Figures in parentheses correspond to the present consumption pattern.
(b) Irrigation pumpsets.
(c) Cooking from animal wastes via biogas.
(d) 17,0 TWh would be required if 7.4 million additional electric pumpsets are installed.
Source: A.K.N. Reddy, 1985a.

The objective therefore of a need-oriented energy strategy should be to electrify *all* homes and provide them with electric lighting. About 21 TWh extra are required by the year 2000 to electrify all the unelectrified homes in the country. But, about 17 TWh of electricity would be saved in the first prong of the strategy by "freezing" the number of electric pumpsets at their present number. Thus, the net increase in electricity demand by electrifying all homes and keeping electric pumpsets at their present number would be about 4 T·Wh in 2000. It is not at all essential that this extra electricity for home electrification should come via grids from centralized power plants; decentralized generation from local sources such as biogas producer gas generator-sets, micro-hydroelectric plants and energy plantations can make a major contribution.

The electrification of homes would make kerosene unnecessary as an illuminant. But, in the first prong of the proposed strategy, it has already been suggested that, by providing biomass-derived fuels, kerosene should be made redundant for cooking. If the rural and urban poor do not need kerosene any more either for cooking or for lighting, then conditions would be established for a synergistic approach to the reduction of diesel consumption in the transport sector. This synergism between the domestic and transport sectors can be brought about in the following way.

The transport sector (Figure 3.15) consumed about 60 percent of the total oil used in the country in 1980-81. Diesel accounts for 78 percent of the oil used in the transport sector, in comparison with gasoline, which is only 12 percent (Figure 3.16). Thus, the pattern of consumption of oil products in India is fundamentally different from that in an industrialized country (Table 3.30).

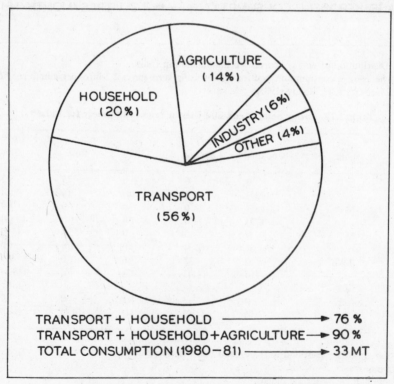

Figure 3.15 Oil consumption by sector in India, 1980-81.
Source: A.K.N. Reddy, 1985a.

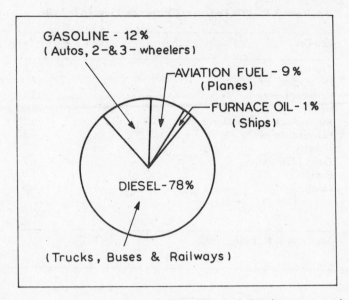

GASOLINE - 12%
(Autos, 2-&3- wheelers)

AVIATION FUEL - 9%
(Planes)

FURNACE OIL - 1%
(Ships)

DIESEL - 78%

(Trucks, Buses & Railways)

Figure 3.16. Oil use in the Indian transport sector by fuel and transport mode.

Table 3.30. 1978 Patterns of Consumption of Petroleum Products in India and USA

Petroleum Product	Percentage Consumption	
	India[a]	USA[b]
Gasoline	6.7	37.4
Diesel	57.7	6.9
Aviation Transport Fuel	5.0	5.8
Kerosene	20.0	0.9
Other	10.6	49.0
	100.0	100.0

(a) *Source:* A.K.N. Reddy, 1985a.
(b) US Energy Information Administration, April, 1981.

India's oil problem, therefore, is in general a problem of the two middle distillates, diesel and kerosene, and in particular, a diesel problem.

The bulk of the diesel consumption in the transport sector is by trucks (Table 3.31 and Figure 3.17).

In 1980-81, trucks consumed 63 percent of the diesel which is about 6 times the diesel consumption of railways. As far as freight is concerned, however, trucks and diesel locomotives haul approximately the same freight. This only confirms the well-known energy inefficiency of truck transport of *high-bulk-density* goods, compared to railway haulage. Despite this, the share of total freight transported by trucks has increased enormously both in relative and absolute terms (Figure 3.18)—in 1950-51, trucks carried 16 percent of the total freight of 105 billion tonne kilometres; in 1976-77 the figure was 33 percent of 233 billion tonne kilometres.

Table 3.31. 1978 End-uses of Petroleum Products in India

End-Use	%
Trucks	27.5
Diesel Pumpsets	14.0
Kerosene Lamps	13,4
Buses	10.0
Automobiles + 2-and 3-wheelers	6.7
Kerosene Stoves	6.6
Industry	6.0
Diesel Locomotives	5.2
Planes	5.0
Other	5.6
	100.0

Source: A.K.N. Reddy, 1985a.

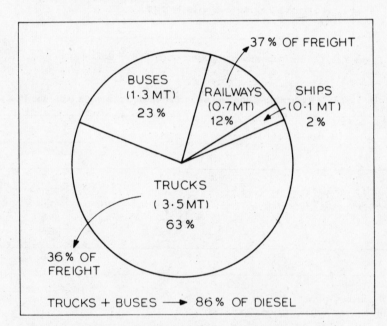

Figure 3.17 Diesel consumption in India by different transport modes.
Source: A.K.N. Reddy, 1985a.

It is well known, however, that the costs of both truck and rail transport (in terms of resources utilized to move a commodity) increase almost linearly with distance, but truck costs, which are lower than rail costs at small distance, rise more rapidly with distance and exceed rail costs at large distances (Figure 3.19). In other words, there are *break-even distances* below which truck transport uses less resources, and above which rail transport is more economical in terms of resource use. These break-even distances decrease as the diesel price increases—but, on the basis of a 50 percent increase over 1979 diesel prices, the break-even distances are between about 100-130 km depending upon the commodity. In fact, the average lead distances over which trucks are

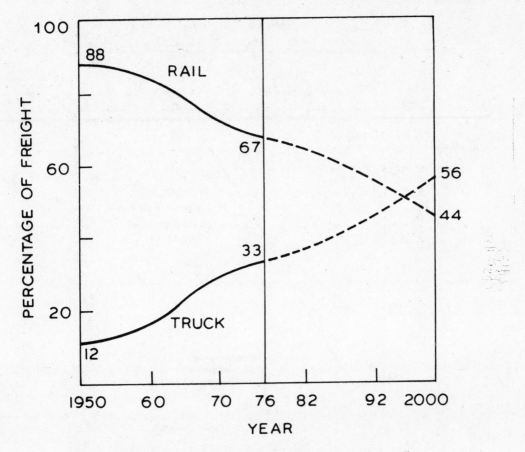

Figure 3.18 Relative contributions of truck and rail to total freight in India.
Source: A.K.N. Reddy, 1985a.

now carrying freight are about three to four times the break-even distances for most commodities (Table 3.32).

Truck operators have been able to achieve average lead distances that are much greater than the break-even distances because their financial costs have been lower than the resource costs to society. This is because the price of diesel in India is subsidized and pegged at the price of kerosene, which is substitutable for diesel in diesel engines. Thus far, diesel prices, could not be increased without roughly equal increases in kerosene prices because, if the kerosene price was much lower than that of diesel, trucks would shift—as shown by past experience—to kerosene fuel. And, hitherto, kerosene prices could not be increased without causing great hardship to the poor because kerosene is used almost wholly in the household sector, about 67 percent for lighting and the rest for cooking. But if, through the strategy proposed here, kerosene becomes unnecessary either as an illuminant or as a cooking fuel, it can either be withdrawn from the market or its price brought on par with that of gasoline. Once this is done, the price of diesel can be raised so that it is equal to that of gasoline, as is the case in the industrialized countries.

The increase in diesel prices would lead to a decrease in break-even distances, but these are, unfortunately, not very sensitive to increases in fuel prices. What would reduce diesel consumption markedly is the shifting to rail of freight traffic allocable to rail, i.e., freight traffic moving

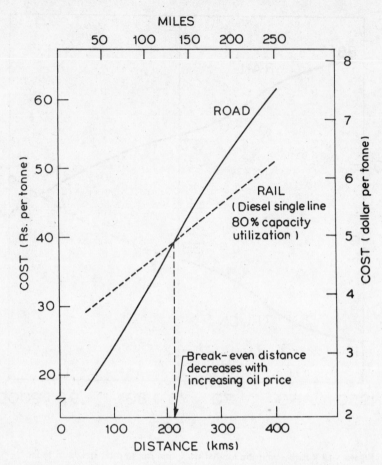

Figure 3.19 Unit costs of rail and road freight transport versus haulage distance in India. *Source:* A.K.N. Reddy, 1985a.

Table 3.32. Originating Freight Traffic, Average Lead and Break-Even Distance for India, 1977-78[a]

Commodity	Originating Traffic (million tonnes)	Average Lead (km)		Break-Even Distance[b] (km)
		Rail	Road	
Coal	69.26	691	408	106
Foodgrains	23.65	1278	277	130
Iron ore	17.23	529	96	106
Iron & Steel	15.21	1101	371	128
Cement	15.20	717	286	116
Fertilizers	11.21	1039	267	107

(a) *Source:* National Transport Policy Committee Report, 1980.
(b) Based on a 50 percent increase over 1979 diesel prices.

by road beyond the break-even distance for road. For instance, if 75 percent and 90 percent of the allocable freight traffic is shifted to rail, the consumption of diesel is reduced from 34 million tonnes to 14.2 million tonnes and 9.6 million tonnes, respectively (Table 3.33).

Still further reductions in diesel consumption can be achieved through changes in the modal mix for passenger traffic. In the case of passenger traffic, railways consume only about 6 percent less fuel per passenger kilometre than buses, in contrast to freight traffic where railways consume 83 percent less fuel per tonne kilometre than trucks. The reason for this difference is that a railway engine has to haul, in addition to the passengers, a great deal of dead weight (coaches and engine)—in fact, 1.07 gross tonnes per passenger. An interesting idea has therefore been proposed to make use of this difference. It is known that about 50 percent of the inter-city rail passenger travel is for distances under 300 km, and that passenger convenience is greater on railways for distances above about 300 km. Hence, if under 300-km passenger traffic is diverted from railways to buses, then the rail haulage capacity which is released can be used for freight, where it is more efficient than truck transport. It turns out that every million passenger-kilometres per year that are diverted from rail to buses results in a net saving of 19.5 tonnes of diesel. Hence, the diversion of 90 billion passenger-kilometres, i.e., 50 percent of half the inter-city rail passenger traffic of 358 billion passenger-kilometres, will result in a net saving of 1.8 million tonnes of diesel (Table 3.33).

The combination of all these strategies can therefore reduce the year 2000 diesel demand in the transport sector from about 34 million tonnes to about 8 million tonnes (Table 3.33).

Even this major reduction in oil consumption may not be adequate because

(1) intra-regional or short-haul traffic is best achieved by road transport, and
(2) for a sustainable future, the dependence of road transport must shift to fuels derived from the fuelwood that can be saved by adopting the first prong of the strategy proposed here.

If diesel fuel in trucks and buses is replaced with methanol, then the only diesel demand from the transport sector will come from the railways, i.e., 2.5 million tonnes.

The two-pronged energy strategy proposed above shows that by the provision of electric lighting and efficient cooking fuels/devices to all homes, India can move towards a virtually oil-free road transport system and a dramatic improvement in the quality of life for its people. And if development is viewed as a process of satisfying basic needs (starting with the needs of the neediest), of strengthening self-reliance, and of ensuring harmony with the environment, then this strategy promotes genuine development of the country.

The proposal which has just been described emphasizes the importance of reorienting energy planning. In particular, it suggests that the energy problem in India is unlikely to be resolved if it is viewed purely as a supply problem to sustain economic growth. It is crucial to scrutinize energy demand from a development perspective, and to assign the highest importance to energy needs associated with developmental priorities. As much attention must be focused on energy *end-uses* as upon energy sources, and both the consumption and generation aspects must be examined. In contrast to conventional energy planning, which has been largely restricted to *sectoral* analyses, what is required is a multi-sectoral approach in which all the sectors of the energy system, i.e., transport, household, agriculture, industry, etc., are considered in their interdependence. It follows that the present patterns of energy generation and consumption cannot be accepted as given and immutable. They must be questioned and new strategies must be explored because a multi-sectoral approach can reveal the possibilities of synergisms between sectors. Moreover, a constant preoccupation of energy planning should be with improvements in the efficiencies of generation, distribution, and consumption of energy.

Table 3.33. Estimates of Diesel Consumption (for the year 2000) based on Various Modal Mixes

Modal Mix	Diesel Consumption in the Year 2000 (Million Tonnes)						Total
	Rail			Road			
	Freight	Passenger	Total	Freight	Passenger	Total	
1. If present trends continue[a]	0.957	1.254	2.211	25.662	6.427	32.089	34.300[b]
2. 75% of allocable freight traffic[c] shifted to rail	1.138	0.880	2.018	8.065	4.137	12.202	14.220
3. 90% of allocable freight traffic[d] shifted to rail	1.364	0.880	2.244	3.220	4.137	7.357	9.601
4. 50% of short-distance rail passenger traffic shifted to buses	1.481	1.023	2.504	0.743	4.602	5.345	7.849
5. Trucks and buses on Producer Gas/Methanol	1.481	1.023	2.504	—	—	—	2.504

(a) Reference Level Forecast of Energy Policy Working Group.

(b) Differs from number given in EPWGR by 4 percent because of difference in norms.

(c) National Transport Policy Committee Report—Assumption IV (b), i.e., 75:25 shares of rail:road freight traffic; Diesel price = 50 percent increase above 1979 diesel price and 75 percent of allocable traffic now moving by road shifted to rail.

(d) Allocable traffic = Freight moving by road beyond break-even distance for road.

Source: A.K.N. Reddy, 1985a.

Detailed need-oriented thinking and end-use focused approaches imply, however, that energy strategies must be locale-specific and country-specific. There are universal principles and methodologies, but there are no strategies valid for all locations, even for developing countries as a category.

As in the case of economic growth, energy planning is meaningless unless one asks: ''energy for whom?'' and ''energy for what?'' In the case of India, it appears that the country has been engulfed by grave oil and firewood crises because it has ignored two crucial basic needs of poor households: efficient energy sources for lighting and for cooking.

Since the strategy outlined above was first proposed in 1981, there has been a spectacular increase in oil production in India, largely from off-shore sources. As a result, the picture regarding oil imports has changed radically—from being an importer of about two-thirds of its oil requirements in 1980-81, India in 1983-84 is likely to import only about 40 percent of its oil. This remarkable success in increasing oil supplies through indigenous production may raise doubts regarding the necessity and value of the needs-oriented strategy described here.

A very basic issue has arisen: Should India try to reduce oil consumption by initiating, implementing and coordinating all the separate efforts in the domestic, agricultural and transport sectors constituting the alternative strategy described above? Or should it persist in its thus-far successful strategy of trying to increase indigenous oil production? It may be noted that this issue is only one more version of the ubiquitous choice in energy management and planning: should energy supplies be increased to maintain (or improve) energy services, or should the same (or greater) services be provided with reduced energy consumption by better demand management (including more energy-efficient technologies)?

Since the present pattern of energy consumption in India (and most other developing countries) is skewed largely towards the affluent consumption patterns of the elite rather than the basic human needs of the poor, the choice is whether the overall objective is to preserve an iniquitous dual society or move towards a just equitable society. If the main outcome of increasing indigenous oil production is to tide over current oil import problems only to carry on *business-as-usual* and prolong the energy *status quo*, then the purely supply-oriented approach may *solve* the energy problem related to oil imports, but it will not advance genuine need-oriented development. Even this *solution* is temporary because the reserves that are currently being used are known to have a limited capacity peaking after about 10 to 20 years (if Britain's experience is any guide—Figure 3.20) during which time the developmental tasks will only multiply unless they are tackled immediately.

A development-oriented energy strategy would attempt simultaneous solutions to the fuelwood and oil crises. Such a strategy would emphasize the use of energy for satisfying basic needs, particularly of the poorer sections, but would simultaneously reduce oil consumption and the burdens of oil imports. In effect, what it would attempt is to store the finite oil reserves as undergound ''reserves'' for the future when its price will become very high in a situation of extreme scarcity.

However, the implementation of such a strategy would require major investments and foreign loans. If funding from abroad is only forthcoming for increasing indigenous oil production—but not for measures to reduce oil consumption—then the country has no choice except to increase oil production. But, it can certainly use the proceeds of this extra oil to finance the development-oriented energy strategy. Thereby, oil would be used to promote development rather than to maintain waste and inequality.

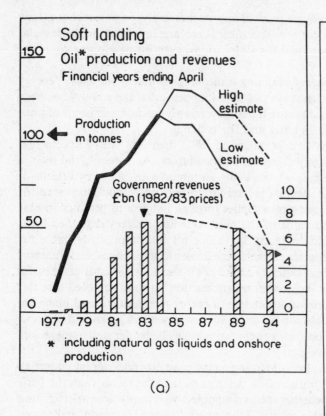

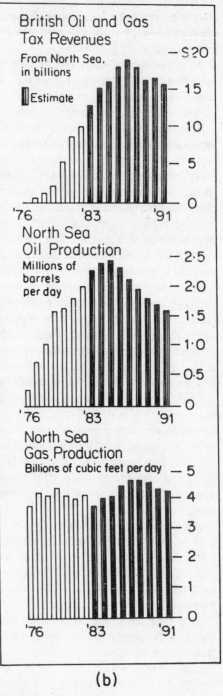

Figure 3.20   (a) Britain's North Sea oil production and revenues. (b) British North Sea oil and gas tax revenues, and oil and gas production.
Source: (a) The Economist, 1984.

(b) The New York Times, 1983.

3.5.6 *The Limits of the Domestic-Agriculture-Transport Synergism*

The question arises at this stage whether the strategy proposed for India is useful for other developing countries.

The second prong of the strategy involves the recapture of freight traffic by the railways and thereby the reduction of oil consumption through the removal of the subsidy on diesel and kerosene prices. For this prong to be applicable to an oil-importing developing country, it is necessary that:

(1) kerosene is used extensively as an illuminant,
(2) kerosene prices are subsidized so that they are within the reach of the poor (Table 3.34).
(3) diesel prices are subsidized so that they are pegged at the level of kerosene prices (Table 3.34).
(4) diesel consumption is a major fraction of oil imports,
(5) trucks account for the bulk of diesel consumption, and
(6) there is a widespread railway network capable of carrying a predominant share of the freight traffic.

Table 3.34. Petroleum Product Price Indicators

Country	Domestic retail prices as a percentage of diesel oil retail prices, 1980			
	Premium gasoline	Kerosene	Diesel oil	Heavy fuel oil
Burundi	148	94	100	—
Chile	122	76	100	59
Dominican Republic	223	84	100	63
Equador	182	55	100	64
Egypt	507	121	100	29
El Salvador	151	93	100	—
Ethiopia	238	121	100	—
Ghana	144	61	100	38
Greece	212	128	100	48
India	213	61	100	75
Indonesia	416	72	100	84
Ivory Coast	146	63	100	67
Jamaica	161	56	100	60
Kenya	154	67	100	39
Korea, Rep. of	380	105	100	76
Malawi	109	76	100	—
Malaysia	257	100	100	—
Mexico	688	150	100	44
Morocco	188	100	100	46
Pakistan	185	90	100	41
Paraguay	259	121	100	76
Philippines	169	101	100	67
Sri Lanka	168	66	100	60
Sudan	371	174	100	61
Tanzania	239	87	100	—
Thailand	161	83	100	59
Uganda	166	87	100	55

Source: The World Bank

The first five of these characteristics can be found in a large number of developing countries, and steps can be taken to replace kerosene as an illuminant by means of home electrification. But, in many of these countries, the last condition pertaining to the railway network may not be satisfied. This is because the railway systems in most developing countries was largely evolved in the heyday of colonial rule to transport goods in and out of the country from and to the imperial metropolis, and after independence, most of these countries under-emphasised the expansion of their railway systems in favour of road transport.

In such countries, there will be few opportunities for reducing truck transport of goods, unless water-borne transport is an attractive alternative, which may be the case with Bangladesh's riverine network. For example, simple escalations in the price of diesel to bring it on par with the price of gasoline may not serve the purpose. In fact, if diesel is not subsidized, increases in its price are invariably passed on to the consumers of the transported commodities. And, price increases of essential commodities such as salt inevitably have an adverse impact on the poor.

Thus, in countries where diesel consumption cannot be reduced through (transport) modal shifts away from truck transport, an important possibility (for cutting down diesel consumption) is through fuel substitution with alternative renewable fuels.

It is in this context that the first prong of the strategy proposed for India becomes relevant to all those developing countries in which non-commercial energy, and fuelwood in particular, is presently being used extensively in their domestic sectors for cooking. The implementation of high-efficiency fuels/devices would lead to the saving of biomass which then can be converted into biomass-derived fuels such as methanol or ethanol.

Whether saved fuelwood is used for the production of biomass-derived fuels or good agricultural land is used—as in Brazil—to grow fuel crops such as sugarcane depends, of course, on whether land is a constraint (Section 3.5.3). Implicit in this statement is the fact that there can not only be synergisms between sectors but also *conflicts* between, say, food and fuel. Actually, there is a much broader range of interactions which are relevant to development objectives, and these interactions will now be considered to prevent a preoccupation with energy from having adverse impacts on other goals.

3.5.7 *Food, Forest/Fuel, Fibre, Fodder, Fertilizer and Foreign-Exchange Interactions*

One fundamental factor in all interactions between energy and non-energy activities is the availability of land. Land is a crucial constraint in all developing countries except in countries like Brazil where the population density is low. The source of the constraint is that land has to serve multiple uses and the possibility of conflicts between these uses is accentuated in developing countries with high population densities.

In the first place, land has to be used for food production, but there are competing demands which can generate conflicts. This competition often arises from the need to produce fuel, fibre (e.g., cotton or jute) and fodder (for livestock).

An idea of this competition can be obtained from the land-use pattern in India (Table 3.35) which has three important features:

- the ratio of pasture land (including fallow land and uncultivable wasteland) to cropland is about 37 percent, which is what is expected if livestock get only 40 percent of their fodder needs from pasture land and the remainder from crop residues;
- fibre crops account for about 5 percent of the net cropped area;
- forests occupy about 20 percent of the total geographical area in contrast to the cropped area and the barren land which are about 43 percent and 7 percent, respectively.

Table 3.35 Land-use Pattern in India 1975-76

Category	Area (Million hectares)	%
1. Net Cropped Area	142.2	43.3
(i) Foodgrains	(106.9)	(32.5)
(ii) Fruits & Vegetables	(3.5)	(1.1)
(iii) Oilseeds	(12.5)	(3.8)
(iv) Fibre Crops	(6.7)	(2.0)
(v) Sugar & Spices	(3.8)	(1.2)
(vi) Other	(8.8)	(2.7)
2. Forests	66.4	20.2
3. Not available for cultivation	39.5	12.0
(i) Area under non-agricultural uses	(17.4)	(5.3)
(ii) Barren & uncultivable land	(22.1)	(6.7)
4. Fallow	22.0	6.7
5. Cultivable Wasteland	17.6	5.3
6. Permanent Pastures	12.6	3.8
7. Tree Crops	4.0	1.2
8. Unclassified	24.4	7.4
Total Geographical Area	328.8	100

Source: Basic Statistics relating to the Indian Economy, 1980.

A direct competition for cropland would arise if cropland is used for growing fuel forests or fuel crops. It follows that simple expansion of forest areas can be achieved mainly in barren land—this would correspond to an approximately 30 percent increase in forest area. Further expansion is likely to affect adversely the production of food (if cropland is used for fuel production) or of fodder (if pasture land is used for forestry). Fortunately, the fuel-fodder conflict can be resolved, for instance, by growing two-tier forests in pasture land with the shorter plants producing fodder and the taller trees, fuel. If this approach is adopted, forest area can be increased to about 29 percent of the total geographical area, i.e., approximately a 50 percent increase.

Apart from the food-fuel-fodder interactions mentioned above, the dedication of part of the cropland to the growth of agricultural products for export purposes is a more serious matter. In effect, this is equivalent to an export of cropland.

One example of this competition for arable land from foreign exchange-earning agricultural crops is the land devoted to export-oriented crops in Brazil. In 1982, about 31 percent of the land cultivated in Brazil was devoted to export crops (16 percent for soyabean), compared to 63 percent for food crops and 3 percent for sugarcane for ethanol (Table 3.24). Jute in Bangladesh is another example. An area of about 0.627 million hectares, or 7.5 percent of the net cropped land, was used in 1976 for jute production for export, even though this area could have yielded about 1.16 million tonnes of rice paddy, or 6.3 percent of the year's rice production.

Foreign-exchange-earning crops can also compete with forest land. For example, Tanzanian production of tobacco is based on clearing of forests and shifting cultivation with yields in the neighbourhood of 0.5 tonnes per hectare. But, tobacco is cured in Tanzania with heat energy obtained from the combustion of fuelwood—about 100 m³ of fuelwood per tonne of tobacco. It turns out that one hectare of forest has to be cleared annually to obtain the fuelwood to cure the tobacco grown on one hectare of forest. The area used for growing tobacco is only about

30,000 hectares or about 0.6 percent of the agricultural land, but it could yield on a renewable basis—at 15 tonnes fuelwood per hectare and 55 percent conversion efficiency into methanol—enough methanol to replace about 0.10 million tonnes of diesel in comparison with the 0.16 million tonnes of diesel and gasoline used in Tanzania in 1979.

These indicative examples of the demands on land for food, fuel, fibre and other crops suggest the importance of establishing a set of decision rules for the use of land in developing countries with land constraints. A tentative set of such decision rules is shown in Figure 3.21. The essential principle underlying this set is the use of arable land only for food and fibre crops and of non-arable land for forests.

The non-ligneous portion of crop residues can be used for fodder for livestock in livestock-rearing economies. In turn, livestock can yield

- food in the form of milk, meat, etc.,
- wastes which can be fermented anaerobically to generate biogas fuel, and
- draught power for agriculture, transport and rural industry where the use of draught animals is customary.

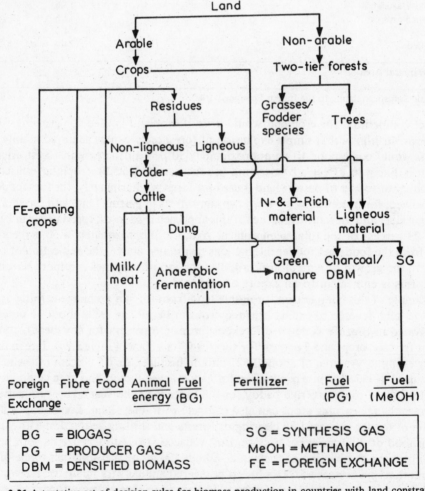

Figure 3.21 A tentative set of decision rules for biomass production in countries with land constraints.
Source: A.K.N. Reddy, 1985a.

If, instead, a part of the arable land is set aside for pasture, then a situation can arise where there is competition between land for food for human beings and land for fodder for livestock. In this context, the replacement of draught animals with mechanical power sources, for example, tractors, lessens the demand for fodder, and therefore pasture land.

The forests on non-arable land can be of two types: (i) two-tier fodder-cum-fuel forests on pasture land, and (ii) fuelwood forests. From the trees in the forests of both types, the non-ligneous portions that are rich in nitrogen and phosphorous must be returned to the land as organic fertilizer, either directly as green manure or indirectly as biogas sludge after being processed along with livestock wastes in biogas plants.

It is the ligneous fraction either from trees or from crop residues which should be turned into fuels for engines. In some situations, e.g., for irrigation pumpsets, it may be advantageous to gasify the wood, perhaps after densification, and utilize the resulting producer gas to run engines within ranges close to the biomass production site. In general, however, it may be preferable to proceed from wood gasification to the production of methanol.

The establishment and use of such decision rules designed to suit the specific conditions of different countries would help to optimize the land-use pattern.

3.5.8 *Synergisms involving the Agricultural, Industrial and Transport Sectors*

The possibilities for synergisms between the transport sector and the agricultural and/or industrial sector can be discovered by scrutinizing the patterns of inter-regional freight traffic. The magnitude of freight traffic is often accentuated by one or more of the following three factors:

- non-uniform development of the various parts of the country;
- production of goods and services for consumption at distant locations; and
- inadequate utilization of local resources.

Consider first that in India (Table 3.36) there is an approximately 1000-kilometre average lead distance for foodgrains, which account for 12.5 percent of the total freight (in tonne kilometres). This long lead distance is a reflection of the fact that there are distinct and widely separated food-surplus and food-deficit regions in the country. In general, the flow of foodgrains is from the North of India to the South. If, however, the development of agriculture in the country had been more uniform, a large fraction of this traffic in foodgrains could have been avoided, and a proportional reduction in energy consumption associated with this transport achieved.

For processed agricultural and industrial products the trend towards production for consumption in distant markets implies not just added energy requirements for transport but also extra energy for processing, packaging, distribution, and storage. For example, the trend towards the development of a nation-wide food production-transport-distribution-storage consumption system in India is indicated by the 500-odd kilometre lead distance for fruits and vegetables (Table 3.36). This change-over from decentralized production for small, local markets to centralized production for nation-wide, or even world-wide markets, is inevitably associated with increasing off-the-farm energy expenditures on food processing, packaging, transport, distribution, and storage. Today, this off-the-farm energy expenditure accounts for about 75 percent of the total energy consumption of the industrial food system of the United States. This energy price will have to be paid if developing countries evolve food systems patterned after those in the industrialized countries.

Further, the added energy and other costs for processing, packaging, transport, distribution, and storage associated with centralized production for distant markets invariably increase the

Table 3.36. Commodity Flows of Inter-Regional Traffic for 1977-78 in India[a]

Commodity	Freight Traffic				Average Lead
	MT	%	MTKMS	%	(KMS)
1. Coal	69.3	25.3	46,332	25.6	669
2. Foodgrains	23.7	8.6	22,681	12.5	957
3. Iron & Steel	15.2	5.5	12,778	7.1	841
4. Mineral Oil	17.9	6.5	9,676	5.4	541
5. Cement	15.4	5.6	9,452	5.2	614
6. Fertilizers	11.6	4.2	9,213	5.1	794
7. Iron Ore	17.2	6.3	9,052	5.0	526
8. Fruits & Vegetables	7.9	2.9	4,206	2.3	532
9. Building Materials	14.6	5.3	3,848	2.1	242
10. Household Items	6.9	2.5	3,826	2.1	554
11. Wood & Timber	6.6	2.4	3,723	2.1	564
12. Limestone & Dolomite	6.9	2.5	2,516	1.4	365
13. Others	61.8	22.4	43,465	24.1	703
Total	275.0	100	180,879	100	657

(a) *Source:* National Transport Policy Committee Report, 1980.

prices for the products so produced beyond the means of the poorest consumers. This situation could be avoided if, instead, more emphasis were given to decentralized production for local consumption, which will often be a more appropriate way to satisfy basic needs.

The approximately 600 kilometre lead distance for the 15 million tonnes of cement (which accounts for about 5 percent of the total freight traffic) highlights a third factor affecting freight demand—the inadequate use of local resources. If cement and building materials are considered together, construction materials account for 7.3 percent of the total freight traffic. In this context, traditional architecture and construction suggests an important guideline for energy strategies: the maximum economical use of local resources.

Thus, there are important synergisms whereby interventions in the agricultural and industrial sectors can affect energy consumption in the transport sector. These synergisms involve

- a more uniform development of all parts of a country,
- production for local consumption, and
- increased use of local sources.

3.6 Energy and Agriculture

In Chapter 1 (Section 1.3.10), attention was directed to the FAO projection that, in order to double food production by 2000, commercial energy use for agriculture in developing countries would have to increase at an average annual rate of 8 percent, from 0.74 million barrels per day of oil equivalent in 1980 to 3.5 million barrels per day in 2000.[58] This projection implies a major expansion of energy-intensive modern agricultural practices in developing countries. But such a shift to energy-intensive agricultural practices is the central issue of the widely debated "green revolution", and it makes the FAO projection somewhat controversial.

It is useful, therefore, to recapitulate some of the more important criticisms levelled against the energy-intensive approach to agricultural development.

The first criticism is that the shift to energy-intensive, highly mechanized agriculture leads to reduced labour demand and thus exacerbates the rural unemployment problem.

The second criticism is that modern agricultural technologies have mainly benefitted the affluent farmers, and thus aggravated an already skewed income distribution. The spread of high-yielding seed varieties and advanced technologies depends upon access to credit and the ability to make the required investments, and it is mainly the larger, more prosperous farmers who are creditworthy and able to invest. One manifestation of the inequitable impact of the green revolution is that even though some improvements have been achieved in the seeds of millet and sorghum, the staple foods for poor people in many countries, in most parts of the world these crops have not shown the major increases in yield attained by crops such as wheat or rice.[59] This "failure" is largely because the poor farmers who grow crops like millet and sorghum cannot afford and therefore do not provide the increased supplies of fertilizer and irrigation water to make the most of the improved seed.

The third criticism is that a stress on energy-intensive agricultural production methods in oil-importing developing countries would generate a demand for extra oil imports. These further increases in oil imports would be intolerable because oil import bills are already so large. [60] Further, the foreign exchange necessary to pay for these imports may have to be diverted from other important development programmes.

Notwithstanding these criticisms, the issues are a good deal more complex than is suggested by a simplistic debate that crudely pits modern against traditional agricultural development strategies. Unfortunately, a major obstacle clarifying these issues is the conventional approach of treating agricultural crop production as a "black box" in which what economists call inputs— capital, labour, energy, etc.—are somehow converted into crop outputs.

An alternative approach is to view crop production as a sequential combination of individual agricultural operations, and for each operation, one option (technique) is chosen out of a set of possible technical options (a menu of techniques). [61] It follows from this that as many technologies of crop production are possible as there are feasible combinations of technical options (techniques). Thus, by scrutinizing the individual operations which are in fact the end-uses of agriculture and considering alternative technical options (techniques) for every operation, it becomes obvious that *agricultural modernization is not a unique possibility, but a generic process involving the replacement in one or more number of operations of the traditional technique for an agricultural operation with a modern technique for that operation.* * Thus, there can be many versions or brands or variants of modern agricultural technology for crop production. In contrast, the conventional approach may lead to a picture of modern agricultural technology as consisting of a total replacement of all the traditional techniques for the individual operations with modern techniques.

The multiplicity of technological possibilities can be revealed through an illustrative example of *rain-fed rice production*.[61, 62] This crop is of major importance because rice constitutes more than 50 percent of the diet of about 1.6 billion people, uses about 11 percent (145 million hectares) of the world's arable land, and is produced in large amounts (411 million tonnes in 1982).[63] Further, roughly half of Asian rice production is rain-fed, involving "flooded wetland systems that receive all moisture from accumulated rainfall".[62]

---

*Of course, the techniques chosen for the various operations must be compatible with each other. For example, the choice of hybrid seeds may compel the use of modern fertilizers, pesticides, and herbicides for the crop-care operation.

Four technologies of rain-fed rice production can be considered:

(1) the truly *traditional* technology of rice production which does not use hybrid seeds, chemical fertilizer, pesticides, herbicides, and either transport vehicles or draught-power sources running on oil-derivatives, but instead depends upon draught animals for land preparation, traditional seed varieties, manual transplanting, organic fertilizer, and manual harvesting, threshing and winnowing;

(2) *Variant 1* of modern rice technology which differs from traditional technology in that it uses modern biological-chemical inputs (hybrid seeds + fertilizer + insecticide + herbicide);

(3) Variant 2 which differs from Variant 1 by using tractors for ploughing and mechanically driven vehicles for transport while still retaining draught animals for harrowing;

(4) Variant 3 of modern rice technology which uses power threshers and replaces the tractors of Variant 2 with power tillers ("walking tractors") for ploughing *and* harrowing.

The human labour, animal labour, direct inanimate energy, indirect inanimate energy, fixed capital costs and paddy output for the four technologies considered above can be computed operation by operation. In a sense, the results represent the "unpackaging" of the "black boxes" constituting the patterns of rice production selected for discussion.[61] Though this end-use approach to agricultural crop production makes it possible to understand *inter alia* which operations of each technology are consuming the labour, animal labour, energy (direct and indirect) and costs, the main focus in the discussion here will be on the overall "first-approximation" implications of these technologies (Table 3.17). Some of the main "first-approximation" conclusions that may be drawn from the consolidated Table 3.37 are discussed below.

Table 3.37. The Energy (Direct and Indirect), Human Labour, Animal Labour, Fixed Capital Costs and Paddy Output of a Few Selected Technologies of Rice Production[a]

Technology	Inanimate Energy (MJ per hectare)		Human Labour (hours/ hectare)	Animal Labour (hours/ hectare)	Fixed Capital Costs ($/hectare)	Approximate Paddy Production (kgs/ hectare)
	Direct	Indirect				
Traditional[b]	0	331	725	342	10	1860
Variant 1[c]	0	4570	983	440	10	3500
Variant 2[d]	1256	4501	905	78	24	3500
Variant 3[e]	3112	5012	652	0	30	3500

(a) Based on A.K.N. Reddy, 1985b.

(b) *Traditional Technology:* No hybrid seeds, chemical fertilizer, pesticides, or herbicides; no use of oil for transport vehicles or land preparation. Instead, traditional seed varieties and organic fertilizers are used, with draught animals used for land preparation, manual transplanting, and manual harvesting, threshing, and winnowing.

(c) *Variant 1:* Differs from Traditional Technology in that it uses modern biological-chemical inputs (hybrid seeds + fertilizer + insecticide + herbicide).

(d) *Variant 2:* Differs from Variant 1 by using tractors for ploughing and mechanical/driven vehicles for transport, while retaining draught animals for harrowing.

(e) *Variant 3:* Utilizes power threshers and replaces the tractors of Variant 2 with power tillers ("walking tractors") for ploughing *and* harrowing.

(1) Variant 1 of modern technology—it will be recalled—is the replacement of traditional seeds with the hybrid varieties and by the addition of packages of chemical inputs (fertilizer, insecticide and herbicide) for the crop-care sub-operations. This technology has the potential of increasing the paddy output (Figure 3.22a) by approximately 90 percent compared to traditional technology.

Despite this potential increase in output which is realizable in practice, there can be a 35 percent increase in labour requirements (Figure 3.22b) because of the greater rice yield and therefore greater work, particularly in the operations commencing with and subsequent to harvesting. With the bulk of the draught power still coming from draught animals, the animal labour requirements show an increase of about 29 percent (Figure 3.22c). There is still no demand for direct sources of inanimate energy (Figure 3.22d), and the fixed capital costs (Figure 3.22f) remain the same.

The big change, however, is in the indirect energy (Figure 3.22e) embodied in the hybrid seeds, fertilizers, pesticides and herbicides—it is approximately 14 times that associated with traditional technology. These escalations in indirect energy use are, of course, associated with increases in variable costs with the purchase of inputs that are intensive in indirect energy. This indirect energy is also a proxy for the dependence on external agencies that have to be established for the supply of seeds, fertilizer, pesticides, herbicides, etc., the urban manufacture of which acquires the character of an industry servicing the agricultural sector.

(2) Variant 2 of modern technology—when compared to Variant 1—shows the same paddy output (Figure 3.22a) and roughly the same amount of indirect energy. But, there is a major decrease in animal labour (Figure 3.22c) along with a tremendous increase in direct inanimate energy (Figure 3.22d) as well as an increase by a factor of about 2.5 (Figure 3.22f) in the fixed capital costs. All these differences arise because Variant 2 of modern technology basically involves the replacement of draught animal power for ploughing and animal-drawn vehicles with the mechanical power of the fuel-consuming tractors and vehicles. However, some animal labour requirements remain because animal labour is used for the harrowing operation. Of course, as the ploughing rate increases significantly with the use of machines, the labour time required for ploughing decreases, and therefore there is a small decrease of about 8 percent in the labour requirements of this technology when it is compared to Variant 1. However, the human labour required is still higher than in traditional technology to the extent of about 25 percent.

(3) Variant 3 of modern technology deals the *coup de grace* to draught animal power (Figure 3.22c) which is replaced not only in the ploughing and transport operations but also in the harrowing operation. If these were the only input-replacements, there would have been little impact on human labour requirements, but Variant 3 also includes the replacement of manual threshing with a power threshing. This replacement not only reduces the labour requirements for threshing but also the labour required for the traditional winnowing operation which is made redundant. All told, there is an approximate 10 percent reduction in the labour traditionally required. These changes are associated with a further increase in the fixed capital costs (Figure 3.22f) and 2.5-fold increase in the direct energy consumption (Figure 3.22d) relative to that with Variant 2 (tractor) technology.

(4) The "mechanized" variants of modern technologies of rice production, i.e., Variant 2 and Variant 3 (minus its power thresher), do *not* increase the paddy output of a single crop; hence, they are most probably undertaken for reasons other than increasing the yield.[64, 65] One possibility is that draught animal power is a constraint in many parts of the developing world because of the scarcity of pasture land and fodder, or simply because of the unpleasant arduousness and drudgery of using animal power. It can also be that the management of large

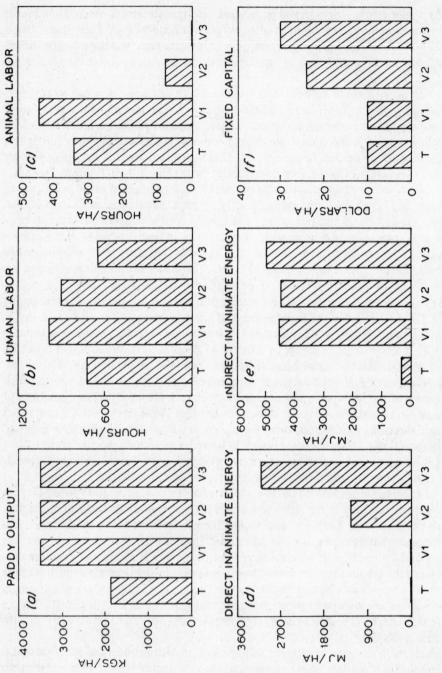

Figure 3.22 Characteristics of four technologies of rain-fed rice production.
Source: A.K.N. Reddy, 1985b.

holdings necessitates the management of large numbers of cattle and cattle-herds, and this may not be as economical as in the case of small holdings. For whichever one or more of these reasons, the impetus to replace draught animals with tractors and power tillers can lead to "mechanized" land preparation and transport.

(5) The replacement of manual threshing with a power thresher in Variant 3 of modern rice technology is another matter; this replacement leads to a significant reduction in labour requirements which must be seen against the reduction in losses in grain incurred during traditional threshing. The question is whether such labour-saving input-replacements are introduced in conjunction with the rising demand for labour in urban and/or rural off-farm industry and therefore with the decreasing availability of agricultural labour, or whether they are brought in because of price distortions embodied in or backed by associated subsidies, incentives, etc.

(6) Of the many conclusions that follow from this simple version of the implications of a limited set of examples of modern technologies of crop production, perhaps the most important is that there are many variants of modern rain-fed rice production technologies, i.e., agricultural modernization does not consist of a unique package which must either be accepted *in toto* or not at all. In particular, it appears that the modernization of rice technologies can be accomplished in three phases based on the introduction of:

- *Phase 1:* output-increasing, land-saving biological-chemical inputs corresponding to what may be called "a green revolution without either tractors or machines",
- *Phase 2:* draught-animal-replacing techniques—"a green revolution with tractors/power tillers only"—which may, in addition, save the land required to grow the fodder for the draught animals,
- *Phase 3:* human-labour-replacing techniques corresponding to complete modernization on the same lines as the industrialized countries—"industrialized-country-style green revolution" with all biological/chemical inputs, tractors/power tillers, and all machines such as combine harvesters, threshers, etc.

(7) The above conclusions are for a single crop, but the introduction of mechanized operations can open up the possibility of multi-cropping in the following way. Whereas the duration of a crop is largely governed by crop physiology, the overall time taken for the completion of all the crop operations can be much greater. If, therefore, a short-duration crop variety is chosen, and the mechanical operations of land preparation, harvesting, threshing, etc., are speeded up, it may become feasible to go through another crop cycle within the same agricultural year. In such a case, the energy, labour, animal labour, capital costs and output would have to be multiplied by as many crops as there are in a year. For instance, if Variant 2 of modern technology permits two rice crops per year, and traditional technology only one, the annual labour requirement per hectare is 1810 hours/hectare which is more than the 725 hours/hectare of traditional technology. In short, the comparison of technologies must be done on a *per year* basis, and not simply on a *per crop* basis.

(8) The elimination of the winnowing operation by the introduction of power threshers has a further implication that can be revealed by the end-use analysis presented here. If human labour is further disaggregated into male and female labour then it becomes clear that since the winnowing operation is largely carried out by women, there will be selective reduction in the employment of women with all the attendant impacts on their incomes and status.[66] Similarly, the likely effect of introducing mechanical transplanters and harvesters merits examination by considering further variants of modern rice technology. Thus, agricultural modernization can not only lead to a displacement of human labour, but there can be a gender bias in this displacement with women bearing the brunt of the process.

(9) Rain-fed rice production is widespread in character—50 percent of rice production in South Asia and South-east Asia is rain-fed and does not involve lift-irrigation; "another 25-35 percent is irrigated for at least one crop, with the remainder categorized as dryland in which water is not impounded".[62] But, the restriction of the above discussion to rain-fed rice was not only for this reason but also because it reflects the tremendous importance of water as a limiting factor in crop production. "Controlled water supplies make all the difference between a good crop and no crop even during the dry season. The majority of the irrigation system in Asia draw water, by gravity, from reservoirs or diversion structures. An increasing number of farmers are, however, turning to powered pumpsets for a small, low-lift irrigation system—studies indicate that supplemental irrigation during the wet season...can increase yields by as much as 0.35 t/hectare.* Irrigation during the dry season can *double* the annual yield compared to the rain-fed water regime which permits only one crop per year. Therefore, irrigation can provide large increases in annual production with relatively small decreases in energy output-input efficiency."[62] It is clear that energy for water-lifting (provided that water is available in subterranean sources) becomes crucial for the achievement of multiple-cropping, and thereby increased crop output and employment.

(10) The discussion here has been devoted to sketching the implications of agricultural modernization because they are the "raw material" and database for the choice of technologies. But it has become obvious that in this domain of decision-making, energy consumption cannot be the sole determinant in the choice of agricultural technologies; several factors such as agricultural productivity, employment generation and investment costs have to be taken into account.

The possibility of choosing from many variants of modern agricultural technology for a particular crop facilitates the advancement of specific objectives such as agricultural productivity, employment generation, energy conservation, etc. When this is done and consideration is given to the inanimate and animate (human and animal) energy, productivity, employment and cost implications of different variants of the modernization of agricultural technology, it turns out that there may be ways of surmounting the standard criticisms against modern agricultural technologies.

(i) *The Rural Employment Issue*: In view of the overriding importance of raising agricultural productivity to feed the growing populations of developing countries, agricultural modernization is imperative. However, its extent and pace are crucial issues for decision-making. A broad guideline would be that the pace of displacement of labour has to be geared to the pace of employment generation in rural and urban industries. It is unlikely that the pace of industrialization can be stepped up to absorb all the labour displaced by a wholesale shift to modern techniques for all agricultural operations. In order to provide sufficient employment for the rapidly growing labour force, this shift can be done selectively for certain operations and at the same time many new jobs can be created in the agricultural sector.

The FAO Agriculture: Toward 2000 projections requires worldwide (Table 3.38) about the same share of human labour in 2000 as in 1974-76 to produce the targeted doubling of food production. What is envisaged is a generalization of Variant 2 of modern rice technology involving much more a substitution of tractors and power tillers for draught animals than the replacement

---

* This corresponds to a 10 percent increase over un-irrigated crop production by means of Variant 1, 2 or 3 of modern rice technology.

Table 3.38. Power Used for Crop and Livestock Production, Developing Countries, Average 1974-76 and Projections for 2000[a]

	Share of total power supplied by					
	Human labour		Draught animals		Tractors	
	1974-76	2000	1974-76	2000	1974-76	2000
Africa	83	82	13	10	4	.8
Far East	64	66	33	25	3	9
Latin America	56	37	20	9	24	54
Near East	63	55	19	9	18	36
Developing Countries	66	63	27	18	7	19

(a) Based on data from revised normative high scenario of FAO, 1981, aimed at doubling food production by the year 2000 in the 90 countries accounting for 97 percent of the population of the developing world outside China.

of people with machines. In particular, the use of tractors and power tillers for land preparation may enable the achievement of more than one crop per year. That is, the slightly increased labour requirements reflect in large part the emphasis given in the FAO projection to double- or multiple-cropping.

(ii) *The Inequality Issue:* While some modern agricultural practices do favour the rich over the poor, this is not necessarily always the case. For example, an emphasis on multiple cropping may have just the opposite effect:
"[With multiple cropping] yields per acre are usually far higher on small landholdings than they are on larger farms, since families use their many hands to exploit what we have described as the intensive margin. With appropriate support facilities and appropriate technology land distribution to favour small-holder agriculture has already played a large role, and it can play a much larger one in improving both agricultural production and rural income distribution."[59]

(iii) *The Energy Consumption Issue:* While the FAO projection calls for a rapid rate of increase in the amount of commercial energy used for agriculture, the projected increase is modest in an absolute sense. Crop and livestock production in 1972/73 accounted for only 3.5 percent of total world commercial energy use.[67] Moreover, only a small fraction of oil use in developing countries is today accounted for by agriculture (Table 3.39). Instead, the lion's share of oil use is accounted for by transportation—mainly for truck and automobile use. But there are major technological opportunities for achieving a doubling in the fuel economy of trucks and at least a two- to four-fold improvement in the fuel economy of automobiles (Section 2.5.3.3). Such opportunities for the improvement of transport energy efficiency could release substantial oil resources for use in agriculture. Certainly, on a global basis, resource scarcity should not in principle be a limiting factor regarding the targeted increased energy requirements for agriculture in developing countries. *The increased requirements for the year 2000 projected by the FAO, some 2.8 million barrels per day of oil equivalent, is equal to the amount of oil saved by the U.S. alone between 1978 and 1981.*

Table 3.39. Distribution of Petroleum Consumption by End Use

	Ethiopia, 1979[a]	India, 1978[b]
Agriculture	5.5	13.8
Industry	12.3	6.5
Transport	64.7	55.3
Household	0.7	20.3
Commercial	3.0	
Mining	0.5	
Road Construction	5.3	4.1
Electricity Production	5.6	
Other	2.4	
	100	100

(a) *Source:* R. Hosier *et al.*, 1982.
(b) *Source:* A.K.N. Reddy, 1982.

Moreover, a couple of decades from now energy used in modern agriculture need not be largely petroleum-based. Energy requirements for agriculture provide a powerful motivation for developing synthetic fuels from biomass—such as methanol derived through thermochemical processes from organic residues or wood. Methanol in the amount of 2.8 million barrels per day oil equivalent could be produced at a 50 percent overall conversion efficiency, using 40 percent of the present level of organic wastes (crop residues, animal manure, and food processing wastes) in developing countries. Alternatively, the methanol could be produced from wood grown in energy plantations. At a yield of 10 tonnes/hectare/year, some 66 million hectares would be required—about 3 percent of the forest area in developing countries. Thus, energy supplies need not be a fundamental constraint on the FAO food production strategy.

(iv) *The Technological Alternatives Issue*: The green revolution debate involves a number of major concerns other than just energy supply availability. It is these other concerns which will probably decide the debate. It is clear, however, that if the details of food production and use are closely examined, it becomes obvious that posing the choices for the future in terms of the traditional versus the modern path is a gross oversimplification. The choices are not so restricted. Instead, there is a broad spectrum of options for resolving the issues in the green revolution debate. Some of these options are indicated below.

(a) The role of good husbanding, with timely planning, planting and weeding, is not to be underestimated.[68, 69] It has been argued that a 145 percent increase in yield is possible by shifting from the combination of bad husbandry, local seed and no fertilizer to the combination of good husbandry, local seed and no fertilizer. In comparison, only a 60 percent increase in yield may be achieved if, instead, the shift is to the combination of bad husbandry, hybrid seed, fertilizer (Table 3.40). Of course, when all three factors are improved simultaneously, the greatest gains are realized.

(b) Another agricultural development option is to make only a *partial* shift to energy-intensive practices—for example, replacing draught animals with liquid fuel-based tillers and tractors. One estimate is that there are 250 million draught animals in developing countries, each producing about 100 kWh/year of useful work (0.5 kW for 200 hours).[70] While these animals work only about 2-3 percent of the time, they must be fed all year round, so that the overall "fodder-to-

Table 3.40. The Effects of Husbandry and Capital Inputs on Agricultural Yields[a]

	2.0 tonnes/ha	3.2 tonnes/ha	4.9 tonnes/ha	8.0 tonne/ha
Bad husbandry	×	×		
Good husbandry			×	×
Local Seeds	×		×	
Hybrid Seeds		×		×
Fertilizer		×		×
No Fertilizer	×		×	
Yield	2.0 tonnes/ha	3.2 tonnes/ha	4.9 tonnes/ha	8.0 tonne/ha

(a) These data are based on surveys in Kenya. *Source:* A.Y. Allen, 1968.

useful-work-efficiency'' is low—a fraction of a percent. Replacing them all with machinery at 20 percent efficiency would require only 200,000 barrels of oil per day and would free up for human use large amounts of land, now used for fodder production.[68] In the United Kingdom, the complete substitution of about 700,000 farm horses by some 250,000 tractors between 1945 and 1960 released 30 percent of the total farm area for food production.[71]

(c) One approach to the problem of high-cost fertilizer is to focus on the potential of biological nitrogen fixation to help boost crop yields without the intensive use of fertilizers. Research activity along these lines is being carried out at a number of the International Agricultural Research Centers.[72]

(d) Another approach to the problem of high-cost fertilizer that offers a number of additional potential benefits in tropical areas is "agro-forestry", involving the coproduction of crops and/or fodder with the leguminous tree crop *Leucaena*. *Leucaena* helps to enrich soil and aid neighbouring plants because its foliage rivals manure in nitrogen content. Recent experiments have shown that if the foliage is harvested and placed around nearby crop plants they can respond with yield increases approaching those affected by commercial fertilizer. Moreover, leucaena foliage provides a palatable, digestible, and protein-rich forage suitable for cattle, water buffalo, and goats. and the fast-growing wood produced in leucaena agro-forestry plots can be used directly as a household or industrial fuel or as a source of fibre or as a feedstock for the production of synthetic gaseous or liquid fuels (producer gas or methanol), for agricultural or other purposes.[73]

(e) Finally, much more attention must be given to the entire post-harvest system, involving the storage, transport, processing, and final consumption of agricultural products. Concerning storage, it has been pointed out that:

"...food losses in developing countries often exceed 20 percent for cereals and legumes because of rodents, insects and moles. The losses for fruits and vegetables in tropical countries are likely to be more than twice as high. Badly needed are new techniques of food storage and preservation designed for poor households and villages."[59]

Even at the point of final consumption, food conservation can be achieved by ensuring that people are basically in good health. The frequent episodes of diarrhoea and other infectious diseases and the high prevalence of intestinal parasites in poor countries reduce the efficiency of food uptake.

(v) *Conclusion:* This discussion shows that there are no significant energy constraints inhibiting the goal of meeting future food requirements in developing countries and that there is an abundance of technological possibilities for achieving this goal in ways that are responsive to concerns about the technology of modern agriculture.

Realizing these opportunities will not be easy, however. Agricultural investment in developing countries will have to increase from about 50 billion dollars in 1980 to more than twice that much by the year 2000.[58] And the leadership in developing countries must have the political will to make it happen.

Further, innovations that bring big returns to large agribusiness groups may even reduce availability of food in a country. In some areas of Mexico, for example, the production of fresh vegetables for export to the U.S. worsens the food situation when the more profitable new crops take over land formerly used to grow staple foods—this is another example where crop exports virtually exports cropland (Section 3.5.7).[58]

3.7 Generic Strategies for the Industrial Sectors in Developing Countries

Most developing countries share a common history of having been subject to colonial rule. This ex-colonial status has resulted in a similarity in the process of industrialization of developing countries. This similarity is important to understand because it can suggest the possibility of certain generic energy strategies for their industrial sectors.

In the first place, the initial steps in the industrialization of most developing countries were taken in the heyday of the imperial powers. These steps involved the establishment of raw-materials-extraction industries with imported equipment and of a transportation system to haul these raw materials to the port cities of the colony, and out from there to the processing and manufacturing industries located in the industrialized countries.

But industrialization could not easily be restricted to the raw materials extraction industries. Expatriates from the imperial countries and the natives who worked for them (or with them) developed consumption patterns that replicated the patterns in the great capitals of the industrialized world. Since the numbers of the expatriates and associated locals were initially quite small, their requirements for goods could be imported. However, as more and more of the local people became part of the growing "elite", the sheer volume of imports became too large, and indigenous manufacturing started to develop.

The local industries in almost all cases were established on the basis of technology imported from the industrialized countries. Further, most of this imported technology was not the latest. Very often, plants which were becoming obsolescent in the advanced countries were shipped out to the developing countries.

But the inefficiency of these industries mattered little because they very rarely had to face competition. The local industrialists (whether they are from the multinational companies or they are nationals) protected themselves from external competition by influencing the national government to enact import restrictions and erect tariff barriers. In this way, economic inefficiency, often reflected in the inefficient use of energy, was able to continue for some time without too much pressure for improvement.

A continuation of this course is no longer viable, however. For the future economic efficiency generally and energy efficiency in particular have assumed vital importance in the industrial sector.

The small size of the local market is a crucial factor in determining the need to break with the past. Local production is primarily geared to satisfying the needs of the elite, i.e., a mere 10-20 percent of the population. In the case of Brazil, for example, its industry has a size which is compatible with its 20 million elite rather than with the total country's population of 120 million people. Once the small elite market is saturated, there are two possibilities for breaking out of the severe industrial stagnation—either a determined bid can be made to capture export markets, or the purchasing power of the masses can be increased through income-generating and poverty-eradication programmes, in which case the general rise in incomes constitutes a tremendous expansion of the national market. In either case what is needed is a modern, economically efficient industrial structure—in the former case in order to be able to compete in world markets and in the latter case in order to promote rapid growth with low product prices. The need to stress economic efficiency in turn implies that energy efficiency is important, in light of both the high cost of energy from new sources and the high costs and foreign exchange requirements of purchasing imported oil.

The modernization of industry requires both the use of best-available existing technologies and also the introduction of new technologies that are compatible within the broad development goals of employment generation, self-reliance, etc. In this modernization process particular attention should be given to new technologies.

Many new inventions that would radically change industrial processes, industrial products, or consumer products offer the opportunity to do more with less, thereby assisting the development process by speeding up the rate of economic growth. While the adoption of radical innovations involves significant economic risks and can also be very disruptive of the "technological status quo," modern societies throughout history have adopted those inventions that offered a sufficient number of process or product improvements that the benefits were judged to outweigh the risks and inevitable dislocations involved.

In the area of energy-using technologies in particular there are many opportunities for introducing advanced technologies that lead not only to energy savings but often to reduced capital and material requirements, reduced environmental damage, etc., as well (Section 2.5.3). Such technologies have taken on new significance in the new era of high-cost energy.

Despite the attractions of advanced technologies, it is generally thought that the process of commercializing advanced technologies is an inappropriate activity for developing countries, which do not have the infrastructure needed for technological innovation and which cannot afford the risk-taking of innovation when there are so many pressing developmental needs to attend to. The conventional wisdom is that developing countries should avoid technological risk-taking and base their development only on technologies that are already well-established.

While less risk would be involved if the technologies adopted by developing countries were those which have been already brought to commercialization in industrialized countries, several considerations weigh against this course of action. In some instances it would be desirable for developing countries to bring to commercialization promising advanced technologies before these technologies are introduced in the already industrialized countries, as we will now argue for the case of energy technologies.

First, many of the industrial technologies being brought to commercial readiness today in the North are capital-intensive and labour-saving—characteristics which are not well-suited to industrial activity in most of the South, where labour is cheap and abundant, while capital is costly and scarce.

Second, the comparative advantages in natural resources are often quite different for many countries of the South from those of the North. In the case of energy, many developing countries are blessed with largely undeveloped and relatively low-cost hydroelectric resources, while most industrialized countries must turn to more costly thermal power sources for increased electrical capacity. Similarly, biomass is a promising source of chemical fuels for many developing countries, requiring decentralized development strategies quite unlike the centralized strategies that have been pursued by the fossil fuel-rich countries.

Third, human needs are quite different in the South from those of the North, because of climatological differences, because of different cultural aspirations, and especially because the satisfaction of basic human needs and infrastructure-building must be given prominent focus in the economic planning of the South, dictating patterns of production which are markedly different from those of the North.

In particular, developing and industrialized countries are out of phase in their industrial development. In the industrialized countries, the areas where most growth and innovation are taking place are electronics, information technology, communications, medical technology, etc.— generally areas involving high value-added fabrication and finishing activities. Both a shift in consumer preferences away from material-intensive goods and the oil price shocks of the 1970's have curbed growth in the demand for basic materials in industrialized countries (Section 2.4.2). This stagnation in demand provides a poor climate for innovation in the basic materials industries, even though the existing capital stocks have been made largely obsolete by the energy price increases of the last decade. Thus, in industries of crucial importance for infrastructure-building, the North is not innovating at a pace sufficient to provide for the needs of the South.

Finally, the potential for rapid growth in the demand for basic materials in the South suggests that some countries of the South may be more promising theatres for innovation in these areas than the countries of the North, where demand is stagnating.

For all of these reasons, developing countries should not retrace the development path of the North but should pursue new directions and assume the risks of innovating in some areas of potentially high pay-off.

Not only are there many theoretical arguments favouring the initial introduction of new technologies in developing countries, but also the historical record shows examples of attempts at such "technological leapfrogging"—some of which have been successful, some not.

One important success story illustrating what is possible is Brazil's sugarcane-based alcohol programme, which has grown rapidly to provide a renewable alternative to imported oil (Section 3.5.2). Another concerns the development of modern charcoal-based steel-making in Brazil (Box 3.7).

What is particularly noteworthy about the charcoal-based steel example is that in this instance blast furnaces processing hundreds of tonnes of steel per day can produce steel that is competitive with steel derived from conventional coke-based furnaces processing thousands of tonnes per day.

The charcoal-based steel example is of great importance to developing countries generally. The technology is labour-intensive; it is well-matched to the resource base of the many biomass-rich, fossil-fuel poor developing countries; and the scale of its installations often provides more appropriate increments in productive capacity in relation to the size of local markets than giant coke-based facilities. It is a dramatic illustration of how a biomass-based, labour-intensive, relatively small-scale industry can at the same time be modern and highly competitive in world markets.

Leapfrogging should not be regarded as a universal strategy for industrialization. It is appropriate only where a unique set of circumstances and capabilities offer great enough benefits to justify the risk-taking that would be involved. The challenge to planners in developing countries and to the international assistance community is to identify and facilitate the exploitation of promising opportunities for leapfrogging and to support the development of the infrastructure needed for innovation.

Promoting innovation this way could greatly broaden the technological choices available to developing countries generally in meeting their development goals. And while a technological success in one developing country will not necessarily be relevant to all developing countries,

Box 3.7: Charcoal-Based Steel-Making in Brazil

In the industrialized countries coke began to replace charcoal as a reducing agent for iron-making in the middle of the 18th century, as a response to rising charcoal costs at that time.[78] The shift to coke led to much larger scale iron-making blast furnaces than was possible with charcoal, as coke has much greater mechanical strength to resist crushing under the load of the blast furnace charge.

While most of the world's steel industry is now based on coke, 37 percent of Brazilian steel production (i.e., 4.9 million tonnes) was based on charcoal in 1983. This anomaly reflects the scarcity of high quality coking coal in Brazil. Coke-based steel production in Brazil is based on a mix of 80 percent imported coal and 20 percent high ash content domestically produced coking coal. Though charcoal-based steel production is widely viewed as an anachronistic way of making steel, charcoal-based steel is a better quality steel, because charcoal has less impurities than coke. And Brazilian charcoal-based steel is competitive in world markets. This is because the industry is far advanced in relation to the ancient charcoal-based industry abandoned by the now industrialized countries some 200 years ago.

The major cost item in charcoal-based iron-making is charcoal, which accounts for 65 percent of the total pig-iron production cost in Brazil. Charcoal for steel-making is produced mainly from planted forests, primarily of eucalyptus and pine. The total plantation area exceeded 5 million hectares in 1983.

Technical developments relating to charcoal-based steel production are advancing at a rapid rate, as indicated in Table B3. 7.1. There has been roughly a doubling of plantation yield over the last decade, and a further 50 percent increase in yield is expected. Over the last decade charcoal yields from wood have improved about 20 percent, and still another 10 percent increase is expected for the near term. At the same time charcoal requirements for ironmaking have also been declining. The net result of all these improvements is that in the near term future, the land area required for a given level of steel production is expected to fall to just 1/5 of what was required in the 1970's (Table B3.7.1).

Table B3.7.1. Parameters Relating to the Annual Production of One Million Tonnes of Steel Based on Charcoal in Brazil[79]

	Wood Yield on Plantations[a] (tonnes/ha/yr)	Wood-to-Charcoal Conversion Rate (m³/tonne)	Specific Charcoal Consumption (m³/tonne pig iron)	Required Area for Plantations (1000 ha)	Investment Required to Establish Forest (million US$)
1970's	12.5	0.67	3.5	336	201.6
1980's	25	0.80	3.2	128	76.8
Near future	37.5	0.87	2.9	71	42.6

(a) Air dry tonnes (25 percent moisture).

on balance it is likely that innovations generated via leapfrogging will be more widely applicable in the South than innovations coming from the North, since the countries of the South are often more similar to each other in terms of their needs and capabilities than they are to countries of the North.

References

Section 3.1
1.   Ismail-Sabri Abdalla, "Heterogeneity and Differentiation—the End of the Third World?" *Development Dialogue,* No. 2, 1978.
2.   "Population Projections by Country and Region," *World Development Report,* International Bank for Reconstruction and Development, Washington, D.C., 1983.
3.   Paul Streeten, "Approaches to a New International Economic Order," *World Development,* Vol. 10, No. 1, pp. 1-17, 1982.

Section 3.3
4.   ASTRA, "Rural Energy Consumption Patterns—A Field Study," *Biomass,* Vol. 2, pp. 255-280, 1982.
5.   R.P. Moss and W.B. Morgan, "Fuelwood and Rural Energy Production and Supply in the Humid Tropics," Natural Resources and Environment Series, Vol. 4, U.N. University Press, Tokyo.
6.   National Council of Applied Economic Research, "Survey of Rural Energy Consumption in Northern India," 1978.
7.   D.O. Hall, G.W. Barnard and P.A. Moss, *Biomass for Energy in the Developing Countries,* Pergamon Press, Oxford, 1982.

Section 3.4
8.   V. Bravo, G. Gallo Mendoza, J. Legisa, C.E. Juarez and I. Zynierman, "Estudio sobre Requerimientos Futuros de Fuentes No Convencionales de Energia en America Latina," Fundacion Bariloche, Buenos Aires, Argentina, 1979.
9.   M.D. Morris and F.B. Liser, "The PQLI: Measuring Progress in Meeting Human Needs," Communique on Development Issues No. 32, Overseas Development Council, Washington, D.C., 1977.
10. P.F. Palmedo *et al.,* "Energy Needs and Resources in Developing Countries," Brookhaven National Laboratory Report No. BLN 50784, March 1978.
11. Amilcar O. Herrera, Hugo D. Scolnik, Graciela Chichilinisky, Gilberto C. Gallopin, Jorge E. Hardoy, Diana Mosovich, Enrique Oteiza, Gilda L. de Romero Brest, Carlos E. Suarez and Luis Talavera, *Catastrophe or New Society? A Latin American World Model,* International Development Research Centre, Ottawa, 1976.
12. N.H. Ravindranath, H.I. Somasekhar, R. Ramesh, Amala Reddy, K. Venkatram and A.K.N. Reddy, "The Design of a Rural Energy Centre for Pura Village: Part I. Its Present Pattern of Energy Consumption," *Employment Expansion in Indian Agriculture,* ILO, Bangkok, pp. 171, 1979.
13. ASTRA, "Rural Energy Consumption Patterns—A Field Study," Report for the Indian Council for Social Science Research, 1981, published in *Biomass,* Vol. 2., No. 4, 1982.
14. Howard S. Geller, "Cooking in the Ungra Area: Fuel Efficiency, Energy Losses, and Opportunities for Reducing Firewood Consumption," *Biomass,* Vol. 2., pp. 83-101, 1982.
15. V.R. Vanin, G.M.G. Graca and J. Goldemberg, "Padroes de Consumo de Energia Brazil 1970," *Ciencia e Cultura,* Vol. 33, No. 4, p. 477, April 1981.
16. Gladys Sumithra, "An Analysis of 1977-78 National Sample Survey on Energy Consumption (32nd Round)— Karnataka," Perspective Planning Division, Planning Department, Government of Karnataka, July 1981.
17. D.P. Sen Gupta, "Energy Planning for Karnataka State," Report of the Karnataka State Council for Science and Technology, 1977.
18. Government of Karnataka, "Report of the Working Group Constituted for Advance Planning for Utilization of Power in Karnataka," 1982.
19. Hans Binswanger, *The Economics of Tractors in South Asia: An Analytical Review,* Agricultural Development Council, New York, and International Crop Research Institute for Semi-Arid Tropics, Hyderabad.
20. Amartya Sen, *Employment, Technology and Development,* Clarendon Press, Oxford, 1975.
21. K.N. Raj, "Investment in Livestock in Agrarian Economies,; *Indian Economic Review,* IV, 256, 1969.
22. Amulya Kumar N. Reddy and K. Krishna Prasad, "Technological Alternatives and the Indian Energy Crisis," *Economic and Political Weekly,* Vol. XII, pp. 1465-1502, 1977.
23. Stuart Wilson, "Bicycle Technology," *Scientific American,* pp. 81-91, March 1973.
24. N.H. Ravindranath, Amulya Kumar N. Reddy, R. Ravindranath, and H.I. Somasekhar, "Draught Power Availability and Requirement for Double Cropping in Dry-land Agriculture," in *Proceedings of the ASTRA Seminar,* pp. 56-58, September 1983.
25. Jose Goldemberg, "Energy Policies in Brazil," *Economic and Political Weekly,* Vol. XVIII, pp. 305-314, February 26, 1983.

26. Amulya Kumar N. Reddy, "A Strategy for Resolving India's Oil Crisis," *Current Science,* Vol. 50, pp. 50-53, 1981. Reprinted in *Bulletin of the Atomic Scientists,* Vol. 38, pp. 47-49, 1982, and as "India—A Good Life without Oil," in *New Scientist,* Vol. 91, pp. 93-95, 1982.

27. M.R. Yogananda and K.S. Jagadish, "Lime-Pozzolana Cements," in *Proceedings of the ASTRA Seminar,* pp. 83-85, September 1983.

28. U. Shrinivasa and H.S. Mukunda, "Wood Gas Generators for Small Power ($\sim$ 5hp) Requirements," *Current Science,* Vol. 52, No. 23, 1094-1098, 1983.

29. P. Raghavendran, *Economic Times,* p. 5, August 28, 1981.

30. Solar Energy Research Institute, *Generator Gas: The Swedish Experience,* T.B. Reed, ed., 1981.

31. G. Foley and G. Barnard, "Biomass Gasification in Developing Countries," Earthscan Technical Report No. 1, London, 1983.

32. S. Soundaranayagam, "Minihydroelectric Demonstration Unit," 1981-82 Annual Report of the Karnataka State Council for Science and Technology.

33. K.S. Jagadish, S. Saini, A. Rege and B.V. Venkatarama Reddy, "Better Houses with Mud—1," ASTRA, Indian Institute of Science, Bangalore, 1983.

Section 3.5

34. C.R. Prasad, K. Krishna Prasad and Amulya Kumar N. Reddy, "Biogas Plants—Prospects, Problems and Tasks," *Economic and Political Weekly,* Vol. IX, pp. 1347-1364, 1974.

35. David C. Stuckey, "The Integrated Use of Anaerobic Digesters Biogas in Developing Countries—A State-of-the-Art Review," International Reference Center for Waste Disposal, Dubendorf, Switzerland, April 1983.

36. J. Goldemberg, "Global Options for Short Range Alternative Energy Strategies," in *Renewable Energy Prospects,* edited by W. Bach, et al., Pergamon Press, 1980.

37. J. Goldemberg, "Energy Issues and Policies in Brazil," *Annual Review of Energy,* Vol. 7, pp. 139-174, 1982.

38. J. Goldemberg, "Brazil—Energy Options and Current Outlook," *Science,* Vol. 200, p. 158, 1978.

39. Howard S. Geller, "Ethanol Fuel from Sugarcane in Brazil," *Annual Review of Energy,* Vol. 10, pp. 135-164, 1985.

40. "Assessment of Brazil's National Alcohol Programme," Ministry of Industry and Commerce—Secretariat of Industrial Technology, Brasilia, 1981.

41. J.R. Moreira and J. Goldemberg, "Alcohol—Its Use, Energy, and Economics: A Brazilian Outlook," *Resource Management and Optimization,* Volume 1, No. 3, p. 231, 1981.

42. Armand Pereira, "Ethanol, Employment and Development: Lessons from Brazil", International Labour Organization, 1986.

43. "Modelo Energetico Brasileiro—version II," Ministerio de Minas e Energia, Brasilia, May 1981.

44. G.M. Graca and J. Goldemberg, unpublished, 1982.

45. J. Goldemberg, "A Centralised 'Soft' Energy Path," pp. 187-200, in *World Energy Production and Productivity,* edited by Robert A. Bohm, Lillian A. Clinard and Mary R. English, Ballinger Publishing Company, Cambridge, Mass. USA. 1981.

46. J.R. Moreira and J. Goldemberg, "Solving the Energy Problem in Latin America," International Conference on Energy Use and Management-III Berlin, 1981.

47. G.S. Dutt, "Reducing Cooking Energy Use in Rural India", Report PU/CEES 74, Centre for Energy and Environmental Studies, Princeton University, Princeton, New Jersey, USA, 1978.

48. Amulya Kumar N. Reddy, "Rural Fuelwood: Significant Relationships," in *Wood Fuel Surveys,* pp. 29-52, Food and Agriculture Organization, Rome, Italy, 1983.

49. *Wood Heat for Cooking,* K. Krishna Prasad and P. Verhaart, Eds., Indian Academy of Sciences, Bangalore, India, 255 pp., 1983.

50. S.S. Lokras, D.S. Sudhakar Babu, Svati Bhogle, K.S. Jagadish and R. Kumar, "Development of an Improved Three-Pan Cookstove Stove," in *Proceedings of the ASTRA Seminar,* pp. 13-17, September 1983.

51. S. Baldwin, G. Dutt, H.S. Geller and N.H. Ravindranath, "Improved Woodburning Cookstoves: Signs of Success," *Ambio,* Vol. XIV, No. 4-5, pp. 280-287, 1985.

52. Amulya Kumar N. Reddy, "Energy for Development in India," Workshop on End-Use Focussed Global Energy Strategies, Princeton, April 21-29, 1982.

53. Planning Commission, Government of India, "Report of the Working Group on Energy Policy," 1979.

54. "Current Energy Scene in India," Centre for Monitoring Indian Economy, Bombay, India, 1983.

55. Amulya Kumar N. Reddy and B. Sudhakar Reddy, "Energy in a Stratified Society: Case Study of Firewood in Bangalore," *Economic and Political Weekly,* Vol. XVIII, No. 41, pp. 1757-1770, October 8, 1983.

56. ''Report of the National Transport Policy Committee,'' Planning Commission, Government of India, New Delhi, 1980.
57. Centre for Monitoring Indian Economy, ''Current Energy Scene in India,'' Bombay, May 1984.

Section 3.6
58. Food and Agriculture Organization, *Agriculture: Toward 2000,* Rome, 1981.
59. N.S. Scrimshaw and L. Taylor, ''Food,'' *Scientific American,* pp. 74-84, September 1980.
60. R. Hosier, P. O'Keefe, B. Wisner, D. Weiner, and D. Shakow, ''Energy Planning in Developing Countries: Blunt Axe in a Forest of Problems?'' *Ambio,* Vol. 11, No. 4, 1982.
61. Amulya Kumar N. Reddy, ''The Energy and Economic Implications of Agricultural Technologies: An approch based on the Technical Options for the Operations of Crop Production,'' Working Paper WEP 2-22/WP 169, Technology and Employment Branch, ILO, Geneva, Switzerland, 1985.
62. D.O. Kuether and J.B. Duff, ''Energy Requirements for Alternative Rice Production Systems in the Tropics,'' IRRI Research Paper Series, No. 59, IRRI, Manila, Philippines, 1981.
63. M.S. Swaminathan, ''Rice,'' *Scientific American,* Vol. 250, No. 1, pp. 80-93, 1984.
64. Iftikhar Ahmed, ''Green Revolution: With or Without Tractors,'' *Agricultural Mechanization in Asia,* Vol. 6., No. 2, pp. 86-87, 1975.
65. ''Mechanization and Employment: Case Studies from Four Continents,'' ILO, Geneva, Switzerland, 1973.
66. Bina Agarwal, ''Agricultural Modernization and Third World Women: Pointers from the Literature and an Empirical Analysis,'' WEP 10/WP 21 Working Paper, ILO, Geneva, Switzerland, 1981.
67. Food and Agriculture Organization, *The State of Food and Energy—1976,* Rome, 1977.
68. Gerald Leach, ''Energy for Agriculture,'' paper prepared for the Workshop on an End Use Focussed Global Energy Strategy, Princeton Univeristy, April 21-28, 1982.
69. A.Y. Allen, ''Maize Diamonds,'' *Kenya Farmer,* January 1968.
70. B.A. Stout *et al., Energy for World-Wide Agriculture,* April 1977 draft manuscript.
71. Gerald Leach, *Energy and Food Production,* IPC Science and Technology Press, Guilford, England, 1976.
72. D.L. Plucknett and N.J.M. Smith, ''Agricultural Research and Third World Food Production,'' *Science,* pp. 215-220, July 16, 1982.
73. National Academy of Sciences, *Leucaena: Promising Forage and Tree Crop for the Tropics,* Washington D.C. 1977.

Boxes for Chapter 3
74. ''Standard of Living of the Indian People,'' Centre for Monitoring Indian Economy, Bombay (1979).
75. ''Basic Statistics: All India,'' Centre for Monitoring Indian Economy, Bombay, 1982.
76. Amulya Kumar N. Reddy and B. Sudhakar Reddy, ''Energy in a Stratified Society—Case Study of Firewood in Bangalore,'' *Economic and Political Weekly,* Vol. XVIII, No. 41, 1757-1770, October 8, 1983.
77. Anon., ''Report of the Working Group on Energy Policy,'' Section 5.3, page 28, Planning Commission, Government of India, New Delhi, 1979.
78. Ch. K. Hyde *Technological Change and the British Iron Industry: 1700-1870,* Princeton University Press, Princeton, New Jersey, 1977.
79. R. de Almeida (Florestal Acesita, S.A., Belo Horizonte, Brazil), presentation at the *Second End-Use Oriented Global Energy Workshop,* Sao Paulo, Brazil, June 1984.

References for the Figure Captions, Tables and Notes

A.Y. Allen, ''Maize Diamonds,'' *Kenya Farmer,* January 1968.
Th. R. Beck, ''Improvements in Energy Efficiency of Industrial Electrochemical Processes'', Report prepared for the Office of Electrochemical Project Management, Argonne National Laboratory, ANL/OEPM-77-2, January, 1977.
BEN, 1985.
V. Bravo, G. Gallo Mendoza, J. Legisa, C.E. Juarez, and I. Zynierman, *Estudio sobre Requerimientos Futuros de Fuentes No Convencionales de Energia en America Latina* (Project RLA/74/030, Foundation Bariloche), Report to United Nations Development Programme, Buenos Aires, Argentina, 1979.
CNP/PETROBAS, *Suma Estatistica,* Fev/1985.
The Economist, June 9, 1984.
Energy Information Administration, U.S. department of Energy, *1980 Annual Report to Congress, Vol. Two: Data,* DOE/EIA-0173(80)/2, April 13, 1981.

Food and Agriculture Organizatioan (FAO), *1977 FAO Production Yearbook 31,* (FAO Statistics Series), Rome, 1978.

Food and Agriculture Organization (FAO), *1978 FAO Production Yearbook 32,* Rome, 1979.

Food and Agriculture Organization (FAO), *FAO Production Yearbook 184/185, Rome, 1984.*

Food and Agriculture Organization, *Agriculture: Toward 2000,* Rome, 1981.

Food and Agriculture Organization (FAO), "Energy for World Agriculture," (Agriculture Series: 7), B.A. Stout, ed., Rome, 1979.

Howard S. Geller, "Ethanol Fuel from Sugar Cane in Brazil," *Annual Review of Energy,* Vol. 10, pp. 135-164, 1985.

J. Goldemberg, "Energy Issues and Policies in Brazil," *Ann. Rev. of Energy,* Vol. 7, pp. 139-174, 1982.

J. Goldemberg, *et al.,* "Country Study—Brazil, A Study on End-Use Energy Strategy," presented at the Global Workshop on End-Use Oriented Energy Strategy, Sao Paulo, Brazil, June 1984.

J. Goldemberg, T.B. Johansson, A.K.N. Reddy, and R.H. Williams, "Basic Needs and Much More with One Kilowatt per Capita," *Ambio* Vol. 14, No. 4-5, pp. 190-200, 1985.

Government of the State of Karnataka, India, "Report of the Working Group Constituted for Advance Planning for Utilization of Power in Karnataka," 1982.

D.O. Hall, G.W. Barnard, and P.A. Moss, *Biomass for Energy in the Developing Countries,* Pergamon Press, Oxford, 1982.

A.O. Herrera, H.D. Scolnik, G. Chichilinisky, G.C. Gallopin, J.E. Hardoy, D. Mosovich, E. Oteiza, G.L. de Romero Brest, C.E. Suarez, and L. Talavera, *Catastrophe or New Society? A Latin American World Model,* A Report of the Fundacion Bariloche, Buenos Aires, International Development Research Centre, Ottawa, 1976.

F. von Hippel, "US Transportation Energy Demand," PU/CEES Report No. 111, Center for Energy and Environmental Studies, Princeton University, Princeton, N.J., 1981.

R. Hosier, P. O'Keefe, B. Wisner, D. Weiner, and D. Shakow "Energy Planning in Developing Countries: Blunt Axe in a Forest of Problems?" *Ambio,* Vol. 11, No. 4, 1982.

International Monetary Fund, *International Financial Statistics Yearbook 1985,* Vol. 38, Washington D.C., 1985.

A.M. Khan and A. Holzl, *Evolution of Future Energy Demands till 2030 in Different World Regions: An Assessment Made for the Two IIASA Scenarios,* International Institute for Applied Systems Analysis Report RR-82-14, Laxenburg, Austria, April, 1982.

A. Makhijani and A. Poole, *Energy and Agriculture in the Third World,* Ballinger Publ. Co., Cambridge, Mass., 1975.

M.J. Mwandosya and M.L.P. Luhanga, "Energy Use Patterns in Tanzania," *Ambio,* Vol. 14, pp 237-241, 1985.

The New York Times, Dec. 25, 1983.

OTA, Office of Technological Assessment, United States Congress, *Energy from Biological Processes,* Washington, 1980.

P.F. Palmedo *et al., Energy Needs and Resources in Developing Countries,* Brookhaven National Laboratory Report No. BNL 50784, March, 1978.

Planning Commission, Government of India, *National Transport Policy Committee Report,* New Delhi, 1980.

N.H. Ravindranath, *et al.,* "An Indian Village Agricultural Ecosystem—A Case Study of Ungra Village," *Biomass,* Vol. 1, 1980.

A.K.N. Reddy, *Energy for Development in India,* Workshop on End-Use Focussed Global Energy Strategies, Center for Energy and Environmental Studies, Princeton University, Princeton, N.J., April 21-29, 1982.

A.K.N. Reddy, "An End-Use Methodology for Development-Oriented Energy Planning in Developing Countries, with India as a Case Study," PU/CEES Report No. 181, Center for Energy and Environmental Studies, Princeton University, Princeton, N.J., 1985a.

Amulya Kumar N. Reddy, "The Energy and Economic Implications of Agricultural Technologies: An Approach based on the Technical Options for the Operations of Crop Production," Working Paper WEP2-22/WP 169, Technology Branch, ILO, Geneva, Switzerland, 1985b.

A.K.N. Reddy and D.K. Subramanian, "The Design of Rural Energy Centres," *Proc. Ind. Acad. Sci.,* Vol. 2, part 3, 1979.

Republica Federative do Brasil Ministerio das Minas e Energia, *Balanco Energetico National,* Brasilia, Brazil, 1983.

R. Revelle, "Energy Use in Rural India, *Science,* Vol. 192, pp. 969-975, June 4, 1976.

Shell Briefing Service, "Energy in the Developing Countries", January, 1980.

K.C. Shukla and J.R. Hurley, "Development of an Efficient, Low NO_x Domestic Gas Range Cook Top," Report prepared by Thermoelectron Corporation, Waltham, Massachusetts, for the Gas Research Institute, July, 1983.

Solar Energy Research Institute, *A. New Prosperity: Building a Sustainable Energy Future—the SERI Solar/Conservation Study,* Brickhouse, Andover, Massachusetts, USA, 1981.

P. Steen, T.B. Johansson, R. Fredriksson, and E. Bogren, *Energy—for What and How Much?* Liber Forlag, Stockholm, 1981 (in Swedish). Summarized in T.B. Johnsson, P. Steen, E. Bogren, and R. Fredriksson, *Science, 219* (1983), p. 355.

G. Sumithra, "An Analysis of 1977-78 National Sample Survey on Energy Consumption (32nd round)—Karnataka, "Perspective Planning Division, Planning Department, Government of Karnataka, July, 1981.

United Nations, *1979/80 Statistical Yearbook,* United Nations, New York, 1981.

United Nations, *Recent Trends and Future Prospects,* ECOSOC Committee on Natural Resources, Report of the Secretary General, (Tables 8 and 9, pp. 31-32), New York, April 1977.

D.A. Waitzman *et al.,* "Fertilizer from Coal," Paper prepared by the Division of Chemical Development, Tennessee Valley Authority, Muscle Shoals, Alabama, and presented at the Faculty Institute on Coal Production Technology and Utilization, Oak Ridge Associated Universities, Oak Ridge, Tennessee, July 21—August 11, 1978.

4. Global Energy Demand and Supply

4.1 The Evolutionary End-Use Approach to Global Energy Demand/Supply

In this chapter, we explore a long-run global energy demand/supply strategy which is based on the end-use approach and is compatible with the solutions of other important national, regional, and global problems.

Most other global energy studies attempt to forecast what future energy demand will be on the basis of extrapolating historical trends and to match projected demand levels with a mix of energy supplies that is determined to a large extent by relatively narrow engineering criteria (Section 1.2). This conventional approach to the energy problem usually leads to energy strategies which conflict with the solutions of other important social problems (Section 1.3).

The analysis of the previous chapters, however, provides ground for optimism that long-term energy strategies can be developed that are consistent with the solutions of these other problems and that *the future of energy demand/supply is more a matter of choice than of prediction.*

This optimism stems from our findings that radical improvements in the living standards of the poor majorities of developing countries and substantial further improvements in amenities in industrialized countries are possible at much lower levels of energy demand than has been projected in conventional energy supply-oriented analyses. These lower levels of demand make possible an energy supply mix in which reliance on the more troublesome options can be avoided or kept to low levels. The lower the demand level, the greater the number of supply options available to meet the demand. At high demand levels, as in the WEC and IIASA projections for the period beyond the turn of the century, essentially all available supply options must be pushed to their limits.

4.2 A Base Case Energy Demand/Supply Scenario

A Base Case global energy demand/supply scenario will be described here. It illustrates that the prospects for identifying a long-range energy future which is both compatible with the achievement of a sustainable world and technically and economically feasible are hopeful.

The scenario is not a forecast but a normatively constructed energy future which could evolve with appropriate public policies. The rationale underlying the construction of the scenario is summarized schematically in Figure 4.1, which highlights explicitly the normative aspects of the analysis.

The first step is to understand present and future needs for energy services—requirements for cooking, lighting, domestic hot water, passenger and freight transport, basic industrial materials, etc. Fortunately, most energy use is concentrated in just a few activities in each energy-using sector (residential, commercial, transportation, and industry), so that the list of important end-use activities that must be scrutinized is manageable in most instances. Estimated levels of energy services associated with alternative economic development paths can be based on

	DEMAND		SUPPLY
INDUSTRIALIZED COUNTRIES	Energy Use Per Capita	Future Activity Levels Based on Projections of Trends	Normative Supply Choices. Within Constraint of Technoeconomic Feasibility Avoiding Over Dependence on Oil, Fossil Fuels, Nuclear Power
		Future Energy Intensities Based on Normative Choice of Best/Advanced Technologies Which Are Technoeconomically Feasible	
	Population	U.N. Projection	
DEVELOPING COUNTRIES	Energy Use Per Capita	Future Activity Levels Based on Normative Assumption of Large Increase in Amenities, Ranging Up to W. European Level of 1970's	Normative Supply Choices, Within Constraint of Tech., Econ. Feasibility Modernize Bioenergy, Promote Self-Reliance
		Future Energy Intensities Based on Normative Choice of Improved Efficiencies, Ranging up to Efficiencies of Best/Advanced Technologies	
	Population	U.N. Projection	

Figure 4.1 Schematic summary of the methodology used to develop the Base Case global energy demand/supply scenario.

extrapolations of historical trends, taking into account ongoing shifts and saturation effects, as in the cases of the energy scenarios presented for Sweden (Section 2.6.1) and the United States (Section 2.6.2), or departures from historical trends can be specified to conform to feasible societal goals, e.g., shifts from historical trends in production and consumption may be desirable in particular developing countries to ensure that basic human needs are satisfied.

With energy service levels specified, the next task is to obtain estimates of the energy intensities for these service activities, that is, the energy required per unit of service provided, e.g., kJ of kerosene per passenger-kilometer of air travel. Both potential improvements in energy efficiency and the use of alternative energy carriers are taken into account, where they are technically feasible and cost-effective.

With these assumptions, future aggregate demand estimates are obtained by summing (over all activities) the products of the activity levels for energy services and the corresponding energy intensities. The demand levels so obtained can then be matched to estimated available energy supplies. Ideally, a comprehensive perspective on future energy demand/supply should be constructed "from the bottom up" in order to reflect differences in resources, climatic and cultural conditions from one country to another.

It is not yet possible to put together a global energy scenario according to this idealized prescription, however, because detailed country studies and country strategies have been developed through the end-use approach for only a few countries. It is nonetheless feasible to formulate a preliminary global energy perspective based on extrapolating to the global situation what is known for a few countries and thereby get an indication of the prospects in achieving the long-term goals for global energy planning described in Chapter 1. This preliminary perspective can also be helpful in formulating long-term energy strategies for countries and regions, to help ensure that these strategies are compatible with global objectives.

Our scenario analysis focuses on the year 2020. This date is sufficiently far in the future to allow ample time to carry out programmes aimed at satisfying basic human needs in developing countries and to achieve considerable improvements in living standards beyond that minimum goal. There is also ample time for the widespread adoption of the kinds of energy-sufficient, end-use technologies identified in this book. Moreover, the world should be well into the transition to the Post-Petroleum Era by 2020. And, if conventional energy strategies are pursued between now and then, the carbon dioxide problem associated with continued reliance on fossil fuels will have reached a critical stage by that time. So, too, will have the nuclear weapons proliferation problem associated with large scale expansion of nuclear power. Yet 2020 is sufficiently close that it has an important bearing on long-range energy planning today.

This analysis is a first step toward determining whether it is feasible to put together an energy supply mix compatible with the achievement of a sustainable world at the projected energy demand levels. But because the energy demand/supply balances arrived at here are highly aggregated at the global level, the analysis necessarily neglects regional variations in energy demand and supply availability, and hence, sheds no light on the problems of how particular energy resources concentrated in particular locations might be made more widely available geographically. Detailed country and regional studies are needed to deal with such problems.

Ultimately, of course, the long-term energy strategies that will be adopted will be determined in the political process. The purpose of the present analysis is to provide decisionmakers in the political process with a more informed basis for making their decisions, especially in terms of the choices which are available to them. Past energy analyses have, we think, given decisionmakers a too narrow view of the energy landscape ahead. Our analysis clearly indicates that decision-makers definitely do *not* have to accept as a given, before formulating long-term energy strategies, that they have no choice but to push conventional energy supply sources to the limit. In fact, to do so, will cause them and the rest of the residents of earth no end of trouble.

4.2.1 *Energy Demand*

On the demand side, the Base Case global energy scenario is obtained by extrapolating to industrialized countries in aggregate the results of the analyses for Sweden and the US (Section 2.6) and to developing countries in aggregate the 1 kW per capita analysis for a hypothetical developing country (Section 3.4.1).

4.2.1.1 *Future Per Capita Energy Demand in Industrialized Countries:* The major responsibility for the global risks associated with conventional energy supplies described in Section 1.3 lies with the industrialized countries, on account of their heavy appetite for energy. With only ¼ of the world's population, these countries account for ⅔ of world energy use.

Fortunately, both ongoing structural changes and major opportunities for energy efficiency improvements suggest that a large reduction in the energy intensity of economic activity is feasible in the industrialized world.

One important factor bearing on future energy demand is the phenomenon that the industrialized world is making a transition to the Post-Industrial Era, which is characterized by growth in industrial activities such as finishing and fabricating, which are inherently less energy-intensive than the processing of basic materials, the industrial activities which dominated growth in the Industrial Era (Section 2.4). Super-imposed on this changing demand structure is the potential for radically improving the efficiency of energy use (Section 2.5.3).

A Base Case energy demand scenario for the industrialized world which takes into account such factors will now be presented, extrapolating from the detailed results obtained for Sweden and the United States.

The results of the Swedish case study were presented in Section 2.6.1. While per capita GDP in Sweden is comparable to that in the United States, final energy use per capita is only about 3/5 as much—averaging 5.4 kW per capita in 1975. Indeed, Sweden is generally looked to as a model "energy-conserving" society. Yet, the analysis presented for Sweden shows major opportunities for energy savings. Two different levels of technological improvement were considered: *best available technology* (that which is or is judged to be economic *today*) and *advanced technology* (for which some success is assumed in ongoing R&D efforts to improve energy efficiency, and for which costs are in the range of interest). Our analysis indicates that by using best available technology, Sweden's per capita final energy use would be reduced to about 3.5 kW if its consumption of goods and services increases 50 percent, while it would be reduced to 4.2 kW if goods and services consumption increases 100 percent. Assuming the use of advanced technology, which may be more appropriate when looking as far ahead as 2020, Sweden's per capita consumption declines to 2.7 kW or 3.3 kW, depending on whether the consumption of goods and services increases 50 or 100 percent.

The U.S. country study, the details of which are presented in Section 2.6.2, is also based on the wide use of cost-effective, energy-using technologies. By the year 2020, it would be feasible to reduce the United States' per capita final energy use to 4.3 kW from 9.0 kW in 1980, with a 50 percent increase in the per capita consumption of goods and services, or to 4.6 kW with a 100 percent increase in goods and services per capita consumption. Because U.S. energy use is presently so high, the projected *reduction* in aggregate U.S. energy use between 1980 and 2020 is very large—some 0.71 to 0.87 TW—equal to 25-30 percent of all energy use in developing countries in 1980.

In light of these results for Sweden and the U.S. and the broad applicability of the technologies involved in these analyses, it is reasonable to expect that a 50 percent average reduction in per capita final energy use could be achieved in all industrialized countries between 1980 and 2020 through energy efficiency improvements, along with continuing and significant improvements in standard of living, reducing per capita final energy use from 4.9 kW to 2.5 kW. This is comparable to what could be realized in Sweden with "advanced technology" and a 50 percent increase in the per capita consumption of goods and services over the 1975 level.* Since Sweden's 1975 per capita GDP was some 75 percent higher than the average for all industrialized countries in that year (Note 4.2.1-A), an energy demand level of 2.5 kW could be associated with an increase in the average per capita GDP for all industrialized countries which is considerably greater than 50 percent.

Although this level of future per capita energy demand in industrialized countries is radically different from that envisaged in previous major global energy studies, this scenario, on the basis of what has emerged from the U.S. and Sweden case studies, appears to be technically feasible and economical—i.e., no more costly and perhaps less costly than an equivalent amount of the energy supply expansion.

While this scenario appears to be feasible, it is very unlikely that it would evolve unless the obstacles to the exploitation of energy efficiency opportunities are removed. The needed public policy measures include, among others, elimination of subsidies for energy supply expansion, marginal cost pricing of energy, and shifting energy utilities from being merely purveyors of energy supplies to providers of energy services (Chapter 5).

A major uncertainty concerning this scenario is the extent to which opportunities for energy efficiency improvements, identified mainly in OECD countries, would be seized upon and pursued in CMEA countries, which to date have made little headway in adopting energy efficiency improvements (Figure 2.4).

4.2.1.2 *Future Per Capita Energy Demand in Developing Countries:* The energy demand situation is completely different for the three-quarters of the world's population who live in developing countries and account for only ⅓ of world energy use.

At present, per capita energy use in developing countries averages about 0.9 kW per capita, of which some 0.4 kW per capita is non-commercial energy consumed largely by the two-thirds of the population who live in rural areas, isolated for the most part from market economies.

The challenge for energy planning in developing countries is to assure that the needed energy services are available—for satisfying the basic human needs of the poor; for providing food for a growing population; and for generally bringing about a much higher standard of living than at present for the population as a whole—in environmentally sound and sustainable ways that promote self-reliance and peace.

---

* The reduction in per capita energy demand could vary from country to country and could be quite small for low-income industrialized countries, without affecting very much the overall average level of future energy demand in industrialized countries. To illustrate this point suppose that for the six industrialized countries with the lowest per capita energy use (Greece, Ireland, Portugal, Spain, Albania, and Hungary) average per capita energy use in 2020 were at the 1980 level (2.2 kW). Then in 2020 the average level of per capita energy use for all industrialized countries would be 2.58 kW instead of 2.5 kW.

How much energy is needed in developing countries in the decades ahead to meet these goals? At this time, a definitive estimate cannot be made of the energy required to meet development goals over the next 40 years. There are two major reasons for this. The present patterns of energy end-use in developing countries are not nearly so well understood as they are in industrialized countries—both because available data are not very comprehensive and because to date much less end-use analysis has been done for developing countries than for industrialized countries. And secondly, it is inherently harder to make meaningful long-range projections of future energy demand for rapidly industrializing countries than for mature industrialized countries.

Due to these problems, we have pursued a different approach for estimating long-term energy requirements for developing countries. In Section 3.4.1, a scenario was presented for an hypothetical future developing country with amenities comparable to those in the WE/JANZ region in the mid- to late-1970s but employing energy efficient end-use technologies for major energy-intensive activities (either the "best available" or "advanced" end-use technologies). The scenario showed that this high level of amenities—corresponding to roughly a tenfold increase in per capita GDP—could be achieved with essentially the same total level of final energy use per capita as at present, although energy would be supplied primarily through high quality energy carriers (electricity, gases, liquids, and high quality solid fuels).

Attaining the high levels of average performance in end-use technology characteristic of this scenario, could in principle be achieved more quickly in developing countries than in already industrialized countries. In the developing world there is such a large potential demand for new energy-using capital stock that the rate of introducing new efficient technology is not limited by the turnover rate of the existing stock, as would be the case in industrialized countries.

The set of activity levels assumed in this scenario should not be construed as targets for developing countries to be achieved by 2020 or any other date. The appropriate mix and levels of activities for, say 2020, may well have to be different to be consistent with overall development goals. But this analysis suggests that it is possible to provide a standard of living in developing countries anywhere along a continuum from the present one up to a level of amenities typical of Western Europe today without departing significantly from the average per capita energy use level in developing countries today, depending on the level of energy efficiency that is emphasized.

That development goals could be achieved with little change in the overall per capita level of energy use should not obscure the challenge of bringing such a future about, however. As in the case of development generally, large amounts of capital would be required to bring about shifts to modern energy carriers and to efficient end-use technology. But while improved end-use devices generally cost more than conventional units, for a wide range of plausible sets of activity levels and for a wide range of end-use technologies it would be less costly to provide energy services using the more efficient end-use technologies than to provide the same energy services with conventional, less-efficient end-use technologies and increased energy supplies. This was shown for various illustrative industrialized country examples in Section 2.5.3 and appears to hold for a wide range of circumstances in developing countries as well (Note 4.2.1-B). The economic resources freed up from pursuing improvements in end-use technologies could be spent for other development purposes.

On the basis of these considerations, 1.0 kW is chosen here as the per capita energy use level in developing countries for the Base Case global energy scenario in 2020—a level of energy use which, with emphasis on energy efficiency improvement and modern energy carriers, would be adequate both to ensure that basic human needs are satisfied and to allow for considerable further improvements in living standards.

4.2.1.3 *Total Global Energy Demand:* The results presented above for industrialized and developing countries can be combined with a Low Variant U.N. population projection* to obtain a global energy demand projection, as summarized schematically in Figure 4.2. Assuming a 50 percent reduction in the per capita final energy use of industrialized countries, 1980-2020, total final energy use by industrialized countries would be reduced from 5.5 TW in 1980 to 2.9 TW in 2020 (Note 4.2.1-C). Assuming for developing countries a per capita final energy use level of 1.0 kW, total final energy use there, 1980-2020, would nearly double (Note 4.2.1-C).

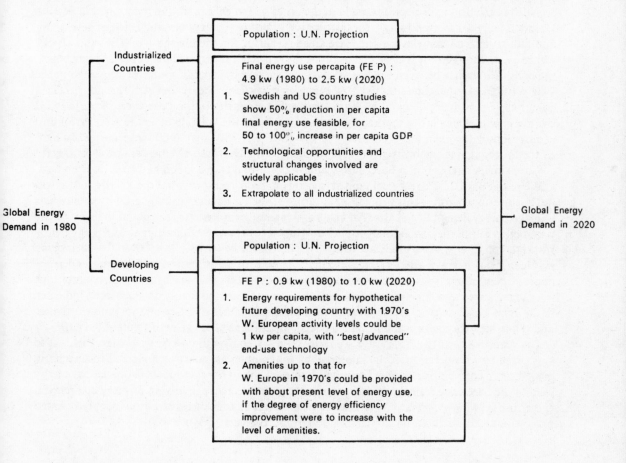

Figure 4.2 Schematic summary of the assumptions underlying the Base Case global energy demand scenario.

---

* Whereas population is usually treated as an exogenous variable in conventional energy projections, it is reasonable to expect that population growth would be slower with than without a basic human needs policy, because the economics of large families tend to be favourable to the poor.[1] Since a quantification of the impact of a basic human needs policy on population has not been carried out, this effect is reflected here by the adoption of the U.N.'s 1980 Low Variant Projection of world population—7.0 billion people by 2020 (vs. 7.8 billion for the Medium Variant).[2] The significance of higher population growth rates is explored through sensitivity analysis in Section 4.3).

Overall global final energy demand would be about the same in 2020 as in 1980, and the developing country share would increase from ⅓ to ⅔ of the total (Note 4.2.1-C). While the global level of energy demand would not change much between 1980 and 2000 in the Base Case scenario, there would be a marked shift to higher quality energy carriers. A continuation of the ongoing "electrification" of the global energy economy is assumed, so that the electricity share of final demand would increase from 10 percent in 1980 to 18 percent in 2020 (Note 4.2.1-C). This increase in the share of electricity in final demand implies a doubling of electricity production.

Figure 4.3 (top) shows the global energy demand level for 2020 projected in the Base Case scenario in relation to projections made in the other global energy studies mentioned in Section 1.2, and Figure 4.3 (bottom) shows the distribution of primary energy demand between developing and industrialized countries for the Base Case scenario, for the alternative WEC and IIASA scenarios, and for 1980. It is clear that as far as the projected energy demand levels are concerned, the major difference between the Base Case scenario and the energy scenarios presented in the 1982 WEC energy study and the 1981 IIASA energy study lies in the treatment of the industrialized countries. The primary demand level for developing countries in the Base Case scenario is 1.2 kW per capita, which is the same as the average value for the WEC high and low scenarios and is only slightly less than the average of 1.5 kW for the IIASA high and low scenarios. In light of the much greater emphasis given here to energy efficiency improvements, however, the 1.2 kW of primary energy in the Base Case scenario corresponds to a much higher living standard.

While the Base Case energy demand scenario is not a forecast of what the energy future will be, it is useful to express this projection in terms of the parameters usually used by economists to make energy demand forecasts. An analysis expressing future energy demand in terms of the parameters of the energy-economy model developed for the U.S. Department of Energy by the Institute for Energy Analysis/Oak Ridge Associated Universities (IEA/ORAU)—an income elasticity,* an average energy price elasticity,** and a nonprice induced energy efficiency improvement rate***—shows that, with public policy efforts that are neither overly ambitious nor radical aimed at redirecting the course of the energy future the energy demand levels indicated in the Base Case scenario are consistent with plausible values of price and income elasticities and plausible expectations about future energy prices and GDP growth (Note 4.2.1-D).

The plausibility of the Base Case scenario vis a vis conventional energy projections is also supported by consideration of the implications for commercial energy demand of this Scenario in relation to conventional energy forecasts. Table 4.1 shows the evolution of commercial energy in both the Base Case scenario and an alternative global scenario involving primary commercial energy use growing at a rate of 2 percent per year—a rate characteristic of various supply-oriented projections such as those of the IIASA and WEC forecasts (Figure 4.3).

---

* The income elasticity is the percentage increase in energy demand associated with each percent increase in GDP, for constant energy prices and no non-price induced efficiency improvement.
** The energy price elasticity is the percent increase in energy demand associated with each percent increase in energy price, for constant GDP and no non-price induced energy efficiency improvement.
*** The non-price induced energy efficiency improvement rate is the annual rate of decline in the energy intensity of the economy due to non-price induced technological innovations. It is used in the IEA/ORAU model to take into account the phenomenon that in the industrial sector energy efficiency improvements have taken place even in times of declining energy prices. In the present analysis, this parameter can also be used as a measure of the public policy effort required to bring about energy efficiency improvements, beyond those that would be induced by energy price increases.

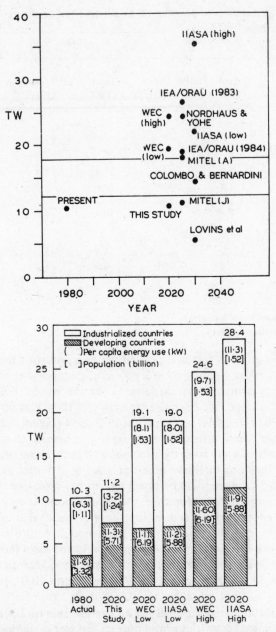

Figure 4.3 Alternative projections of global primary energy demand, in TW (TW-years per year).
The Figure on the top shows the the Base Case primary energy demand scenario of this study in relation to the actual rate of primary energy use in 1980 and recent projections of global energy use reviewed in J. Goldemberg, T.B. Johanson, A.K.N. Reddy, and R.H. Williams, 1985. See the caption to Figure 1.7 for the references to the other global energy studies indicated here.

The figure at the bottom shows projections of global primary energy use by region for 2020 for the IIASA study, the 1982 WEC study, and the Base Case scenario of this study, in relation to the actual rate of primary energy use in 1980.

Table 4.1. Comparison of Scenarios for Commercial Primary Energy Use

	1980		2020		Average Growth Rates, 1980-2020 (% Per Year)	
	Total (TW)	Per Capita (kW)	Total (TW)	Per Capita (kW)	Total	Per Capita
Scenario with Global Energy Growth @ 2 Percent per Year and Constant Per Capita Energy Use in Industrial Countries						
Industrialized Countries	7.0	6.3	7.8	6.3	0.3	0.0
Developing Countries	1.8	0.54	11.6	2.0	4.8	3.3
World	8.8[a]	2.0	19.4	2.8	2.0	0.8
Base Case Scenario						
Industrialized Countries	7.0	6.3	4.3	3.5	−1.2	−1.5
Developing Countries	1.8	0.54	6.9	1.2	3.4	2.0
World	8.8[a]	2.0	11.2[b]	1.6	0.6	−0.5

(a) Excludes 1.5 TW of non-commercial bioenergy.
(b) Includes 1.6 TW of commercial bioenergy.

If global energy were to grow as fast as 2 percent per year, most of the growth would take place in the developing countries, because there is powerful evidence that even under "business-as-usual" conditions energy demand in industrialized countries would probably grow no faster than population or would decline. We have already seen that there was no net increase in total primary energy use in OECD countries in the period 1973-1984 (Figure 2.2)—corresponding to a 6 percent reduction in per capita primary energy use in this period. In addition, the ongoing shift in industrialized countries away from the processing of basic materials toward services and toward fabrication- and finishing-intensive goods-producing activities means that the future economies of industrialized countries will be much less energy-intensive than at present, even if energy efficiency is not emphasized. Moreover, the industrialized energy economies have not yet fully adjusted to the energy price increases that have already taken place, so that even if there are no more energy price increases or only very modest future price increases, these economies will become more energy-efficient. A quantification of these factors using the above-described IEA/ORNL energy-economy model shows that under a wide range of plausible business-as-usual conditions and a 50 to 100 percent growth in per capita GDP, per capita energy use in industrialized countries will not be higher in 2020 than in 1980 (Note 4.2.1-E).

If commercial energy demand grows globally at an average rate of 2 percent per year, while energy demand in industrialized countries grows only as fast as population, energy demand in developing countries would have to grow at an average rate of 4.8 percent per year and per capita commercial energy use at an average rate of 3.0 percent per year (Table 4.1).

It is unlikely that adequate capital would be available to expand conventional energy supplies at such a rate. As pointed out in Section 1.3.3, the World Bank estimated that increasing per capita commercial energy use in developing countries at an average rate of 2.5 percent per year, 1980-1995, would require capital investments in energy supply averaging $130 billion per year, 1982-1992. To meet this target the Bank estimated that real foreign exchange allocations to energy supply expansion would have to grow in real terms at 15 percent per year in this period—a wholly unrealistic target.

Such considerations show that 2 percent average growth in global energy use is wholly implausible. The energy supply challenges for the Base Case scenario, in comparison would be far less formidable. While commercial energy use per capita would still grow in developing countries at a non-trivial 2 percent annual rate (Table 4.1), the dramatic reduction of energy demand in industrialized countries in this period would tend to greatly ease upward pressure on energy prices, making this expansion more affordable than it would be under business-as-usual conditions. In addition, emphasis on energy carriers such as biomass-derived producer gas for cooking and decentralized power generation can be expected to result in lower energy supply expansion costs than would be the case where only centralized energy sources are pursued.

4.2.2 *Energy Supply*

In the discussion in Chapter 1 of the global risks associated with energy production and use, three risks were singled out as being especially troublesome: the threat of climatic change associated with the atmospheric build up of CO_2 due to the burning of fossil fuels; the global security risk arising from heavy dependence of the industrialized market economies on Middle-Eastern oil; and the risk of nuclear weapons proliferation associated with the widespread use of nuclear power. These risks are unique among the wide range of risks associated with energy production and use in that they are "irreducible," i.e., they defy technical solution. Society must either "learn to live with" these risks or move away from dependence on oil, fossil fuels, and nuclear energy sources. The conventional wisdom regarding these problems is that society has no choice but to accommodate to these risks because all major energy sources must be expanded to meet society's needs. But this view does not take into account the great potential for decoupling energy and economic growth.

4.2.2.1 *Atmospheric Carbon Dioxide and the Burning of Fossil Fuels:* The prospect that the atmospheric carbon dioxide level would double in the latter half of the next century, triggering major changes in the global climate in this period, if fossil fuel use expands as projected in the WEC and IIASA studies, has led to a rapidly growing literature on accommodation strategies for dealing with the carbon dioxide problem. For example, one of the major recommendations of the 1983 U.S. Environmental Protection Agency report on the carbon dioxide problem is to accelerate and expand research on improving our ability to adapt to a warmer climate.[3]

An alternative response to the CO_2 problem is to pursue an end-use oriented energy strategy which results in a level of fossil fuel use that is far below what has been forecast.

To illustrate what might be achieved by making only relatively minor changes in the present energy supply mix (but major changes in the energy supply "requirements" which have been forecasted) it is assumed for the Base Case scenario that the total fossil fuel use is the same in 2020 as in 1980, and there is a modest shift away from dependence on coal.

The mix of fossil fuels use is important first because CO_2 emissions differ from one fossil fuel to another—with coal being more of a poroblem than either natural gas or oil. Per unit of fossil fuel energy released, coal generates 1.2 times as much as oil and 1.8 times as much as natural gas (Note 4.2.2-A). Moreover, the world's remaining coal resourcers are vast—much greater than the remaining gas and oil resources. Using up all the remaining ultimately recoverable oil and gas resources (with no further use of coal) would lead to an atmospheric CO_2 level only 1.5 times the pre-industrial level (440 ppm), but using up half of the coal left in the ground would increase the CO_2 level 4 times over the pre-industrial level (Note 4.2.2-A).

To relate future levels of fossil fuel use to constraints on the ultimate level of CO_2 in the atmosphere, it is assumed that: (1) half of the released carbon dioxide remains in the atmosphere; (2) all estimated ultimately recoverable oil and gas resources are eventually used up, so that concerns about carbon dioxide are reflected as constraints on coal production; and (3) coal production falls exponentially over time. Under these assumptions, the rate of decline of coal use in the future depends on the CO_2 ceiling level (Note 4.2.2-B and Figure 4.4). If the ultimte ceiling were as low as 1.5 times the pre-industrial level, coal would have to be phased out very rapidly, falling to one-half the present level before the turn of the century, but this is not a practical target. On the other hand, if the allowable level were 2.0 times the preindustrial level, then coal production could be phased out extremely slow, falling to half the 1980 level only after 230 years.

While major climatic changes are expected if the atmospheric CO_2 level doubles, it is impossible to say what an "acceptable" CO_2 ceiling should be. For the purposes of the present analysis, a ceiling of 1.7 times the pre-industrial level (490 ppm) is arbitrarily selected for the Base Case scenario. This means that coal use would have to fall to half its present level only after 100 years, and coal use would decline 20 percent between 1980 and 2020.

In the Base Case scenario, the reduction in coal use from the present level is relatively modest,

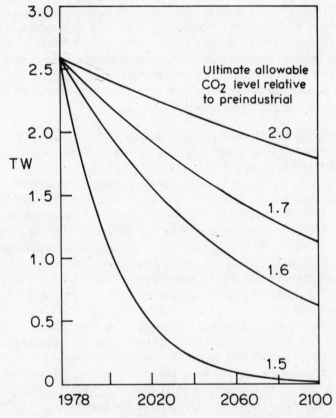

Figure 4.4 Constrained annual coal production as a function of the allowable ultimate atmospheric carbon dioxide level, assuming that the carbon dioxide constraint is reflected entirely as a constraint on the use of coal, along the lines discussed in the text.

but coal use would be only 20 to 40 percent as high in 2020 as in the WEC and IIASA scenarios (Figure 4.5). And emissions of CO_2 in 2020 from the burning of all fossil fuels would be only 40 to 60 percent as large as in the IIASA and WEC scenarios. In the Base Case scenario, CO_2 level in 2020 would be about 1.3 times the pre-industrial level (380 ppm).

Despite the slow rate at which coal would be phased out under this scenario, the notion of controlling the atmospheric carbon dioxide buildup by restricting coal use poses major economic challenges. The assumed CO_2 ceiling would require limiting coal use in the long run to about ¼ of the amount of coal available at prices less than ½ the world oil price in 1982 (Note 4.2.2-C).

But coal is a dirty fuel, the use of which in environmentally acceptable ways requires considerable capital investment, hence its use is not nearly so competitive as a comparison of fuel prices would indicate. The economic challenge of shifting from coal could be eased if oil and gas prices did not rise too much in the coming decades, as may be feasible in an energy-efficient world (Section 4.2.2.2). In such a world, it may prove to be easier during the transition to the "Post Petroleum Era" to make do with less coal and rely relatively more on natural gas, which is clean and convenient to use.

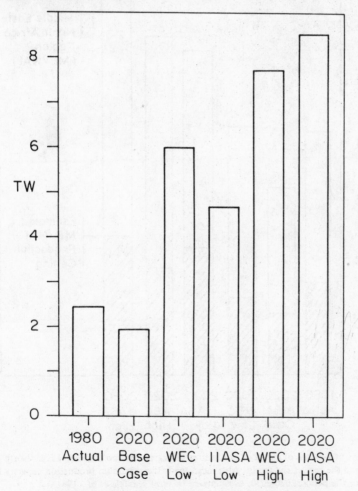

Figure 4.5 Alternative projections of global coal production.

Moreover, the fact that almost 90 percent of the coal left in the ground is concentrated in just three countries—the U.S.S.R., the U.S. and China (Note 4.2.2-A)—makes the prospect of control easier than would be the case if the resources were distributed widely throughout the world and agreement among many countries were required.

Even with the reduced coal use in the Base Case scenario, coal would remain a major energy resource for a long time to come; the cumulative allowable production is equivalent to a 150-year supply at the present rate of usage.

4.2.2.2 *The World Oil Problem:* Most global energy studies envisage that future oil demand will be so high that the balance of oil market power must again shift back to the Persian Gulf in the period before 2020, as illustrated in Figure 4.6. In the IIASA energy scenarios and in the WEC high scenario, the Middle East-North, Africa (ME/NAf) region in 2020 would be

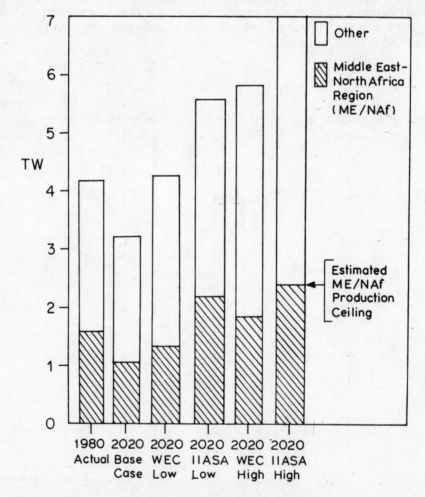

Figure 4.6 Alternative projections of global oil production, disaggregated into the Middle East/North Africa (ME/NAf) Region and the rest of the world. Also indicated is the maximum production capacity for the ME/NAf region for the period 2020-2030, as estimated in Wolf Haefele *et al.*, 1981.

required to produce oil at or near capacity levels—a situation similar to that in 1979, when world oil supplies were so tight that the Iranian revolution triggered the second oil price shock (Note 4.2.2-D).

It is possible, though, to avoid this tight oil supply situation for the entire period out to 2020 by implementing end-use energy strategies, in conjunction with efforts to shift the mix of oil and gas use to give greater emphasis on natural gas.

There is about as much natural gas as oil left in the world. But the gas resource is much greater in relation to current consumption. The amount of remaining gas judged ultimately recoverable at the global level is about a 200-year supply, compared to a 100-year supply for oil; outside the ME/NAf region and countries with centrally planned economies, there is about a 100-year supply of gas, compared to a 40-year supply of oil, at the current consumption rate (Note 4.2.2-E).

The prospects for fuel substitution as well as resources limitations suggest greater emphasis should be given to natural gas. Natural gas can be readily substituted for oil not only in stationary applications but in vehicles as well (Note 4.2.2-F).

In the Base Case scenario, it is assumed that by 2020 gas and oil production rates become equal (see Figure 4.7). Together with the assumptions about overall fossil fuel use and coal use, this implies that gas use would increase 1.85 times, but global oil use would decline from 63 to 45 million barrels per day, 1980-2020 (Note 4.2.2-D). At this lower level of world oil demand, there may well be adequate oil supplies available outside the ME/NAf region at production costs lower than $30 per barrel (in 1982$) to keep demand for ME/NAf region oil down to 15 million barrels per day—the 1983 "world oil glut level" (Note 4.2.2-G).

Such a scenario implies much greater global security and far lower oil prices than in the WEC and IIASA scenarios—and perhaps even stable oil prices for the entire period out to 2020. The prospect of more secure oil supplies and stable oil prices means that oil would be a more dependable and affordable energy source during the transition to the Post-Petroleum Era, and, in particular, oil would be more available for essential development purposes in developing countries during this critical transition period.

4.2.2.3 *Nuclear Weapons Proliferation and Nuclear Power:* Regarding the nuclear power-nucelar weapons connection, it is often said that the "genie is out of the bottle," so that we must learn to live with the risks of a proliferated world. The genie certainly would be out of the bottle if the WEC or IIASA projections become reality—projections that nuclear power will grow to levels 10 to 30 times as high as in 1980 (Figure 4.8). But these projections are not necessarily destiny. The implementation of end-use-oriented energy strategies, together with the poor economic prospects of those nuclear power technologies which offer the greatest proliferation risk, provides a basis for a less grim outlook.

As noted in section 1.3, by avoiding nuclear power technologies involving the reprocessing of spent nuclear fuel, the risk of nuclear weapons proliferations could be reduced and, in particular, the risk of *latent proliferation* could be greatly reduced.

Even a ban on reprocessing technologies would not *prevent* proliferation by means of the recovery of plutonium or other weapons-usable material from spent fuel—the risk of which increases with the extent of world-wide nuclear power development. Thus nuclear power should be regarded as an energy technology of last resort, limited to those situations where viable alternative energy technologies are not available.

The Base Case global energy scenario reflects this perspective on nuclear power. For the year

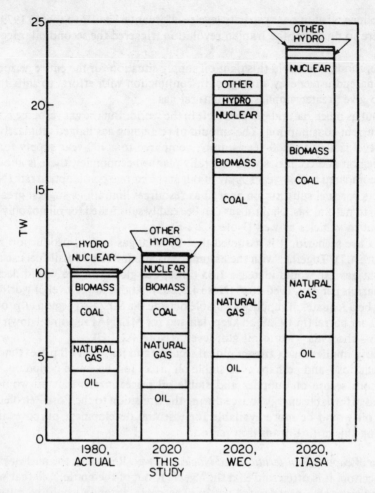

Figure 4.7 Primary energy use scenarios for 2020 by energy source. Averages of the low and high growth WEC and IIASA scenarios are compared to the Base Case of this study and to the actual use levels of 1980.

2020, this scenario involves nuclear power production amounting to 0.3 TW,* which represents official judgements in the early 1980's as to the worldwide production level for nuclear power by the year 2000 (Note 4.2.2-H). While this would be a fourfold increase in nuclear power production over the 1980 level, it implies essentially no net nuclear capacity increment beyond what is already planned for the year 2000, so that beyond the turn of the century the only new nuclear power plants that would be built are those that would replace retired units.

At this level of nuclear power development (and even considerably higher levels), there would be an abundance of low-cost uranium in the world for a long time. The economics of nuclear fuel reprocessing are not favourable today and would be only marginally favourable at very high uranium prices (Note 4.2.2-I). It has been shown, for example, that U.S. uranium resources are large enough to make reprocessing and the breeder reactor uneconomical for a period of at least 100 years even if there is rapid growth in nuclear power in the United States, i.e., U.S.

---

* Assuming an average capacity factor of 65 percent, this corresponds to an installed nuclear capacity of 460 GW (e).

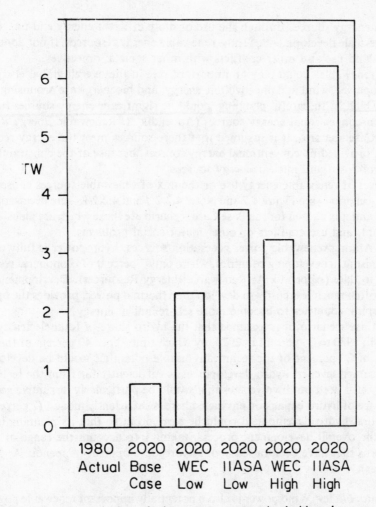

Figure 4.8 Alternative projections of primary energy use associated with nuclear power generation.

nuclear power output grows to 0.4 TW by 2020[4]. The same analysis showed that even uranium-poor nations could afford to forego reprocessing and the breeder reactor and could achieve uranium supply security simply by stockpiling uranium. (Because uranium, unlike oil, is very cheap, the cost of stockpiling is very low—on the order of 1 percent of the cost of producing electricity). Thus, the economics of policies aimed at avoiding reprocessing due to the proliferation problem are quite favourable.

Similarly, the more favourable economics of investment in energy efficiency improvement relative to investments in energy supply expansion generally also work in support of efforts to make nuclear fission power the energy technology of last resort.

4.2.2.4 *Roles for Renewable Resources:* One way to cope with the global risks posed by dependence on oil, fossil fuels generally, and nuclear power is to shift to greater reliance on less troublesome renewable energy sources. The prospects for major shifts to renewable resources, however, are far more speculative at the present time than the prospects for reduced dependence on conventional energy sources through the use of more efficient energy end-use technology. In addition, large-scale development of some renewable energy resources, if not done carefully, could also pose land use and other conflicts with other societal objectives.

Nevertheless, renewables could play an important role in the overall global energy mix. In particular, hydropower, wind and photovoltaic energy, and bioenergy are promising renewable energy sources which, with careful planning, could be significant energy sources that are less troublesome than conventional energy sources (Appendix A: Renewable Energy Resources).

For the Base Case scenario, it is assumed that these sources meet the energy requirements which cannot be provided by conventional energy sources, because of the constraints assumed to limit the use of fossil and nuclear energy sources.

Neither the level of renewable energy use or the mix of renewable sources chosen for 2020 in the Base Case scenario—see Figure 4.7 and Notes 4.2.2-J and 4.2.2-K—are necessarily optimal. Rather, the level and mix chosen for the Base Case scenario are those which are plausible without causing significant land use conflicts or other major social problems.

Hydropower: Among renewable power generation sources, hydropower is fully proven and is especially promising in developing countries, where only 7 percent of economical reserves have been developed to date (Appendix A; Renewable Energy Resources). Development of hydro power, which is often much less costly to develop than thermal power, provides the opportunity for many developing countries to become more self-reliant in energy.

For the Base Case scenario, it is assumed that the hydro share of total electricity increases from 20 percent in 1980 to 25 percent in 2020, by which time about 40 percent of the economic hydro potential, or 20 percent of the technically usable potential, would be developed (Note 4.2.2-J). The assumed level of hydro development is sufficiently far from the technical limit of the resource that it need not involve sites that would be particularly disruptive ecologically. Moreover, the pace of hydro expansion envisaged here is sufficiently modest (2 percent average annual growth) that hydro development could be carried out so that its planning is carefully integrated into the overall development process, taking into account the range of social and ecological concerns that have been raised about hydro development (Appendix A: Renewable Energy Resources).

Wind and Photovoltaics: Wind power is also a potentially important renewable power source, for which large, mass-produced wind machines would appear to be competitive in windy areas with conventional sources of electric power (Appendix A: Renewable Energy Resources). The most promising wind regimes are in industrialized countries, and commercial wind energy systems are currently being built in several of them. To achieve a high level of wind energy development would probably require emphasizing large (2 to 3 MW) windmills that are integrated into the utility grid system in ways that would improve the reliability of wind electricity. There are many promising ways to do this, but the present prospects for large-scale wind development are still uncertain.

Though not yet commercially established, photovoltaic technology (especially involving innovations such as amorphous silicon solar cells) holds forth great promise as a power generating source application in a wide range of circumstances (Appendix A: Renewable Energy Resources). Owing to the commercial uncertainties surrounding photovoltaics, however, no separate estimate is made of the level of use of this technology but instead wind and photovoltaic technologies are considered together, and it is assumed that these sources together account for 5 percent of total electricity use in 2020.

In the event that photovoltaics technology is not commercialized, all of this electricity would be provided by wind. This would require a level of wind energy development which is only a tiny fraction of the estimated wind energy potential (Note 4.2.2-J). On the other hand, if the promise of photovoltaic technology is realized, the photovoltaic contribution could be considerable, with perhaps less emphasis on wind.

Fuel-Fired Thermal Power Plants: It is assumed that electricity requirements in excess of what is provided by nuclear power, hydropower, wind and photovoltaics, would be provided by fuel-fired thermal power plants—either cogeneration plants or conventional central station thermal power plants. Assuming that 15 percent of the electrical demand could be met by cogeneration—a percentage which reflects the analysis of the industrial cogeneration potential in the United States (Section 2.6.2)—this implies that the amount of electricity required from fuel-fired central station thermal power plants would be the same in 2020 as in 1980, despite a doubling of the overall use of electricity (Note 4.2.2-J).

Biomass Energy Sources: Biomass is often regarded as "poor people's energy"—an energy source unfit for a modern society. Indeed, the industrialized world made a transition a century ago from wood to coal and other fossil fuels. Hence, it's difficult for many people to take seriously any notions of giving greater emphasis to bioenergy sources. Yet bioenergy sources have many attractive aspects (Appendix A: Renewable Energy Sources).

Used directly, biomass is a fuel that is usually less costly than oil and is often competitive with coal. And for the production of synthetic gaseous or liquid fuels biomass is in many ways superior to coal: it has much less sulfur than coal, little ash, and, because of its looser molecular structure, it can be gasified at a lower temperature than coal.

An important advantage of biofuels from the societal perspective is that, grown on a renewable basis, biomass becomes a source of chemical fuels, the production and use of which leads to no net increase in the atmospheric carbon dioxide level.

Using biomass sources to produce modern gaseous and liquid energy carriers also enables fossil fuel-poor but biomass-rich countries to become more self-reliant. Moreover, there is already a rapidly growing interest in modern biomass-derived fuels in some countries as an alternative to continued reliance on oil imports. For example, in a very short period of time, Brazil was able to establish a large-scale ethanol programme based on sugarcane. While sugarcane-based ethanol may be relevant only to a few parts of the world where fertile and is abundantly available, methanol production could be based on wood grown on land poorly suited for food production in many parts of the world.

For the Base Case scenario, it is assumed that half of all cogenerated electricity, e.g., cogeneration in the forest products and agricultural processing industries, is based on biomass—requiring some 0.2 TW of primary biomass energy (Note 4.2.2-K). Providing direct fuel needs in 2020 that could not be met with fossil fuels, owing to the assumed constraint on the overall fossil fuel supply, would require an additional 1 TW of biomass derived fuels (Notes 4.2.1-C and 4.2.2-K). In other words, 1/8 of all direct fuel use would have to come from biomass solids, liquids, and gases. Assuming that biomass feedstocks are converted to useful solid, liquid, and gaseous energy carriers at an average conversion efficiency of 70 percent,* some 1.4 TW of

---

* This is the approximate average conversion efficiency which would result if the mix of energy carriers were ½ biogas and methanol and ½ synthesis gas and solid fuel.

primary biomass energy would be required to make these fuels, bringing the total primary biomass energy requirements for the Base Case scenario to 1.6 TW, which is only slightly higher than the use of bioenergy sources in 1980 (Figure 4.7 and Note 4.2.2-K).

We assume that attempts would be made to utilize bioenergy on a renewal basis in order to reverse the ongoing process of deforestation associated with the current harvesting of biomass. Toward this end, there are two biomass sources, which if used carefully could help reverse deforestation: organic wastes and biomass grown for energy purposes.

On a global scale, there is an enormous production of organic wastes—forest product industry wastes, crop residues, manure, and urban refuse. In 1980, it amounted to about 2.8 TW—which is some two-thirds as large as world oil production. If the production of organic wastes increases simply in proportion to population, it would reach 4.1 TW by 2020 (Note 4.2.2-L). Because of competition for these wastes for other purposes (Note 4.2.2-M), however, it is assumed for the Base Case scenario that only about 0.8 TW(1/5 of the total) of organic wastes is recovered for energy purposes. With this level of organic waste use, an equal amount of biomass would have to be provided by growing biomass for energy purposes.

Because of the relatively low productivities of existing forests, the alternative of managed biomass production from bioenergy "plantations" or "farms" or "woodlots" for energy use or for multiple purposes is emphasized here. With managed biomass production, productivities can be much higher than in natural forests. In the following discussion, the word "plantations" refers to such managed production—whether on a small, medium, or large-scale basis.

For plantations to provide 0.8 TW of biomass for energy purposes would require the annual production of some 1.4 billion tonnes of dry biomass per year. The amount of land needed to grow this much biomass depends on the biomass productivity. For the purposes of the present analysis, it is assumed that the mean recoverable biomass productivity is 10 dry tonnes per hectare per year. While this is far less than what has been achieved under very favourable circumstances, there is not enough good long-term data available to justify higher productivities overall (Note 4.2.2-N). At 10 dry tonnes per hectare per year, the plantation land area required by 2020 would be some 140 million hectares, or roughly 4 percent of the world's forest area.

In the cases of both organic waste use and bioenergy from plantations, it is clear that the demands on the biomass system are rather modest and would not seem to be limited by any significant land-use or other constraints.

4.2.2.5 *Overview of Energy Supply for the Base Case Scenario:*

We have not attempted to put together an energy supply mix that is optimal in an economic sense. Rather, the supply mix chosen (Figure 4.7) is intended to be a plausible one which satisfies the condition that it be compatible with solutions to other important global problems having strong links to energy.

Our analysis indicates that it is both technically and economically feasible to find energy strategies compatible with and supportive of the solutions to other important global problems, and that these strategies do *not* represent a radical departure from the present. It is the "conventional" WEC and IIASA energy supply scenarios which are, in fact, radically different from the present situation and which would pose formidable economic and institutional challenges to bring about. The Base Case scenario involves no major energy supply changes from the present situation—the overall level of fossil fuel use is unchanged, the renewable share is up only modestly—from 16 percent in 1980 to 19 percent in 2020—and no exotic energy sources would be required. The energy supply problem would be quite manageable, in our view.

4.3 Sensitivity Analysis

The preceding analysis indicates that for the assumptions underlying the Base Case scenario a plausible energy future can be described which is compatible with and supportive of the solutions of other important global problems. But how dependent is this outcome on the various assumptions?

To address this issue, a sensitivity analysis will be presented. The analysis focusses on variations in the most important parameters underlying the Base Case scenario, and it considers first biomass and then oil and natural gas as "swing energy sources"—that is, the levels of these energy supplies for the year 2020 are adjusted to bring energy supply and demand into balance as variations are made in the scenario assumptions.

4.3.1 *Biomass as the Swing Energy Source*

Biomass is a potentially important swing energy source because: its use on a renewable basis would not aggravate the global carbon dioxide problem; it is a widely available renewable energy source that can be readily utilized with technologies which are available or can be brought to commercialization with little developmental effort; and the planting and careful management of biomass for energy purposes would be an important part of a global effort of afforestation.

As in the Base Case scenario, the two sources of biomass considered are organic wastes and the managed production of biomass for energy purposes on energy woodlots, farms, or plantations.

Because of the uncertainties surrounding the degree to which it will be possible to utilize organic wastes for energy purposes, it is assumed in the sensitivity analysis that the use of organic wastes for energy in 2020 cannot exceed 1.0 TW, or 1/4 of the estimated total organic waste generation rate (Note 4.2.2-L). It is also assumed that plantations and organic wastes make equal contributions to the total biomass supply until the organic waste use level reaches 1.0 TW and that beyond this level all extra biomass must be biomass grown for energy purposes.

An index which illustrates the required biomass production effort is the average rate of starting new plantations (in million hectares per year) needed to assure that enough biomass is available for harvesting in 2020 to meet the projected biomass supply level; it will be referred to as the plantation expansion rate, or PER.* For the Base Case scenario this rate would be 4.6 million hectares per year—which is comparable to the present PER in developing countries (Note 4.3.1-A).

The following discussion examines how the demand for biomass is affected by changes in each of several important parameters—the population growth rate, per capita energy use rates, the allowed ultimte atmospheric carbon dioxide level, and the growth of the nuclear power.**

4.3.1.1 *Population:* The global population in 2020 would be 13 percent higher in 2020 if the U.N. medium variant projection were realized or 23 percent higher if its high variant were realized, instead of the low variant (Note 4.3.1-B). If the extra energy requirements of a larger population were met entirely with biomass as the swing energy source, the relatively small percentage increases

---

Here it is assumed that the average period of rotation for the plantations is five years, so that the planting for supplies to be harvested in 2020 must be carried out in the period 1985 to 2015.

** For all the alternatives to the Base Case scenario, it is assumed that: (1) electricity accounts for 18 percent of final energy demand; (2) hydro (wind and photovoltaics) accounts for 25 percent (5 percent) of electricity production; (3) cogeneration accounts for 15 percent of electricity production, with a 50/50 mix of fossil fuels and biomass; (4) the nuclear electricity level is fixed @ 0.3 TW unless specified otherwise; (5) the remainder of electricity production is by fossil fuels in central station plants.

in population would be associated with quite large increments in the PER. As shown in Figure 4.9, the PER would increase from less than 5 to 14 or 22 million hectares per year as the population increased from the low to the medium or high U.N. variant for 2020.

The biomass effort clearly becomes more much difficult at the higher population growth rates, and this underscores the importance of solving the population problem as means of dealing with global resource management problems generally. But even these more ambitious bioenergy development efforts may be not only feasible but desirable. The PER value associated with the medium population variant is comparable to the 11 million hectares per year expansion rate for agriculture in the period 1950-1975.[5] Moreover, even with the high population variant, the required PER is still low in relation to the global deforestation rate. According to FAO statistics, the world's forest area decreased from 4400 million hectares in.1952 to 3800 million hectares in 1972, or at an average rate of 30 million hectares per year in this period.[5] A large-scale reforestation effort will be needed to reverse this process, and the development of bioenergy plantations could be a major part of such an effort.

Of course at the higher PER values, the issue of competition between bioenergy and food production must be dealt with very carefully. However, the best use of land may be to pursue complementary rather than competitive strategies for bioenergy and food production. It may be both feasible and desirable to restrict agricultural production to the better lands, increasing productivity there with inputs of energy—tractors, fertilizers, irrigation, etc.—and to grow biomass for energy purposes on the more marginal lands, using some of the energy produced this way to support more modern, more energy-intensive agricultural methods.[6-7] Tree farms with rotations of 5-10 years or more would probably be a more sustainable use of marginal lands than attempts at agricultural production.

4.3.1.2 *Per Capita Energy Use:* Figure 4.9 illustrates the impacts on the PER of alternative per capita energy demand levels—specifically the impacts of alternative combinations (x, y) where

x = per capita final energy use rate (in kW) for developing countries.

y = per capita energy use rate for industrialized countries.

For developing countries, a final energy use rate of 1.3 kW per capita is considered as an alternative to the Base Case value of 1.0 kW. This alternative value involves a 50 percent increase in per capita final energy use, 1980-2020. This higher rate of final energy use corresponds to a primry energy use rate of about 1.7 kW, which is near the high end of the WEC and IIASA projections for 2020. [The per capita primary energy use rate for the WEC (IIASA) scenarios in 2020 is in the range 1.1. to 1.6 kW (1.2 to 1.9 kW)]. This alternative energy use rate would imply only modest improvements in the efficiency of energy-using technologies in developing countries and certainly no significant "technological leap-frogging" to energy-efficient, advanced energy-intensive processes.

For industrialized countries, per capita final energy use levels of 3.3 kW and 4.9 kW per capita are considered as alternatives to the Base Case final energy use rate of 2.5 kW, representing considerably less ambitious energy conservation efforts. The 3.3 kW case involves per capita final energy use in 2020 which is one-third less than in 1980 and is equal to the average 1980 final energy use level for the WE/JANZ region. In the 4.9 kW case, energy efficiency would improve, 1980 to 2020, just enough to offset per capita economic growth.

Figure 4.9 illustrates the impacts of alternative assumptions about per capita energy demand on the requirements for bioenergy development, assuming the U.N. low population variant and biomass as the swing fuel. This figure shows that while in the Base Case (1.0, 2.5) the PER

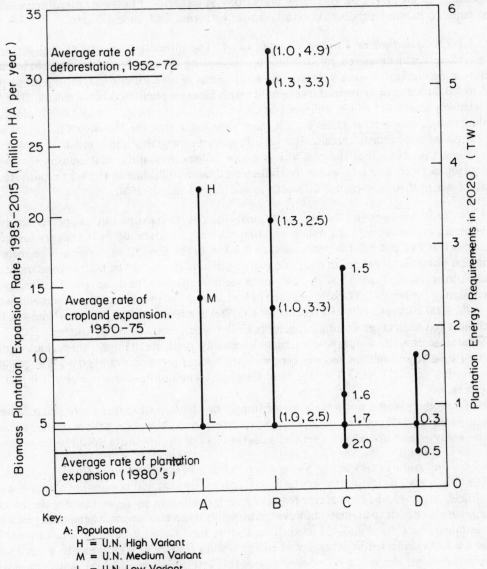

IMPACT OF ALTERNATIVE SCENARIO
ASSUMPTIONS ON PLANTATION ENERGY REQUIREMENTS

Figure 4.9 The results of sensitivity analysis, in which biomass is the "swing energy source" to bring energy supply and demand into balance as various scenario assumptions are altered. Shown here are the biomass requirements from energy plantations, farms, and/or woodlots—both in terms of the required global energy production rate in 2020 (in TW) and in terms of the average plantation expansion rate (PER) in the period 1985-2015 that is required to ensure that enough biomass is available for harvesting 2020 to meet global bioenergy requirements. See text for detailed asssumptions.

value is much less than the average rate of cropland expansion, 1950-1975, the PER is comparable to the cropland expansion rate for the (1.0, 3.3) scenario, and it is 2 to 3 times the rate of cropland expansion for the (1.3, 2.5), (1.3, 3.3), and (1.0, 4.9) scenarios. For these latter three scenarios the required biomass production efforts would be formidable undertakings.

4.3.1.3 *The Atmospheric Carbon Dioxide Level:* The ultimate level of CO_2 buildup in the atmosphere could be varied by speeding up or slowing down the rate of coal phase-out by substituting biomass. Figure 4.9 shows how variations in the ultimate CO_2 ceiling in the range 1.5 to 2.0 times the preindustrial level would affect biomass plantation requirements. The PER is relatively insensitive to the atmospheric CO_2 ceilings in the range 1.6 to 2.0. To reduce the ceiling to 1.5, however, requires a PER more than triple that for the Base Case.

The major impediment to realizing a 1.5 ceiling level is probably not so much its implications for biomass production as the prospect of major dislocations in the coal industry that would arise from a rapid coal phase-out—coal use would have to decline at the very rapid rate of 4 percent per year to a level in 2020 which is just 1/5 of that in 1980.

4.3.1.4 *Nuclear Power:* In the sensitivity analysis, two alternative nuclear power levels are considered, consistent with the philosophy that nuclear power should be the energy sources of last resort: 0 TW and 0.5 TW, compared to 0.3 TW in the Base Case scenario. The former is a nuclear phase-out scenario, in which no more nuclear plants would be built beyond those now under construction. The 0.5 TW nuclear power scenario involves the same rate of nuclear power expansion in the period 2000-2020 as in the period 1980-2000; in this scenario new plants would be built, 2000-2020, at a rate of 25 plants (of 1 GW(e) average capacity) per year (Note 4.3.1-C), with installed nuclear generating capacity reaching nearly 800 GW(e) by the year 2020.

What is noteworthy about these alternative scenarios is the small impact these changes make on the PER—from 3 million hectares per year if the nuclear power use rate is 0.5 TW to 5 million hectares per year if it is 0.3 TW (the Base Case) or 10 million hectares per year if it is 0 TW (Figure 4.9).

Nuclear power would not have a major impact on the overall global energy picture unless it were widely used throughout the world at levels several times higher than the maximum assumed in this analysis, and under such circumstances the risks of proliferation would be large as well.

4.3.2 *Oil and Natural Gas as the Swing Fuels When Energy Demand is High*

The above analysis shows that at the higher energy demand and population levels, it would be difficult to provide the extra energy required from biomass sources alone. Limiting the marginal energy supply options to biomass, however, is probably unnecessarily constraining. In particular, the assumed levels of oil and gas production for the Base Case scenario are much lower than what has been projected in other global energy studies (Figure 4.7) and may be lower than is necessary to keep the world oil price at reasonable levels and to make the rest of the world reasonably invulnerable to disruptions of oil supplies.

To examine this issue, a High Demand scenario was constructed, involving the highest aggregate energy demand for the cases considered above—1.3 kW per capita in DC's and 4.9 kW per capita in IC's (Note 4.3.2-A). For the supply mix, the same conditions as for the cases in which biomass is the swing fuel are assumed, except that:

- the biomass supply is limited to 3 TW or twice the 1980 level—1/3 which would come from organic wastes (1/4 of all organic wastes) and 2/3 from plantations, corresponding to a PER of 12 million hectares per year, 1985-2015, approximately the rate of cropland expansion, 1950-1975;

- the extra demand not met by biomass is met instead by increased oil and natural gas production, the levels of which are assumed to be equal by 2020;
- oil production outside the ME/NAf region is again assumed to be increased enough to limit the need for oil from the ME/NAf region to the 1983 glut level of 15 million barrels per day (Note 4.3.2-B).

The detailed energy supply features of this alternative scenario are described in Notes 4.3.2-C and 4.3.2-D. Here are some of the highlights:

- Primary energy demand would increase from 10 TW in 1980 to 15 TW in 2020, a level which is still only about ⅔ as large as in the averages for the WEC and IIASA scenarios (Figure 4.7).
- Electricity demand would still grow at less than half the 5 percent per year average growth rate of the 1970s (Note 4.3.2-D).
- While oil use would be about the same as in 1980, the assumed expansion in natural gas production implies that the sum of oil and gas use would increase from 6 TW in 1980 to 9 TW in 2020 (Note 4.3.2-C), compared to averages of 9 and 10 TW for the WEC and IIASA scenarios, respectively (Figure 4.7).
- It may even be feasible under this scenario to maintain the world oil price at or near the present level through 2020, since the requirements for oil production outside the ME/NAf region are comparable to the estimated remaining supplies of oil outside the ME/NAf region with production costs less than $30 per barrel (Note 4.3.2-D).
- The ultimate level of carbon dioxide in the atmosphere would be the same as in the Base Case scenario, since only the rate of using up oil and gas resources would be increased. Moreover, total fossil fuel emissions in 2020 would still be only 60 percent as large as in the WEC and IIASA scenarios in 2020 (Note 4.3.2-F).
- The level of nuclear power development would be the same as in the Base Case, so that proliferation concerns would not be exacerbated.
- The level of hydropower development for 2020 would amount to only about half of the 1976 WEC estimate of the extent of hydro resources that can be economically developed.

It would seem from these numbers that most of our concerns relating to energy supply that motivated the Base Case scenario could be dealt with in the High Demand scenario as well. Therefore, a critical policy issue arises: Why should energy planners seek to attain the lower energy demand levels associated with the Base Case scenario if the same objectives could be mostly realized at the higher demand level?

There are two reasons. First, it is probably cheaper at today's energy prices to provide energy services with the higher efficiency end-use technologies which characterize the lower demand scenario, than to provide the same energy services with less efficient end-use technologies but more energy supplies. In addition, the lower energy demand level would provide a significant margin for error in planning. Flexibility to contend with planning errors decreases, the higher the energy demand level.

The possibility of maintaining a relatively stable world oil price until 2020 with the demand levels in the High Demand scenario depends sensitively on the assumptions that: (1) natural gas production can be expanded 2.5 times, 1980-2020, to a level comparable to that projected in the WEC and IIASA high scenarios—an endeavour that may be feasible but nonetheless would be difficult; and (2) low cost oil resources outside the ME/NAf regions are indeed as large as estimated in the IIASA study, so that production of such resources would be effective in keeping ME/NAf oil output at the relatively low, post-1980 oil glut level (Note 4.3.2-B). Moreover, if world oil demand were higher than the level assumed for this scenario, it is very likely that most of the extra demand would have to be met by ME/NAf producers. If world oil demand were just ¼ higher than in the High Demand scenario (corresponding to a mere 7 percent increase in world energy use) and if this incremental demand had to be met through increased ME/NAf production, that region would have to produce at or near capacity levels—a condition that would undoubtedly mean a much higher oil price (Note 4.3.2-G) and greatly reduced global security.

The incremental energy demand associated with the High Demand scenario could of course be met by expanding coal and nuclear power use instead, but the incremental supplies needed from these sources in order to provide reasonable assurances of a secure oil supply situation would greatly exacerbate the problems of atmospheric carbon dioxide and nuclear weapons proliferation.

Of course the commercial success of new technologies such as amorphous silicon solar cells might enable expanded energy demand without jeopardizing the global goals identified for this analysis, but such technologies cannot be counted on at this time.

4.4 Conclusion

It appears both technically and economically feasible: (1) to meet basic human needs and to improve living standards considerably beyond the satisfaction of basic human needs in developing countries without increasing per capita energy use above the present average level—by taking advantage of opportunities to make much more efficient use of energy in both the modern and traditional sectors of these countries; and (2) to provide a continuing improvement in living standards in industrialized countries while cutting per capita energy use in half over the period 1980 to 2020—a possibility that arises as a consequence of both the ongoing structural changes in the economies of industrialized countries as well as the many opportunities for making more efficient use of energy.

From the global perspective, these findings mean that it is feasible to meet ambitious economic goals in the period 1980-2020 without increasing the overall level of primary energy requirements. A continuing shift to higher quality energy carriers will be necessary, however. For developing countries, this means a major transition from the present situation—where nearly half of primary energy requirements are provided by fuelwood, used largely for cooking—to wide use of modern solid, liquid, and gaseous fuels. For developing and industrialized countries alike, it means a much greater degree of electrification.

As long as global energy demand is not too large, there can be considerable flexibility in the choice of energy supplies, the mix of which can be adjusted so as to develop a long-run energy strategy which is consistent with the solutions of other major global problems.

The sensitivity analysis undertaken here shows that while it would not be necessary to achieve the precise energy demand levels of the Base Case scenario to bring about an energy future consistent with the solutions of other global problems with strong links to energy, the strong emphasis given to energy efficiency improvement in the Base Case scenario is a sensible approach to dealing with these global issues. At the lower energy demand levels, there is room for mistakes and miscalculations. If demand were in the neighbourhood of that for the High Demand scenario however, there would be little room for manoeuvre.

Overall, our analysis highlights the importance of trying to pursue a global energy course in which total global energy use changes little over the next several decades. Energy efficiency improvements would offset the expanded use of energy services that would arise from population and economic growth. Our analysis stops short, though, of identifying a truly sustainable long-run energy future, because it does not extend beyond the year 2020 and in the period out to 2020 no major shift to renewable resources is envisaged. We think it less important at this juncture to be able to assess the prospects for the period beyond the year 2020 than to understand the opportunities for getting just that far. Before 2020, much new technology will emerge that will make long-range problems even more manageable.* Energy planners need not strive to solve the energy problem for all time but rather to pursue an evolutionary strategy consistent with and supportive of the achievement of a sustainable world.

It will not be easy to bring about a fundamental reordering of the energy problem from preoccupation with supply expansion to concern for the most effective ways to provide energy services. It will require a readjustment of energy-related public policies. But no other energy course for the future will be easy either. The energy supply approach to the energy problem is foundering. The last several years have witnessed some of the most costly dry holes in the history of petroleum exploration. At the same time, the falling world oil price brought on by the world oil glut beginning in 1980 has stymied efforts to develop supply alternatives to oil—e.g., in the United States the massive attempts to launch a synthetic fuels industry have collapsed. The very survival of the nuclear power industry is threatened in many parts of the world. And numerous electric utilities are now financially strapped by the overinvestments they made in the 1970's in big new generating facilities, especially nuclear power plants. In relation to such problems, the challenge of creating an effective energy service industry does not seem so formidable.

If the end-use approach really catches on, it may actually prove to be easier to pursue than indicated here. This analysis has been restricted to technologies that are either already commercialized or are in such an advanced state of development that energy performance characteristics can be described fairly well. No attempt has been made to "guess" wholly new end-use technologies or to try to identify technological limits for future improvements in energy efficiency. Yet, it is clear that there is room for much more innovation than what we have described. The technologies emphasized have energy performances that are still far from thermodynamic limits. It must be remembered that the era of concern for high-cost energy is only a decade old. We are only now beginning to see the fruits of R&D efforts on end-use technologies that were initiated in the aftermath of the first oil crisis.

To sum up, the end-use strategy described here, which involves emphasis on energy efficiency

---

* To illustrate the long run possibilities, consider the carbon dioxide problem. In the analysis presented here, it was assumed that eventually all oil and gas supplies would be used up. But there are conceivable circumstances where this assumption could be relaxed. One of the most promising technologies in the offing for producing chemical fuels involves the use of amorphous silicon solar cells (perhaps laid out flat on the deserts on simple support structures) to produce hydrogen by means of electrolysis. A preliminary analysis indicates that if present industry expectations regarding the costs of amorphous silicon solar cell panels are realized, it may be feasible to produce hydrogen at costs competitive with the costs of oil and natural gas (Appendix A: Renewable Energy Resources). Suppose that this technology is successfully developed and that it proves to be feasible to substitute the produced hydrogen for all fossil fuels in the period 2020 to 2050. The result is that the ultimate level of atmospheric carbon dioxide could be limited to 1.4 times the pre-industrial level or 400 ppm (Note 4.4-A). The land area required for solar cells would be of the order of half the state of Texas or about 2 percent of the world's warm deserts (Note 4.4.-B).

improvements in industrialized and developing countries alike, is not dependent on technological breakthroughs. The strategy is economically feasible in the sense that investments in energy efficiency often involve direct costs that are less than the costs of investments in the equivalent amount of energy supply. There is no question that the degree to which the world's oil, atmospheric buildup of CO_2, and nuclear weapons proliferation problems can be solved will be greatly affected by the extent to which the impediments to more efficient energy use can be overcome by new public policy initiatives.

References for the Text of Chapter 4

1.   The World Bank, *World Development Report 1984,* Oxford University Press, New York, 1984.
2.   United Nations, *World Population Prospects as Assessed in 1980,* New York, 1981.
3.   Stephen Seidel and Dale Keyes, *Can We Delay a Greenhouse Warming? The Effectiveness and Feasibility of Options to Slow a Build-Up of Carbon Dioxide in the Atmosphere,* a Report of the Strategic Studies Staff of the Office of Policy and Resources Management of the US Environmental Protection Agency, September, 1983.
4.   Harold A. Feiveson, Frank von Hippel, and Robert H. Williams, "Fission Power, an Evolutionary Strategy," *Science,* pp. 330-337, January 26, 1979.
5.   B. Bolin, E.T. Degens, S. Kempe, and P. Ketner, *The Global Carbon Cycle,* a report of the Scientific Committee on Problems of the Environment (SCOPE) and the International Council of Scientific Unions (ICSU), John Wiley & Sons, Chichester, 1979.
6.   D. Pimentel *et al.,* "Deforestation: Interdependency of Fuelwood and Agriculture," draft report, New York State College of Agriculture and Life Sciences, Cornell University, Ithaca, New York, 1985.
7.   Robert H. Williams, "Potential Roles for Bioenergy in an Energy Efficient World," *Ambio,* vol. XIV, No. 4-5, pp. 201-209, 1985.

References to the Figure Captions, Tables, and Notes of Chapter 4

American Paper Institute, *Statistics of Paper, Paperboard, and Woodpulp,* New York, 1981.

Ellis L. Armstrong, "Hydraulic Resources," in *Renewable Energy Resources: the Full Reports to the Conservation Commission of the World Energy Conference,* IPC Science and Technology Press, 1978.

B. Bolin, E.T. Degens, S. Kempe, and P. Ketner, *The Global Carbon Cycle,* a report of the Scientific Committee on Problems of the Environment (SCOPE) and the International Council of Scientific Unions (ICSU), John Wiley & Sons, Chichester, 1979.

The British Petroleum Company, *BP Statistical Review of World Energy,* 1982.

Bureau of the Census, U.S. Department of Commerce, *Statistical Abstract of the United States 1984,* U.S. Government Printing Office, Washington DC, 1984.

E.A. DeMeo and R.W. Taylor, "Solar Photovoltaic Power Systems: an Electric Utility R&D Perspective," *Science,* pp. 245-251, April 20, 1984.

J. Edmonds and J. Reilly, "A Long-Term Global Energy-Economic Model of Carbon Dioxide Release from Fossil Fuel Use," *Energy Economics,* vol. 5, pp. 74-88, 1983a.

J. Edmonds and J. Reilly, "Global Energy Production and Use to the Year 2050," *Energy, the International Journal,* vol. 8, pp. 419-432, 1983b.

J. Edmonds, J. Reilly, J.R. Trabalka, and D.E. Reichele, "An Analysis of Possible Future Atmospheric Retention of Fossil Fuel CO_2, DOE Report DOE/OR/21400-1, Washington DC, 1984.

Energy Information Administration, U.S. Dept. of Energy, *Estimates of US Wood Consumption from 1949 to 1981,* US Government Printing Office, Washington DC, August 1982.

Energy Modeling Forum, *Aggregate Elasticity of Energy Demand, Vol. 1,* EMF Report No. 4, Stanford University, Stanford, California, 50 pp., 1980.

J.R. Frisch, *Energy 2000-2020...Where Are We Going? Regional Stresses...,* Conservation Commission of the World Energy Conference, London, 1982.

H.S. Geller, "The Potential for Electricity Conservation in Brazil," Companhia Energetica de Sao Paulo, Sao Paulo, Brazil, April 1985.

J. Goldemberg, T.B. Johansson, A.K.N. Reddy, and R.H. Williams, "An End-Use Oriented Global Energy Strategy," *Annual Review of Energy,* vol. 10, pp. 613-666, 1985.

J. Goldemberg and R.H. Williams, "The Economics of Energy Conservation in Developing Countries: a Case Study for the Electrical Sector in Brazil," in *Energy Sources: Conservation and Renewables,* D. Hafmeister, H. Kelly, and B. Levi, eds., American Institute of Physics Conference Proceedings, No. 135, American Institute of Physics, New York, 1985.

Wolf Haefele *et al.,* Energy Systems Program Group, International Institute for Applied Systems Analysis, *Energy in a Finite World, A Global Systems Analysis,* 837 pp., Ballinger, Cambridge, 1981.

D.O. Hall, G.W. Barnard, and P.A. Moss, *Biomass for Energy in the Developing Countries,* Pergamon Press, Oxford, 1982.

International Energy Agency, *World Energy Outlook,* Organization for Economic Cooperation and Development, Paris, 1982a.

International Energy Agency, *Energy Balances of OECD Countries, 1976/1980,* Paris, 1982b.

International Energy Agency, *OECD Energy Balances 1970-1982,* OECD, Paris, 1984.

A.M. Khan and A. Holzl, *Evolution of Future Energy Demands till 2030 in Different World Regions: An Assessment Made for the Two IIASA Scenarios,* International Institute for Applied Systems Analysis Report RR-82-14, Laxenburg, Austria, April 1982.

W.D. Nordhaus, "The Demand for Energy: An International Perspective," in *International Studies of the Demand for Energy,* W.D. Nordhaus, ed., North-Holland, Amsterdam, 1977.

Office of Policy, Planning, and Analysis, U.S. Department of Energy, *Energy Projections to the Year 2010,* Technical Report in Support of the National Energy Policy Plan (DOE/PE-0029/2), October 1983.

Office of Technology Assessment (U.S. Congress), *Energy from Biological Processes, Volume II-Technical and Environmental Analyses,* 1980.

W. Palz and P. Chartier, *Energy from Biomass in Europe,* Applied Science Publishers, Ltd., London, 1980.

R.O. Sandberg (Bechtel) and C. Braun (EPRI), "Economics of Reprocessing—U.S. Context," paper presented at the American Nuclear Society's Topical Meeting on Financial and Economic Bases for Nuclear Power, Washington DC, April 8-11, 1984.

T.B. Taylor, R.P. Taylor, and Steven Weiss, "Worldwide Data Related to Potentials for Widescale Use of Renewable Energy" PU/CEES Report No. 132, Center for Energy and Environmental Studies, Princeton University, Princeton, New Jersey, March 1, 1982.

Gordon Thompson, "The Prospects for Wind and Wave Power in North America," PU/CEES Report No. 117, Center for Energy and Environmental Studies, Princeton University, June 1981.

U.S. Environmental Protection Agency, "Refuse-Fired Energy Systems in Europe: An Evaluation of Design Practices. Executive Summary," Publication No. SW 771 of the Office of Water and Waste Management, November 1979.

United Nations, *1979 Yearbook of World Energy Statistics,* New York, 1981b.

United Nations, *1979/80 Statistical Yearbook,* New York, 1981c.

United Nations, *1981 Yearbook of World Energy Statistics,* New York, 1983.

J.P. West and L.G. Brown, *Compressed Natural Gas,* Publication P14 of the New Zealand Energy Research and Development Committee, Auckland, New Zealand, April 1979.

World Energy Conference, *World Energy Resources 1985-2000,* IPC Science and Technology Press, 1978.

World Energy Conference, *Survey of Energy Resources,* 1980, prepared for the 11th World Energy Conference, September 8-12, 1980, Munich, 1980.

Bin Zhu, "Afforestation and Energy for China," draft report prepared at the Center for Energy and Environmental Studies, Princeton University April 1982.

5. Policies for Implementing Energy Strategies for a Sustainable World

5.1 Introduction

In this chapter, we will suggest ways in which alternative policy instruments might be used to formulate policies for implementing the energy strategies described in detail in chapters 2-4, which were constructed with the objective of serving the basic social goals of equity, economic efficiency, environmental soundness, self-reliance, long-term viability and peace (Figure 5.1).

The policy discussion is organized to conform to major features of our end-use oriented strategies described in Chapter 1, emphasizing:

- Satisfying basic human needs,
- Meeting the energy needs of the poor and disadvantaged,
- Creating fair economic competition between all energy sources and end-uses,
- Promoting energy efficiency improvements,
- Beginning a transition to renewable energy resources,

GOAL	=	AN OBJECTIVE TO BE ACHIEVED
STRATEGY	=	A BROAD PLAN TO ACHIEVE GOAL
POLICY	=	A SPECIFIC COURSE OF ACTION TO IMPLEMENT A STRATEGY
POLICY INSTRUMENT	=	A PARTICULAR MECHANISM FOR INITIATING AND MAINTAINING A POLICY
POLICY AGENT	=	AN AGENT WHO WIELDS A POLICY INSTRUMENT

Figure 5.1 Definition of terms.

- Generating new knowledge and technological advances,
- Promoting national self-reliance,
- Being compatible with the solutions to other global problems.

Both national and global policies are presented here. In Chapter 6 we will discuss the context in which these policies would be implemented, i.e., the political economy of energy.

5.2 National Policy Instruments

A variety of policy instruments are available to policymakers for implementing various energy policies. They include:

- reliance on market forces to determine energy demand levels and the appropriate mix of energy production and end-use technologies.
- administrative allocation of energy carriers, capital, and/or technology
- subsidies
- regulation
- taxes/duties
- government-sponsored competitions
- administrative generation of data and dissemination of information
- administrative setting of research and development priorities and support of R&D
- creation of appropriate policy agents

We will examine the strengths and weaknesses of each of these instruments.

5.2.1 The Market

It is within the economic self-interest of a producer to expand his output of goods or services by one unit if the price he receives for this extra unit is greater than or equal the added cost of producing it, the so-called marginal cost; if the price is less than his marginal cost, it would not be worthwhile to produce more.

For the economy as a whole, economic theory tells us that only when the prices of goods or services are equal to marginal costs is an economy squeezing from its scarce resources and limited technical knowledge the maximum of outputs. In a perfectly competitive free market economy, where no single producer can influence the market price, prices will automatically seek the level of the marginal costs.

5.2.1.1 *The Attractiveness of the Market Mechanism:* Reliance on market forces to allocate resources—national resources, technology and capital—may have intrinsic appeal, especially in terms of implementing many elements of an end-use oriented energy strategy. The use of energy throughout an economy is determined by a complexity of factors—including user preferences and income, the resourcefulness and technical know-how of those who produce and sell energy and energy-using technologies, etc. The technological options for providing particular energy-related services are varied, and they are always changing as new technologies are developed. In the market, these complexities are dealt with by those with intimate knowledge of what they need and can afford, i.e., the buyer, and by those very well acquainted with what the energy or energy-using device costs to produce, i.e., the seller. And it is within both's self-interest to keep costs down. Hence, the efficient allocation of resources results, theoretically at least.

The alternative is to allocate resources by administration. The difficulty with this is that bureaucracies are notoriously ineffectual when it comes to keeping track of the needs and preferences of a multitude of users and in replacing the myriad interactions of the market with their own procedures and rules—which is why black markets often spring up when bureaucracies take over the allocation and distribution of resources. Moreover, when in charge of production, public bureaucracies do not seem to be as productive, cost-conscious or innovative as producers in the market, perhaps because they do not directly benefit if output per hour increases or costs are cut.

5.2.1.2 *Market Imperfections:* Despite its theoretical advantages, the market has many well-known shortcomings.

Perfectly competitive free market conditions exist only in isolated instances in capitalist economies, e.g. in a few agricultural industries where many producers of identical products conduct marketing by auction. The energy industry in particular typicallly involves monopoly or oligopoly (few seller) situations.

Because of inappropriate market signals to producers and consumers, markets are not efficiently allocating resources in the energy sector. Moreover, for a variety of reasons typical individual energy consumers tend not to behave as free market economic theory would have them behave, i.e., evaluating lifecycle costs of alternative options, using the market rate of interest to put costs and benefits in different time periods on a comparable basis. Instead, consumer decisions relating to energy investment and purchase tend to be oriented to keeping first costs low.

While eliminating these market imperfections would lead to economically efficient decision-making, such an achievement would still leave major challenges for energy policymakers, because there are problems that the market is inherently incapable of handling: poverty (an equity issue); environmental and security problems (externalities); and the need for research and development (to provide adequately for the future).

And finally, the market is ineffective in generating and communicating to consumers and policymakers the kinds of information needed to provide a basis for decisions aimed at eliminating or compensating for market imperfections.

Inappropriate market signals: Historically, governments in both industrialized and developing countries have provided subsidies to encourage energy production and consumption. Supply subsidies are of course popular with energy producers; and lower prices are always more popular with consumers than high prices. The appeal to policymakers of producer tax breaks or price controls or price subsidies arises from both the political support such policies evoke from the beneficiaries and the administrative simplicity of such policies.

Consider first producer subsidies. There is a long tradition in many countries for governments to give more favorable tax treatment or other direct subsidies to the energy supply industries, e.g., oil and gas producers, energy utilities, etc., than to other industries (Note 5.2-A). The energy industry and its supporters generally argue that the oil industry "needs" special incentives to make sure that enough drilling is done to find more oil and that tax breaks or other special treatment is needed to make sure that enough new power plants are built to meet future consumer demands (Note 5.2-B).

These subsidies to producers lead to energy prices that are below true marginal costs. Perhaps more important, these producer subsidies draw large amounts of capital to the energy supply industries, thus helping make capital resources more scarce for both investments in energy efficiency and for other economic activities. In the United States, for example, nearly 40 percent of all new plant and equipment expenditures in 1982 were accounted for by the energy industry (Figure. 2.3). Furthermore, subsidies to energy producers may not even lead to increases in energy production. Baumol and Wolff have shown that unless energy supply subsidies are devoted to research activities, with no prospect of direct commercial payoff, they have an inherent tendency to *decrease* net energy yields (Note 5.2-C).[1]

In addition to providing subsidies to the energy supply industries, governments have also imposed price controls to protect consumers of particular energy carriers or particular classes of consumers. For example, in many developing countries the price of kerosene is pegged well below the price this energy carrier should have in relation to the world oil price (Chapter 3). While the intent is laudable, to keep this fuel affordable to poor people, a major economic inefficiency results from keeping the kerosene price artificially low. The price of diesel fuel, which is similar to kerosene, must also be kept low, with the result that truck freight expands rapidly at the expense of rail freight, thereby increasing the nation's oil import bill and increasing the foreign exchange burden of paying for the extra oil.

Major inefficiencies are also inherent in the rate structures of energy utilities. In many parts of the world, for example, electricity prices are based on average costs. This means that the electricity price is equal to the total cost of production divided by the total electricity sales, with rates adjusted for different consumer groups to reflect in some approximate way the variations in the costs of providing service to different groups. The problem with this policy of "average cost pricing" is that in many areas new electricity supplies must come from sources—e.g., new coal or nuclear thermal power plants or marginal hydroelectric sources—for which the costs are often considerably higher than the costs of producing electricity from existing sources (see, for example, Note 2.1-E). As these new power plants are brought into operation, consumers usually experience in their utility bills only a very diluted effect of these higher costs, which are "rolled in" to an average cost together with the costs of older, much cheaper electricity sources. Accordingly, consumers tend to consume more electricity than they would if prices were more appropriately based on marginal costs.

The situation is often made worse by rate structures that further insulate particular groups of consumers from marginal costs. One such practice has been to give large volume consumers enormous discounts (see Figure 2.28, which illustrates the situation for U.S. industry). This practice can attract large-scale, energy-intensive industry such as primary aluminium production. Offering discount rates is also an easy way for the electricity producer to "smooth out" the electricity supply/demand imbalance that would otherwise result when a large increment of new electricity supply is added to the system—an especially serious problem in many developing countries, where a single new power plant often increases the installed capacity by a large percentage. But again major economic inefficiencies can arise from the practice of offering discounted rates to large volume consumers. We have shown in Chapter 3, for example, how the opportunity costs of these practices can be high in developing countries. Electricity-intensive industries tend to provide relatively little employment. In circumstances where unemployment is a serious problem, the same amount of electricity supply could be used to create many more jobs if used in a large number of employment-intensive industries instead of in a small number of electricity-intensive industries.

Still another rate structure problem is that some utilities, e.g., in Sweden, have rate structures such that a large fraction of the consumer's bill is constant, independent of the consumption level. The rationale for this fixed charge is that it is needed to cover the utility's fixed costs, which it incurs, independent of its production level. But the effect is to discourage energy efficiency measures, because utility bills are thereby reduced much less than in proportion to the savings.

High Consumer Discount Rates: Whereas investments in new energy supplies are made by energy companies, investments in energy efficiency improvement are made instead by many consumers of energy in buildings, transportation, and industry.

In making investments in energy-using technologies, the rational choice among alternative options for providing a given energy service (space heating, refrigeration, cooking, travel, etc.) would be the option with the least lifecycle cost, where costs and benefits are discounted at the consumer's market rate of interest.

In practice, though, this often does not happen—especially when the individual faces investment decisions relating to his own energy use. Many analyses have been done in recent years which show that the discount rates implicit in actual consumer purchases that involve choices among options of different energy efficiency are far in excess of market interest rates—often in the range of several tens to several hundred percent, compared to (inflation-corrected) market interest rates of 10 to 15 percent or less.[2-6]

Consumer discount rates relating to investments in energy efficiency are high for a variety of reasons: inadequate or unreliable information about opportunities for and the potential dollar savings from investments in energy efficiency; the "hassle" of making such investments; the inadequate availability of capital resources for financing these investments, and uncertainties about future energy prices. In addition, some energy using devices such as appliances and furnaces and air conditioners may be purchased by landlords and builders who have no interest in the operating costs which tenants or homebuyers will be shouldering (Notes 5.2-D, 5.2-E, 5.2-F, and 5.2-G).

The failure to deal effectively with poverty: Even if they worked perfectly markets would not be able to deal effectively with the problem of poverty, which concerns equity, not economic efficiency.

The best that can be expected from the workings of an efficient market is more rapid economic growth than would otherwise be the case. It has often been argued that while the rich get richer with more rapid growth, the poor get richer too. But, as we have pointed out, the "trickle down" approach to the development of developing countries has not been very effective in addressing the problem of poverty.

In industrialized countries, the high energy prices brought on by the oil shocks of the 1970's, though effective in making middle and upper income households more energy-efficient, have created severe economic hardships for poor people, who have been unable to afford investments in more efficient cars, appliances, or in thermally tighter homes.

The failure to address externalities: Decisions determined in free markets do not reflect social costs that are not accounted for in market prices. These so-called "external" social costs were discussed in Chapter 1. They include: the loss of self-reliance in market power and in foreign policy because of over-dependence on oil imports; the risk of war, even nuclear war, as a consequence of the dangerous dependence of the industrialized market economies on Persian Gulf oil; the acid rain problem associated with fossil fuel use in stationary power plants and automobiles; the risk of nuclear weapons proliferation associated with the availability of weapons-usable materials from nuclear power fuel cycles; the risk of climatic change associated with the atmospheric build-up of carbon dioxide from the burning of fossil fuels; etc.

Inadequate emphasis on research and development: Research and development efforts by the private sector tend to concentrate largely on opportunities for near-term improvements in existing products and processes. The lack of private sector interest in basic and applied research reflects the inadequacy of the market in dealing with the long term. Private sector R&D efforts are inherently weak in this area because the potential payback extends beyond business planning horizons. Another factor is that private firms cannot capture the benefits of R&D that provides only generic information. And there is little incentive for private firms to pursue research and development in areas where much of the potential pay-off involves broad societal benefits—e.g., research aimed at understanding better the problems of the poor and other disadvantaged groups and how these problems might be alleviated, research on the carbon dioxide problem, etc.

Inadequacies in generating and disseminating information: Consumers need reliable and readily understandable information about the energy performance, cost-effectiveness, and side-effects of alternative energy end-use technologies. Planners also need such information, plus detailed, highly disaggregated information about patterns of energy end-use, plus general information about possibilities for technological innovation as a basis for policymaking aimed at eliminating or compensating for market imperfections.

The market does not generate the needed detailed information on patterns of energy use or on the prospects for technical changes. And much of the information generated in the market on particular technologies is often not standardized or is biased in favor of particular products. Such information provides a poor basis for comparing alternative technologies.

5.2.1.3 *Conclusion:* Given its advantages, the market should be relied upon as an instrument for implementing end-use oriented energy strategies, wherever possible. And policies which impede the efficient functioning of the market, e.g., price controls, average cost pricing, and producer subsidies, should be eliminated.

However, and this is crucial, the public sector has to intervene in the energy market to eliminate market distortions, to lower consumer discount rates, and to deal with the problems of poverty, externalities, the need for research and development, and scarcity of needed information—problems that the market is intrinsically incapable of dealing with adequately.

5.2.2 *Administrative Allocation of Energy Carriers, Capital, and/or Technology*

National, state and local governments can allocate resources adminstratively as a substitute for or supplement to the market. In the case of energy this can involve the administrative allocation to non-governmental groups of energy carriers, capital, technology or energy services. And the government can administratively allocate resources to ·meet its own needs by means of procurement.

5.2.2.1 *Rationing:* The rationing of energy carriers such as gasoline and LPG has frequently been resorted to by governments in times of scarcity as a more equitable means of allocating resources than by price. Rationing was widespread during World War II, but exists even now in many countries.

Rationing ensures that the poor are able to get needed supplies in times of scarcity. As a means of allocating resources in more normal circumstances, however, rationing is a cumbersome policy instrument, as it involves a great deal of administrative work and also lends itself easily to irregularities. Moreover, even in times of scarcity there may be less cumbersome methods of ensuring a fair allocation of resources.

Schelling has argued in the case of gasoline rationing, for example, that rationing through marketable coupons is better than through non-transferable coupons, because those who need more than their "fair share" can always get more by buying coupons that other users don't need—a far simpler process than appealing to a rationing board. And those who don't need their full allocated share can sell their unwanted coupons to generate income for more pressing needs. He further points out that if one accepts this principle, then it would be preferable instead to ration with a tax, the revenues from which could be fully rebated to consumers in the same manner that the coupons would have been distributed.[7] The latter would allocate resources the same way as with marketable coupons but would avoid the need to produce ration coupons, the risk of counterfeiting, and other administrative burdens.

More generally, energy policy should be oriented to avoiding the supply shortage situations that make rationing necessary or desirable.

5.2.2.2 *Allocation of Energy Supply Access:* In the case of electric or gas utilities an important form of administrative allocation involves connecting consumers to the utility grid systems, thereby providing them with access to electricity or gas supplies.

Privately owned utilities in the United States and some other industrialized countries are granted service area franchises in exchange for an agreement to be regulated as "natural monopolies." A condition of the agreement has often been that all those seeking service in the franchise territory must be hooked up. Such arrangements have been successful in providing broad access to energy supplies.

5.2.2.3 *Capital Allocation:* Capital resources are very often administratively allocated to energy systems. The tax breaks and other subsidies to energy supplies described above are examples of generic schemes for allocating capital to particular classes of energy suppliers. In the case of publicly-owned utilities, capital is also allocated administratively to particular activities or projects. And development aid is typically allocated by the funding agencies to particular energy supply projects e.g., hydroelectric projects.

Capital resources could also be administratively allocated to investments in energy efficiency improvement, as a means of dealing with the capital scarcity problem relating to such investments.

5.2.2.4 *Allocation of Technology/Energy Services:* In some instances, it may be desirable to allocate administratively equipment, components of equipment, or improved energy services to particular groups.

The market cannot allocate resources among the poor in developing countries, whose activities lie largely outside the market economy. This shortcoming might be compensated for by administrative allocation of technology. To bring about the wide use of improved wood stoves in developing countries might involve, for example, distributing the grates which have proved to be crucial to the achievement of high efficiencies.

Even in situations where the poor do participate in markets, such schemes may be desirable. For example, one way to deal with the high heating fuel bills of the poor in Northern climates would be for the utility or government to "retrofit" the houses of the poor with energy efficiency improvements as a public service activity.

5.2.2.5 *Government Procurement:* When a new technology is introduced there are often institutional obstacles to the creation of new markets. Producers may be reluctant to enter new markets if the burdens of convincing enough consumers to buy the new products are too great. The problem is compounded by the tendency of consumers to be unwilling to take the large risks associated with purchases of "first-of-a-kind" products. Moreover, initial costs of new products tend to be high, owing to the need to recover developmental costs, even if there may be good prospects of much lower costs later on, realized through the economies of mass production, "learning curve" experience, and competition.

Government can help create new markets for socially desirable new products by procuring specified quantities of the products for meeting its own needs. Examples relevant to an end-use oriented energy strategy might include the purchase of energy-efficient light bulbs for government buildings, or automobiles or trucks of high fuel economy for government motor vehicle fleets.

Government procurement could help reduce both the risks to producers of market development and the risks to consumers of being first-of-a-kind purchasers, through the government's demonstration of the attractions or shortcomings of the new products.

The drawbacks of procurement as an instrument for helping to create new markets are the risks that economically inefficient industrial production will thereby be sustained, that competition will be constrained, and that innovations beyond the initial product designs will be retarded.

Such problems could be minimized if procurement were carried out through competitive bidding, if the government specified minimum product performance and cost targets, and if procurement programs were designed to be phased out in a timely manner.

5.2.2.6 *Conclusion:* While the administrative allocation of resources can be an effective way of dealing with some particularly troublesome problems, the use of this instrument as an across-the-board allocative mechanism would be bureaucratically burdensome. The use of this instrument should be reserved for implementing policies where market mechanisms or "quasi-market" mechanisms (e.g., energy taxes) would not be very effective. An excellent example of such a case is the providing the poor in developing countries with efficient woodstroves.

5.2.3 *Subsidies*

If markets are distorted by large subsidies that favor the development of conventional energy supplies, it is natural to consider neutralizing this effect by offering subsidies for energy productivity improvement and alternative supplies (e.g. solar energy) as well. While such subsidies may lead to a situation that is more economically efficient than the status quo, there are major problems with the targeted subsidy that limit its utility as an instrument of pubic policy.

The use of subsidies to promote a particular course of action can tend to discourage innovation and creative uses of technology by directing capital resources to well-defined (and hence not innovative) technologies. Both the bureaucratic burden of matching the subsidy to the most deserving cases and the political factors that influence the allocation process can limit the economic efficiency of the targetted subsidy as a policy instrument. When subsidies are applied to many and diverse energy end-use situations, the administrative burden of measuring the effectiveness of targetted subsidies can be formidable. Such shortcomings of subsidies to promote energy end-use strategies are illustrated by a case study of tax credits for residential and industrial investments in energy efficiency improvement and solar energy in the United States (Note 5.2-H).

An important reason for being cautious about proposals to introduce new subsidies to create in the marketplace a more even-handed treatment of investments in conventional and alternative energy sources is that the need for subsidies to achieve this objective has not been established. A 1982 review of U.S. utility programs to finance energy conservation investments based on consumer surveys showed no clear need for financial incentives to motivate investments in energy efficiency.[8] Factors such as advertising effectiveness, consumer protection, and convenience features seem to have at least as strong a bearing on consumer participation in these programs as the financial incentives involved.

Despite these problems with the subsidy as an instrument to promote end-use energy strategies, there are instances where the costs of subsidies can be justified or where alternative instruments would be ineffective in implementing goals.

As in the case of the administrative allocation of resources, targetted subsidies can be an effective way of dealing with some particularly troublesome problems. But it should be reserved for situations where alternative instruments cannot be used more efficiently.

5.2.4 *Regulation*

Regulation can be a powerful instrument for implementing public policy; for a given policy, regulation reduces the uncertainty of the outcome perhaps more than any other instrument can.

There are four generic types of regulation of importance to energy policy: economic regulation, regulation aimed at reducing external social costs, regulation to improve the flow of information, and regulation of energy performance.

5.2.4.1 *Economic Regulation* Economic regulation can be used to determine energy prices, to keep in check anti-competitive tendencies in nominally competitive energy markets, and to frame the charters and operating rules for public energy utilities.

In centrally planned economies prices are determined administratively rather than through market forces. Similarly, in many mixed economies certain prices are also fixed administratively— because competitive conditions do not exist or to protect especially vulnerable consumer groups from the burden of high prices. These price controls give rise to economic inefficiencies, making very costly the achievement of the underlying goals.

Where more economically efficient means of protecting vulnerable consumer groups are available—e.g., allocation of capital or more efficient energy-using equipment to poor groups— they should be deployed instead of keeping energy prices below marginal costs.

Economic regulations aimed at promoting competitive conditions in energy markets, e.g., anti-trust regulations, can be effective in promoting economic efficiency.

Utility charters define the allowable activities of utilities. In the past these charters defined allowable activities as the narrow task of providing particular energy supplies in well-defined service areas. The focus on the provision of energy supplies was a natural result of the past preoccupation of energy planners with energy supplies. Now these charters and operating rules could be broadened to require that publicly-regulated energy companies become involved not only in the supplying of energy but also in providing consumers with a service—helping them use energy more efficiently—and to make incentives available to the companies for doing so. However, care should be exercised to ensure that the existence of these regulated "energy service companies" does not inhibit the development of energy service companies in the unregulated private sector, where economic conditions are such that private companies might play a useful role in the economy.

5.2.4.2 *Regulation to Reduce Externalities*

The wide range of regulations of air pollution, water pollution, land use, use of toxic materials, etc., which have evolved over the last 15 to 20 years in many parts of the world indicate how external social costs can be reduced through regulation.

Regulations could also be used to limit the external energy-related costs emphasized in this book—namely, acid rain, climatic change associated with the burning of fossil fuels, global insecurity associated with dependence of the industrialized market economies on Persian Gulf oil, nuclear weapons proliferation associated with widespread development of nuclear power, deforestation associated with non-renewable use of biomass.

Here the crucial question is: Is regulation the most economically efficient way of reducing these external social costs? An alternative approach would be to tax the offending activity (see below).

5.2.4.3 *Regulation to Improve the Flow of Information:*

Ignorance about the opportunities for energy efficiency improvement or for using alternative energy sources is a major obstacle to rational decision-making in energy-related purchases. Hence, regulations that serve to improve the flow of information can be very effective policy measures.

The mandated labeling of the fuel economy of new automobiles sold in the U.S. illustrates the value of labeling laws that indicate energy performance based on standardized tests. The mpg index effectively communicates energy performance information in a manner that is readily understood by consumers. While the mpg rating was determined in testing procedures that may not accurately reflect actual on-the-road conditons, it usually ranks the fuel economies of different cars in the correct order. Labeling energy performance of an energy-using product is applicable to those energy end-uses where energy performance can be well-defined by a simple index that can be readily understood by consumers. Refrigerators are a good example of such an energy-using product.

For a wide range of other energy-using activities that would fall under the purview of publicly-regulated energy companies, information flow could also be improved by requiring these companies to provide their customers in all sectors information regarding the energy performance and cost-effectiveness of alternative investments on the customer's side of the meter.

5.2.4.4 *Regulation of Energy Performance*

As an instrument of public policy the regulation of energy performance has the advantage that it reduces the uncertainty about future energy use perhaps more than any other approach. Thus, such regulations should be particularly helpful to energy planners who would like to count on energy efficiency improvements as a major source of "equivalent energy supply."

In fact, there have been some major success stories involving the regulation of energy performance. Particularly noteworthy is the 1975 law passed by the U.S. Congress mandating that auto producers double the corporate average fuel economies of the cars they produce by 1985. This law will bring the energy efficiency of the U.S. automobile up to the level of most of the rest of the world.

But there are limits as to what can be achieved with regulations of energy performance. In practice, such regulations can be applied only to a limited set of energy-using activities, and they tend to be weak stimuli for promoting technological innovation (Note 5.2-I). Energy performance regulations are best suited for:

- that subset of energy using activities that can be characterized by an energy performance index that it is easily measurable, readily understood, and widely applicable.
- those situations in which it is worthwhile to clear the market of "energy guzzling devices."
- problems that defy alternative, administratively simpler solutions.

5.2.4.5 *Conclusion*

Regulations of all four types described here can have important roles in implementing end-use oriented energy policies. Economic regulations of regulated energy companies can be used to create an infrastructure for implementing energy efficiency opportunities. Regulation of offending activities is one way to deal with externalities associated with energy production and use. Regulation can be used to improve the flow of information to provide a sounder basis for market decisions relating to energy purchases and investments. And, for a limited but often important set of energy-using activities, energy performance regulation can be effective in encouraging the production and purchase of energy efficient products such as automobiles.

5.2.5 *Taxes/Duties*

The levy of taxes or duties on energy carriers or energy-using devices is a market-like instrument for allocating resources among energy producing and consuming activities that discourages the production or purchase of the activities or items being taxed. Taxation can be used to promote either broad social goals or narrow purposes.

Taxes might be used in a developing country, for example, to help shift the distribution of production away from activities that serve mainly the desires of the elites to activities that would better meet the needs of the poor majority. Taxes might be levied on the automobile or on gasoline, for example, to limit the growth of the automotive sector, thereby freeing up investment resources for other purposes. This is already being done in many countries. The high gasoline taxes and automobile taxes in Western Europe, for example, arose from the view that these were luxury items that should be limited through taxation.

A more narrow role for taxes would involve their use as instruments for reducing the social costs associated with particular energy production and use activities. Taxation would be administratively simpler than regulation and would give the offender wider choice in how to reduce the offending activity. Thus, taxation would tend to be more conducive to innovation in control technology than would prescriptive regulations which specify both the degree of control required and prescribe or limit the choices for control technology. Taxation brings the advantages of price as an allocator of resources to the problem of controlling externalities. When taxes are used to control externalities, investments in less damaging alternative energy supply technologies and energy efficiency improvements become more economically attractive.

A major shortcoming of taxation is that typically external social costs are difficult or impossible to quantify in dollar terms. Hence, the tax level must be, to a large degree, arbitrary. Also, the outcome of a tax-based control strategy will tend to be less predictable than that of a regulatory approach.

A further disadvantage of imposing taxes on energy carriers is that they are reflected in energy price increases, which are regressive. For this reason, tax measures must be accompanied by policies that compensate for this inequity. One way to deal with the equity problem would be to use some of the revenues generated by the taxes to assist groups like the poor that are especially vulnerable to price increases. Economists often argue that the most efficient approach would be to deal with the problem of poverty as a separate issue and merely add the energy tax revenues to the general revenue pot. The problem with this strategy is that unless the tax legislation ties some of the revenues to helping the vulnerable groups there might be no compensating assistance for these groups.

To conclude, taxation as a policy-making instrument has a special appeal for implementing end-use oriented energy strategies where the diversity of end-uses and end-use technologies involved makes administrative solutions exceedingly difficult. While in some areas taxes will be less effective than regulations or other administrative solutions, taxation is well-suited to the wide range of energy activities that are not readily amenable to administrative solutions. Taxation can have an important role in an end-use oriented energy policy, wherever effective ways can be devised to deal with the equity problems that arise with consumption taxes (Section 5.3.2.1).

5.2.6 *Government-Sponsored Competitions*

A major challenge for energy policy is how to encourage technological leaps forward, as distinct from incremental changes in technology. We have shown in Chapter 2, for example, that for many energy-intensive activities such as space heating and the automobile, it is technically and economically feasible to improve energy performance for many important energy-using systems 5- or even 10-fold. Most of the policy instruments we have discussed so far are not likely to be especially effective in promoting such innovations. Regulations of energy performance are useful for clearing the market of "energy-guzzling" technologies but not for promoting highly energy-efficient technologies. Similarly, energy taxes, unless very large (and thus politically difficult to introduce) would not be powerful instruments for promoting technological leaps forward.

A mechanism which could be effective in promoting radical innovations would be government-sponsored design competitions, which would provide especially innovative producers either direct monetary rewards or indirect rewards through the publicity generated for winning designs. A good model of a successful, relatively low-cost venture of this sort was a design competition for the construction of passive solar homes carried out in the United States in 1980 by the Solar Energy Research Institute with the Denver Home Builders Association in the state of Colorado. At a very low cost to the government, this program attracted considerable publicity and may have had a major impact in changing the way the local construction industry builds houses (Note 5.2-J).

5.2.7 *Administrative Setting of R&D Priorities and Support of R&D*

The various failings of the market for motivating the private sector to do the R&D needed in support of society's long term goals (Section 5.2.1.2) indicate the importance of an active government role in R&D. It is important for the government to set R&D priorities in areas where private sector efforts are weak and to provide support for R&D that could be carried out in government laboratories, universities, and in the private sector.

5.2.8 *Administrative Generation and Dissemination of Information*

For the implementation of end-use strategy, a crucial element is good information on energy end-use—on how energy is being used (disaggregated by end-uses), on the extent to which energy is being provided in ways that further societal goals, on alternative technical means of providing desired energy services, on the cost-effectiveness of alternative end-use technologies, on the efficacy of alternative institutional mechanisms for providing desired energy services, and on opportunities for making future energy performance gains through technological innovation.

Because the information needed by policymakers and consumers is often not generated by market forces (Section 5.2.1.2), an active government role is called for in data collection and analysis, data base maintenance, and the dissemination of information to appropriate user groups.

These information production and dissemination roles for government are activities that offer the potential for large payoffs at relatively low cost. Information generation often requires the cooperation of energy-users, however, and for some users (e.g., small industrial energy users) the paperwork could be unduly burdensome, if government requests for information were not carefully designed to avoid such problems.

5.2.9 *Creation of Appropriate Policy Agents*

While many government bodies and private firms exist for promoting the expansion of energy supplies, the institutional infrastructure needed to implement end-use oriented energy strategies is still in its infancy in most parts of the world. One instrument that could be used to implement end-use oriented energy policies would be to create, through legislative or executive action, the needed new institutions that would be the homes of the needed policy agents. The needed institutions could be in the public sector—e.g., specialized centralized agencies or departments for dealing with various aspects of energy end-use, bodies for coordinating various energy supply-related and demand-related efforts being carried out in specialized agencies, decentralized extension centers, etc.—or the private sector—e.g., new industries that market energy services instead of energy supplies. The creation of new institutions in the private sector would be accomplished through incentives that would make their creation attractive and, as already noted, by broadening the charter of publicly-regulated energy utilities.

5.3 National Policies

We have organized the discussion of energy policies around the key features of the energy strategies we have proposed. Some policies are effective in implementing several parts of these strategies and are thus discussed more than once. This repetition might be avoided in a "laundry list" approach, but it is unavoidable in a holistic approach such as ours, which stresses the inter-relationships among policies and examines some policies in several contexts.

5.3.1 *Policies Relating to the Poor and Other Disadvantaged Groups in Developing Countries.*

5.3.1.1 *Energy Supplies for the Satisfaction of Basic Human Needs:* To bring about the satisfaction of basic human needs in an energy strategy requires a specific energy policy that makes basic human needs a direct, immediate target.

Concrete and specific courses of action are needed in developing countries that would channel energy supplies *on a priority basis* towards: the production, processing and storage of food;

the cooking of food; the manufacture of building materials; the use of these building materials for the construction of homes and public and commercial buildings; the maintenance of adequate comfort levels of lighting, heating/cooling in the interior of these buildings; the production of fibers, fabrics and clothing; the supply of potable water and the removal of human and other wastes through proper sanitation systems; the maintenance of a safe and healthy environment; the construction and operation of educational establishments; the running of various modes of transportation and channels of communication; and so on.

Special emphasis should be placed upon cooking, the lighting of homes, and the supply of sufficient and safe water for domestic use—end-uses that are normally ignored in energy planning.

Each of these end-uses warrants its own package of detailed policies. Policies regarding cooking should deal with choices for both fuels and stoves, promoting not only the sustainable supply of cooking fuels but also the distribution of stoves that are far more efficient than traditional stoves. Policies regarding lighting should give particular attention to the electrification of homes in efforts to bring electrification to settlements (cities, towns, villages); preferably these policies should involve mechanisms for overcoming the obstacles that inhibit the hook-up of some households in settlement electrification efforts, where this has been a problem. Policies regarding safe drinking water should aim to ensure that the availability of energy is not a constraint on the boring of wells or the pumping of water.

5.3.1.2 *Energy Supplies for the Energy Needs of the Poor and Other Disadvantaged Groups:*

A policy of priority allocation of energy for basic human needs in developing countries is relevant to the entire population. In theory, such a policy would ensure that the energy needs of the poor and other disadvantaged groups are met *along with the rest of the population.* But in practice, what invariably happens in dual societies is that the allocation of resources in a resource-scarcity situation starts with the wants of the privileged elite. Thus, it is best to make doubly sure that the energy needs of the poor and other disadvantaged groups are met on a priority basis by undertaking specific courses of action, instead of relying on macroeconomic policies, to provide energy services for the poor, women and other disadvantaged groups.

Equity is an important goal here, but not the only one. For people living at low levels of nutrition, reducing the energy expended on avoidable chores is a way of improving nutritional status and increasing labour productivity. It is important to examine how inputs of inanimate energy can alleviate the situation of the poor, particularly poor women. If such possibilities are found, supplying the requisite energy inputs should be given high priority.

The problems of poverty, oppression and exploitation cannot be solved through energy strategies and policies alone. But because of the ubiquitous role of energy in society, strategies and policies aimed at reducing these evils will have major energy aspects. Many energy strategies and policies have a direct bearing on the condition of the poor.

5.3.1.3 *Utilizing Energy for Generating Employment:*

Most developing countries are burdened with severe unemployment problems. Generating employment should be a major objective of the development process in general and in industrialization in particular, since employment is one of the most important basic needs. Employment generates purchasing power, which enables the employed to gain access to the means to satisfy other basic needs.

Making energy supplies available can help create employment, but the employment potential per unit of energy or power provided can vary markedly from one industry to another. The number of jobs generated per unit of installed electric power, for example, depends in a significant way upon the particular industry. There are highly energy-intensive industries where energy is virtually a major "raw material", e.g., aluminium smelting, which create very little employment per kW. And there are non-energy-intensive, labour-intensive industries where energy is mainly an amplifier of labour or controller of information.

It follows that if the bulk of energy is diverted to energy-intensive industries, then the production and distribution of energy may make little contribution to employment generation, and therefore to development. This is not simply a theoretical possibility. A recent report pertaining to the state of Karnataka in India (where the industrial sector utilized 74 percent of the state's annual electricity output of 6.2 TWh) showed that 18 major electro-metallurgical firms consume two-thirds of the industrial electricity and provide direct employment to 4000 people, while 1200 other firms which use the remaining one-third of the electricity provide employment for 250,000 people.[9] Of course, a rigorous analysis should also take into account the down-stream employment (e.g., the employment provided to those who manufacture aluminium conductor cables) as well as the direct employment.

Thus, energy should be used as an instrument for employment generation in countries beset with unemployment. Specifically, this means that considerations of energy and employment intensity should be taken into account in allocating to new industries energy supplies, particularly electricity.

5.3.2 *Policies for Dealing with Poverty in Industrialized Countries*

While middle and upper income households in industrialized countries can afford to make some adjustments to higher energy prices—by insulating their homes, by buying more fuel efficient new cars, etc.—the poor especially in the United States (Section 1.3.5), with no or little capital resources and credit, often continue to live in poorly insulated, drafty houses built before the oil crises and to have to depend on the behemoths discarded by the better-off for transportation

The burdens of high energy costs would be reduced for the poor through the implementation of some general energy efficiency policies—such as energy performance standards for appliances and for automobiles (Section 5.3.5.2). But these measures alone are not likely to adequate to prevent economic hardship for the poor. Moreover, other policies aimed at raising energy prices to reflect the marginal cost of production (Section 5.3.4.2) will aggravate the suffering of the poor if not carefully crafted.

5.3.2.1 *Energy Pricing Reform and the Poor:* The suffering of the poor has often been given as a reason to keep energy prices low. We contend, however, that instead of relying on economically inefficient systems of price controls to protect the poor, it is far preferable to complement an economically efficient pricing system with programs specifically targeted at the needs of the poor.

In adjusting utility rates to reflect marginal rather than average costs, one way to ease the burden on the poor would be to keep the overall revenue generation rate fixed but to allow the rate charged for energy to vary with the consumption level—rising from a rate below the present price for low levels of consumption to marginal costs for high consumption rates. Alternatively, rates could be raised across-the-board to marginal cost levels, while compensating measures are enacted to protect the poor. When rates are raised to marginal costs, utilities would realize windfall profits proportional to the difference between the marginal and the average costs. These windfall profits should either be taxed away or rebated to consumers in ways not directly related to energy consumption. In either case part of the excess revenues could be allocated to subsidies for the poor.

Similarly, if new taxes are to be levied on energy carriers, the regressive nature of the tax might be compensated for by creative use of the revenues generated by the tax. For example, energy tax revenues might be used to offset some existing tax which is regressive—such as the social security tax in the United States.[10] Or energy tax revenues might be used to help finance energy efficiency investments by the poor.[11]

5.3.2.2 *Energy Subsidies for the Poor:* It will not always be possible to help the poor through creative pricing reform proposals. In many instances, it will be desirable to enact programs to help the poor in more direct ways.

Already spending such a large fraction of income on energy expenditures (Section 1.3.5) and generally having poor credit ratings, the poor can benefit only marginally from many of the measures aimed at general consumers (Section 5.3.5). Some kind of direct subsidy to poor households is generally needed. But what should be the nature of this subsidy?

In the United States at the present time, modest federal assistance is provided to help pay for the fuel bills of the poor, and an even more modest effort is committed to providing support for winterizing the homes of the poor. Far more aggressive programs are called for. While assistance in paying fuel bills is needed to reduce near term suffering, the greatest need is subsidies for *investments* to reduce fuel bills—especially for space heating. Because of the inferior quality of much of the housing of the poor, e.g. in the United States, it is desirable to coordinate the provision of energy efficiency improvements where necessary with programs aimed at general improvement of the housing stock of the poor.[11]

Large one-time investment subsidies to the poor for retrofitting their homes would be more economically efficient than continually subsidizing their fuel bill subsidies and would serve to break the demoralizing effect of continuing dependency associated with all assistance programs.

5.3.2.3 *Employment Aspects of End-Use Oriented Energy Policies:* In the long run, changes in the structure of employment associated with implementing an energy efficiency strategy may well be more important to low income people in industrialized countries than the changes experienced by them as consumers. Essentially, all econometric studies show that energy and labor are substitutable inputs in the overall economy and that rising energy prices will lead to substitution of labor for energy.[12] Thus energy taxes in particular should tend to generate employment. More specifically, economic production associated with energy efficiency devices and low energy-intensive products tends to be characterized by higher employment levels per dollar of economic activity than would the production that was replaced.[13]

Employment associated with end-use efficiency would also tend to be less specialized. For example, a case study of reusable vs. throwaway bottles showed that with the less energy-intensive reusable bottle option, new jobs in retail stores, local delivery, and bottle cleaning would replace a smaller number of jobs lost in basic materials processing assembly line operations and long distance shipping associated with the throwaway option.[14] Similarly, the demand for efficiency services, such as repair work on durable products and construction and renovation of buildings to improve energy performance, would create new jobs requiring rather general skills. This increase in less specialized employment could be a very favourable side effect of energy conservation, in that it would tend to reduce structural unemployment, which has proved to be one of the most intransigent poverty problems confronting industrialized countries.

5.3.3 *Policies for Priority Allocations of Energy to Vulnerable Sectors in Developing Countries*

If there are energy policies for the satisfaction of basic needs and for the energy needs of the poor and other disadvantaged groups, it might seem redundant to reflect the same concerns through sectoral allocations of energy. But this is not the case. Allocation of energy—electricity, oil, coal, etc.,—are not made either according to basic needs or according to income groups; *they are made on a sectoral basis.* Furthermore, government ministries and departments are generally (but not exclusively) constituted on sectoral lines—industry, agriculture, transport, etc. As a result, the energy policies concerning basic human needs and the poor and other disadvantaged groups have to be incorporated into sectoral allocations.

5.3.3.1 *Emphasis on Energy Services for the Domestic Sector:* If energy policies are to protect the poor and other disadvantaged groups and be biased towards their basic needs, it is implicit that there should be a special emphasis on energy services for the domestic sector. In particular, the energy services of cooking, lighting and space heating/cooling, and the energy inputs associated with maintaining a safe and healthy environment should receive high priority. Indeed, if it is not assigned top priority in the allocation of energy supplies, the domestic sector will usually be by-passed in the planning process, in favour of other sectors with more powerful lobbies. For example, the rural electrification programmes in developing countries have until recently neglected the electrification of all homes.

5.3.3.2 *Priority Allocation for the Agricultural Sector:* Though the rich farmer's lobby in developing countries is usually powerful enough to secure adequate supplies of, for instance, electricity for irrigation pumpsets, the utilities are usually reluctant to erect the transmission-distribution lines to remote scattered settlements and farms because of the high costs of supplying load centers with low load factors and small loads. However, in view of the vital importance of food production to most developing countries, allocation policies should aim to ensure that energy supply is not a constraint on the productivity and output of the agricultural sector.

Fortunately, providing energy for agriculture does not require centralized generation nor does the energy carrier have to be electricity—decentralized generation may prove to be more economical beyond certain break-even distances from the centralized source, and alternative sources for example, pumpsets fueled by producer-gas derived from wood, may prove to be more economical. These alternative modes of generation and energy carriers would be given fair consideration if the utilities were mandated the responsibility of providing the energy service of lifting groundwater for irrigation or of *energizing* irrigation pumpsets rather than being restricted to *electrifying* them from the grid.

5.3.4 *Policies to Promote Fair Competition Among All Energy Sources and Between Energy Supply and Energy Efficiency Technologies*

Defects in energy markets created by subsidies and the failure of energy prices and utility rates to properly reflect the costs of bringing forth new energy supplies (Section 5.2.1.2) are not inherent market imperfections but distortions that can and should be removed in the interest of promoting economic efficiency. Without these distortions, alternative energy supply and end-use efficiency investments would be treated much more evenhandedly in the market.

For this reason, it would be wise to move toward eliminating subsidies to energy producers and toward marginal cost pricing in markets where prices are controlled by government policy. Such policies should be carried out in conjunction with complementary policies needed to implement an end-use oriented energy policy.

Also, a good, highly disaggregated energy demand/supply database is essential as an aid to policymakers in bringing about an evenhanded treatment of alternative energy options and in coordinating these actions with other energy policy objectives.

5.3.4.1 *Eliminating Energy Supply Subsidies:* Despite the distortions posed by the existing programs of subsidies to energy supply industries, it is often regarded as politically infeasible to eliminate these subsidies. A popular response to the existing system of subsidies for conventional energy supplies is to propose similar subsidies for alternative investments in energy productivity improvement and solar energy.

As noted in an earlier discussion, such alternative subsidies also pose serious problems (Section 5.2.3). It is generally very difficult to neutralize one market distortion with another. The compensating measure will tend to only partially offset the effects of the original distortion and it may do so in unintended, undesirable ways. Moreover, it is by no means clear that there is any need to introduce subsidies in order to make alternative technologies associated with end-use oriented energy strategies economically competitive.

A more appropriate policy would be to move away from reliance on targeted subsidies. Instead of introducing new subsidies to promote energy productivity improvement, "fairness" in energy markets should be promoted by the more politically challenging approach of eliminating existing subsidies for energy supplies. This approach should be coupled to public policies that would lead to lifecycle cost minimization in market decisions relating to energy (Section 5.3.5).

In general, energy subsidies should be restricted to solving the problems which the marketplace cannot solve, such as the energy problems of the poor and the promotion of R&D.

5.3.4.2 *Rationalizing Energy Prices:* Economic efficiency will be enhanced by the elimination of energy price controls that keep prices below marginal costs and by the redesign of utility rate structures so that consumers have a greater incentive to use energy more efficiently.

It is essential that policies enacted to bring energy prices in line with marginal costs must be carried out in conjunction with policies that would address the problems that the original pricing policies were attempting to solve.

Consider first the situation in developing countries, where the price of kerosene is often set at a low level to protect the poor. Raising the price of kerosene and its sister diesel fuel to market clearing levels should be done only in conjunction with programs that would provide the poor with alternatives to kerosene for lighting and cooking—e.g., rural electrification for lighting and biogas or producer-gas for cooking. The excess revenues from kerosene and diesel fuel sales might be used to help finance such programs.

This example illustrates the general principle that the excess revenues which would arise from adopting a marginal cost pricing policy could be used constructively for complementary purposes. One possibility would be for the government to tax these excess revenues, perhaps using them to offset other taxes that are regressive or otherwise unpopular. Another possibility would be to return the excess revenues to consumers through rebates in ways not directly related to consumption, perhaps serving some other purposes. There are countless ways this might be done. In the case of residential customers, the rebates might be on a per capita or per household basis. In countries with serious unemployment problems, rebates to industrial customers might be set in proportion to the number of employees in the firm.

Rebates may not be practical in some situations. For example, it may not be practical to rebate excess revenues to domestic customers in many developing country situations because many low income households are not well integrated into the market economy and hence the administration of rebates would be difficult.

An alternative to rebating the excess revenues would be to use them to help finance investments in energy productivity improvement or alternative supplies, with priority given to the groups most severely affected—e.g., the poor.

For utilities, as noted previously, an alternative would be to keep the overall level of revenue collection the same but to allow the rate to vary with the consumption level—rising from a rate below the present price for low consumption levels (so as to protect the poor) to the marginal cost for a sufficiently high consumption level. This strategy could also be used to convert present utility rates that involve a large fixed cost component (and thereby discourage efficiency) into rates that would provide a powerful incentive to use energy more efficiently.

There are of course many more ways besides these that would enable a transition to marginal cost pricing. Whatever the path, however, there would likely be strong vested interests opposed to moves in this direction, making such moves politically very challenging.

5.3.4.3 *Database Generation and Maintenance:* To try to bring about an evenhanded treatment of technologies aimed at reducing energy demand and alternative energy supply options in ways that are consistent with other energy policy objectives requires that decision-makers know much more about the existing energy system than do decision-makers who implement conventional energy supply strategies. Specifically, energy planners need detailed information on overall requirements and consumer prices paid for various energy carriers, energy requirements and expenditures for different sectors and for various subgroups within sectors, patterns of energy consumption disaggregated by end-uses, locally available energy resources, e.g., biomass, etc. And they need demographic and economic data to complement this energy data.

To meet these energy planning needs, detailed and highly disaggregated energy demand/supply databases should be developed and maintained.

5.3.5 *Policies Relating to Energy Efficiency Improvement*

While a rationalized energy price environment would lead to more investments in energy productivity improvements, the level of such investments would still be much lower than would take place if lifecycle costs were minimized using the market rate of interest as a discount rate. For the many reasons we discussed earlier (Section 5.2.1.2), there are numerous institutional obstacles to such consumer decision-making, and the high consumer discount rates implicit in typical consumer decisions relating to investments in energy efficiency are the result. Both the prospect of economic gains and the various external societal benefits associated with low energy demand futures justify new public policies aimed at promoting cost-effective investments in energy efficiency improvement—i.e., investments which lead to reduced lifecycle costs when the discount rate is near the level of market rate of interest.*

---

* In comparing alternative investments, the discount rate used in each case should reflect the relatives risks involved. Thus, in comparing an investment in new energy supply with an investment in energy efficiency improvement, the discount rates used in each case to estimate lifecycle cost may be different. One cannot easily generalize that energy supply investments are more risky than investments in energy efficiency or vice versa. For the purpose of the present discussion, we shall assume that such differences in risk are small enough to justify saying that cost-effective investments in energy efficiency improvement would involve using discount rates near the market interest rate.

There are a number of actions that can be taken to promote cost effective investments in energy efficiency:

- Improving the flow of information.
- Targeting energy performance improvements.
- Making capital more readily available.
- Promoting comprehensive energy service delivery.
- Promoting cogeneration.

5.3.5.1 *Improving the Flow of Information:* Since lack of good information about energy saving opportunities is a major impediment to investments in cost-effective investments in energy efficiency improvements, government can play an important role in improving market performance by various measures that would improve the flow of information to consumers. This can be accomplished in many ways.

Generic information might be made available through community-based "energy extension services," akin to the agricultural extension services, that have successfully provided farmers with information about new agricultural techniques. Such extension services could be useful in industrialized and developing countries alike. In the latter, for example, such centers might be involved in showing how to make and properly use efficient cooking stoves.

Also energy utilities might be required to offer advice that reflects customers' unique needs, as determined by audits of their premises. Such audits for residential customers are required of most large gas and electric utilities in the United States. The challenge here is to ensure that high-quality and useful information in transmitted to the customer. Care must be taken with regard to the audit design (Note 5.3.5-A), and the utilities involved must be sufficiently motivated to do a good job.

Still another way to improve the flow of information would be to require the labeling at the time of sale of the energy performance of various energy-intensive products. Labeling would be relevant for those products for which there exists an energy performance index that can be readily measured (for some prescribed duty cycle), that is relatively unambiguous, and that is easily understood. Candidates for such labels are automobiles (a label showing 1/100 km or mpg), various household appliances, and even houses (a label giving some indication of heating requirements). [11]

It is not always necessary to improve the flow of information dramatically in order to overcome the market barrier posed by the high cost of information acquisition, however. An alternative strategy would be to encourage the development of companies offering energy services, instead of particular energy technologies. These companies could relieve the individual users, in part or in full, of the burden of information acquisition by selling energy services instead of devices (see below).

5.3.5.2 *Targeting Energy Performance Improvements:* For many of the same energy-using activities that are suitable for labeling energy performance, market interventions that go beyond mere labeling may have to be considered, because: (1) lack of information is just one factor that has led to high consumer discount rates (Section 5.2.1.2); (2) consumer purchases of high efficiency devices that lead to reduced lifecycle costs will help promote overall improved economic efficiency (Section 2.5.3); and (3) there are sometimes important external societal benefits that result from a shift to high energy performance devices at little incremental direct cost—a point that warrants some elaboration.

An example relating to the automobile illustrates this point. With present technology, it is technically feasible to improve the fuel economy of automobiles from the present global average of 13 1/100 km (18 mpg) to some 3 1/100 km (80 mpg) or better (Section 2.5.3.3). The economics of such improvements are indicated graphically in Figure 2.10, which shows the total cost of owning and operating a car per km (mile) of driving as a function of the fuel economy—it indicates that up to fuel economies of some 8 1/100 km (30 mpg) the total cost declines rapidly with fuel economy but thereafter remains roughly constant. The consumer has no incentive to see higher fuel economies because the savings due to reduced fuel requirements are approximately offset by the car's higher first cost associated with its improved fuel economy, and because at high fuel economy levels fuel accounts for such a small fraction of the total cost of owning and operating a car. While the individual consumer would be no better off in dollar terms having a car with a fuel economy falling on the higher efficiency side of this curve, he would be no worse off either. But if most consumers had such efficient cars, society would be enormously better off, because of the reduced oil import requirements and the lower world oil price. The prospect of generating such enormous societal benefits without burdening the consumer provides a powerful rationale for some kind of market intervention to promote high fuel economy cars.

This finding, that the total lifecycle cost varies only marginally over a wide range of efficiency improvements, is not restricted to the automobile. There are numerous others. For example, in a study of space heating systems for new gas heated houses in the United States. It was found that the total lifecycle cost for space heating varied by only a few percent with various combinations of heating systems and levels of insulation, while fuel use varied by factor of three.[11]

There are various policy instruments that could be used to bring about market shifts to high efficiency products. The regulation of energy performance might be used to promote high automotive fuel economy or high efficiency appliances. Or taxes might be used to accomplish much the same goal—e.g., devices performing worse than the average for new devices might be taxed at the time of purchase, with the penalty increasing in proportion to expected extra lifecycle energy requirements.

In the case of household appliances utilities might be required to offer rebates to customers who purchase appliances more efficient than the average for new appliances, with the rebates increasing in proportion to the expected energy savings. To the extent such rebates allow the deferral of more costly new supplies, the consumer, the utility, and its other rate payers would all benefit. A growing number of utilities in the United States offer such rebates for certain appliances.

One way to promote market entry of highly efficient energy performers would be for the government to sponsor design competitions, as has been done in various instances for new houses (see, for example, Section 5.2.6) and has been proposed in the United States for automobiles (Note 5.3.5-B). Design competitions could in principle be applied for any end-use device, even cooking stoves in developing countries, which can be well characterized by a suitable energy performance index.

5.3.5.3 *Making Capital More Readily Available for Investments in Energy Efficiency:* The problem of capital scarcity for investments in energy productivity improvement can be dealt with through both general tax reform measures and measures that direct capital resources to specific applications.

Since so much capital has in the past been directed to energy supply investments, elimination of subsidies to new supplies should help make capital more available for purposes other than energy production. The question is whether more than this should be done to promote the availability of capital for investments in energy efficiency by energy-using businesses, other institutions, and individual consumers?

Consider first the energy-intensive, basic materials processing industries of industrialized countries. The slow demand growth in many of these industries (Section 2.4) provides such a poor climate for innovation that many have argued for some kind of government assistance that would accelerate modernization of these industries in order to ensure their continued competitiveness in world markets and safeguard national security. But if more capital is made available to these industries then there would be less capital available for other industries, which may have better growth prospects. Questions relating to government assistance for such industries are difficult and politically charged ones, for which there are no easy answers. Perhaps the most promising way to save the stagnating basic materials processing industries in the North (or some parts of them) is to leave them on their own, bringing the full pressure of foreign competition to bear upon them and forcing them to innovate. According to Harvard economist Harvey Lieberstein, this aspect of foreign competition—the spur it gives to innovation—is a far more important reason for a country not to erect barriers to free trade than the classical argument that free trade enables countries to best exploit their comparative advantages.[15]

The problem of capital scarcity for the individual consumer is more clear-cut. In many instances, capital resources should be made more readily available than they now are in many instances. This could be accomplished in several different ways. For example, mortgage laws could be changed so that in determining the size of the allowable mortgage, consideration would be given to the household's increased ability to make loan repayments if the house is more energy efficient. Or special funds could be established by government to finance qualifying energy efficiency investments. Or energy utilities could be encouraged to make capital available for energy efficiency investments on its customer's premises.

It may not be necessary to make this capital available at below market interest rates, though, in order to promote effectively investments in energy efficiency (Section 5.2.3). Providing such subsidies limits the overall capital supply. It may often be the case that simply making more capital available at market interest rates, in conjunction with improved overall delivery of energy services is sufficient inducement for consumers to invest in energy efficient improvements. Capital subsidies should be reserved for the situations where less costly alternatives cannot work—e.g., investments in energy efficiency by the poor (Sections 5.3.1 and 5.3.2).

5.3.5.4 Promoting Comprehensive Energy Service Delivery: It is easy for a consumer to purchase specified quantities of particular energy carriers like oil or electricity. Well-tested economic systems exist for making such transactions; the quantities exchanged are easy to measure; and both producer and consumer understand the values of the commodities involved.

This is not the case for investments in energy savings. The marketing of energy end-use efficiency is inherently more complicated than the marketing of energy supplies or conventional end-use technologies. One must be concerned not just with producing new energy-efficient devices, but with the full spectrum of relatively novel marketing problems; (1) diagnosis of the individual consumer's needs for obtaining energy services in the most cost-effective manner—which will often lead to the identification of needed technical changes; (2) consumer education as to the need to make these changes—a task made difficult because the expected savings is often ambiguous; (3) the financing of any new devices or contractor work that may be required—a problem that arises because energy efficiency improvements usually requires increased first costs; (4) after-purchase servicing; (5) monitoring of performance in the field to ascertain actual savings, with feedback that can be used to modify energy saving strategies as needed in light of this experience.

To effectively promote energy end-use efficiency across a broad range of energy services—from cooking with fuelwood stoves in developing countries to space heating in industrialized countries—policies should address all such aspects of the marketing of conservation—i.e., they should be concerned not just with the production of the hardware involved but with all the needed supporting "software" as well.

There are various ways such comprehensive marketing of energy conservation might be achieved. One possibility, as noted earlier, would be to convert energy utilities into energy service companies, that is, companies that market *energy services* (heating, cooling, lighting, etc.) in much the same way they today market energy carriers. If utilities were to evolve in this way, they would come to play a role originally envisioned for them by Thomas Edison when he invented the incandescent bulb; he proposed that utilities sell *illumination,* thereby giving them a financial interest to provide this illumination in the most cost-effective way.

The electric and gas utilities are good candidates for marketing the services required for such an effort. Already a number of the more progressive utilities in the United States have initiated energy conservation programs that include: (1) providing advice on investments in energy efficiency; (2) offering to arrange for contractors to carry out such work; (3) financing such investments with low or zero interest loans; and (4) providing rebates to consumers for the purchase of energy efficient appliances and/or to appliance dealers for promoting their sales.[11]

Accustomed to accumulating large quantities of capital, utilities are well-positioned to direct these resources to energy efficiency investments. Also they have an administrative structure for channeling the capital to essentially all households and businesses. Moreover, the utility billing system offers the opportunity for customers to pay "life cycle cost bills" as alternatives to fuel or electricity bills, if they were to receive loans from the utility which they could pay off on their utility bills.

Finally, utilities are in a good position to deal with a number of particularly difficult tasks, such as the retrofitting of existing buildings with energy efficiency improvements. The barriers that now inhibit many householders and small businesses from making improvements in energy efficiency could be greatly reduced if not overcome if utilities were to offer a comprehensive "one stop retrofit service," involving: audit and post-retrofit inspection, coordinating the contractor work, and arranging long-term financing.

Where utilities are privately-owned and publicly-regulated, utility regulators could play an important role in creating effective programs. Instead of simply requiring the utility to establish energy efficiency programs, regulators should consider modifying the reward structure to give the utility a financial stake in exploiting opportunities for cost-effective improvements in energy efficiency. One promising approach would be for the public utility commission to allow the utility to treat program costs as operating expenses and grant financial rewards for energy savings actually realized or financial penalties for failure to meet established goals (Note 5.3.5-C).

In some instance utilities will be unable or unwilling to create needed energy conservation programs. For example, electric or gas utilities may not wish to offer a retrofit service for oil-heated homes. Or a utility may have so much excess generating capacity that it sees no need to help its customers use energy more efficiently. In such circumstances, government could stimulate the creation of independent new companies that would market energy efficiency improvements, e.g., by making loan or grant assistance available to customers of firms that provide such a service.

If neither utility nor private sector endeavors are feasible or practical for marketing comprehensive energy conservation, local governments could assume the responsibility.

5.3.5.5. *Promoting Cogeneration:* Cogeneration, the combined production of electricity and useful heat in small-scale installations located at or near the point where the heat would be used, is an important energy-saving concept (Section 2.5.3.4) which warrants special attention from policy makers.

Cogeneration has long been discouraged throughout the world by electric utilities who have seen cogeneration as a threat to their businesses. Utility resistance to cogeneration has been manifest in policies ranging from refusal to allow cogenerators to hook-up to the utility grid to unfavorable pricing arrangements for cogenerators that are allowed to hook-up—charging cogenerators extraordinarily high rates for back-up power and offering extraordinarily low rates for the purchase of power generated in excess of the cogenerator's needs.

Policies providing fair treatment in the market for cogenerators are needed. One possibility is along the lines of a law passed in the United States in 1978, which requires utilities to hook-up cogenerators to the utility grid and to pay them for the electricity they want to sell to the utility at a price equal to the cost the utility could avoid by not having to provide this power (Note 5.3.5-D). The law has had a tremendous impact in creating a cogeneration industry. As a result, cogeneration's share of U.S. electricity supply is expected to increase from 3 percent in 1980 to some 15 percent in 2000. Additionally, incentives might be directed to getting electric utilities or special subsidiaries of utilities directly involved in cogeneration projects, a possible new role of utilities as "energy service companies" (Note 5.3.5-E).

5.3.5.6 *Scorekeeping:* Above all, policymakers need to know the efficacy of using alternative policy instruments in promoting various energy saving goals. Such information cannot generally be known without empirical investigation. To help policymakers better understand what works and what does not and why, requires good "scorekeeping." Scorekeeping efforts which monitor actual savings realized in ongoing energy savings programmes and provide timely feedback to planners could be used to provide a basis for continually improving these programs, and they would help protect consumers against fraudulent or incompetent energy service firms.

5.3.6. *Policies Relating to Renewable Resources*

An emphasis on renewables in R&D (Section 5.3.7) is needed to enhance the prospects of making a timely shift to dependence on renewable energy resources. Such a shift is essential in the long run in order to bring about an energy future consistent with the achievement of a sustainable world.

But even in the near term, renewable energy in the forms of hydropower and biomass can make significant contributions to energy supply and contribute to energy self-reliance.

5.3.6.1 *Hydroelectric Policies:* While undeveloped hydroelectric resources are enormous in many developing countries (Appendix: Renewable Energy Resources), the successful development of these resources requires extensive advanced study to identify potential major ecological and social problems—such as the opportunity costs of diverting large land areas to hydroelectric facilities; the consumptive water losses associated with the creation of large artificial reservoirs in arid regions; the potential for sudden dam failure, with catastrophic loss of life and property; the disruption of fisheries and the spread of certain water-borne diseases such as schistosomiasis, malaria, and sleeping sickness; the displacement of populations.

In some areas, where such costs may outweigh the potential benefits, no dams should be built. Where dams are to be built, the planning should aim to avoid or minimize the potential adverse impacts and be carefully integrated into overall development planning. Emphasis on energy end-use efficiency makes more time available for such integrated planning to take place that would be the case with more rapid electricity demand growth.

5.3.6.2 *Bioenergy Policies:* Biomass is a widely available feedstock for energy use. And, like hydroelectricity, biomass is already widely used for energy purposes, but inefficiently and often in non-sustainable ways (Chapter 3). For bioenergy to make a signficant contribution in the meeting of energy needs, it will be necessary to adopt policies that promote both the sustainable production of biomass and a shift to modern, efficient biomass energy conversion technologies.

Policies to promote the sustainable production of biomass should be carefully integrated with forestry policies and more general land-use policies. Because agricultural expansion onto marginal lands is a major contributor to deforestation, particular attention should be given to agriculture-forestry land use conflicts.[18] One option would be to concentrate future expanded agricultural production on the better lands, using modern production techniques, and simultaneously to commit the more marginal lands to energy production on biomass farms or plantations.[18, 19] Another option would be to promote multiple use of the more marginal lands through agro-forestry production techniques.[20-21]

A major challenge in bringing about the sustainable development is providing adequate incentives for the production of biomass for energy purposes as an alternative to the overuse of natural forests. The present low biomass price has been cited as a formidable obstacle to halting and reversing the process of deforestation.[22] But experience to date with biomass plantations has demonstrated that such plantations in many developing countries can be profitable at biomass prices far less than what is needed for the biomass to be competitive with imported oil (Note: 5.3.6-A). The challenge is to make biomass available in forms and in markets where it can be compete with imported oil.

If raw biomass were converted to high quality solid, gaseous, liquid, and electrical energy carriers, bioenergy could often compete with imported oil, and the present problem of a low biomass price would probably be largely resolved.[19] Emphasis should be given in public policy therefore to commercialization efforts based on present technology and to R&D efforts that would promote this modernization of the bioenergy resource.

Efforts to modernize bioenergy must be carefully coordinated with efforts to resolve the fuelwood problem. Efforts to raise the biomass price could have a devastating effect on the poor who depend on low cost fuelwood for cooking, unless measures are simultaneously made to improve the efficiencies of biomass cookstoves and to shift to more modern energy carriers (e.g., biogas or producer-gas derived from biomass).

5.3.7 *Policies to Promote the Generation of New Knowledge and Technological Advances*
The generation of new knowledge and new technology could facilitate the achievement of all the goals associated with end-use energy strategies.

5.3.7.1 *Priority Areas for R&D in Support of End-Use Oriented Energy Strategies:* Research is needed on both social and technical aspects of energy end-use, on renewable energy, and on the side effects of energy production and use.

Social science research is needed: (1) to better understand the particular problems of the poor and other disadvantaged groups and how the needs of these groups might be effectively addressed; (2) to improve techniques for measuring the success of conservation programs; (3) to understand the reasons for successes and failures in such programs; and (4) to understand how to motivate consumers to base their energy decisions on consideration of lifecycle costs, or equivalently, how to design conservation programs so as to make cost-effective energy saving options attractive to consumers.

While major improvements in energy productivity can be made with end-use technologies that are already commercially available, there remains a considerable potential for further improvement that could be realized with the appropriate R&D effort. The era of energy demand consciousness is barely a decade old, which means that the state of end-use technology is 'low on the learning curve.' This point can be illustrated with reference to Figure 2.6, which shows the history the efficiency of central station thermal power plants—a curve which might be regarded as representing a 'learning curve' for electricity production technology. The eight-fold increase in efficiency in the period 1900-1960 brought the electric utility industry from a low point on this learning curve to a high point, beyond which further improvement was only marginal. If energy efficient end-use technology were to be placed on such a learning curve, it would be located where the thermal electric generating technology was at the turn of the century. Energy efficient end-use technology is still in its infancy.

More generally, R&D is needed to provide a continuing basis for 'technological leaps forward,' in which many characteristics of the system under change are improved at the same that its energy characteristics are improved. Innovations of this type, made repeatedly in modern history, have been major contributors to economic progress (Section 2.5.3.4).

And finally, continuing research is needed on the external social costs of energy production and use, to provide a better understanding of these problems and what to do about them.

In all these cases, the private sector cannot be relied on to generate the needed R&D (Section 5.2.1.2), so government action is needed.

5.3.7.2 *The Dimensions of Present R&D Efforts:* Despite the long-run importance to end-use oriented energy strategies of research on both the social and technical aspects of energy end uses, and on renewable energy sources, these efforts play only minor roles in present government sponsored R&D programmes.

Table 5.1 shows government R&D expenditures for 1981 and 1983 in the International Energy Agency (IEA) countries. Overall expenditures declined. In 1981, only a minor share of government R&D resources was committed to energy conservation and renewables—about 6 percent and 13 percent, respectively. Total government expenditures in these areas was only-one-third of that spent on nuclear energy or one-fourth of that spent on nuclear and fossil energy sources combined. By 1983, overall government R&D expenditures had declined by 20 percent, and the emphasis on nuclear energy was strengthened. In the United States under the Reagan Administration, R&D expenditures on energy conservation and solar energy were reduced nearly six-fold while nuclear R&D expenditures remained essentially unchanged (Figure 5.2). These data show that there is a clear need for a major restructuring of energy research and development efforts to carry out an end-use oriented approach in industrialized countries.

Among the efforts that have been made by aid agencies to support the development of a technological capability within developing countries are those of agencies such as UNIDO (United

Table 5.1. Distribution of 1981 and 1983 Government Energy R&D Budgets in IEA countries (in percent)

	1981	1983
Total expenditure in *current* prices ($ millions)	8,356	6,632
Conservation	5.9	5.9
Oil and Gas	3.6	4.5
Coal	14.1	8.0
Nuclear (non-breeder)	27.8	36.6
Advanced nuclear	26.0	31.2
New Energy Sources (Solar, Wind, Ocean, Biomass, Geothermal)	12.9	8.8
Other Sources and New Vectors	0.7	0.4
Supporting Technologies	9.1	4.6
	100	100

Sources: OECD, 1982; OECD, 1984.

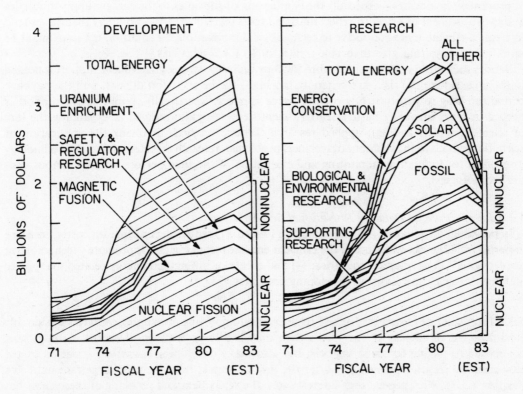

Figure 5.2 Expenditures of the United States federal government on energy research and development. *Source:* National Science Foundation, 1983.

Nations Industrial Development Organization), which has a charter specifically designed to support industrialization and which recently became active in the area of energy.

UNIDO sponsors studies and international conferences in addition to acting as a clearinghouse for information exchange, working rather closely with governments and technological institutions. Despite such linkages, the effectiveness of this organization in identifying and encouraging new directions for technological development has not been outstanding—perhaps because its role has been limited to advising and counselling.

While organizations such as UNIDO have a potentially important role as clearinghouses for technological information, what is needed are more structured programs that aim not just at providing information but at creating and developing new technologies relevant to meeting broad development goals. Linkages of R&D efforts to the development planning process are needed to assure both that R&D is in line with development goals and that planners are aware of the long term prospects for new technology.

5.3.7.3 *Using Various Policy Instruments to Redirect R&D:*

The role of government and aid agencies in relation to the R&D needed to support end-use oriented energy strategies should be to help create an economic climate conducive to greater private sector R&D in this area and to sponsor promising R&D activities in areas where the private sector effort is inadequate.

The policies outlined above for improving the climate for investments in energy efficiency improvement in industry—especially the elimination of subsidies to the energy supply industries and the adjustment of prices to reflect marginal costs (Section 5.3.4)—would by themselves tend to create a climate more conducive to technological innovation. In addition, tax laws might be modified in ways that give favorable treatment to R&D activities.[23]

Besides such efforts to encourage more R&D by industry, strong government support is needed in the areas where the risks are generally too high, the benefits too diffuse, and the payback period too long to justify strong private sector support. Specifically, public sector support is needed in areas involving basic research (scientific investigation aimed at expanding the lore of scientific understanding), applied research (generic research involving the application of scientific methods to the solutions of technical problems), research on externalities, and technology assessment (including the monitoring and evaluation of the field performance of various end-use technologies).

5.3.8 Energy Policies Relating to National Self-Reliance

In pursuing self-reliance, policies should be seek to: (1) reduce dependence on costly, insecure imported oil; (2) make wider use of the efficient end-use technologies; (3) promote "technological leapfrogging" in developing countries; (4) and create or strengthen energy-related manpower and institutional capabilities in developing countries.

5.3.8.1 *Reducing Oil Imports:*

The self-reliance of oil-importing developing countries was doubly eroded by the oil shocks of the 1970's. Not only is a large fraction of export earnings now committed to paying for these imports, but also many developing countries, attracted by the easy credit of recycled petrodollars after the first oil shock, have incurred large external debts (Section 1.3.3). With present high interest rates, the repayment and servicing of these debts has proved such a major burden that the debtor countries have often altered their internal and foreign policies according to the desires of lending institutions and of oil-exporting countries—an adjustment that is hardly consistent with their self-reliance.

The industrialized countries may not have suffered economically to the same extent, but they too have been forced to use foreign policy and trade (including arms trade) measures to ensure secure supplies. They have displayed a strategic dependence on volatile parts of the world and have felt compelled to get embroiled in the conflicts there.

In the case of developing and industrialized countries alike, self-reliance would be enhanced by reduced dependence on oil imports. And developing countries would be in a better position to pay their debts. Moreover, both developing and industrialized countries would have greater flexibility in foreign policy.

5.3.8.2 *Promoting the Wide Use of Efficient End-Use Technologies in Developing Countries:*
All of the policies aimed at promoting efficient end-use technologies (Section 5.3.5), promoting sustainable indigenous energy production (Section 5.3.6), and creating marketplace conditions that give an even-handed treatment of various energy supply and conservation technologies (Section 5.3.4) would contribute to energy self-reliance.

But the question often arises as to whether the energy self-reliance that would result if developing countries adopt efficient end-use technologies would lead to a net increase in national self-reliance or instead simply shift dependency to another area. This question revolves around two issues: the present absence of a manufacturing capability for many energy-efficient end-use technologies in developing countries and the extra capital required for investments in end-use efficiency.

To address such issues it is essential for developing countries to carry out detailed assessments of alternative end-use technologies, to determine the net likely contribution to self-reliance of adopting more energy-efficient end-use technologies—looking at technology transfer issues, the prospects for domestic manufacture, net capital requirements, etc.

It is very likely that many efficient end-use technologies would be identified that would satisfy the criterion of making a net contribution to self-reliance. One study of a wide range of possible electricity-saving technologies for Brazil found that most could be manufactured within a few years time in Brazil, if manufacturers felt there would be sufficient demand.[24] And the extra first costs to the consumer associated with choosing a more energy efficient end-use technology are often more than offset by the national capital savings resulting from the extra energy supplies that would thereby not be needed. This is illustrated in Figure 5-3, which shows the results of an "acid test" for energy efficiency in developing countries—a comparison of the total capital costs for efficient but expensive lightbulbs with that for the equivalent amount of new hydropower supplies in Brazil.[25] The large capital savings potential to the country as whole associated with a shift to such bulbs provides a strong motivation for programs that would promote their use—e.g., making the bulbs available from the utility on the "instalment plan," where the consumer makes monthly amortization payments on his utility bill.

5.3.8.3 *Promoting Technological Leapfrogging in Developing Countries:*
While much energy efficiency improvement can be achieved with technology that is already available, even more dramatic improvements could be made with technologies still under development (Section 2.5). In some instances, it may make sense to commercialize promising advanced technologies first in developing countries. In the case of many basic materials processing industries, for example, demand growth for basic materials in many industrialized countries is so stagnant (Section 2.4) that many of the "smokestack industries" involved are in a depression. Fighting for their very

survival, these industries are not strong candidates for making fundamental innovations. This is in sharp contrast to the situation in developing countries, where the demand for basic materials is far from saturation.

If a developing country were to be the first to introduce a new energy-efficient technology, self-reliance could be enhanced not only from the energy savings that would arise, but also from the prospect of moving to a stronger manufacturing position in the global marketplace, because

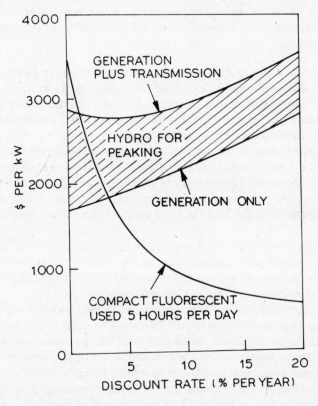

Figure 5.3. The discounted present values of the cost of peaking electricity produced via a hydroelectric power system and the cost of saving electricity via the installation of energy-efficient compact flourescent lightbulbs.

The costs are for electricity delivered to or saved at the household. All capital investments over the estimated 50 year life of the hydroelectric power plant are included.

The assumed hydroelectric power costs are characteristic of new plants that would be built in Brazil. Specifically, it is assumed that the "overnight" construction cost of the hydroelectric facility is $1170 per kW of installed capacity, that the plant is constructed over a six year period, and that the plant is paid for in six equal instalments over the construction period. To provide 1 kW of peak demand requires 1.16 kW of installed capacity to allow for a 16 percent reserve margin. The total installed capacity required to provide 1 kW of residential peak demand is 1.45 kW, when allowance is made for 20 percent transmission and distribution losses.

It is assumed that transmission facilities lasting 30 years cost $710 per kW. Some of this transmission cost could be avoided if new generating capacity could be deferred via investments in more efficient light bulbs.

It is assumed that 13 Watt compact fluorescent lightbulbs lasting 6,000 hours and costing $9.20 each replace 40 Watt incandescent bulbs lasting 1000 hours and costing $0.50 each. It is assumed that the light bulbs are used five hours a day, including the early evening hours, so that they contribute to the peak electricity demand.

Source: J. Goldemberg and R. Williams, 1985.

such innovations tend to improve many factors of production simultaneously, thereby lowering production costs (Section 2.5.3.4).

But such "technological leapfrogging" should be approached cautiously. First, governments need to determine whether in fact, for cases involving foreign technology, net increases in self-reliance would result. Clearly, the answers would depend on the technology transfer agreements reached. Second, the country must be economically strong enough to assume the risk of failure associated with any technological innovation. And finally, governments need to determine whether introducing the new technology would be compatible with other development goals, e.g., reducing unemployment.

While there is much potential gain to be realized with technological leapfrogging, in terms of speeding up the development process, making the right choices requires technologically sophisticated manpower and a strong institutional infrastructure for managing new technologies.

Efforts should be made to encourage the development of the needed manpower capabilities and an industrial technological infrastructure capable of supporting "technological leapfrogging ventures".

5.3.8.4 *Initiating and/or Increasing National Energy-Related Manpower Capabilities in Developing Countries:* Technical manpower is needed to implement any energy strategy—for the analysis of future needs, for technology assessment, for R&D, and for planning and management.

An important public policy issue is whether or not to foster the development of this capability in developing countries. It is often believed that it is less time-consuming and more cost-effective to use the capability that is found in industrialized countries to perform these energy-related tasks *for* developing countries. Hence, there has been a mushrooming of university groups and consulting firms in industrialized countries to carry out the energy work for developing countries. Even if such work makes major contributions to the solution of the energy problems of developing countries, it certainly undermines the national self-reliance of these countries and contributes to their lack of confidence in their ability to tackle their own problems.

Moreover, it is doubtful whether end-use oriented energy strategies could be implemented without intimate knowledge of the locale, without access to detailed information on end-uses, without an understanding of all the socio-economic factors associated with the use of end-use technologies by consumers, and without a sustained commitment to formulate, implement and live through energy solutions. In general, it is mainly the residents of a country who qualify on all counts for the tasks associated with implementing end-use energy strategies.

For these reasons, a policy aimed at initiating and/or increasing the energy-related technical manpower capabilities in developing countries through education and technical training is needed.

5.3.8.5 *Building and/or Strengthening National Energy-Related Institutions:* If the energy-related manpower capability created in developing countries by deliberate policies is to be sustained, then it should be developed in association with the building and/or strengthening of energy-related institutions in these countries. Moreover, institutions capable of implementing end-use oriented energy strategies should have broad enough mandates so that they can provide a wide range of energy services, instead of just providing particular energy supplies, which is the case with most present energy-related institutions. Specific policies are required for this purpose.

5.3.9 *Policies Relating to Other Global Problems*
The global environmental and security problems posed by the production and consumption

of oil, fossil fuels generally, and nuclear power (see Section 1.3) can be mitigated by policy initiatives relating to both energy demand and energy supply.

5.3.9.1 *Reducing Oil Imports:* Policies that promote oil import reduction through energy efficiency improvement (Section 5.3.5) and fuel substitution (Section 5.3.6) would contribute not only to national self-reliance (Section 5.3.8.1) but also to reduced global insecurity caused by overdependence of the industrialized market economies on Persian Gulf oil.

5.3.9.2 *Reducing Emissions from Fossil Fuel Combustion:* Despite the uncertainties associated with establishing clear causal relationships among lake and forest damage, acid deposition, and air pollution emissions from the combustion of fossil fuels, decisive action cannot await the generation of the new knowledge needed to clarify unequivocally these relationships.

As pointed out earlier (Section 1.3.6), the problem of uncertainty might be dealt with by seeking out emissions control strategies that make sense for a wide range of reasons besides the direct benefits of associated with reduced acid deposition. One such control strategy would involve adoption of fluidized bed combustion techniques for coal-burning; another, the use of low sulfur instead of high sulfur coal in stationary combustion applications, while shifting high sulfur coal away from direct combustion to the production of methanol.[26] The production of methanol from coal would lead not only to greatly reduced sulfur oxide emissions but to reduced nitrogen oxide emissions from motor vehicles as well, since nitrogen oxide emissions can be kept to low levels with methanol as a motor vehicle fuel. Both fluidized bed combustion technologies and methanol transport fuel are desirable for many reasons besides the air pollution reduction advantages they would bring.

The policy instruments of greatest relevance for encouraging such technological innovations are regulations and taxes on pollutant emissions.

If the regulatory approach were taken it would be desirable to frame the regulations in ways that give the polluter flexibility in the choice of controls, instead of mandating the use of particular control technologies, e.g., flue gas desulfurization systems. Taxes on emissions—making the polluter pay for polluting—would automatically have this effect.

Whatever instruments are used, international agreements are needed in some regions, because of the migration of pollutants from the emissions sources across international borders.

5.3.9.3 *Reducing Fossil Fuel Consumption to Reduce the Atmospheric Build-up of Carbon Dioxide:* As there appear to be no viable emissions control strategies to deal with the potential for climatic change associated with the atmospheric build-up of CO_2 from the combustion of fossil fuels (Section 1.3.8), avoidance of major climatic change seems to require reduced dependence on fossil fuels.

Policies we have described to eliminate energy supply subsidies and move to energy prices based on marginal costs (Section 5.3.4), to promote energy conservation (Section 5.3.5), and to promote bioenergy as a chemical fuel (Section 5.3.6) would all be effective in reducing dependence on fossil fuels.

But it may be necessary to go further than such policies and introduce incentives specifically tailored to reduce fossil fuel use, e.g., a carbon emissions tax on fossil fuels. Moreover, since this is a global problem, efforts to reduce fossil fuel use may require cooperative international agreements (Section 5.4.1.2), a truly formidable challenge.

5.3.9.4 *Making Nuclear Power the Energy Source of Last Resort:* In section 1.3 we singled out two important problems posed by nuclear power development: the disposal of radioactive wastes and nuclear weapons proliferation.

The radioactive waste problem, like the acid rain problem, is a problem soluble via technical fixes, at least *in principle,* although a solution is not yet at hand (Section 1.3.6). The wisdom of proceeding with nuclear power development before solving this problem is dubious.

The proliferation problem is more troubling. The nuclear weapons connection to nuclear power is inherent in the technology and is not amenable to a technical fix (Section 1.3.10). The nuclear weapons proliferation risk increases, dramatically if there is a shift from current once-through nuclear fuel cycles—as a result of increased demand for uranium—to ones involving fuel reprocessing and the recycle of the recovered plutonium in fresh fuel.

The proliferation risk can be greatly reduced if reprocessing technologies are avoided and if nuclear power is regarded as the energy technology of last resort. Making nuclear power the energy technology of last resort would deal effectively with the proliferation risk, while leaving the nuclear option open to those countries that would have difficulty finding viable alternatives to nuclear power.

The global aspects of this policy will be discussed in Section 5.4.1.3. Here we just briefly describe supporting efforts at the national level.

Both marginal cost pricing (Section 5.3.4) and the policies we have described to promote energy end-use efficiency (Section 5.3.5) would be effective in supporting this nuclear power policy. The removal of subsidies to the nuclear power industry, including subsidies to promote nuclear power exports would be especially effective in limiting the role of nuclear power, both because of the poor economics of nuclear fuel reprocessing relative to once-through fuel cycles (Note 4.2-L) and because even with once-through fuel cycles nuclear power has proved to be much more costly than originally (Note 2.2-B).

Further constraints on nuclear power, e.g., a formal ban on reprocessing, might also be considered, both unilaterally and in the context of negotiated international agreements.

5.4 Global Energy Policies

It is in the nature of end-use oriented energy strategies that most of the actions have to be taken at the local, state and national levels. However, some global energy-related problems require collective global efforts and the coordination of national energy policies. In addition, national policies are often strengthened by a climate of international support.

In the following discussion, no attempt is made to be comprehensive. The objective is to illustrate the types of global energy policies that would have to be considered, consistent with the achievement of a sustainable world.

5.4.1 *Global Policies to Address Global Problems*

Three global energy problems have been singled out in this book as warranting particular attention: (1) the need to smooth the global transition from oil, (2) the need to limit the atmosphere build-up of CO_2 in the atmosphere from the burning of fossil fuels, and (3) the need to denuclearize the world.

5.4.1.1 *Promoting a Smooth Transition from Oil:* Over the period of the next several decades, the world must begin to make a transition away from oil, because the remaining world oil resources are limited. Continuing the global trend of recent years to reduced oil use would also contribute to enhanced global security by reducing the dependence of industrialized market economies on Persian Gulf oil; and it would contribute to improved economic efficiency, because of the reduced expenditures on oil imports and the lower world oil price arising from reduced oil demand.

Today, however, uncertainties about future oil prices discourage investments in energy efficiency and supply alternatives. In a situation with major fluctuations in oil prices, corresponding to shock-glut-shock-glut conditions, the onset of a glut following each shock gives investors the misleading impression that the oil crisis is over and encourages increased oil consumption. This sets the world on a course leading to still another shock. Moreover, the glut periods, as we have seen in mid 1980's diminish the sense of urgency that is needed to induce oil importers to reduce their long-run vulnerability to supply disruptions.

The levels of oil prices reached in the second oil price shock of 1979 stimulated major reductions in oil use rates. Between 1979 and 1983, OECD countries reduced oil use by 18 percent and net oil imports by 39 percent, even though aggregate GDP increased 5 percent in this same period. But this trend may be reversed by the rapid drop in the world oil price that has taken place since 1981. In fact the downward trend in oil use by OECD countries actually reversed, increasing by 2 percent, 1983-84, after four years of decline from the peak consumption level of 1979.

The rational approach to a future of dwindling oil resources involves global oil policy aimed at:

- Minimizing the possibility of oil-price shocks by reducing dependence on oil and oil imports, thereby reducing the vulnerability of oil-importing nations to supply disruptions and enhancing global security.
- Enabling a shift of economic resources from expenditures on oil imports to other developmental purposes, as a result of a lower world oil price.
- Fostering a more stable environment for investments in efficient use of energy and renewable energy sources.

Reducing the possibilities of oil-price shocks and the vulnerability of oil-importing countries to supply disruptions: The oil-price shocks of the 1970's came about because as the industrialized market economies increased their demand for oil, they had to rely increasingly on oil imports from Persian Gulf producers.

It is possible, of course, to cushion the impact of future shocks by means of stockpiling and building enough flexibility into the energy system to shift to alternative fuels during emergencies. These are the solutions usually considered to deal with this global security problem. A more fundamental approach would be to try to avoid future shocks by avoiding the conditions that give rise to them. The combinations of conditions needed to avoid future shocks appears to be a level of world oil demand sufficiently low that dependence on Middle East producers is not too great and a world oil price sufficiently high that marginal producers outside the Middle East are adequately motivated to produce.

One oil demand scenario that might lead to these conditions was presented in Chapter 4 in connection with our base case global energy demand scenario, involving an average annual rate of decline in global oil use of 0.7 percent per year, 1980-2020. Under this scenario, it may be feasible to keep demand for Middle East oil 2020 at about the "glut" level of the early 1980's—some 15 million barrels per day, or 3/5 of the 1979 peak production level—with remaining demand met by other oil suppliers, if the world oil price did not fall too far.

Enabling a shift of economic resources from oil purchases to meeting other needs: A successful policy of reducing oil demand would lead to reduced payments for oil imports not only because of the reduced import level but also because the world oil price would be lower. At higher demand levels, the need to operate the world oil production system near full capacity or to exploit especially costly marginal oil supply sources would drive up the world oil price.

The group of oil importing developing countries, which has perhaps been hit the hardest by oil-price hikes, would probably benefit the most from reduced oil import bills. They would be able to divert valuable economic resources now committed to paying for oil imports to a variety of development projects. These countries in fact would probably gain more from lowered oil prices than from direct OPEC aid—the total oil import bill of oil-importing developing countries was $60 billion* in 1980, while direct OPEC aid to them totalled about $6.5 billion in that year.

Fostering a more stable environment for investments in energy efficiency and renewable energy: Oil import reduction targets could be met in a manner consistent with the other goals we have articulated for the global energy system with a variety of policies, including the elimination of oil industry subsidies and oil price controls (Section 5.3.4) and various measures to promote energy end-use efficiency (Section 5.3.5) and bioenergy development (Section 5.3.6.2). However, a important element of any policy package should be provisions to eliminate or at least greatly damp out fluctuations in the consumer oil price, in order to create a stable economic climate for investments in efficiency improvements and alternative energy sources. This might be accomplished by means of taxes on oil products or oil import tariffs that vary with the world oil price in such a way that the real price seen by consumers would be constant or perhaps slowly increasing.

While such taxes or tariffs could be levied by individual countries acting on their own, it may well be desirable for importers to carry out an oil demand reduction effort cooperatively, perhaps under the terms of an *international oil-import reduction agreement.* A particular country might be hesitant to restrict or tax the use of oil by its own industry unless other countries did the same, out of fear that its industry would become less competitive in the world market. In contrast, each country participating in a cooperative effort would benefit from every other country's efforts, since the world-market price of oil is likely to go down with the aggregation of reduced demands from each country.[28] Even oil exporters could benefit from such planning in the long run. The reduced uncertainties about the future demand for oil would allow them to plan more efficiently investments in new productive capacity.

The cooperative tariff should be high enough to keep oil demand under control but not so high that it would force the world oil price so low that marginal oil producers would drop out of the market. At very low world oil prices, the bulk of world oil production would shift back to the Persian Gulf leaving oil importers once more vulnerable to the prospect of supply disruption.

Despite the obvious mutual benefits that would arise from the levy of a cooperative tariff, achieving the required cooperation would be an heroic political accomplishment.

5.4.1.2 *Limiting the Build-up of Atmospheric Carbon Dioxide:* The build-up of CO_2 in the atmosphere can lead to major global climatic changes in a period of decades, and only limitations on CO_2 emissions can improve the situation; no viable method is known for scrubbing out carbon dioxide.

---

* Imports of oil by oil-importing developing countries amounted to 4.8 million barrels per day in 1980.[29] Here a world oil price of $34 a barrel is assumed for 1980.

There is no scientific *certainty* about the rate and impacts of the atmospheric CO^2 build-up. The only certainty will come after the effects are registered. It is prudent and both technically and economically feasible, however, to limit greatly the extent of climatic change by reducing dependence on fossil fuel use.

Any attempt to limit the atmospheric build up of CO_2 focus on limiting the use of coal, which is far more abundant than oil or gas. The need to limit coal use is reflected in the construction of the energy supply mix for our base case global energy scenario described in Chapter 4. In this scenario coal use declines at an average rate of 0.6 percent per year, leading to an atmospheric CO_2 level in 2020 of about 1.3 times the pre-industrial level. If this rate of coal phase-out were to persist after 2020, the carbon dioxide level would eventually reach 1.7 times the pre-industrial level, assuming all oil and gas resources were eventually used up—much less than the doubling of the CO_2 level in the mid 21st century associated with most supply-oriented energy strategies.

Even lower CO_2 levels could be achieved if alternative energy supplies, e.g., electricity and hydrogen from amorphous silicon solar cells (Appendix on Renewable Energy), could be developed, permitting an even more rapid phasing out of fossil fuels. However, it is clear that for the next several decades the major challenge in dealing with the CO_2 problem will be to prevent the greatly expanded coal production envisaged for the coming decades in most supply-oriented global energy futures.

Limiting coal use will be one of the most formidable challenges of the energy strategy proposed in this book. However, synergisms associated with the energy strategy articulated here would facilitate a transition from coal. While coal is a much cheaper fuel than oil, it is a dirty fuel requiring much more capital investment than either oil or gas to use in environmentally acceptable ways. Increasing concerns about acid rain that are reflected in requirements to burn coal in ways that greatly reduce sulfur oxide and nitrogen oxide emissions would tend to make coal use even more costly than it is already. At the same time, the strategy described above for limiting global oil use in the decades ahead would tend to keep both oil and natural gas prices from rising, making these fuels relatively more attractive in the transition period of the next several decades.

Despite these hopeful prospects for dealing with the carbon dioxide problem, it might be necessary to go beyond measures that force the clean burning of coal and stabilize oil prices. It might be necessary to attempt to limit the use of coal directly, through the levy of carbon dioxide emission taxes or the use of some other direct control mechanisms.

The fact that nearly 90 percent of the world's estimated geological coal resources lie in just three countries—the U.S.S.R. the U.S. and China—indicates that the fate of the CO_2 problem rests largely in the hands of these countries. Agreements on limiting coal production in these countries and leadership by these countries in the pursuit of non-coal intensive energy futures could be effective in dealing with the CO_2 problem. At the same time, countries not having indigenous coal resources should explore alternatives to importing coal.

5.4.1.3 Denuclearizing the World: The main institutional deterrent today to the horizontal proliferation of nuclear weapons is the Non-Proliferation Treaty (NPT).

Unfortunately, despite its virtues, the NPT has two fundamental shortcomings: it has failed to attract several states into the Treaty, and it does not forbid activities that will bring nations increasingly closer to nuclear weapons capability.

The states most closely allied to the U.S. and Soviet Union appear to have accepted the two caste system underlying the NPT, in which the world is formally divided into weapons and non-weapon states.

But many of the states choosing or forced to tread a more independent path, e.g., Argentina, Brazil, India, Iran, Israel, Pakistan, South Africa, South Korea, Taiwan, Yugoslavia, have opposed a formal rejection of the nuclear option and have mostly refused to ratify the Treaty. Most of these countries have both a reasonable technological basis for and experience in nuclear energy; hence, in all likelihood, they could build nuclear weapons if they desired to do so, and for this reason they are considered "threshold" states.

There seems little likelihood that groups within these "threshold" nuclear countries will oppose the further proliferation of nuclear weapons or be willing to support far-reaching measures of international control over nuclear power. These moves have been effectively inhibited by the discriminatory character of the current non-proliferation system. Under this system, making a formal renunciation of nuclear weapons can be seen as equivalent to accepting a fundamental restriction on the political independence of the non-nuclear country, condemning it to a neo-colonial status with respect to the superpowers. As a result, under the present ground rules, there seems no prospect of widening support for the NPT.

Even more serious is the fact that the groundwork for reprocessing and recycling plutonium in the industrialized countries has already created demand for these technologies in Argentina, Brazil, Pakistan, South Korea, Taiwan, and elsewhere, including several countries which have joined the NPT. Such a demand is inspired by a combination of technical considerations, prestige, economics, and weapons motives. Indeed, it is precisely the ambiguous character of the "plutonium fuel cycle" technologies that is troublesome. With these technologies, states will be able to move step by step toward a weapons capability without having to announce or decide their ultimate intentions in advance. It is this process of "latent proliferation" which is undermining the effectiveness of present nuclear safeguards arrangements.

The process is not significantly constrained by the NPT because the Treaty permits the development of all types of civilian nuclear power "without discrimination." Indeed, in Article IV, parties to the Treaty undertake to facilitate the "fullest possible" exchange of equipment, materials, and information for the peaceful uses of nuclear energy. Even countries which have ratified the NPT are moving closer to a technical capability to produce nuclear weapons. And under extraordinary circumstances, they could withdraw from the Treaty altogether at relatively short notice.

As long as plutonium remains in spent fuel, it is protected against diversion by the intense radiation from the fuel elements. Plutonium can be separated only through reprocessing spent fuel, but there are no sound reasons for doing so at present. It is uneconomical to reprocess spent fuel to recover the plutonium for recycle in present day reactors (Note 4.2-L). Moreover, there is no need to reprocess spent fuel for breeder reactors (which require plutonium as fuel). The nuclear power systems in the world will not be constrained by limited access to uranium for at least fifty years, even with moderately high projections of nuclear power growth.[29] The need for breeder reactors would be far less with the nuclear power scenario presented in Chapter 4. Finally, the argument that the disposal of nuclear power wastes requires reprocessing also does stand up to critical analysis. It appears to be no less cumbersome to dispose of high-level waste than spent fuel elements.[30] No scheme has been worked out satisfactorily for any complete waste handing system, but there appears to be no new inherent problems with a direct spent-fuel disposal scheme.

An important policy towards reducing the risk of proliferation would therefore consist of avoiding the reprocessing of spent fuel. Imposing such a constraint on non-nuclear countries, whether they have participated in the NPT or not, would be possible only if the nuclear weapons states accepted parallel obligations—certainly on their civilian power programs but perhaps also on their weapons programs. In other words, *the issues of horizontal and vertical proliferation of nuclear weapons are inevitably, intimately and inextricably linked.* A global policy to avoid the reprocessing of spent fuel from civilian power programs (implemented through an international agreement) may have to be supplemented by a policy (agreed to by the nuclear-weapons states) not to reprocess spent fuel to produce plutonium for weapons purposes. In order to secure the support of the "threshold" states for effective non-proliferation conditions, international safeguards obligations should be as nondiscriminatory as possible between nuclear and non-nuclear weapons states. This symmetry of obligations could be embodied in a *Denuclearization Treaty* that might replace the present Non-Proliferation Treaty.[31]

Such a treaty would require non-nuclear states to accept the present restrictions imposed on them by the NPT and necessary additional constraints on reprocessing, but it would also require nuclear states to submit their programs to the strictest international safeguards. The "threshold states" would presumably support the treaty because it would mean a significant overall decline in the number of nuclear weapons around the world and reduce the chance of a nuclear holocaust.

A denuclearization treaty is not a complete solution to the problem of proliferation—it becomes increasingly incomplete as the number of reactors increases. Hence, it is important that simultaneously there are global policies to supplement national policies to make nuclear power an energy sources of the last resort. Perhaps one step in this direction would be to broaden the charter of the International Atomic Energy Agency to cover all energy sources.

5.4.2 Strengthening National Energy-Related Capabilities and Efforts in Developing Countries

Because of the scarcity of their economic and technical resources, most developing countries have had to depend on international aid for energy-related activities.

The total expenditure on the energy programs of the multilateral and bilateral aid agencies since 1972 has been approximately $14 billion (Table 5.2). The bulk of this expenditure has come from the large development banks, but bilateral aid has accounted for 30 percent of the total.

Most of the present energy aid programs were established under the general umbrella of the United Nations and were designed specifically to protect developing countries from encroachment by industrialized countries trying to enlarge their markets or to otherwise advance their national intetrests. The UNDP (United Nations Development Program) has funded hundreds of projects with special emphasis on assistance for energy planning but has no funding for the execution of projects. The United Nations Center for Natural Resources, Energy and Transport (CNRET) is another small U.N. agency working in assistance for energy planning.

By far the most important agency engaged in multilateral aid is the World Bank. This is the largest of the international development banks; its actions influence the constellation of regional and national development banks: the Inter-American Development Bank (IDB), the Asian Development Bank, the European Development Fund, the African Development Bank, the OPEC Fund and others.

The very large sums of money lent by the World Bank ($12.2 billion in 1980 in all areas)— two-thirds of which was below commercial rates (approximately 8 percent interest and a 15-20 years payback period)—make this source of money very attractive. Of the remainder, one-third of the loans are concessional with practically no interest and at least a 40-year payback period;

Table 5.2. Expenditures of Multilateral and Bilateral Aid Agencies in the Energy Area (in millions of current dollars)

	Conventional Power Generation (Hydro, Nuclear, Thermal), Transmission, Distribution; Power Sector Studies	Fossil Fuels Recovery (includes studies and training)	New and Renewables (includes Geothermal, Fuelwood)	Tech. Assistance, Energy Planning, Other	Total Energy Aid
MULTILATERAL AID					
1. World Bank (FY '72-Dec. '78)	5,210	305	170	—	5,686
2. Inter-American Development Bank (FY '72-FY '78)	2,596	158	4	—	2,758
3. Asian Development Bank (FY '72-FY '78)	1,183	21	0	—	1,204
4. European Development Fund (to May 1978)	141	—	9	—	150
5. UN Development Programme (to Jan. 1979)	72	23	29	13	137
6. UN Center for Natural Resources, Energy and Transport (to Jan. 1979)	3	5	4	5	17
Subtotal	9,205	512	216	18	9,952
BILATERAL AID					
1. French Aid (1976-1979)	229	16	30	5	280
2. Canadian Int'l Devel. Agency (1978/79, 1979/80)	88	0	2	1	91
3. German Aid (1970-present)	1,925	41	81	48	2,095
4. Kuwait Fund (FY '73-FY '78)	437	99	1	—	536
5. Swedish International Development Agency	?	?	?	?	?
6. Netherlands—Dutch Development Cooperation (1970-present)	119	71	7	2	198

7. UK Overseas Development Administration (1973-present)	146	1	3	—	149
8. US AID (FY '78-FY '80)	403	2	96	46	546
GRAND TOTALS	12,719	757	437	121	14,033
% in each sector	91	5	3	1	100

Source: T. Hoffman and B. Johnson, *The World Energy Triangle*, Ballinger Publishing Co., Cambridge, Massachusetts, 1981.

these loans go only to very poor countries. Energy accounts for approximately 25 percent of the World Bank's lending.

Over the past decade, most industrialized countries have tried to limit their general UN-related efforts and to give more emphasis to bilateral aid programs. All industrialized countries operate bilateral aid programs, with West Germany, the United States, and Kuwait providing significant support for energy-related activities (Tables 5.2 and 5.3).

Table 5.3. Bilateral Energy Aid Programs

1. *Canada* (Canadian International Development Agency, CIDA). Concentrates activities in energy production and distribution, evaluation of resources and energy management. Invested approximately $610 million, 1972-1982.
2. *France* Aid for Africa and Francophone countries is administered by the Ministry of Cooperation. Aid for Fast Asia and Latin America is handled by the Foreign Ministry. Emphasis was always given in the aid programs to solar energy (biomass, solar coolers and refrigeration, wind, photovoltaics, etc.). The aid is well coordinated with French industry (mainly the French company Sofretes).
   Between 1976 and 1979 approximately $280 million were committed to energy programs, with an emphasis on hydropower and rural electrification; non-conventional sources accounted for less than 10 percent of the total.
3. *West Germany* (Ministry of Economic Cooperation-DMZ). In the past most of the energy programs were handled by capital loans for large projects. More recently the Agency for Technical Cooperation (GTZ) which handles technical assistance has established a German Appropriate Technology Exchange (GATE). From 1970-present approximately $2.1 billion were invested, mainly in power generation and distribution.
4. *Kuwait* Established in 1962, the Kuwait Fund is the largest Arab aid agency. From 1973-78, approximately $340 million were invested, mainly in conventional power projects. Almost 50 percent was spent outside the Arab World.
5. *Holland* [Directorate General in International Cooperation (DGIC), Ministry of Foreign Affairs]. Emphasis of the program is on training, assessment, and investments in hydropower, although work in forestry is on the rise. From 1970-1980 approximately $200 million were invested in energy programs.
6. *United Kingdom* (Overseas Development Administration-ODA). ODA has provided foreign exchange for energy projects and more recently has shown interest in the "basic needs approach" to energy. From 1973 to present it spent almost $150 million in energy projects.
7. *United States* (Agency for International Development-AID). USAID's energy budget has been approximately $150 million per year with an increasing emphasis on renewables, which in 1980 accounted 25 percent of the budget. Large scale power generation is being downgraded in USAID funds but not rural electrification. Main areas of concern in renewables are energy production (charcoal, village, woodlots, etc.), energy resource evaluation, and energy policy studies and sector management.
8. *Sweden* (Swedish International Development Authroity-SIDA). SIDA has concentrated its efforts in a few countries in order to achieve maximum impact. In the past, it has been active in large hydro projects, primarily in Botswana, Kenya, Tanzania, Tunisia and Turkey. From 1979 on, four areas of priority were established: hydroelectric projects (conventional and mini-hydro), forestry, energy planning, and other (biogas, geothermal, etc.).
9. *Venezuela* (Venezuelan Investment Fund-VIF). VIF has large assets coming from petroleum revenues; in 1975 VIF resources were already of the order of $5 billion. It acts through regional and national development banks, such as the Inter-American Development Bank, the Andean Development Corporation, the Caribbean Development Bank, the National Bank of Panama and others. The Charter of VIF prohibits grants.

The supply aspects of energy have clearly dominated all energy aid programs (Table 5.2). And among the various supply options, over 90 percent of the investments have been for large systems for the generation, transmission and distribution of electricity (mainly hydropower). Fossil fuel exploration has accounted for about 5 percent, and new and renewable energy sources about 3 percent of energy aid. Efforts on the demand aspects, directed mainly at conservation in the industrial sector, have accounted for less than 1 percent of total expenditures. Demand management, including interfuel substitution and the reorganisation of consumption in the different sectors of the economy or across them, has received very little attention.

It is unclear, at this stage, how much energy-related aid is necessary to implement the energy strategies suggested in this book. Many of the technologies which have been emphasized, however, such as improved cooking stoves, biogas plants, producer gas generators and engines, biomass-fired gas turbine cogeneration systems, wood-to-methanol plants, more efficient light bulbs, more efficient engines for motor vehicles, etc., involve relatively modest "front end" costs.

The $2 billion that is required to finance the construction of a single nuclear power plant could be used instead to finance the construction of 200-400 million energy-efficient cooking stoves at $5-10 per stove—far more than enough for all poor urban and rural households in India, for example—200 million energy-efficient compact fluorescent lightbulbs, etc. By emphasizing more efficient end-use technologies, less net capital is usually required to provide a given level of energy service than with an equivalent amount of a conventional energy supply (see, for example, Figure 5.3).

The challenge in implementing end-use strategies lies not so much with the quantity of capital required as with the institutional hurdles involved. Once a decision has been made to build a $2 billion power plant, the subsequent construction is a relatively straightforward task that can be carried out by a relatively small and disciplined team. But the number of factors involved in spending $2 billion on end-use technologies is likely to be large and the activity far more difficult to organize.

To help implement an end-use oriented energy strategies in developing countries, the international aid community should consider some fundamental restructuring of its aid programs. The elements of a more end-use oriented aid policy include:
- Shifting the emphasis from project to program support,
- Building and strengthening energy-aid-receiving indigenous institutions.
- Strengthening indigenous energy-related technical capability, and
- Supporting "technological leapfrogging" efforts.

5.4.2.1 *Shifting the Emphasis from Project to Program Support:* Financial support for energy activities from aid agencies and financial institutions like the World Bank has invariably been project-oriented, typically directed to large, supply-biased projects, e.g., the construction of hydroelectric facilities. One reason perhaps for this tendency is that the bureaucratic paper-work to process a small project is scarcely different from that for a large project, and therefore the disbursal of huge grants can be quickly completed by approving a very small number of massive projects.

Aid in the project mode may be appropriate for supply-oriented energy strategies where the preoccupation is with massive energy-supply plants. But this approach is inadequate for implementing end-use-oriented energy strategies which emphasize diverse and often small-scale technologies tailored to regional and local conditions. The implementation of a large number of small projects is impractical with project-type support, in which the disbursal of funds is closely administered by the funding agency. Accordingly, it is desirable to reorient aid from specific projects to broad programs for which the detailed allocation of program resources is largely the responsibility of locally based institutions, in accordance with the overall program objectives.

A drawback of this policy of shifting from project to program support is that many developing countries may not have the technological and management institutions and expertise to plan and administer such programs. In fact, this is another reason why aid support has not emphasized programs but instead has supported projects that are closely and narrowly defined at the proposal stage so that the aid agencies do not have to rely much on local institutions and capabilities.

The only way of overcoming this weakness is to devote efforts to building institutions and strengthening indigenous capability. Though this is a time-consuming and often frustrating task, the long-term pay-offs are so sure and enormous that aid agencies would do well to resist the temptation of achieving "quickie successes" and undermining self-reliance in the process.

5.4.2.2 *Building and Strengthening Energy-Aid-Receiving Indigenous Institutions:* First, a portion of overall aid should be directed to building the necessary energy-related institutions, perhaps in much the same way that the Rockefeller Foundation in the early years of the 20th century successfully contributed to institution-building in the medical area.

Aid resources should be directed in ways that would enable energy-related institutions to support staff who are familiar with local development problems and who are capable of carrying out the needed technology assessments, formulating the appropriate programs, monitoring these efforts, and improving programs in light of field experience.

The large utility companies of developing countries are particularly attractive candidates for "institutional renovation." In this instance, what is needed is reorientation of technically competent staff from preoccupation with energy supply expansion to the administration of broad end-use-oriented energy strategies.

As this institutional capability is developed, a greater and greater shift from project support to program support could take place.

5.4.2.3 *Strengthening Indigenous Energy-related Technical Capability:* Traditionally, aid has not been very effective in directly fostering local technical capability. In part, this has been due to the emphasis on large projects for which highly specialized support services are required. The result has been that procurement and consulting arrangements are frequently left to foreign companies who become better and better at providing these specialized skills. But another, and perhaps more important, reason is that most of the large loans and grants managed by international or bilateral organizations are given specifically to cover expenses involving foreign currency. Local expenditures are usually not covered by the loans. A typical loan covers about one-third of the overall project cost. *The aid money is therefore spent mainly on consultancy and engineering services and on machinery imported from abroad.* Often, a sizeable fraction of bilateral support must be spent in the donor (lending) country and/or on its personnel.

These practices, which are *de facto* methods of recycling the aid back to the donor country, are not consistent with facilitating self-reliant development. Much more in the interest of the aided country is a policy of strengthening indigenous technical capability, one which stiputes that

- Before foreign consultancy services are recruited, it be proved that they are both essential and unavoidable, and when they are hired, measures be taken to associate local groups with the projects/programs.
- A significant fraction of the aid be spent domestically in the recipient countries, so that it is able to contribute to building the local technical capability.

5.4.2.4 *Promoting Technological Leapfrogging:* Industrialized countries could play important roles in promoting "technological leapfrogging" in developing countries by supporting research and development efforts in both their own countries and in developing countries on scientific and technical problems relevant to the needs of developing countries.

A particularly important focus for industrialized country efforts is basic research on problems relevant to the energy needs of developing countries in areas where developing country capabilities are weak. For example, the production of producer gas from wood is a promising technology for rural development. While the use of wood as a feedstock for gasification offers the prospect of much more efficient utilization of the biomass resource than is possible with charcoal (owing to the energy losses of charcoal production), wood gasification is plagued by problems of tar formation that complicate the operation of producer gas engines—a problem that does not exist for charcoal-based systems. A better fundamental understanding of the gasification process might lead to gasifier design strategies that result in low levels of tar formation requiring little or no messy gas clean-up operations for producer gas engine systems.[32]

The needed basic biomass research is not likely to be done in developing countries in the near term, for two reasons. First, developing country scientists who work on producer gas problems are typically under great administrative pressure to do applications-oriented research aimed at getting technologies into the field quickly; there is little support available for exploring basic scientific quesitons. Second, even if there were support for such activity, few developing country laboratories are adequately equipped with the fairly sophisticated laboratory equipment that is often needed to facilitate path-breaking work. Such equipment is readily available, however, in the various combustion laboratories at universities and government laboratories in industrialized countries, where it is used primarily for aerospace research and advanced fossil fuel research applications. Directing even a tiny fraction of this space age analytical capability to fundamental scientific problems relating to biomass could be of enormous benefit to developing countries.

An institutional constraint on such efforts, however, is the view held by most international aid agencies that basic research on problems relevant to developing countries should be done only in developing countries. While this view may be meaningful in the area of *technology development,* to ensure that new technologies are indeed well-suited to local needs, its application to *basic research* implies that important opportunities for supporting technological leapfrogging may be lost.

In some areas industrialized country efforts to promote technological innovation to meet their own needs might be oriented so that they would simultaneously serve developing country needs as well. For example, it turns out that the technology for gasifying urban refuse, a large fraction of which is paper, i.e. cellulose, is very similar to the technology required to gasify wood and other biomass forms. In fact, a number of gasifiers being developed for biomass could also be used for urban refuse applications, where they may offer environmental and perhaps also cost advantages (through the sale of the produced gas or gas-derivative products such as methanol or cogenerated steam and electricity) over urban refuse incineration technologies.[33] In other words, the pursuit of gasification technologies that could be important in solving the urban refuse disposal crisis facing many industrialized countries could also provide the opportunity for speeding up the commercialization of needed advanced biomass gasification technologies.

Industrialized countries can also help promote technological leapfrogging both by helping to build the capabilities for doing research and development in developing countries and by supporting efforts to commercialize attractive advanced technologies.

The latter strategy is of course contrary to the tradition in aid programs of directing support to technologies which have been "proven" in the North. Yet it is desirable that some aid be used to support first-of-a-kind commercial demonstrations in rapidly industrializing countries that provide a favourable climate for innovation. Particular attention should be given to promosing technologies that would exploit the comparative advantages of the aid-receiving countries.

A list of candidates might include:

- Fuel-saving, aluminium-intensive, lightweight cars in a hydro-rich countries that produce both aluminium and automobiles.

- Plasmasmelt steelmaking technology in hydro-rich, coal-poor, rapidly industrializing countries (Section 2.5.3.4).
- Wood-to-methanol plants in biomass-rich, oil-importing countries heavily dependent on highway transportation.
- The fuel-efficient direct-injection stratified-charge (DISC) engine with multi-fuel capability for highway transportation in an oil-importing, automobile-producing country with alternative fuel capability.
- Bagasse-fired gas turbine cogeneration systems in sugar or alcohol producing countries.[19]

While there is a risk of failure in all these cases, the potential pay-offs with successful commercialization would be enormous. Moreover, the risk involved should be put in perspective. Unlike the situation with, say, nuclear fuel reprocessing technology, where a single production facility would represent a billion dollar and very uncertain investment (in light of the trouble this industry has had to date), all of the above technologies involve much smaller scale front-end investments. (Even a plasmasmelt steel plant would be a much smaller-scale production facility than a conventional steel plant). The cost of failing is typically much smaller for new technologies associated with an end-use-oriented energy strategy than it is with new centralized energy supply production technologies.

But a developing country should carefully assess the new technology, the indigenous capability for managing it and developing it further, and the overall implications for development of its adoption before getting involved in a "technological leapfrogging" venture.

If one of the ventures were to involve a bilateral effort with an industrialized country which has developed the technology, it is important that the developing country secure an agreement ensuring that it would not just serve as the field-station for the developer who would subsequently make all the profits as the technology catches on but would share with that country the rewards of success.

References

1. William J. Baumol and Edward N. Wolff, "Subsidies to New Energy Sources: Do They Add to Energy Stocks?" *Journal of Political Economy,* vol. 89, No. 5, pp. 891-913, October 1981.
2. Harry Chernoff, "Individual Purchase Criteria for Energy-Related Durables: the Misuse of Lifecycle Cost." *The Energy Journal,* Vol. 4, no. 4, pp. 81-86, 1983.
3. D. Gately, *"Individual Discount Rates and the Purchase and Utilization of Energy-Using Durables: Comment",* *The Bell Journal of Economics,* Vol. 11, Spring, pp. 373-374, 1980.
4. J. Hausman, "Individual Discount Rates and the Purchase and Utilization of Energy-Using Durables," *The Bell Journal of Economics,* Vol. 10, pp. 33-54, 1976.
5. J.E. McMahon and M.D. Levine, "Cost/Efficiency Tradeoffs in the Residential Appliance Marketplace," *Proceedings of the American Council for an Energy Efficient Economy's 1982 Summer Study of Energy Efficient Buildings, Santa Cruz, California, August 1982,* Energy Information Center, New York, 1983.
6. Alan K. Meier and Jack Whittier, "Consumer Discount Rates Implied by Consumer Purchases of Energy-Efficient Refrigerators," *Energy, The International Journal,* Vol. 8, No. 12, pp. 957-962, 1983.
7. Thomas C. Schelling, "Alternatives for Gasoline," in *The Dependence Dilemma: Gasoline Consumption and America's Security,* Daniel Yergin, ed., Center for International Affairs, Harvard University, Cambridge, Massachusetts, 1980.
8. Linda Berry, "The Role of Financial Incentives in Utility-Sponsored Residential Conservation Programs; a Review of Customer Surveys," Oak Ridge National Laboratory Report No. ORNL/CON-102, 1982.
9. Government of Karnataka, India. "Report of the Working Group constituted for Advance Planning for Utilization of Power in Karnataka," 1982.

10. Marc H. Ross and Robert H. Williams, *Our Energy: Regaining Control,* McGraw Hill, New York, 1981.

11. Robert H. Williams, Gautam S. Dutt, and Howard S. Geller, "Future Energy Savings in U.S. Housing," *Annual Review of Energy,* vol. 8, pp. 269-332, 1983.

12. E.R. Berndt, "Aggregate Energy, Efficiency, and Productivity Measurement", *Annual Review of Energy,* vol. 3, 1978.

13. Clark W. Bullard III, "Energy and Employment Impacts of Policy Alternatives", in *Energy Analysis, a New Public Policy Tool,* Martha W. Guilliland, ed., AAAS Selected Symposium 9, Westview Press, Boulder, Colorado, 1978.

14. Bruce Hannon, "System Energy and Recycling: a Study of the Beverage Industry", Center for Advanced Computation, University of Illinois, Urbana, March 1973.

15. Harvey Lieberstein, "Microeconomics and x-efficiency theory: If there is no crisis, there ought to be," in *The Crisis in Economic Theory,* Daniel Bell and Irving Kristol, eds., Basic Books, New York, 1982.

16. H.S. Geller, *Energy Efficient Appliances,* a report of the American Council for an Energy Efficient Economy and the Energy Conservation Coalition, Washington DC, 1983.

17. Stuart Diamond, "Cogeneration Jars the Power Industry," *The New York Times,* June 10, 1984.

18. David Pimentel, Wen Dazhong, Sanford Eigenbrode, Helen Lang, David Emerson, and Myra Karasik, "Deforestation: Interdependency of Fuelwood and Agriculture," draft report, New York State College of Agriculture, 1985.

19. Robert H. Williams, "Potential Roles for Bioenergy in an Energy Efficient World," *Ambio,* vol. XIV, No. 4-5, pp. 201-209, 1985.

20. J.B. Raintree and B. Lundgren, "Agroforestry Potentials for Biomass Production in Integrated Land Use Systems," paper presented at the Symposium on Biomass Energy Systems: Building Blocks for Sustainable Agriculture, Airlie House, Virginia, January 29-February 1, 1985.

21. Ray Wijewardene and Parakrama Waidyanatha, *Conservation Farming for Small Farmers in the Humid Tropics,* Department of Agriculture, Sri Lanka, 1984.

22. David French, "The Economics of Reforestation in Developing Countries," paper presented at the Symposium on Biomass Energy Systems: Building Blocks for Sustainable Agriculture, Airlie House, Virginia, January 29-February 1, 1985.

23. Joseph Cordes, *The Impact of Tax and Financial Regulatory Policies on Industrial Innovation,* National Academy of Sciences, Washington DC, 1980.

24. H.S. Geller, "The Potential for Electricity Conservation in Brazil," Companhia Energetica de Sao Paulo, Sao Paulo, Brazil, April 1985.

25. J. Goldemberg and R.H. Williams, "The Economics of Energy Conservation in Developing Countries: A Case Study for the Electrical Sector in Brazil," in *Energy Sources: Conservation and Renewables,* D. Hafmeister, H. Kelly, and B. Levi, eds., American Institute of Physics Conference Proceedings No. 135, American Institute of Physics, New York, 1985.

26. Charles L. Gray and Jeff Alson, *The Optimum Solution to the Acid Rain and Energy Problems: Methanol—the Transportation Fuel of the Future,* The University of Michigan Press, 1985.

27. The World Bank, *The Energy Transition in Developing Countries,* Washington, DC, 1983.

28. Hung-Po Chao and Stephen Peck, "Coordination of OECD Import Policies: A Gaming Approach," *Energy,* vol. 7, No. 2, pp. 213-220, 1982.

29. Harold A. Feiveson, Frank von Hippel, and Robert H. Williams, "Fission Power: an Evolutionary Strategy," *Science,* Vol. 203, pp. 330-337, January 26, 1979.

30. Hartmut Krugmann and Frank von Hippel, "Radioactive Waste: the Problem of Plutonium," *Science,* Vol. 210, pp. 319-321, October 17, 1981.

31. Harold A. Feiveson and Jose Goldemberg, "Denuclearization," *Economic and Political Weekly,* Vol. XV, No. 37, pp. 1546-1548, September 13, 1980.

32. Eric D. Larson, "Producer Gas, Economic Development, and the Role of Research," PU/CEES Report No. 187, April 1985.

33. Birgir Arnason, "Methanol from Biomass and Urban Refuse: Prospects and Opportunities," M.S. Thesis, Mechanical and Aerospace Engineering Department, Princeton University, Princeton New Jersey, 1983.

References for the Figure Captions, Tables, and Notes

Section 5.2
Alliance to Save Energy, *Industrial Investment in Energy Efficiency: Opportunities, Management Practices, and Tax Incentives (Summary)* Washington DC, 1983.

William J. Baumol and Edward N. Wolff, "Subsidies to New Energy Sources: Do They Add to Energy Stocks?" *Journal of Political Economy,* Vol. 89, No. 5, pp. 891-913, October 1981.

D.B. Goldstein, "Refrigerator Reform: Guidelines for Energy Gluttons", *Technology Review,* Vol. 86, No. 2, pp. 36-46, 1983.

E. Hirst, R. Goeltz, and H. Manning, "Household Retrofit Expenditures and the Federal Energy Conservation Tax Credit", Oak Ridge National Laboratory Report No. ORNL/CON-95, 1982.

D. MacMillan, "Denver Metro Homebuilders Program," American Council for an Energy Efficient Economy Report, Washington DC, 1982.

Leslie K. Norford, *An Analysis of Energy Use in Office Buildings: the Case of ENERPLEX,* Ph. D. thesis, Department of Aerospace and Mechanical Engineering, Princeton University, Princeton, New Jersey, June 1984.

K.H. Smith, "Denver: Trendsetter in Solar Homes," *Solar Age,* vol. 9, no. 9. pp. 24-29, 1982.

Section 5.3.5

Alvin L. Alm and Kathryn L. Mowry, "PURPA: Purpose and Prospects," Energy and Environmental Policy Center Report E-83-03, John F. Kennedy School of Government, Harvard University, 1983.

California Public Utilities Commission, *Decision No. 91107,* 243 pp., San Franscisco, California, 1979.

California Public Utilities Commission, *Decision No. 82-10-021,* 17 pp., San Franscisco, California, 1982.

Stuart Diamond, "Cogeneration Jars the Power Industry," *The New York Times, June 10, 1984.*

Marc H. Ross and Robert H. Williams, *Our Energy: Regaining Control,* McGraw Hill, New York, 1981.

Robert H. Williams, Gautam S. Dutt, and Howard S. Geller, "Future Energy Savings in US Housing," *Annual Review of Energy,* vol. 8, pp. 269-332, 1983.

Section 5.3.7

OECD, "Energy Research Development and Demonstration in the IEA countries. 1981 Review of National Programmes", OECD, Paris, 1982.

OECD, "Energy Research Development and Demonstration in the IEA countries. 1983 Review of National Programmes", OECD, Paris, 1984.

National Science Foundation, 1983.

Section 5.3.8

J. Goldemberg and R.H. Williams, "The Economics of Energy Conservation in Developing Countries: A Case Study for the Electrical Sector in Brazil," in *Energy Sources: Conservation and Renewables,* D. Hafmeister, H. Kelly, and B. Levi, eds., American Institute of Physics Conference Proceedings No. 135, American Institute of Physics, New York, 1985.

Section 5.4

T. Hoffman and B. Johnson, *The World Energy Triangle,* Ballinger Publishing Company, Cambridge, Mass., 1981.

6. The Political Economy of End-Use Energy Strategies

6.1 Introduction

We have shown that if there were a shift of emphasis in energy planning from supply expansion to improvements at the point of energy use, global energy strategies could be shaped that are compatible with the achievement of a sustainable world. The end-use approach we propose represents a dramatic departure from the conventional wisdosm and will require some public sector interventions in the market to adjust the course of the evolving political economy of energy. But we live in political times in which fiscal austerity and *laissez faire* reign, so the question arises: Is the change of course we advocate at all realistic?

We will explore this question in this chapter. And we will set forth our reasoning in concluding that the prospects for the end-use approach to energy are much brighter than one might expect from a quick read of the current political situation and in fact they are quite good.

6.2 The Unrealism of the Conventional Wisdom

The energy future envisioned in the conventional, supply-oriented approaches to the global energy problem, upon close analysis, looks utterly unrealistic.

In terms of energy demand, most conventional energy forecasts envisage future global energy growth to be much less than the long term historical growth rate (3.6 percent per year for commercial energy, 1925-1972), but they are consistent with the global energy growth rates since the early 1970s—1.9 percent per year, 1972-1983.

But continued growth in global energy demand at rates of the order of 2 percent per year is highly unlikely. As we noted in Chapter 4, given the ongoing structural changes in the economies of the industrialized market countries and the expected continued market responses to energy price increases which have already occurred, per capita energy use will probably not grow and may even continue the ongoing decline, even if their public sectors do nothing to encourage energy efficiency. This means that to bring about global growth at a rate of some 2 percent per year would require that nearly all the growth come from developing countries. But as we showed, this would be financially impossible—it would take too much capital.

And in terms of energy supply, the conventional approaches to the global energy problem are positively horrific. The capital costs and institutional stresses of having to expand energy production capacity at the rate envisaged—e.g. one new 1 GW(e) nuclear power plant every 4 to 6 days for the next 50 years, etc., would be staggering. The expansion of the industrialized market economies' demand for oil imports would lead inevitably to their becoming overdependent once again on oil exports from the politically volatile Middle East, thereby increasing the chances of a superpower confrontation and perhaps even a nuclear war. It would also mean higher world oil prices and the approaching exhaustion of low-cost oil resources outside the Middle East—developments that would seriously constrain the economic development of oil-importing poor countries in the critical decades ahead. The expanded use of fossil fuels generally could lead to climatic changes requiring formidable adjustments in a few decades time, a mere instant in geological time. The envisaged expansion of nuclear energy would lead to so much weapons-usable material being discharged from nuclear reactors and fuel recycling facilities around the world that it would be literally impossible to control the proliferation of nuclear weapons. And the neglect of the deforestation problem in conventional energy strategies could mean irreversible loss of already rapidly dwindling world forest resources.

But, fortunately, such a course is not inevitable.

6.3 Satisfying Broad Social Goals

A fundamental weakness of the conventional approaches is that they are too narrrowly focussed on the requirements of the energy system itself. An energy system can be sustained over the long run, in an engineering sense, while it is eroding the long-term sustainability of the world. Our end-use energy strategies and the policies to implement them are designed to satisfy broad social goals—equity, economic efficiency, environmental soundness, long-terms viability, self-reliance and peace.

Since the political economy of energy does not exist in a social vacuum, we thought it a necessity that the solution to the energy problem be compatible with the solutions to other problems. And an added advantage of broadly-based strategies is that they tend to have strong political appeal over the long run. The reasons for this are fairly obvious. The broader base attracts the support of a wider range of constituencies within the political process. Also, it offers decision makers the chance of killing two, three or four birds with one stone. As we saw in the case of the Brazilian alcohol program, for example, multiple benefits can flow from a country's effort to become less dependent on oil imports and more energy self-reliant—that program is a good employment generator, it frees up foreign exchange earnings for capital investment within the country, it stimulates indigenous technological development, and so on. A tariff on oil imports aimed at stabilizing the consumer oil price would have similar multiple benefits—it would enhance global security, it would create a more stable economic climate for a wide range of investments in energy efficiency and supply alternatives to imported oil, it would lead to a lower world oil price, and some of the revenues so generated might be targeted to meeting the needs of the poor (e.g. energy for basic human needs in developing countries) and/or to promoting self-reliance in energy (e.g. investments in energy efficiency or bioenergy development).

Another appealing aspect of the end-use approach is that it gives decision makers an opportunity to do something about particularly intransigent problems such as deforestation or desertification or the debt crisis while also doing something about the energy problem. Not that implementation of energy end-use strategies alone will solve such problems—it will not. But it could make a dent in them. Just the relatively straightforward matter of providing every poor family in a developing country with an efficient wood-burning or gas stove—and, as we have seen, you can do that for less than the cost of one new nuclear power plant, even in a country as populous as India—would have a real and an immediate impact on deforestation and desertification problems. So would efforts to promote the replacement of imported oil with modern energy carriers derived from biomass, by helping to support biomass prices that would make it profitable to grow biomass on plantations or farms as an alternative to the irreversible cutting down of natural forests.

At the global level, end-use energy strategies offer decision-makers in aid organizations and international institutions a similar opportunity. They can do something positive about nuclear weapons proliferation by deferring reprocessing of spent fuel because any economic rationale for plutonium recycle would be transferred to the remote, indefinite future. By promoting the technology and institutional changes needed for more efficient energy use they can even do something about such formidable problems as the carbon dioxide build-up in the atmosphere, because there would no longer be compelling economic reasons for a major shift to coal.

6.4 Why Intervention is Needed

Since the first oil price shock of 1973, there has been a technological revolution in the technologies of energy use. Everything from home heating systems and automobiles to office buildings and steel mills have become more energy-efficient, and even further efficiency gains are sure to come. In addition, the economies of industrialized countries are undergoing structural changes, away from material-intensive and energy-intensive economic activities. As a result, even without any public sector intervention, energy demand in the industrialized countries will grow very little if at all.

So why should there be market intervention to promote more efficient use of energy? Because per capita consumption in the industrialized countries could be cut by some 50 percent if the public sector intervenes clearing away the institutional obstacles and market imperfections which stand in the way of achieving the full range of energy efficiencies which are technically and economically feasible.

The benefits from capitalizing on these efficiency opportunities would be substantial. Overall expenditures on energy services would be reduced on a lifecycle cost basis. The reduced expenditures would come from both direct energy savings and from the lower energy prices resulting from the reduced need to rely on energy based on high marginal production costs. These reduced expenditures on energy would free up economic resources for other purposes. Equally important, the more efficient use of energy would also enhance global security because it would mean the oil importing countries would be less dependent on Middle Eastern oil, making military intervention in the region as well as oil supply disruptions less likely.

Developing countries too would benefit greatly from the more efficient use of energy in the industrialized countries. It would ease pressure on world oil prices and enable these countries to devote less of their foreign exchange earnings to paying their oil import bills. And hence, those same countries would be better able to repay their debts to the industrialized countries. So it also will help stabilize the world financial system.

In addition, of course, the developing countries would benefit greatly from their own more efficient use of energy. In fact because capital is so scarce in developing countries and since investments in energy efficiency typically reduce a country's total capital requirements, developing countries have an especially strong incentive to make more efficient use of energy. The resulting reduced energy service expenditures would free economic resources for speeding up development. The aggressive pursuit of energy efficiency improvement removes the spectre of inadequate supplies of affordable energy as a constraint on development.

The other reason the public sector has to intervene is a simple one. There are certain tasks which benefit society as a whole but do not get done in the market, because there is no or very little money to be made in doing them. They include: meeting the basic human needs of the poor, protecting the environment, and supporting basic research and development. Governments around the world are involved, to a lesser or greater extent, in these tasks today because in the past they were not getting done. All three are incorporated into the end-use energy strategies we have described in this book.

6.5 Why Intervention will Happen

The present political context does not, on the surface at least, appear all that favorable for interventionist policies. There is no question that the political pendulum has swung away from public sector intervention in the economy. Of course the swing has been more pronounced in some countries, i.e., the United States, Great Britain and Chile, than in others. But even in a capitalist country with a long tradition of public sector guidance of the economy such as Japan or in socialist countries such as China and Hungary, there are unmistakable signs of political movement towards *laissez faire*. The democratic socialists throughout West Europe are busy re-thinking, not their basic values, but their policy agendas, and they are searching for alternatives to their historical approaches—nationalization, price controls, expansion of the welfare state, etc.—of accomplishing their social objectives. In many parts of the world today, terms such as "privatization," "de-regulation," "fiscal belt-tightening," and "free market economics" are *au courant*.

To put all of this into perspective, a few aspects of the current political swing need to be noted. Why has it occurred? The reasons are varied and complex, too varied and complex to go into in any depth here, but three factors do seem to stand out. First there has been widespread disillusionment with publicly-owned enterprises due to their poor performance. Second, many people seem to have grown worried about the escalating costs of the public sector and wary of the measures governments have taken to deal with those rising costs—either raising taxes or printing more money. Third, it is only fair to note that the current fiscal austerity-*laissez faire* mindset in the market countries of the North has been imposed upon many developing countries by the IMF and international banks in New York, London, Paris, Zurich, etc., whether they liked it or not, and in many cases they did not, but they agreed because they needed more credit.

Lastly, we need to remember that the political pendulum has a way of swinging back—i.e., it has a kind of built-in corrective mechanism. All through the 1920's, for example, the pendulum was swinging towards *laissez faire* and austerity; then came 1929, and back it headed quite swiftly.

Of course it is impossible to say exactly where the impetus will come for the next change in direction. Perhaps it will be another oil price shock, sometime in the 1990's. Or perhaps it will be a combination of factors. In many industrialized countries, chronic unemployment and slow productivity growth persist. In addition, the manufactured goods exported from developing countries are becoming increasingly competitive in the markets of industrialized countries. These factors may combine with higher oil prices to induce economic stagnation or worse, and under such circumstances, the public sector is bound to be called on to intervene in order to improve economic peformance. Policies designed to stimulate improved energy productivity would logically play an important role in such an effort.

In developing countries, the impetus for public sector intervention is already very strong. Who else but the public sector can resolve the debt crisis? Who else but the public sector can direct capital to crucial development tasks, such as public health, education, etc. But as we noted in Chapter 1 and 3, most developing countries are two-tiered societies—with a relatively small commercial and bureaucratic elite at the top running things and everyone else below having little say in how things are run. So perhaps a more relevant question, particularly in terms of end-use strategies such as the meeting of basic human needs, is: What is in it for the ruling elites?

Several answers come to mind. For one, the problem of poverty is so dire and pervasive that for the elites to ignore it or to deal with it inadequately, through some variation of the "trickle-down'' approach, is to risk social unheaval. It certainly is in the long-term self-interest of the elites to identify and carry out effective programs for dealing with poverty.

In addition, the prospects for sustained economic growth in many developing countries are not promising with a continuation of the business-as-usual course. The difficulty is that a continued heavy emphasis on exports, which has shaped to a large degree the present mix of economic production, may cause serious problems in the years ahead. While the ongoing shift of production from commodities to the production of manufactured goods has helped protect the economies of developing countries from the vicissitudes of the commodity markets, the outlook for the export of manufactured goods to the North is clouded. Already developing countries are confronting a rising tide of protectionism. Even if protectionist tendencies can be curbed, Northern markets for processed basic goods are approaching saturation and may therefore not be very lucrative for the rapidly growing basic industries of the South. In the face of this export market uncertainty, the alternative of shifting the mix of production to exploit mass markets *within* developing countries is a more promising course for realizing sustained long-term economic growth, and it certainly is thus within the self-interest of the ruling elites.

But for there to be mass markets for manufactured goods in the South, there has to be income. You cannot sell goods to people who have no money. Which brings us back to matters dealt with earlier—meeting the basic human needs of the poor, improving the productivity of agriculture, generating productive employment in new rural industries, etc.

6.6 Why it will not be Difficult

Although the end-use approach does represent a dramatic change in thinking from the conventional wisdom, the means of implementing it are really quite ordinary. Radical changes in political institutions may not be required, and the political instruments for doing the job are really quite familiar. No technological breakthroughs are required. No cost reductions in "alternative" technologies are needed.

Mainly what is needed is a refocussing, and there is plenty of evidence this has already begun. In the OECD countries between 1973 and 1984 per capita primary energy use declined 6 percent while per capita GDP increased 18 percent.

The end-use strategies we have suggested require neither publicly-owned enterprises nor budget-busting public expenditures to implement. Some of the implementing policies, e.g., removal of energy price controls, reduce public sector intervention in the market. Others, such as energy performance standards or gas guzzler taxes for automobiles, would mean more intervention. Some, such as ending the subsidies for energy producers, will save the public sector money. Others, such as subsidizing energy efficiency improvements in the homes of the poor, will cost money. Some will just redirect public sector revenues from one part of the budget to another, e.g., from conventional energy supply R&D to R&D on energy efficiency and renewables.

Meeting the basic human needs of poor people in developing countries will obviously require money. Some might come from revenues generated from an oil import tariff and/or from the reallocation of funds now earmarked for energy supply projects that would not be needed with an emphasis on energy efficiency improvements. However, it may also be necessary to tax the elites. While the elites would thereby have to make short-term sacrifices, they would gain in the long run because redistributing present income this way to target the satisfaction of the basic needs of the poor holds forth the most promising prospect for long term economic growth, through the development of mass domestic markets.

We are optimistic. The end-use approach offers industrialized countries a way of solidifying a technological revolution which has already begun and achieved much. That there are many more technically and economically feasible opportunities for energy efficiency improvements left to exploit is tremendously encouraging. And for developing countries, the end-use approach offers hope. It shows that by emphasizing modern energy carriers and energy efficiency improvements in the pursuit of development goals they could achieve a standard of living up to that of West Europe's in the 1970's with very little increase in per capita energy use. That is great news.

APPENDIX A: Renewable Energy Resources

A.1 Introduction

The purpose of this book is to try to identify global energy strategies that are supportive of or at least consistent with the solutions of major global problems with strong links to energy—in short to explore the prospects for the pursuit of global energy strategies for a sustainable world.

We emphasize energy demand and the prospects for evolving viable global futures with much lower levels of energy demand than the levels projected in most conventional analyses. The low energy future for the world described in Chapter 4 would provide considerable flexibility in energy supply planning, making it possible to avoid many of the serious global problems described in Chapter 1, associated with overdependence on oil, fossil fuels generally, nuclear energy, and the non-renewable use of bioenergy resources.

While the illustrative scenario presented for the year 2020 in Chapter 4 involves only modest shifts from the present distribution of energy sources, an energy future that is truly compatible with the achievement of a sustainable world must, *in the long run,* be based largely on the use of renewable energy resources.

This appendix offers a perspective on the potential for renewable energy. This review provides the analytical basis for the important but limited role assumed for renewable energy in the scenarios of Chapter 4 and a basis for optimism that in the longer term future renewable resources could come to play a considerably larger role in meeting energy needs in an energy-efficient world.

A.2 The Attractions of Solar Energy

Solar energy is high quality energy that is available in abundance (Note A.2-A). The total flow of solar energy through the earth's natural systems is some 10,000 times greater than the present flow of energy through man's machines. Even the one percent of the solar influx that generates the great atmospheric pressure systems which drive the winds, and which in turn generate the waves, is some 180 times as large as man's rate of energy use. And though, on the average, the photosynthetic process is less than 0.2 percent efficient, even photosynthetic production creates 10 times as much energy as man uses.

Not only is solar energy abundant but also many (though certainly not all) of the more promising strategies for tapping the solar energy resource would have much less damaging impacts on the environment, public health, and global security than is the case with fossil and nuclear fuels (Chapter 1). There is little question that humanity should exert its ingenuity to rechannel a small amount of the flow of solar energy and thereby reduce or remove his dependence on these conventional fuels.

Despite the obvious attractions of solar energy there are major hurdles to be overcome before solar energy can become supply source as important as oil or coal. Sunlight is diffuse, so large and costly capital equipment will often be needed to put the solar energy resource to good use; sunlight is intermittent, so costly storage will be necessary to provide energy for when the sun does shine. And, solar energy is not totally free of adverse environmental impacts. Massive hydroelectric projects may disrupt natural flora and fauna and create direct problems for man, such as the increased incidence of schistosomiasis and malaria; extensive bioenergy production may have negative environmental impact by competing with agriculture, forestry, and the preservation of wildlife habitat; vast arrays of windmills may be unsightly and noisy; and exploited on a large scale, other solar technologies may also run up against land-use constraints.

Is it possible to devise solar energy strategies so that they avoid or minimize such adverse impacts? A tentative judgment, supported by the analysis presented here, is affirmative. There is such a rich diversity of solar energy options that the chances are high that for the long run solar paths can be found which are affordable and which can be implemented in a manner consistent with the satisfaction of environmental concerns, as long as overall energy demand is not too high.

As the present analysis is intended to serve as an existence proof rather than an articulation of the preferred solar energy course, we have made no attempt at completeness. Four solar options are discussed; hydroelectricity, bioenergy, wind power, and the production of electricity and/or hydrogen from amorphous silicon solar cells. All of these options could potentially play very large roles in the overall energy economy. The first three classes of technology have been singled out because of their favorable economics and relatively mature state of development. The final option was chosen because progress is especially rapid in the development of this truly new technology and because the commercial success of this technology could radically transform the entire energy supply situation, practically everywhere.

A.3 Hydroelectric Resources

The most familiar renewable resource is hydroelectric energy, which is derived from solar energy by means of the evaporation of water that is subsequently returned to the earth as rainfall.

Hydroelectricity provides an existence proof that solar energy can be developed cost-effectively on a large scale. Indeed, hydroelectricity tends to be cheaper than electricity derived from either fossil fuels or nuclear power. This is possible because mountainsides and streams provide low cost "solar collectors" and the artificial reservoirs created by hydroelectric dams can be relatively cheap solar energy storage systems.

The theoretical availability of the hydroelectric resources is the potential power that can be developed from runoff for each segment of land surface from some average surface elevation above sea level. On a global basis the theoretical potential of hydroelectric power is about 5 TW (44,000 TWh per year) of average power output (Note A.3-Aa). To produce the same amount of electricity with thermal power plants would require a fossil fuel input of 15 TW, or 1.8 times the total global use of fossil fuels in 1980. Of course, much of the theoretical resources would be impractical or uneconomic to develop. Nevertheless, only a small fraction of the hydroelectric resource that is judged to be both technically and economically feasible to develop has been developed to date.

It has been estimated (Note A.3-Aa) that about 40 percent of the theoretical potential is technically usable. Moreover, on the basis of a 1976 survey, the World Energy Conference (WEC) has estimated that about half of this technical potential could be developed economically. The WEC estimate of the economic potential, some 1.1 TW of average power output 9700 TWh/year), is 5.6 times the actual world hydroelectric production in 1980 and about 1.2 times the total world level of electric power generation from all sources in 1980.

The distribution of untapped resources is, however, very uneven among groups of countries (Note A.3-Ab). OECD nations have already exploited about half their economic potential, the U.S.S.R. and Eastern Europe about 1/5, but developing countries, which account for about 2/3 of the untapped potential, have tapped only 7 percent of their potential. Moreover, the economic potential for hydroelectricity in developing countries is about 5 times their current level of power generation from all sources.

The resource estimates for regions do not reflect the large country-to-country variations that occur throughout the world. Table A-1 shows these variations for the 30 countries with the greatest hydro potential. It is noteworthy that 1/3 of these countries have economic hydroelectric resources that are from 12 to 270 times current total electricity consumption and thus are well-positioned to "share" their hydroelectric resources with their neighbours.

Regional hydroelectric strategies can give much greater signficance to the hydroelectric resource than single country estimates—even for some industrialized countries. For example, hydroelectric power is considered to be a relatively minor energy option in the United States, where it accounted for only 12 percent of electricity production in 1980. A doubling of U.S. hydroelectricity production appears to be economical (Table A-1). But a regional North American strategy would double again the hydroelectric potential, because the economic potential for hydroelectric development in Canada (which has 1/10 the population of the U.S.) is comparable to that of the United States. As a consequence, the economic hydroelectric potential for both countries is equivalent to nearly half than total electricity production in 1980.

As large as the untapped hydroelectric resources appears to be (Table A-1 and Note A.3-A), the WEC estimates may in fact be low for many countries—in part because economic conditions have changed dramatically since the 1976 WEC Survey. The subsequent doubling of the real world oil price and the escalation in coal and nuclear power plant capital costs would make development of more of the theoretical hydroelectric power potential economically viable today. Moreover, for many developing countries, hydroelectric resources have not even been adequately surveyed, so that estimates of the resource potential continue to increase as better information is obtained. This is evident from the fact that recent official estimates of the economic hydroelectric potential in both India and Brazil are roughly double the estimate in the 1976 WEC survey. The upward adjustments for these two countries alone are equivalent to the output of 150 large nuclear or coal-fired power plants (Note A.3-B).

Planners must take into account that hydroelectric dams not only generate economic benefits, but also they can create severe social and ecological problems. Among the most frequently discussed problems are the opportunity costs of diverting large land areas to hydroelectric facilities; the consumptive water losses associated with the creation of large artificial reservoirs in arid regions; the potential for sudden dam failure with catastrophic loss of life and property; the disruption of migratory patterns of certain fish species and the destruction of fisheries; and the spread of certain water-borne diseases such as schistosomiasis, malaria, and sleeping sickness. In some areas these costs may outweigh the benefits so that no dams should be built—for example, in areas where free-flowing rivers and bottomland in river valleys are regarded as especially scarce ecological resources. In areas where dams are to be built, hydroelectric development should be carried out in a manner that avoids or minimizes adverse impacts.

Table A-1. The 30 Countries with the Largest Theoretical Hydroelectric Potential[a, b]

	Theoretical Hydroelectric Potential (TWh/year)	Ratio of Economic Hydro Potential to Total 1980 Electricity Production	Ratio of Economic Hydro Potential to Actual 1980 Hydroelectricity Production
USSR	3942	0.9	6.1
Argentina	2432	4.9	13.1
China	1927	4.5	24.4
Zaire	1567	?	?
Brazil	1389	4.0	4.3
Colombia	1290	?	?
United States	1063	0.3	2.3
Burma	821	158	259
Canada	817	1.7	2.4
India	750	1.9	18
Indonesia	667	19	356
Vietnam	556	7.4	41
Norway	500	1.5	1.5
New Zealand	500	1.3	1.7
Venezuela	494	?	?
Turkey	436	3.2	6.6
Italy	341	0.3	1.0
Madagascar	320	271	949
Mexico	280	1.4	5.5
Peru	278	12	16
France	270	0.2	1.0
Equador	258	?	?
Ethiopia	228	82	117
Chile	224	7.7	12
Costa Rica	223	16	17
Bolivia	197	60	85
Popua-New Guinea	197	112	361
Sweden	196	1.0	1.5
Nigeria	193	?	?
Gabon	180	195	250

(a) Estimates of the theoretical and economic hydro-electric potential are the WEC estimats given in Ellis L. Armstrong, 1978.

(b) Production levels for 1980 are from United Nations, 1981.

The importance of careful planning in this regard can be seen by consideration of one serious problem, the widely discussed increased incidence of schistosomiasis, which often results from the construction of large hydroelectric facilities in tropical regions.

Urinary schistosomiasis is a slow-killing disease which results from infection with a worm parasite, for which spined eggs are discharged with the urine. Where unsanitary practices permit urine with eggs to enter a body of water, the life cycle of the disease is established. The eggs hatch and the larvae develop to the infective stage in an intermediate host, a small aquatic snail which prefers quiet waters such as the weedy and shallow littoral areas of artificial lakes. Infection of other persons is through their contact with water containing the infective larvae.

Because unsatisfactory human waste disposal systems among peoples who live along the lake shore and the extensive contact of the people with the water provide the necessary conditions for the increase in schistosomiasis, it is clear that hydroelectric development efforts must be accompanied by the installation of sanitary human waste disposal systems and conveniently placed alternative water supply systems.[1]

In general hydroelectric planning should involve extensive advanced ecological and social study to anticipate and avoid major problems and a continuing effort to monitor potential adverse impacts during construction and after the project is completed. And the overall process should be carried out with the participation of the people in the affected communities.

Here, as with other energy supply development efforts, it is clear that the key to success is not only careful planning that is sensitive to ecological concerns but also the integration of energy planning process into the overall development process.

A.4 Energy from Biomass

A.4.1 *General Considerations*

Biomass energy, like hydroelectric energy, is a form of solar energy that is already in wide use, accounting for some 14 percent of total world energy use in 1980. The majority of the world's people rely for most of their energy needs on biomass, primarily in the form of fuelwood for cooking. The widespread use of bioenergy at present reflects its relatively low cost. In fact, fuelwood is gathered at "zero (private) cost" in most rural areas of the world. Although the photosynthetic process is quite inefficient in converting solar energy into useful energy forms, the produced energy is relatively cheap. "Low cost" solar collectors are provided by the land surface on which plants grow, and the produced biomass stores solar energy for use when the sun does not shine.

Biomass is often regarded as "poor people's energy," and the industrialized world made the transition a century ago from wood to coal and other fossil fuels. These facts make it difficult for many people to take seriously the notion of returning to or maintaining a wood economy. If this means a re-introduction or retention of inefficient technologies, then the objection to a wood economy is valid. But the wood energy economy of the future could be one involving advanced technology geared to meeting energy needs efficiently with modern biomass-derived energy carriers—in both developing and industrialized countries.[2]

Biomass can be utilized in the form of convenient pelletized solid fuels or it can be converted to gaseous or liquid energy carriers or to electricity. In developing countries, biomass-derived gas might be used for cooking with much greater efficiency than is possible today with conventional wood stoves; or the gas might be used in both developing and industrialized countries in high efficiency gas-turbine or diesel-type engines that produce mechanical power or electricity plus heat for industrial process use or residential heating (cogeneration). Or biomass might be used directly to produce electricity. Especially promising in this regard is the prospect of firing low-cost gas turbines directly with biomass.[2,3] New developments in gas turbine technology also offer the prospect of electricity generation or cogeneration with low-cost gas turbine systems at very high thermodynamic efficiency.[4] Biomass-derived liquid fuels such as methanol and ethanol might be used in transport or other applications where there is a need for an easily used and storable liquid fuel.

The manufacture of fluid fuels from biomass is not only feasible but is in some ways preferable to their manufacture from coal because of the unique characteristics of biomass.[5] Biomass typically has less ash and much less sulfur than coal, and its looser molecular structure means it can be gasified at a lower temperature than coal. In short, biomass is an inherently more attractive feedstock for synthetic fuels production than coal.

With the use of biomass energy in the form of advanced energy carriers such as pellets, gaseous and liquid fuels, and electricity, consumers would notice little difference between biomass-derived and conventional energy carriers, but the bioenergy production industry would be quite different from today's energy industry. In photosynthesis, typically less than 1 percent of available sunlight is converted to stored chemical energy. This means that, in contrast to the situation with conventional fossil fuels, which exist in rich concentrations in the earth and thus favor the development of large mine-mouth or well-head energy conversion facilities, biomass must be recovered from wide land areas to meet a given need. The dispersed nature of the biomass resources thus tends to favor a more decentralized energy supply system.

While photosynthesis is not a very efficient way of harnessing solar energy, the overall biomass resources can nevertheless be quite significant—as long as overall energy demand is not too large. One measure of the size of the biomass energy resource base is that in the photosynthetic process some 100 TW-years of chemical energy are stored in plants each year (Note A.4-A), about 10 times man's present energy use. To what extent can man harness or augment this biomass resource base to meet his energy needs?

There are essentially three possibilities: recovery of organic waste resources, increased harvesting of biomass from existing forests, and the intensive management of biomass production on energy "farms" or "plantations."

A.4.2 *Biomass Energy from Organic Wastes*

Biomass is already being extensively used by man for the production of food and fiber. For these activities there are significant production and post-consumption waste streams of biomass that are not being fully utilized. In forestry, the wood energy content of forest residues from logging operations, stand improvement cuttings, and wastes associated with the industrial processing of wood, is comparable in the United States to the total roundwood harvested; crop residues are typically produced in quantities of 1 to 2 tonnes per tonne of harvested product; the energy content of livestock manure is comparable to that of crop residues in both industrialized and developing countries; and finally urban refuse contains large quantities of biomass wastes— largely food wastes and paper and other short lifecycle fiber product wastes.

Today these waste streams contain about 2.8 TW of energy or about ¼ of total world energy consumption (Note A.4-B). Moreover, it is estimated that, in 2020, when the world population would be about 60 percent larger than it is today, these waste streams could contain about 4.1 TW of biomass energy—equivalent in energy content to world oil production in 1980 (Note A.4-B). The enormity of the energy content of these waste streams suggests that attempts be made to try to exploit them for their energy value.

Of course, not all organic wastes are available for exploitation as energy sources. Some forest residues should be left in the forest to maintain nutrient levels; this is especially true of foliage, in which nutrients tend to be concentrated. And it is often desirable to leave some crop residues on the fields for their nutrient and soil conditioning value. Moreover, it is simply uneconomic to try to recover some waste streams. Manure, for example, is most readily recovered only from animals in confinement or kept from grazing widely.

But often the practical availability of the waste stream is very sensitive to the kind of technology involved. If manure and crop residues were gasified in biogas digesters, for example, the nutrients would largely remain concentrated in the residuum discharged from the gasifiers. This residuum could be returned to fields as a high quality fertilizer and soil conditioner. Moreover, if a shift is made in agriculture to no-till practices, which leads to better water management and reduces nitrogen losses, the need to maintain crop residues on the field is reduced. And finally, technological innovations can lead to significant reductions in the costs of waste recovery. Agricultural harvesting equipment, for example, might evolve so as to be able to recover both the food crop and the residues simultaneously, in a single operation.

A.4.3 *Biomass From Existing Forests*

While deforestation is a serious problem in many parts of the developing world, the present rate of forest utilization is a small fraction of the rate of wood production. About 6 TW, or 85 percent of above ground wood production in forests, was not utilized in 1970 (Note A.4-Ca), with roughly a 50/50 split of unused increment between industrialized and developing countries (Note A.4-Cb). To those accustomed to hearing that harvesting rates are approaching biomass production rates, these statistics will be surprising. The numbers come out this way, however, because the biomass resource of interest for energy purposes is not the same as the biomass resource of interest to the forest products industry, for the benefit of whom most forest statistics have been prepared. Conventional forest resources estimates have concentrated on the measurement of the stemwood under bark of mature commercial species and have often excluded biomass production associated with bark, branches, stumps, foliage, dead and diseased, or mishaped trees, noncommercial species, and immature trees of commercial species. These omissions can lead to underestimating total above ground biomass production by factors of up to 3 to 6 (Note A.4-D).

In addition, biomass production in commercial forests in some areas could be increased significantly with better management. The U.S. Congressional Office of Technology Assessment (OTA) estimated in 1980 that just by fully stocking U.S. commercial forests with productive species, their biomass production could be doubled. Thus while U.S. Forest Service data indicate that the productivity of U.S. commercial forests is only about 2 tonnes/hectare/year of *roundwood,* the OTA estimates that total biomass productivity is presently up to twice as large and could be increased to an average of about 9 tonnes per hectare per year with full stocking.[6]

As in the case of forest residues, it appears that economic conditions are favorable in many instances to a much fuller utilization of the forest resource. For example, in the early 1980's the paper industry in Sweden paid for the wood used as feedstock for paper and pulp production a price which is about half that of oil in the early 1980's (Note A.4-E); because wood recovered for energy purposes need not have the select qualities required for pulp and paper production, its price may well be lower than this paper feedstock price.

Despite the large untapped fraction of the forest resources and the potential for increasing productivity in many areas with better management practices, there will always be resistance to *full* exploitation of the forest resource, which many people believe should be left in as pristine a state as possible. Larger forest areas might be left in such a condition if a third approach for exploiting the biomass resource is emphasized—intensive biomass management on "energy farms" or "energy plantations".

A.4.4 *Energy Plantations*

By intensive management of biomass crops (trees, reeds, grasses, etc.) and with careful genetic screening for productive hybrids, fertilization, thinning, etc., it is possible to increase biomass yields several-fold over what they would otherwise be in the natural state (Note A.4-F). Under the favorable conditions of ample precipitation and a long growing season in warm climates, annual yields of especially productive species such as eucalyptus and giant leucaena can be on the order of 40 tonnes per hectare or more. Even in Northern boreal forests (Canada/Scandinavia) yields of 10-15 tonnes per year can be expected with species such as willow, poplar, and alder.

And there are even promising energy plantation species for arid conditions (e.g., Mesquite, Saltbush, Johnson Grass, or Kochia), where conventional agriculture and forestry activities are wholly impractical.

The economics of biomass plantations are much more uncertain than the economics of forest residue recovery, since there has been limited experience with intensive management of biomass for energy. However, for 22 wood plantation projects (financed by the World Bank) in developing countries, which produced 120 million tonnes of fuelwood, the economics were quite favorable—with an average wood selling price of $1 per GJ corresponding to a 27 percent rate of return (Note A.4-G). For plantations in Northern industrialised countries the economics would be less favorable, owing to generally lower yields and higher labor costs, but estimates which have been made indicate a plantation-gate production cost per unit of biomass energy which is still far less than the cost of the energy equivalent in crude oil (Note A.4-H).

The biomass production potential on energy farms or plantations can be quite large on relatively limited areas. For example, the energy equivalent of the 1980 level of world oil production (4.18 TW) could be grown on 380 million hectares, a land area equivalent to 10 percent of the world's forest area, with an average productivity of 18 tonnes/hectare.

A.4.5 *Bioenergy Conflicts*

A number of concerns have been expressed about large scale bioenergy production that may limit the extent to which this energy source is exploited. Some of the more significant concerns are as follows.

Large-scale bioenergy prodution is in basic conflict with food and fiber production, and insofar as it drives up the price of food it is an especially regressive technology. Ethyl alcohol derived from sugar cane, for example, requires high quality agricultural land and thus would be an inappropriate option where agricultural land is scarce.

The intensive use of fertilizers to realize large yields and to compensate for nutrient removal in whole tree harvesting can lead to nutrient leaching and runoff and a serious deterioration of water quality.

The tendency toward monocultures in intensely managed forest projects will reduce ecological diversity and increase vulnerability to pests and disease.

While it is beyond the scope of the present analysis to address in detail these and other concerns about large-scale biomass production, there are so many different ways to cultivate biomass for energy that it may be possible to avoid or at least minimize the most serious problems.

The food/bioenergy conflict should be considered in the context of the ongoing process of deforestation, one cause of which is the conversion of forest lands to agricultural production. Much of this converted land is only marginally suitable for agricultural production. It would often make both economic and ecological sense to shift some lands from marginal agricultural production to the production of bioenergy crops, e.g., short rotation trees and simultaneously increase the agricultural output on the better lands by means of more modern, energy-intensive production techniques—perhaps using some of the energy produced on the bioenergy plantations or farms to increase crop output. Such opportunities could be significant, because on average, only about 40 percent as much fertilizer is used per hectare in developing countries as in industrialized countries, and only about 1/3 of all cropland is farmed with heavy machinery (Note A.4-I).

There are many energy crops that do not require the use of high quality agricultural land. While ethanol produced from sugar cane does require good land, methanol is an alternative high quality liquid fuel which could be produced from trees grown on marginal lands:

"One of the advantages of certain species of fast growing trees is their tolerance of poor soils and their ability—once established—to resist droughts, floods, and other environmental stresses."[7]

Also conflicts can be largely avoided by pursuing complementary, rather than competitive, strategies:

"In cases where land is scarce, *agroforestry* systems have much to offer. Agroforestry involves the intercropping of trees with field crops, a practice which can allow food and fuel to be produced from the same land. The crops must be chosen carefully so that they complement rather than compete with each other, *Eucalyptus* for example, is not a good agroforestry species due to its high moisture demand and its tendency to produce toxic substances. *Leucaena* on the other hand, helps to fertilize its surrounding area since it is a legume and itst open leaf structure means that it does not shade-out ground crops as much as some other trees...

Planting trees which can be used for multiple purposes also improves the economics of tree farming. A number of legume trees, for example, produce leaves which are excellent for animal food and their pods can often be used for human consumption. Other trees yield valuable nuts and fruits, and produce wood which can also be used for building purposes."[7]

One way to ameliorate the problem of nutrient leaching associated with intensive fertilization is to add fertilizer in many small doses over time instead of a few large doses. An alternative especially promising approach to this problem is to select for energy plantations nitrogen-fixing species. Fortunately, there are fast-growing nitrogen-fixing species for a wide range of plantation conditions. Giant leucaena is such a species in warm, moist climates. Alder is a good candidate for temperate climates; and mesquite is a nitrogen-fixing species that shows promise for desert bioenergy farms.

The potential for greatly reduced ecological diversity with biomass plantations warrants close attention. It is not an unknown problem. The ecological balance was radically transformed as large tracts of land were converted from forests to organized farmland over the last century. In making a shift to large-scale cultivation of biomass for energy purposes, however, a substantial loss of species diversity may not be inevitable—if the problem is given proper attention.

Consider, for example, the high correlation in forests between the number of bird species and the percent of the standing wood that is dead (see Figure A-1); in essence many species (such as the woodpecker) require dead wood and the insect population it provides for survival. Note from the figure that while the number of species rises rapidly initially with the dead wood pecentage, it rises only very slowly beyond about 5 percent. One way to help maintain a high degree of species diversity therefore is to let a small fraction of planned energy plantation land remain untouched.[8]

There is no doubt that large-scale bioenergy production will alter the ecological balance; however, it is unclear the extent to which the change will be for better or for worse if intensive management is done carefully, as indicated in a 1980 analysis carried out by the U.S. Office of Technology Assessment (see Box A. 1). And the above discussion suggests that as long as bioenergy production is not pushed to the limits of its technical potential, there are many possibilities for dealing with the concerns raised about bioenergy production. If emphasis is given in public policy to opportunities for energy efficiency improvement, bioenergy can be raised to the status of a major energy source without approaching these limits. Moreover, with this emphasis on energy efficiency, one buys time to achieve a good understanding of potential adverse impacts and to develop strategies for avoiding or minimizing such impacts.

A.5 Wind Energy

Harnessing wind energy involves technology that is nearly as old as civilization itself. Windmills were employed for grinding grains as far back as the seventh century in Iran and were important

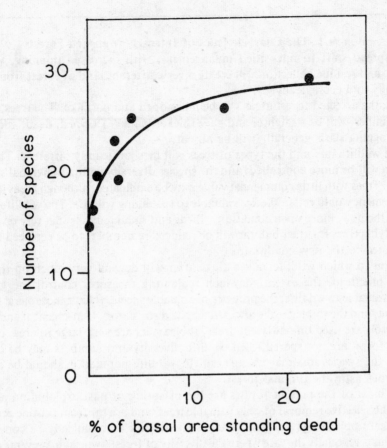

Figure A-1. The relation between bird species richness and proportion of the basal area of the forests standing dead
Source: Sven C. Nilsson, 1979.

in bringing about the first mechanical revolution in Europe in the twelfth and thirteenth centuries.

Wind energy was harnessed early by man because in this form solar energy has already been transformed into high quality mechanical energy, generated by the earth's weather system acting as a "heat engine". Thus, a simple windmill and generator can convert the mechanical energy of the wind to electricity with high efficiency.

Today, wind energy is making a comeback, but not with the windmills of yesteryear. Today's windmill designs reflect the sophistication of modern aeronautics.

Wind energy is being rapidly commercialized at the present time, especially in Denmark, Canada, Hawaii, and California. In Denmark about 1000 small (50-60 kW) units were in operation in early 1984.[9] But California is rapidly becoming the wind energy capital of the world. There the installed capacity grew 10-fold from 1981 to the end of 1984, when the total installed capacity had reached 609 Mw, involving a total of 8400 wind machines;[10] moreover, the California Energy Commission expects installed wind capacity to reach 1000 Mw by 1986.[11]

Wind energy commercialization is taking place so rapidly because of the favorable economics of this technology in areas with good wind resources, assisted by favorable tax treatment in the United States. The rapid growth of the industry has fostered technical advances and competitive

Box A.1: The Characteristics of Intensely Managed Forests

A widespread shift to intensified management, with increased thinning, whole-tree harvesting, and residue collection will create a very different kind of forest from today's, both visually and ecologically.

Visually, the affected forest areas will be more open and parklike. The trees, although fewer in number, will be straighter and have thicker trunks. Downed, dead, and diseased trees and logging slash generally will be absent.

Both the wildlife mix and the types of trees will be significantly different. The type of trees grown will be more controlled, and the species diversity within individual stands will be reduced. Trees with little commercial value may be eliminated, although areas in multiple use management would retain species valuable to sustaining wildlife. The wildlife mix itself will reflect the new, more open conditions. Birds and small animals that rely on slash and dead and dying trees for their habitat will be reduced in number, to be replaced by species better adapted to the new conditions.

The extent to which wildlife values may suffer will depend very much on the type of harvesting practiced, the extent to which replanting measures control the growth of vegetation valuable to wildlife, the presence of valuable species that cannot tolerate intensive management, and the total acreage affected and its distribution. If mechanical and chemical brush controls are used on newly cut areas, if clearcuts are very large in area, or if large pockets of forest are not spared, then wildlife diversity and numbers may be degraded. Otherwise, the species mix may change but the wildlife population should be as diverse and numerous as in the original forest.

Because most of the present forests are the offspring of past exploitation and "high grading" (the selective removal of only valuable trees) and are far from pristine ecosystems, the ecological implications of these changes should not automatically be considered as negative. This is especially the case where the diversity of forest ownership prevents extremely large contiguous areas from being placed in single species management (monoculture).

Managed forests are often described as "healthier" forests than the largely unmanaged forests found in the East. This may be a fair statement from the perspective of measurable economic worth; timber growth will be enhanced, the population of game animals will increase, damage-causing agents such as bark beetles that reside in slash or dead and dying trees will be reduced, and the incidence of forest fires might decrease. However, the effects of intensive management on other components of forest "health" such as long-term stability and resistance to disease epidemics are not as well understood and may be negative in some cases. Also, although large portions of the public may approve the changes in forest appearance and character inherent in an increase in forest management, policymakers should still expect substantial opposition to these changes, especially considering the uncertainties about the potential for long-term soil and productivity effects.

Source: The above is a direct quote from pp. 77-78, vol. I, of ref. 6.

conditions that resulted in significant cost reductions, with the average cost of new installations in California falling from $3100 per kW in 1981 to less than $1900 per kW in 1984,[10] and the unit costs for the least costly units in wind farms approaching $1000 per kW.[11]

The relatively small windmills being sold today in industrialized countries (with rated capacities on the order of 50 kW) can produce electricity at a busbar cost which is comparable to the busbar cost for a new coal-fired, baseload power plant (Note A.5-A). Of course, wind power is not worth as much as baseload electric power, because it is not "firm power." But at the world oil price following the second oil shock, it was cost-effective in displacing the oil still being used to generate electricity (Note A.5-A).

Although the expected cost of electricity from large (2.5 MW) windmills that are being built today is more costly than the electricity from the smaller machines (Note A.5-A), today's large windmills are single first-of-a-kind devices. It is projected, on the basis of economic analysis of the production process, that by the time mass production levels are reached, these multi-MW machines will cost only $1100 to $1300 per kW (Note A.5-B) and be able to produce electricity at a busbar cost which is only 60 to 75 percent as high as that for baseload coal plants (Note A.5-A).[12]

If wind is to provide reliable power, however, storage or back-up power must be provided to compensate for the intermittency of the wind. One of the most promising strategies for doing this involves coupling wind and hydroelectric power sources, using hydroelectric reservoirs for pumped storage. Sorensen has modelled such a strategy for coupling Danish windmills with Norway's hydropower resources.[13] The proposed system would operate as follows: in periods of insufficient wind, additional water would be let down through the Norwegian turbines; if on the other hand, more wind power is being produced than is needed to cover the Danish load, the surplus would be exported to Norway where it would be used instead of hydro-electricity. Sorensen shows that the combined system would be as dependable as the hydro-system alone. Wind energy planning for Sweden involves such a hydro/wind coupling and thereby envisages a wind energy contribution to the Swedish electrical system in excess of 10 percent in the 1990s.[14]

In regions where there is a good wind resource but little hydro-electric generating capacity, alternative storage schemes are needed to make wind power reliable. One possibility is compressed air storage. With this scheme, the electricity generated by the wind machines, when not otherwise needed, is used to run a compressor to compress air for storage in a deep underground reservoir. During peak demand periods, the compressed air is used to produce electricity with gas turbines. (Ordinarily 2/3 of the shift power output of the gas turbine is required to drive the compressor.) Suitable sites for underground compressed air storage are widely available. One 290 MW plant has been successfully operated in Huntorf, West Germany, since 1978, and a 220 MW plant is being built in Decatur, Illinois, in the United States[15]. The major problem with this storage scheme is that it is thermally inefficient. Compression increases the temperature of the air beyond practical limits for storage, so the air must be cooled before storage. The associated heat loss makes the process inefficient. One high efficiency alternative is "adiabatic" compressed air storage, which involves storing the heat generated in compression.

An alternative, large-scale storage scheme which appears to be relatively low-cost and efficient is undergound pumped hydroelectric storage (Figure A-2). With this scheme, water would be pumped during off-peak periods from a reservoir located perhaps a thousand meters underground to a surface reservoir; during peaking periods water would be returned through a turbine to lower reservoir, thus producing power. Underground rock formations suitable for such systems are relatively widely available, and this kind of pumped hydro should be much more environmentally acceptable than schemes involving two surface reservoirs.[16]

No really good estimates of the global wind potential have been made. However, the 1981 IIASA global energy study estimated a global technical potential for wind power of 3 TW—(26,000 TWh per year, the amount of electricity which could be produced with fossil fuel

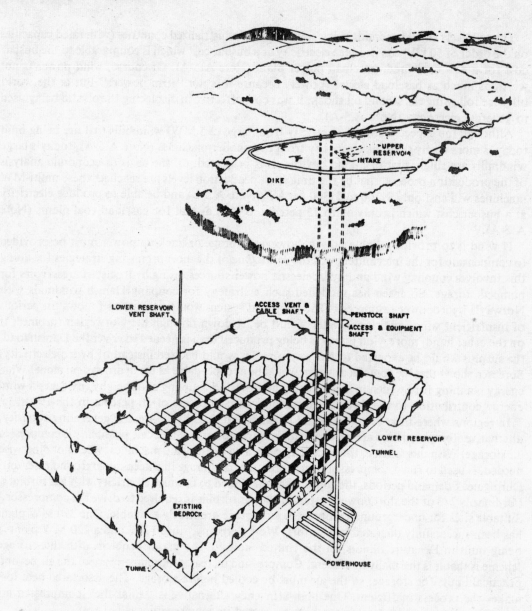

Figure A-2. Typical underground hydroelectric pumped storage.
Source: C.T. Main, Inc., November 1978.

inputs at a rate of 9 TW. (IIASA limited its estimate to continental areas with 1000 km of coasts running from 50 northern degrees to 50 southern degrees latitude.)[17] The IIASA analysts then made the judgment that only 1/3 of this technical potential (1 TW) would be realized because of economic, aesthetic and competing land value constraints.

The IIASA estimate may be low. Thompson has shown[18] that if the wind resource is captured only on sites in the United States and Canada with an annual average wind power density of more than 400 W per square meter of turbine-swept area (exceeded on 16 percent of the land area of the United States and Canada), the average power that could be produced in these two countries alone is 0.9 TW (7900 TWh/year).* Moreover, the best wind conditions in these two countries are in the continental interior (Figure A-3)—regions excluded in the IIASA analysis.

Most estimates of the wind energy potential, are based on examinations of the wind potential in regions where the average wind speed is high. With this constraint, the wind resource would appear to be confined largely to industrialized nations. (see Figure A-4).

However, the role of wind energy in developing nations may be considerably more important than such "global analyses" would indicate. That this may be so is indicated by the results of a detailed analysis of the wind energy resource potential in Karnataka State in South Central India. Even though the annual average wind power density (an average based on measurements at 21 stations in Karnataka state) is only about 25 watts/m² (corresponding to an average wind speed of 2.8 meters/second), the wind energy density is much higher than this when the wind actually blows. About 80 percent of the annual wind energy flux is concentrated in May and the four· monsoon months, June-September. The wind energy resource could be harnessed economically under these conditions if very low cost (and thus relatively inefficient) windmills, were available. Based on the use of simple (10 percent efficient) windmills, it has been estimated that the extractable wind energy potential in Karnataka state is about an order of magnitude larger than the current level of electrical energy consumption.[19] This example illustrates the pitfalls of estimating the potential for wind energy development on the basis of maps giving only regional annual average wind speeds. An adequate assessment requires that local spatial and temporal variations in the wind resource be taken into account and that careful consideration be given to how the available wind resource might be matched to the time-varying patterns of local energy needs.

Despite the shortcomings of the present level of knowledge of the wind energy resource potential in different parts of the world, even the IIASA estimate of the "practical" wind potential is enormous—it is as large as the WEC estimate of the economic potential for hydropower development and roughly equal to the present level of total electric power generation in the world.

The major potential constraint in harnessing wind power appears to be the acceptability to the public of large numbers of wind turbines. Both aesthetic and noise factors will probably determine the extent to which wind power is developed. Because of such considerations, off-shore based systems, which are of considerable interest in Sweden, may prove to be important, particularly in sheltered coastal areas.

A.6 Thin Film Amorphous Silicon Solar Cells

The direct production of electricity from solar energy using silicon solar cells is an ideal way to harness the sun. The basic raw material involved is silicon, which can be derived from silicon dioxide, or common sand. In the energy conversion process, there are no "moving parts" and there is thus nothing to wear out. And photovoltaic PV electricity production with silicon solar cells probably comes as close to being free of adverse environmental effects as is possible with any energy technology.

Up till a few years ago, it was believed that crystalline solar cells were necessary for the direct production of electricity through the photovoltaic effect. There have been significant advances in crystalline solar cell technology in recent years, as system costs have fallen from $20 per peak

---

* This estimate is for an array of Boeing MOD-2 windmills (having a rated capacity of 2.5 MW(e) spaced at intervals of 1.5 km (for a square array). The blade for the MOD-2 is about 100 meters in diameter.

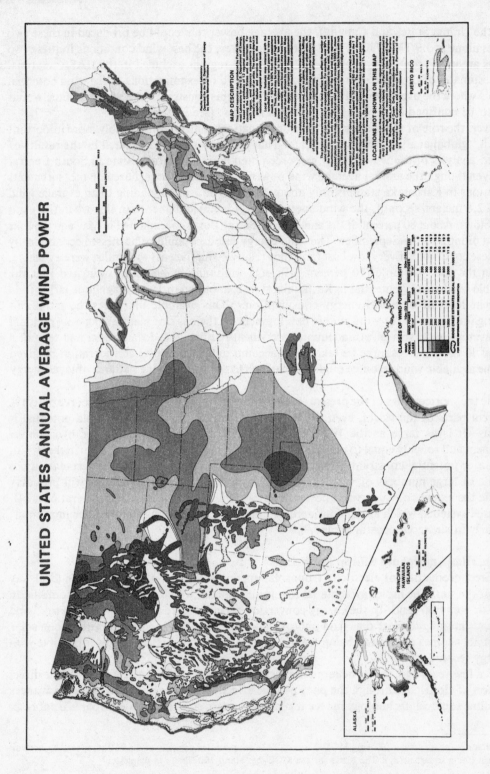

Figure A-3. Estimated annual wind-power density (in Watts per square meter) over North America—Canada (top) at the United States (bottom).

Source: Gordon Thompson, June 1981.

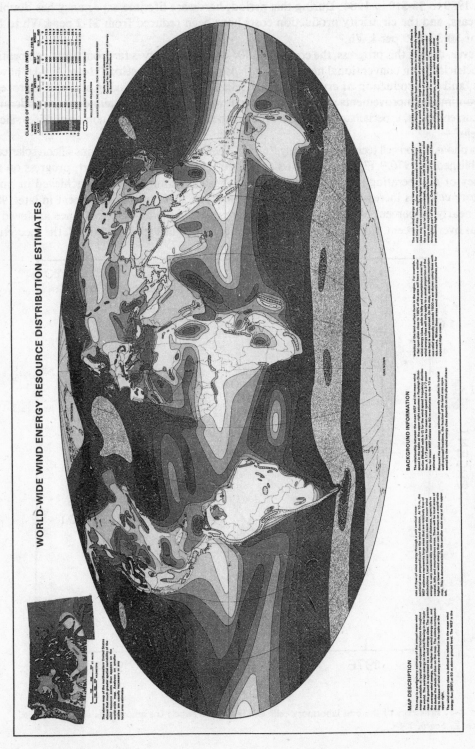

Figure A-4. Maps of wind speed regimes for January (top) and July (bottom) based on inhomogeneous meteorological data, mostly taken at a height of about 10 meters above the ground.

Source: World Meteorological Organization, 1964.

watt in 1980 to $8-$10 in 1984. During this period, hardware lifetime expectancy has doubled to 20 years, and the electricity production costs have been reduced from $1-2 per kWh to the range of $0.30-0.50 per kWh.[20]

However, despite this progress, the cost of this PV-based electricity is far from being competitive with electricity from conventional new peaking (Note A.6-A) and baseload (Note A.5-D) power sources, and the production of crystalline solar cells remains a painstaking and costly process. While continuing improvements are likely to bring down the costs of crystalline cells considerably below current levels, especially promising are the prospects for amorphous (non-crystalline) silicon solar cells.

Amorphous silicon cell technology is truly "new." The first paper on amorphous silicon solar cells was published in 1976.[21] Figure A-5 shows that since the first published report, progress on the efficiency of an operating cell has been dramatic. The maximum efficiency achieved in small laboratory cells has increased from a mere 2.5 percent in 1976 to over 10 percent in late 1982, and to nearly 12 percent by 1984.[22-23] For large-area panels the best performance achieved by 1983 was over 8 percent. Further efficiency improvements can be expected. Indeed, the theoretical

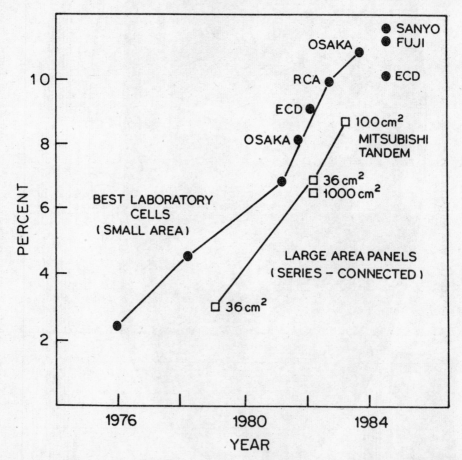

Figure A-5. Efficiency of the best laboratory cells and large area panels for amorphous silicon solar cells. *Source: Z. Smith, 1984.*

limit on efficiency for single layer cells is about 20 percent.[24] Moreover, with multiple layer cells (each layer of which is active for a different part of the solar spectrum) that would be only marginally more costly (per unit area) than single layer cells, theoretical efficiencies as high as 30 percent are possible.[23] A review of such cells by analysts at the Electric Power Research Institute in the United States has concluded that an efficiency of 16 or 17 percent is probably achievable for commercial systems.[26] In 1984 the efficiency of the better commercial amorphous silicon cells was still a more modest 5 percent, however, with R&D targeted to bring the efficiency of commercial cells to the ragne 8-12 percent by 1988.[20]

The promise of this technology is reflected in part by the rapidly growing private sector effort committed to its development. The world's leading manufacturer has been the Japanese firm SANYO, which in 1984 produced 3.3. MW of amorphous silicon cells, largely for consumer product markets.[27] Indeed, Japanese manufacturers generally have shifted their rapidly growing solar cell production to amorphous silicon technology (Figure A-6). In the United States the

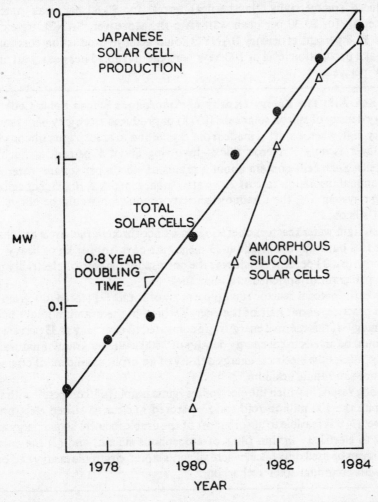

Figure A-6. The trend in the Japanese production of solar cells (total and amorphous silicon). *Source:* Z. Smith, 1984.

leading manufacturer has been Arco Solar, which had a production capacity of 750 kW per year in 1984.[27] Other United States firms with developmental programs that were building manufacturing facilities at the time of this writing were Chronar Corporation (Princeton, NJ), Energy Conversion Devices (Troy, Michigan), and Solarex/AMOCO* (Rockville, Maryland).

The most remarkable features of amorphous silicon solar cell technology are that: (1) the cells can be fabricated by a variety of techniques involving the vapor deposition of thin films of amorphous silicon on a substrate like glass—techniques that are amenable to low cost mass production; and (2) the cells absorb sunlight so effectively that they need be only about 1 micron thick (about 1 percent as thick as present day crystalline cells), so that their energy density (kWh of produced electricity per gram of silicon) is comparable to that for breeder reactors (kWh of produced electricity per gram of uranium)—see Box A.2.

Just how much costs can be expected to fall in the not-to-distant future is suggested in work by RCA researchers in 1983, who projected that 10 percent amorphous silicon modules could be mass-produced commercially "in the near term" for $0.60 per peak watt and "in the intermediate term" for $0.30 per peak watt (A.6-B). Moreover, SANYO researchers in 1981 projected that @ 8 percent efficiency SANYO could achieve production cost targets of $0.25 per peak watt at a production level of 100 MW per year and $0.15 per peak watt at a production level of 1 GW per year.[28]

Box A.2: The Energy Density of Amorphous Silicon Solar Cells

The energy density of silicon solar cells (kWh) of produced electricity per gram of silicon) becomes very high when a shift is made from crystalline to amorphous silicon cells, where the silicon layer is only 1 micron thick—involving about 1 percent a much silicon as crystalline cells. Such cells contain about 3 grams of silicon per square meter. Thus, for an average annual insolation rate of 200 watts/square meter, a 10 percent cell efficiency, and a 20 year system life, the lifetime electricity production would be about 1200 kWh per gram of silicon.

In contrast, light water reactors operating on a once through fuel cycle are characterized by a nuclear fuel burn-up rate of about 33 megawatt-days (th) per kg of heavy metal fuel. For a reactor with a 33 percent efficiency, the corresponding rate of electricity production is 260 kWh per gram of enriched uranium fuel.

In principle, the breeder reactor can utilize most of the 82 GJ stored in 1 gram of natural uranium. In practice, about half of the energy stored in the uranium might be recovered as thermal energy. If this thermal energy were converted to electricity at 33 percent efficiency, the result would be an electrical energy density of 3800 kWh per gram of natural uranium, which is only 3 times the electrical energy density of amorphous silicon; of course the earth has a lot more sand than uranium.

This amazing result, in which the electronic regime is able to "compete" with the nuclear regime, despite the 100 million fold ratio of stored nuclear to stored electronic energy, arises because: (1) it is feasible to absorb most of the usable incident solar energy and convert it efficiently to electricity in thin films of amorphous silicon; and (2) the energy stored in the nucleus can be used only once, where silicon's stored electronic energy can be extracted repeatedly with continual solar recharging.

---

* RCA, a leading developer of amorphous silicon solar cell technology, sold its PV program to Solarex in 1983. Solarex was then bought out by Standard Oil of Indiana (AMOCO).

RCA researchers in 1983 designed an hypothetical 50 MW photovoltaic power plant located near Princeton, N.J., based on their expectations for amorphous silicon cell panel costs. These expectations are for a total system cost of about $1.50 per peak watt "in the near term" and about $0.90 per peak watt "in the intermediate term." The cost of electricity from the "intermediate term" power plant would actually be competitive at the busbar with the cost of electricity from a new baseload coal plant (Notes A.6-C and A.5-D).

A direct comparison with baseload electricity production is somewhat misleading, however, because the photovoltaic power produced is not "firm power." To make a proper comparison with baseload electricity production, electrical storage is necessary to smooth out the electrical output. Alternatively, photovoltaic power plants could be considered for utility peaking applications. This application is particularly interesting because the maximum photovoltaic output will often occur at the same time as the peak demand for summer peaking utilities (Figure A-7). An analysis by RCA and electric utility researchers has concluded that a photovoltaic plant in New Jersey would, in fact, be competitive by the time the total installed cost for the cells falls to $1.1—$1.35 per peak watt.[29]

But the utility peaking market is a relatively small one. For widespread use in utility applications, electrical storage will be necessary. This may be economical "in the intermediate term" in areas with good sunlight. Indeed, the busbar cost of electricity from a photovoltaic power plant could

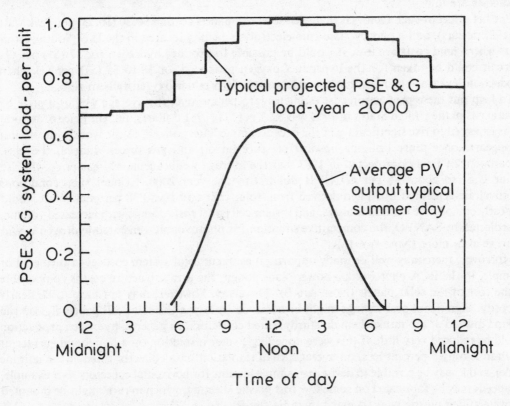

Figure A-7 Shape of the electic power load projected for the Public Service Electric and Gas Company in New Jersey in the year 2000, compared to the electrical output of a photovoltaic system.
Source: W.S. Ku *et al.*, 1983.

be reduced 40 percent simply by moving the plant from New Jersey to United States' sunny Southwest (Note A.6-C). The cost reduction achieved this way provides a considerable cost margin to allow for extra storage costs.

Alternatively, the storage problem could be overcome by generating hydrogen with the produced electricity by means of electrolysis. As a chemical fuel that can be readily stored and distributed through pipelines, hydrogen produced by means of photovoltaic-based electrolysis could serve wide markets in displacing both natural gas and oil, as supplies of these fuels become scarcer and more costly.

Hydrogen production has the advantage that the costly equipment needed to convert into AC power the DC power generated by solar cells could be eliminated, thereby realizing a savings of perhaps 15 percent (Note A.6-C). Hydrogen produced from DC electricity would also be less costly than hydrogen produced from AC electricity because the rectifier would not be necessary for the electrolysis plant (Note A.6-D). Hydrogen also offers the advantage of relatively low cost transportablility. While hydrogen has only 1/3 the heating value of natural gas, it can be produced through efficient electrolysis at high pressure; produced at high pressure, hydrogen can be transported long distances at costs that are comparable to or lower than the costs of transporting natural gas,[30] and much less than the cost of transporting electricity.[31] With cheap transport costs one has the option of siting in areas where land use competition is minimal and land costs are low.

The net effect of such factors is that with solar cell panels costing $0.30 ($0.15) per peak watt (in 1981 dollars), and a photovoltaic-cum-electrolysis facility located in the U.S. Southwest in areas where land costs are low, it would be possible to produce hydrogen for $7 ($5) per GJ if credit could be taken for the byproduct oxygen produced, or $8 to $9 ($7) per GJ when production levels are so high that the value of the byproduct oxygen falls to zero.

To help put these production cost estimates into perspective, the average wellhead price of natural gas in the United States in 1984 was $2.1 per GJ (1981 dollars). But the price of natural gas is expected to rise significantly in the decades immediately ahead, as the worldwide oil glut disappears, once more bringing upward pressure on oil and gas prices. Indeed, the U.S. Department of Energy projected in 1983 that the average wellhead natural gas price will rise in the U.S. to about $6 per GJ (1981 dollars) by the year 2000.[32] Under such conditions photovoltaic-generated hydrogen derived from solar cells costing $0.30 per peak Watt would be nearly competitive with natural gas, and if solar cell panel costs decline with increased volume, as projected by SANYO, the competitive situation for photovoltaic-generated hydrogen would improve even more (Note A.6-E).

Moreover, there may well be many opportunities to cut total system costs even further. For example, in the RCA photovoltaic power plant design, the array structure cost is comparable to the cost of the solar panels (Note A.6-B). The array, however, may not have to be nearly so costly. Significant cost savings might be achieved by laying the solar cell panels flat on the ground instead of mounting them on sturdy, tilted collectors. In general, hydrogen production would be reduced very little if this were done (e.g., solar insolation on a horizontal surface at El Paso is only 8 percent less, on average, than insolation on a collector tilted at the latitude angle), and it may be possible to devise very simple mounts for horizontal collectors. For example, solar cells may be fabricated on some low cost plastic sheeting, which in turn might be mounted on water-filled plastic bags to make them lie flat.[33]

In addition to the potentially attractive economics of photovoltaic-generated hydrogen, this approach to fluid fuels production would put very little pressure on land resources needed for other purposes. To demonstrate this point, a comparison with bioenergy production is instructive. A very productive biomass plantation in the United States might yield 25 tonnes of dry biomass per hectare per year. If this were converted to useful energy forms at an average efficiency of 65 percent, the useful energy produced would amount to about 1 watt per square meter of plantation land area. In contrast, a 15 percent efficient solar cell combined with an 80 percent efficient electrolyzer would yield about 24 watts of hydrogen per square meter in the Southwestern United States. Not only would land requirements be reduced by more than an order of magnitude if photovoltaic-generated hydrogen were produced as an alternative to fluid fuels derived from biomass, but also the needed land could even be low cost desert land.

Even on a grand scale, land-use constraints would not be serious. Replacing the present level of fossil fuel use in the world (some 8.4 TW) with photovoltaic-generated hydrogen would require only 2 percent of the desert area of the world—an area equivalent to half the state of Texas.*

The discussion here has focussed on the great promise of one particular PV technology. In the long run, it may turn out that the direct production of hydrogen by means of hydrogen-evolving solar cells is preferable to the indirect hydrogen production described here, requiring the intermediate steps of electricity production and electrolysis.[34] Or it may turn out that one or more of the strong competitors to amorphous silicon technology will emerge as the dominant PV technology—with high efficiency single crystal systems (both silicon and gallium arsenide) used with solar energy concentrators and flat-plate PV devices made of crystalline silicon sheet material being perhaps the most promising alternatives to amorphous silicon technology.[26] However the technology evolves, progress to date provides strong evidence of a major role for PV technology in the world's future energy economy.

References

Section A.3
1.   Letitia Obeng, "Should Dams Be Built? The Volta Lake Example," *Ambio,* vol. 6, no. 1, pp. 46-50, 1977.

Section A.4
2.   Robert H. Williams, "Potential Roles for Bioenergy in an Energy-Efficient World," *Ambio,* vol. XIV, No. 4-5, pp. 201-209, 1985.
3.   J.T. Hamrick, "Development of Wood-Burning Gas Turbine Systems," *Modern Power Systems,* vol. 4, No. 6, June 1984.
4.   Eric D. Larson and Robert H. Williams, "Steam-Injected Gas Turbines," Pamphlet Paper of the American Society of Mechanical Engineers, presented at the Gas Turbine Conference, Dusseldorf, West Germany, June 1986.
5.   Michael Antal, Jr., *Biomass Energy Enhancement,* a Report to the President's Council on Environmental Quality, July 1978.
6.   Office of Technology Assessment, U.S. Congress, *Energy from Biological Processes,* Washington, DC, 1980.
7.   D.O. Hall, G.W. Barnard and P.A. Moss, *Biomass for Energy in the Developing Countries,* Pergamon Press, New York, 1982.
8.   Ingrid Stjernquist (University of Lund, Sweden), private communication to Robert H. Williams, September 1982.

Section A.5
9.   Ny Teknik 1984: 17, pp. 16-17.
10.   Tony Baer, "Economics Improve at Altamont Wind Farms," *Renewable Energy News,* vol. 8, No. 1, p. 12, February 1985.

---

* Typically the average insolation level in sunny desert regions of the world is well in excess of 200 watts per square meter. Thus with 15 percent efficient solar cells (a realistic target for multiple layer cells[26]) and an 80 percent efficient electrolyzer, 8.4 TW of hydrogen could be produced on some 350 thousand square kilometers.

11. David Hoffman, "Wind Power: Industry Shouldering the Challenge of Commercialization," *Renewable Energy News,* vol. 7, No. 2, p. 10, May 1984.

12. James I. Lerner, "A Status Report on Wind Farm Energy Commercialization in the United States, with Emphasis on California," paper presented at the 4th International Solar Forum, Berlin, Federal Republic of Germany, October 6-9, 1982.

13. Sorensen, "A Regional Wind-Hydro Electricity Supply System," paper presented at the Third International Symposium on Wind Energy Systems, Lyngby, Copenhagen, Denmark, August 26-29, 1980.

14. Namnden for energiproduktions-forskning (National Swedish Board for Energy Source Development), "Wind Energy in the Swedish Power System," NE 1982:12, 79 p., Stockholm, Sweden, 1982.

15. Nadine Lihach, "Breaking New Ground with CAES", *EPRI Journal,* pp. 16-21, October 1982.

16. Gordon Thompson, "Utility-Scale Electrical Storage in the USA: the Prospects of Pumped Hydro, Compressed Air, and Batteries," PU/CEES Report No. 120, Center for Energy and Environmental Studies, Princeton University, Princeton, New Jersey, August 1981.

17. Energy Systems Program Group of the International Institute for Applied Systems Analysis, *Energy in a Finite World: A Global Systems Analysis,* 837 pp., Ballinger, Cambridge, 1981.

18. Gordon Thompson, "The Prospects for Wind and Wave Power in North America," PU/CEES Report No. 117, Center for Energy and Environmental Studies, Princeton University, Princeton, New Jersey, June 1981.

19. U. Shrinivasa, R. Narasimha, and S.P. Govinda Raju, "Prospects for Wind Energy Utilization in Karnataka," *Proceedings of the Indian Academy of Sciences, Section C: Engineering Sciences,* vol. 2, Part 4, December 1979.

Section A.6

20. Christopher Pope, "Photovoltaics: Doubled Shipments Signal Coming of Age," *Renewable Energy News,* vol. 7, No. 2, p. 28, May 1984.

21. D.E. Carlson and C.R. Wronski, *Applied Physics Letters,* vol. 28, p. 671, 1976.

22. D.E. Carlson, *IEEE Transactions,* ED-24, p. 449, 1977.

23. A. Catalano *et al.,* "Attainment of 10% Conversion Efficiency in Amorphous Silicon Solar Cells," RCA reprint, September 1982.

24. D.E. Carlson, "Solar Energy Conversion in a-Si:H," *Topics in Current Physics,* Springer-Verlag, 1983.

25. D.E. Carlson, private communication, October, 1982.

26. Edgar A. DeMeo and Roger W. Taylor, "Solar Photovoltaic Power Systems: An Electric Utility Industry R&D Perspective," *Science,* vol. 224, No. 4646, pp. 245-251, April 20, 1984.

27. Don Best, "Arco, Sanyo Show New Thin-Film PV's," *Renewable Energy News,* vol. 8, No. 1, p. 1, February 1985.

28. Y. Kuwano and M. Ohnishi (of the SANYO Electric Company's Research Center, in Hirakata City, Osaka, Japan), "Industrialization of a-Si Cells," paper presented at the 9th International Conference on Amorphous and Liquid Semiconductors, Grenoble, France, July 2-8, 1981.

29. W.S. Ku, N.E. Nour, T.M. Piascik (of Public Service Electric and Gas Company, Newark, New Jersey), and A.H. Firester, A.J. Stranix, M. Zonis (of RCA Laboratories), "Economic Evaluation of Photovoltaic Generation Applications in a Large Electric Utility System," paper presented at the 1983 IEEE PES Winter Meeting.

30. Diomedes Christodoulou, *Technology and Economics of the Transmission of Gaseous Fuels via Pipelines,* Thesis, Master of Science in Engineering, Department of Mechanical and Aerospace Engineering, School of Engineering and Applied Science, Princeton University, Princeton, New Jersey, April 1984.

31. G. Carleson, "Assessment of the Potential Future Market in Sweden for Hydrogen As an Energy Carrier," *International Journal of Hydrogen Energy,* vol. 7, No. 10, pp. 821-829, 1982.

32. Office of Policy, Planning, and Analysis, U.S. Dept. of Energy, "Energy Projections to the Year 2010: a Technical Report in Support of the National Energy Policy Plan," Report DOE/PE-0029/2, Washington, DC, October, 1983.

33. Theodore B. Taylor, "Solar Fuels: Biomass and Hydrogen Produced by Photovoltaic Cells," a report prepared for the Solar Energy Research Institute at the Center for Energy and Environmental Studies, Princeton University, Princeton, N.J., June 25, 1981.

34. Adam Heller, "Hydrogen-Evolving Solar Cells," *Science,* vol. 223, pp. 1141-1148, 1984.

References for the Figure Captions, Tables, and Notes

Section A.2

M. King Hubbert, "The Energy Resources of the Earth," *Scientific American,* vol. 224, pp. 60-70, 1973.

E.N. Lorenz, *Nature and Theory of the General Circulation of the Atmosphere, World Meteorological Organization ,* Geneva, 1967.

World Meteorological Organization, *Meteorological Aspects of the Utilization of Solar Energy as an Energy Source,* Technical Note No. 172, 1981.

Section A.3

Hartmut Krugman, "Hydroelectric Power in Brazil: Expanding to Meet Increasing Needs," Coppe, Universidade Federal do Rio de Janeiro, May 1982.

M. Schulman (Eletrobras), "O Potencial Hidreletrico Do Brasil," September, 1980.

Ellis L. Armstrong, "Hydraulic Resources," in *Renewable Energy Resources: the Full Report to the Conservation Commission of the World Energy Conference,* IPC Science and Technology Press, 1981.

World Energy Conference, *Survey of Energy Resources,* Munich, September 1980.

Section A.4

Soren Andersson, "Upplandsbranslen," NE 1980:5, a report (in Swedish) of Namnden for energiproduktions-forskning (National Swedish Board for Energy Source Development), Stockholm, Sweden, 1980.

J.S. Bethel *et al., Energy from Wood,* a report to the Office of Technology Assessment prepared by the College of Forest Resources, University of Washington, Seattle, Washington, April 6, 1979.

Sven Bjork and Wilhelm Graneli, "Energy Needs and the Environment," *Ambio,* vol. 7, No. 4, pp. 150-156, 1978.

Richard Davids, "Growing Our Own Fuel," *EXXON USA,* pp. 16-21, First Quarter 1982.

D.E. Earl, Forest Energy and Economic Development, Clarendon Press, Oxford, 1975.

J. Laurence Kulp (Vice President for R&D, Weyerhauser Company), "Biomass from Forests", paper presented at the Inter-American Symposium/Workshop on Biomass Substitutes for Liquid Fuel.

Molokai Study Team (James L. Brewbaker, editor), "Giant Leucaena (Koa Haole) Energy Tree Farm, an Economic Feasibility Analysis for the Island of Molokai, Hawaii," Hawaii Natural Energy Institute, University of Hawaii, Honolulu, Hawaii, September 1980.

Sven G. Nilsson, "Density and Species Richness of Some Forest Bird Communities in South Sweden," *OIKOS,* vol. 33, pp. 392-401, 1979.

Office of Technology Assessment (U.S. Congress), *Energy from Biological Processes, Volume II—Technical and Environmental Analyses,* 1980.

W. Palz and P. Chartier, *Energy from Biomass in Europe,* Applied Science Publishers, Ltd., London, 1980.

D. Pimental *et al.,* "Deforestation: Interdependency of Fuelwood and Agriculture," unpublished report, College of Agriculture and Life Sciences, Cornell University, 1985.

Programplan Energiproduktion 1980-03-12, NE 1980: 4, a report (in Swedish) of Namnden for energiproduktions-forskning (National Swedish Board for Energy Source Development), Stockholm, Sweden, 1980.

United Nations, *1979/1980 Statistical Yearbook,* New York, 1981.

Robert H. Whitaker and Gene E. Likens, "The Biosphere and Man," in *Primary Productivity of the Biosphere,* Helmut Lieth and Robert H. Whitaker, eds., Springer-Verlag, New York, 1975.

Theodore B. Taylor, Robert P. Taylor, and Steven Weiss, "Worldwide Data Related to Potentials for Widescale Use of Renewable Energy," PU/CEES Report No. 132, Center for Energy and Environmental Studies, Princeton University, Princeton, NJ, March 1, 1982.

U.S. Environmental Protection Agency, "Refuse-Fired Energy Systems in Europe: An Evaluation of Design Practices. An Executive Summary," Publication No. SW 771 of the Office of Water and Waste Management, November 1979.

Section A.5

Bechtel Power Corporation, *Coal-Fired Power Plant Capital Cost Estimates,* a report prepared for the Electric Power Research Institute, Report No. PE-1865, May 1981.

Energy Information Administration, *Annual Report to Congress, Volume Three: Projections,* March 1981.

Energy Information Administration, U.S. Dept. of Energy, "Monthly Energy Review," October 1982.

Charles Komanoff, *Power Plant Cost Escalation: Nuclear and Coal Capital Costs, Regulation and Economics,* Komanoff Energy Associates, New York, 1981.

James I. Lerner, "A Review of Large Wind Turbine Systems," September 1982.

C.T. Main, Inc. (Boston, Massachusetts), "Underground Hydroelectric Pumped Storage," report prepared for the U.S. Departments of Energy and Interior, November, 1978.

L.L. Nelson, "Medicine Bow Wind Project," *Proceedings of the Fifth Biennial Wind Energy Conference and Workshop,* sponsored by the U.S. Department of Energy, Washington, D.C., October 5-7, 1981, SERI/CP-635-1340, CONFF-811043, Volume II, p. 557.

Technical Assessment Group of the EPRI Planning Staff, *Technical Assessment Guide,* Electric Power Research Institute Report No. PS-1201-SR, July 1979.

Gordon Thompson, "The Prospects for Wind and Wave Power in North America," PU/CEES Report No. 117, Center for Energy and Environmental Studies, Princeton University, Princeton, New Jersey, June 1981.

World Meteorology Organization, "Paper W/11," in *Proceedings of the U.N. Conference on Alternative Energy Sources, Rome 1961,* United Nations Printing Office, New York, 1964.

Section A.6

Energy Information Administration, Office of Coal, Nuclear, Electric and Alternative Fuels, U.S. Department of Energy, *Cost and Quality of Fuels for Electric Power Plants,* DOE/EIA-0075(81/11), November 1981.

Jan F. Kreider and Frank Kreith, *Solar Heating and Cooling,* second edition, Hemisphere Publishing Corporation, Washington, 1982.

Y. Kuwano and M. Ohnishi, "Industrialization of a-Si Solar Cells," paper presented at the 9th International Conference on Amorphous and Liquid Semiconductors at Grenoble (France), July 2-8, 1981.

W.S. Ku, N.E. Nour, T.M. Piascik (of Public Service Electric and Gas Company, Newark, New Jersey), and A.H. Firester, A.J. Stranix, M. Zonis (of RCA Laboratories), "Economic Evaluation of Photovoltaic Generation Applications in a Large Electric Utility System," paper presented at the 1983 IEEE PES Winter Meeting.

R.L. Leroy and A.K. Stuart, "Unipolar Water Electrolyzers: a Competitive Technology," in *Hydrogen Energy System,* Proceedings of the 2nd World Hydrogen Conference, held in Zurich, Switzerland, 21-24 August 1978, T. Nejat Veziroglu and Walter Seifritz, eds., Volume 1, Pergamon Press, Oxford, 1979.

Sierra Energy and Risk Assessment, Inc., and Robert Foster, Consulting, *Coal Use in California, Volume 1: Appropriate California Coal Applications,* report prepared by under the sponsorship of the California Foundation on the Environment and the Economy, September 1982.

Z. Smith (Department of Electrical Engineering and Computer Sciences, Princeton University) "Solar Electricity from Amorphous Silicon: Japanese Challenge/American Response," seminar at the Center for Energy and Environmental Studies, Princeton University, December 4, 1984.

A.J. Stranix and A.H. Firester, "Conceptual Design of a 50 MW Central Station Photovoltaic Power Plant," presented at the 1983 IEEE PES winter meeting.

Notes

Section A.2

Note A.2-A. Solar Energy Fluxes (TW)

Solar Radiation Intercepted by the Earth[a]	175,000
Solar Radiation Absorbed by the Earth[a]	110,000
Solar Energy Involved in Evaporation[b]	40,000
Solar Energy Used in the Generation of the Atmospheric Pressure System[c]	1,800
Solar Energy Involved in Direct Sensible Heating	68,000
Solar Energy Utilized in Photosynthesis[d]	100
Man's Rate of Energy Use, 1980	10.65

(a)   The solar constant (the intensity of sunlight on the top of the atmosphere) is 1370 watts per square meter and the earth's albedo (the percentage of incident sunlight reflected back into space) is 32 percent (World Meteorological Organization, 1981).

(b)   M. King Hubbert, 1973.

(c)   E.N. Lorenz, 1967.

(d)   From Note A.4-A.

Section A.3

Note A.3-Aa. Summary of Hydro-Electric Resources Potential

	Total Electricity Production in 1980 (TWh)[a]	Hydro Electricity Production in 1980 (TWh)[a]	Theoretical Hydroelectric Potential (TWh)[b]	Technically Usable Hydroelectric Potential (TWh)[b]	Economic Potential (TWh)[d]	Ratio of Hydro Potential to Total 1980 Electricity Production	Economic Hydro Potential to Actual 1980 Hydroelectricity Production
Asia	1313.1	264.2	16,486[c]	5,340[c]	2672	2.0	10.1
South America	261.9	192.0	5,670	3,780	1892	7.2	9.9
Africa	187.0	61.1	10,118	3,140	1569	8.4	25.7
North America	2828.1	551.1	6,150	3,120	1561	0.6	2.8
U.S.S.R.	1295.0	180.0	3,940	2,190	1095	0.9	6.1
Europe	2192.3	462.9	4,360	1,430	714	0.3	1.5
Oceania	120.5	34.2	1,500	390	197	1.6	5.8
TOTAL	8197.9	1745.5	44,280	19,390	9700	1.2	5.6

(a) United Nations, 1981.

(b) World Energy Conference, September 1980.

(c) According to Zhu Xiaozhang (Associate Chief Engineer, Institute of Hydropower, Gansu Province, c/o Minister of Water Conservancy, Beijing, China, and Member of the Technical Panel on Hydropower of the Preparatory Committee for the United Nations Conference on New and Renewable Sources of Energy), this estimate may not include data from China. Mr. Zhu estimates that the theoretical hydroelectric potential for China is 6000 TWh/year and that the technically usable potential is 1900 TWh.

(d) This is the economic potential estimated from the 1976 World Energy Conference Survey of Energy Resources. In light of much higher costs for alternative power generating sources, the author of the report providing these estimates (Ellis L., Armstrong, 1978) expects that the potential would be greater now.

Note A.3-Ab. Summary of Hydro-Electric Resources Potential

	Electricity Production From All Sources in 1980 (TWh)[a]	Hydro-Electricity Production in 1980 (TWh)[a]	Economic Potential (TWh)[b]	Ratio of Economic Hydro Potential to Total 1980 Electricity Production	Ratio of Economic Hydro Potential to Actual 1980 Hydro-Electricity Production
Developing Market Countries	915.9	368.8	4644	5.1	12.6
OECD	5162.7	1062.8	2282	0.4	2.1
China, N. Korea, Vietnam	342.0	79.5	1546	4.6	19.8
U.S.S.R., E. Europe	1777.3	234.4	1199	0.7	5.1
TOTAL	8197.9	1745.5	9700	1.2	5.6

(a)   United Nations, 1981.

(b)   This is the economic potential estimated from the 1976 World Energy Conference Survey of Energy Resources. In light of much higher costs for alternative power generating sources, the author of the report providing these estimates (Ellis L. Armstrong, 1978) expects that the potential would be greater now.

Note A.3-B. Limitations of the WEC Estimates of the Hydroelectric Potential

	(TWh/year)
India	
Economic Potential	
WEC (1978)	221
CWPC (1960)[a]	221
1978 Reassessment[a, b]	396
Total with Micro/Mini[c]	420
Theoretical Potential (WEC, 1978)	750
Brazil	
Economic Potential	
WEC (1978)	548
Schulman (1980)[d]	
Firm Power[e]	933
Average Power[f]	1195
Theoretical Potential[g]	1389-5256

(a)   The first systematic survey of India's hydroelectric potential was made between 1953 and 1960 by the then Central Water and Power Commission. The much higher value for the economic potential given in the 1978 reassessment is due mainly to a more systematic identification of the potential in the Himalayan region.

(b)   This potential is based on water inflows of 90 percent dependability, i.e., inflows likely to be available 90 out of 100 years. The average potential is greater than this.

(c)   The 1978 reassessment value does not include the potential with mini-hydro or micro-hydro schemes, which are (unofficially) estimated to be able to provide an additional 25 TWh/year.

(d)   The latest official projection for Brazil is presented in M. Schulman, 1980.

(e)   Firm power is the output assured at the lowest water flow levels on record, e.g., over the past 40 years for which hydrological data exist.

(f)   The potential with average flow conditions.

(g)   The lower estimate of the theoretical potential is that given by the WEC (Table A-1). The higher estimate is presented in Hartmut Krugman, May 1982.

Section A.4

Note A.4-A. Net Primary Production and Related Characteristics of the Biosphere[a]

	Area (million km²)	Standing Biomass (billion tonnes)	Ratio, Standing Biomass to Net Primary Production (years)	Net Primary Productivity (tonnes/ha/year)	Total Global Production[b] (Terawatts)
Continental					
Tropical Rain Forest	17.0	765	20.5	22.0	22.0
Tropical Seasonal Forest	7.5	260	21.7	16.0	7.1
Temperate Forest					
Evergreen	5.0	175	26.9	13.0	3.8
Deciduous	7.0	210	25.0	12.0	5.0
Boreal Forest	12.0	240	25.0	8.0	5.7
Woodland and Shrubland	8.5	50	8.3	7.0	3.5
Savanna	15.0	60	4.4	9.0	8.0
Temperate Grassland	9.0	14	2.6	6.0	3.2
Tundra and Alpine	8.0	5	4.5	1.4	0.65
Desert and Semi Desert Scrub	18.0	13	8.1	0.9	0.94
Extreme Desert—Rock,					
Sand, Ice	24.0	0.5	7.1	0.03	0.04
Cultivate Land	14.0	14.0	1.5	6.5	5.4
Swamp and Marsh	2.0	30	5.0	30	3.5
Lake and Stream	2.0	0.05	0.06	4.0	0.5
Subtotal	149	1837	15.6	7.8	69.3
Marine					
Open Ocean	332.0	1.0	0.024	1.25	24.5
Upwelling Zones	0.4	0.008	0.04	5.0	0.12
Continental Shelf	26.6	0.27	0.028	3.6	5.66
Algal Beds and Reefs	0.6	1.2	0.75	25.0	0.94
Estuaries (excluding Marsh)	1.4	1.4	0.67	15.0	1.24
Subtotal	361	3.9	0.07	1.55	32.4
Global Totals	510	1841	10.7	3.36	101.7

(a)   Source: Robert H. Whitaker and Gene E. Likens, 1975.

(b)   Here it is assumed that 1 tonne of dry matter has a heating value of 18.6 GJ. Thus 1 tonne/year = 590 watts.

404 *Energy for a Sustainable World*

Note A.4-B. See Note 4.2.2-L.

Note A.4-Ca. The World's Renewable Energy From Forests

	Area (million hectares)	Annual Increment of Wood per Hectare[b] (cubic meters)	Total Increment of Wood [b]	
			(billion cubic meters)	(TW)[c]
Cool Coniferous	800	4.1	3.3	1.29
Temperate Mixed	800	5.5	4.4	1.75
Warm Temperate	200	5.5	1.1	0.46
Equatorial Rain	500	8.3	4.1	1.66
Tropical Moist Deciduous	500	6.9	3.5	1.38
Dry	1000	1.4	1.4	0.55
Totals and Averages	3800	4.7	17.8	7.1

(a) Source: D.E. Earl, 1975.
(b) Estimated to include all wood above ground.
(c) The energy content of the wood is taken to be 12.54 GJ per cubic meter.

Note A.4-Cb. The Utilization of the World's Incremental Forest Resources (1970)

	Annual Increment[b] (billion CM)	Annual Consumption (billion CM)			Unused Increment	
		Industrial	Fuel	Total	(billion CM)	TW[c]
Industrialized Countries	8.8	1.1	0.3	1.4	7.4	3.0
Developing Countries	9.0	0.2	0.8	1.0	8.0	3.2
TOTAL	17.8	1.3	1.1	2.4	15.4	6.2

(a) Source: D.E. Earl, 1975.
(b) Estimated to include all wood above ground.
(c) The energy content of the wood is taken to be 12.54 GJ per cubic meter.

Note A.4-D. Ratio of Stemwood to Total Biomass Production in Forests

Forest Type	Dry Matter Production (tonnes/ha/year)			Ratio of Stemwood to Total Production	
		Steamwood			
	Total	Becking	Weck	Becking	Weck
Tropical High Forest 0-500 m above above sea level	63.0	13.2	8.8	0.21	0.14
500-1500 m above sea level	54.6	11.5	7.6	0.21	0.14
Temperate Rain Forest	31.0	9.3	9.3	0.30	0.30
West European Forest	18.0	5.8	5.8	0.32	0.32

(a) Source: D.E. Earl, 1975.
Note A.4-E. The 1982 market price in Sweden for wood feedstock for paper and pulp was 20 kr per GJ or $2.80 per GJ (1980 dollars).

Note A.4-F. Potential Yields from Biomass Plantations

Location	Species	Yield (tonnes/ha/year)
Sweden	Willow, Poplar[a]	10-15
	Reeds[b]	10
United States		
Washington	Red Alder[c]	22-45
California, Louisiana, Mississippi, Florida	Eucalyptus[c]	27
Wisconsin, Illinois Missouri, New England	Hybrid Poplar[c]	11-34
Georgia	American Sycamore[c]	18-36
North Carolina	Loblolly Pine[d]	11-18
West Texas	Mesquite[e]	7-16
West Texas	Saltbush[e]	>7
West Texas	Johnson Grass[e]	>7
West Texas	Kochia[e]	>7
Brazil		
Aracruz	Eucalyptus[d]	
First Rotation		23
Second Rotation		33
Third Rotation		44
Highest Clone		61
Oahu, Hawaii	Giant Leucaena (f)	21-33

(a) Programplan Energiproduktion, 1980.
(b) Sven Bjork and Wilhelm Graneli, 1978.
(c) James C. Bethel *et al.,* April 6, 1979.
(d) J. Laurence Kulp, 1981.
(e) Richard Davids, 1982.
(f) For tree densities of 5000 to 40,000 per hectare (Molakai Study Team, 1980).

Note A.4-G. See Note 5.3.6-A.

Note A.4-H. The following are estimates of the plantation-gate costs for biomass energy produced on plantations in the U.S., in 1980 dollars per GJ[a,b,c].

	High Productivity Site		Low Productivity Site	
	Min. Cost	Max. Cost	Min. Cost	Max. Cost
Initial Capital Investment[d]				
Land Lease	0.016	0.033	0.016	0.041
Installations	0.012	0.024	0.024	0.049
Engineering & Contingency (30%)	0.017	0.017	0.012	0.027
Interest During Construction (4 years)	0.009	0.019	0.014	0.030
Total Capital Cost	0.044	0.093	0.066	0.147
Annual Operating Cost				
Site Preparation	0.047	0.210	0.098	0.267
Plantation	0.070	0.105	0.140	0.210
Fertilization	0.143	0.229	0.286	0.458
Irrigation	0.222	0.291	0.517	0.656
Weed Control	0.023	0.029	0.034	0.057
Protection	0.052	0.069	0.115	0.172
Harvesting	0.143	0.258	0.344	0.516
General Administration	0.006	0.011	0.011	0.023
Total Operating Cost	0.706	1.202	1.555	2.359
Total Cost	0.75	1.30	1.61	2.51

(a) These U.S. plantation cost estimates, from James S. Bethel *et al.,* April 6, 1979, are for hypothetical plantations with the following characteristics:

Annual Production	207,000 tonnes/year (dry weight)
Biomass Productivity	
Average High Productivity Site	27 tonnes/ha/year
Average Low Productivity Site	13 ½ tonnes/ha/year
Plantation Area Required	
High Productivity Site	7,700 hectares
Low Productivity Site	15,400 hectares
Expected Lifetime	32 years
Rotation Age	16 years
Coppicing	Every 4 years
Terrain	Flat to Gentle

(b) The costs in the original report are converted to 1980 U.S. dollars using the GNP deflator.
(c) Dry wood is assumed to have a lower heating value of 19 GJ per tonne.
(d) A real interest rate of 6 percent per year is assumed.

Note A.4-I. About 33% of all the world's cropland is farmed with high degree of mechanization, and 60% with the intensive use of fertilizers (private communication from David Pimentel, February 19, 1985). However, fertilizers are used much less intensively in developing than in industrialized countries. Averaged over all cropland, the annual fertilizer input rates (for N, P_2O_5, K_2) are 115 kg per hectare in industrialized countries and 48 kg per hectare in developing countries. Moreover, the average is only 11 kg per hectare in Africa. The incremental crop yield per added kg of fertilizer is several times greater in developing countries where the present application rate is low than in industrialized countries (D. Pimentel *et al.,* 1985).

Section A.5
Note A.5-A. Estimates of wind electricity production costs

	1981 cents per kwh
Alternative Wind Systems[a]	
Present day small (50 kW) turbines[b]	4.6
Present day large (2.5 MW) turbines[c]	6.5
Target for mass-produced large turbines[d]	2.8-3.4
Benchmark Costs	
Cost of electricity from a new, coal-fired baseload plant in the U.S.[e]	4.6
Average cost of oil used in electric power generation in the U.S. in 1981[f]	5.5

(a) Utility financing is assumed, for which the appropriate average cost of capital is assumed to be 0.038 (Note A.5-B). This corresponds to a capital recovery factor of 0.072 for a 20 year system life (assumed for small wind turbines) and 0.056 for a 30 year system life (assumed for large wind turbines). In addition the annual insurance cost is assumed to be 0.003 times the initial capital cost. Following James I. Lerner, September 1982, the annual O&M cost is assumed to be 2% of the initial capital cost. The annual average capacity factor is assumed to be 35%.

(b) In James I. Lerner, September 1982, it is estimated that the installed cost for systems being built today is typically about $1500 per kW.

(c) The estimated unit costs for 40 MOD-2 wind turbines for the U.S. Bureau of Reclamation's proposed 100 MW Medicine Bow Wind Project is $2500 per kW (L.L. Nelson, 1981).

(d) It is estimated that in mass production the unit installed cost of the MOD-2 wind turbine will be $1100-1300 per kW. See Note A.5-C.

(e) See Note A.5-D.

(f) For a heat rate of 10,900 KJ per kWh and an average 1981 oil price for electric utilities in the U.S. of $5 per GJ (Energy Information Administration, U.S. Department of Energy, October, 1982).

Note A.5-B. The Capital recovery factor for utilities is

$$\text{CRF }(i,N) = \text{capital recovery factor} = i/[1-(1+i)^{-N}],$$

where i = real cost of capital, and N = expected system life, in years.
The average cost of capital is:

$$i = (a \times cd) + (b \times cp) + (c \times ce),$$

where

cd = debt cost
cp = preferred stock cost
ce = common stock cost

and

a = debt fraction
b = preferred stock fraction
c = common stock fraction.

In Technology Assessment Group, 1979, it is recommended that in evaluating utility investments the appropriate (inflation-corrected) costs of capital from different sources and the mix of capital sources should be the following:

cd = 0.019   a = 0.5
cp = 0.024   b = 0.15
ce = 0.071   c = 0.35

Thus for utility investments, i = 0.038, so that:

CRF (0.038,20) = 0.072.
CRF (0.038,30) = 0.056.

Note A.5-C. Estimated capital cost for mass-producted MOD-2 windmills

	1981 dollars per kW
Site preparation	90.5
Transportation	16.2
Erection and Checkout	76.2
Rotor assembly	182.9
Drive train	210.8
Nacelle subassembly	102.2
Tower subassembly	150.9
Initial spares and maintenance equipment	19.6
Non-recurring costs	19.6
SUBTOTAL	868.9
10% profit	86.9
TOTAL TURNKEY COST	957
Site-related costs	94- 255
Interest during construction	79 – 91
TOTAL INSTALLED COST	1100-1300

(a)   Source: James I. Lerner, September 1982.

(b)   The MOD-2 is a 2.5 MW horizontal axis wind turbine designed and fabricated for the U.S. Department of Energy by the Boeing Engineering and Construction Company.

(c)   The cost estimates presented here are for the 100th unit produced in a dedicated production facility that makes 20 turbines per month.

Note A.5-D. Busbar cost for electricity from a new 500 MW coal-fired baseload power plant.

	1981 cents per kwh
Capital[a,b,c]	1.22
Fuel[d,e]	2.67
O&M[f]	0.76
TOTAL	4.65

(a)   The cost for 2-500 MW units averaged over 15 different U.S. regions has been estimated to be $900 per kW in mid-1978 dollars, which becomes $1175 per kW in 1981 (Bechtel Power Corporation, May 1981).

(b)   For utility financing of a plant with a 30-year life the capital recovery factor is 0.056 (Note A.5-B). In addition, the annual insurance cost is assumed to be 0.003 times the initial investment. All taxes are neglected.

(c)   Over the last decade, the capacity factor averaged about 70% for 300 MW coal plants and 62% for 600 MW units (Charles Komanoff, 1981). On the basis of this experience, a 65% capacity factor is assumed for 500 MW units.

(d)   The real price of coal to utilities in the U.S. increased in the period 1973 to 1982 at an average annual rate of 8.6%, from $0.70 per GJ to $1.50 per GJ, in 1981 dollars. The U.S. Department of Energy has projected that the utility coal price will increase further, to $2.30 per GJ by 1990 and at an average growth rate of 0.9% per year thereafter (Energy Information Administration, March 1981). For a utility discount rate of 3.8% per year this implies a price levelized over the period 1990 to 2020 of $2.60 per GJ.

(e)   For a high efficiency, supercritical steam plant with wet lime/limestone flue gas desulphurization, for which the heat rate would be 10,270 KJ per kWh (Technical Assessment Group, 1979).

(f)   See Technical Assessment Group, 1979.

Section A.6

Note. A.6-A. Busbar cost for electricity from a new peaking power plant

	1981 cents per kwh
Capital[a,b]	0.334/CF
Fuel[c]	8.70
O&M[d]	0.30
TOTAL	9.0 + 0.334/CF

(a) In Sierra Energy and Risk Assessment, Inc., and Robert Foster Consulting, September 1982, a survey of the capital costs for new high efficiency peaking turbines estimated an average capital cost of $356 dollars per kW in 1980 dollars, which becomes $390 per kW in 1981 dollars.

(b) For utility financing of a plant with a 20 year life the capital recovery factor is 0.072 (Note A.5-B). In addition the annual insurance cost is assumed to be 0.003 times the initial investment. All taxes are neglected. Here CF is the annual average capacity factor. In peaking applications CF is on the order of 0.1.

(c) In Sierra Energy and Risk Assessment, Inc., and Robert Foster, Consulting, September 1982, the average heat rate for new efficient combustion turbines was estimated to be 12,900 KJ per kWh. Here the peaking fuel prices is taken to be the average price of distillate fuel for utility peaking applications in the U.S. in 1981, or $6.75 per GJ (U.S. Energy Information Administration, November 1981).

(d) See Sierra Energy and Risk Assessment, Inc., and Robert Foster, Consulting, September 1982.

Note A.6-B. Cost estimates for a 50 MW amorphous silicon PV plant, located in Princeton, New Jersey[a,b,c]

	1981 dollars per peak watt	
	Near Term	Intermediate Term
Wire	0.037	0.026
Site Preparation	0.030	0.015
Electrical Protection	0.009	0.004
Array Structure	0.566	0.354
Power Conditioning	0.141	0.094
Solar Cell Panels	0.585	0.292
Land	0.088	0.088
TOTALS	1.46	0.87

(a)   These estimates were prepared at the RCA Laboratories in Princeton, New Jersey (A.J. Stranix and A.H. Firester, 1983).

(b)   The sources of these cost estimates were the following:

Support structures	Work done by Bechtel
Field wiring and electrical components	Direct quotations from suppliers
Installation of field wiring	Studies by Burt Hill Kosar Rittleman
Field wiring	Associates and private discussions with power engineers
Power conditioning	Personnel at Sandia Laboratories, EPRI, and United Technologies
Solar Cell Panels	RCA's own R&D

(c)   The cost estimates given in A.J. Stranix and A.H. Firester, 1983 were converted to 1981 dollars using the GNP deflator.

(d)   Some of the important parameters for the proposed plant are:

Solar cell efficiency	10%
Plant capacity factor	19.4%
Tilt of solar cell panels	30 degrees
Shadowing losses	< 1%
Wire losses	1.4%
Annual electricity production	85.5 million kWh
Total solar cell area	0.50 million square meters
Total plant site area	94 hectares
Annual O&M costs	$0.45 per square meter
Land cost	$50,000 per hectare

Note A.6-C. Estimated busbar electricity cost for alternative PV power plants

	1981 cents per kwh
AC electricity from a plant located in Princeton, New Jersey [a]	4.1
AC electricity from a plant of the same design located in El Paso, Texas[b]	2.5
AC electricity from the El Paso plant if the land cost were $2500 instead of $50,000 per hectare	2.3
DC electricity from the El Paso plant if the land cost were $2500 per hectare[c]	2.0
DC electricity from the El Paso plant if the land cost were $2500 per hectare and if the solar cell panel costs were reduced from $0.30 to $0.15 per peak watt[d]	1.6

(a) For a capital recovery factor of 0.072 (Note A.5-B), an annual insurance cost equal to 0.003 times the installed capital cost, an installed capital cost of 0.87 dollars per peak watt (the intermediate term estimate in Note A.6-B), and an annual O&M cost of $0.45 per square meter or $0.0026 per kwh (Note A.6-B).

(b) The electricity cost Eep, in El Paso is estimated to be:

$$Eep = Ep \times (In/Iep),$$

where

Ep = The electricity cost in Princeton.

In = The annual average insolation Newyork, N.J., on a collector tilted at the latitude angle.

Iep = The annual average insolation in El Paso on a collector tilted at the latitude angle.

Annual average insolation values for these cities are obtained from Jan F. Kreider and Frank Kreith, 1982.

In = 166 watts per square meter (collector tilted at 40.7 degrees)

Iep = 271 watts per square meter (collector tilted at 31.8 degrees)

(c) The DC power that is produced by a PV array is usually converted to AC power, so that it can be utilized in the existing electrical system. In addition the power level is not constant throughout a day, so that the voltage is usually adjusted to a constant level to assure compatibility with the existing electrical system. For hydrogen production applications, only DC power is needed. Here it is assumed that for hydrogen production applications, the power conditioning costs (accounting for 11% of the total capital cost, as shown in Note A.6-B) can be neglected.

(d) In 1981 researchers at the SANYO Electric Co. Research Center in Hirakata City, Osaka, Japan, projected that at 10% efficiency SANYO could produce cells for $0.15 per peak watt by the time the annual production volume reached about 350 MW per year (Y. Kuwano and M. Ohnishi, 1981). From Note A.6-B it can be seen that a 50% reduction in the solar cell panel cost would lead to a 21% reduction in the total installed capital cost.

Note A.6-D. Capital and operating cost for an electrolyzer without a rectifier, for operation at 25% capacity factor

	Dollars per GJ
Total cost of hydrogen produced from AC electricity costing 1.5 cents/kwh @ 25% capacity factor[a]	9.80
Less cost of electricity[b]	−5.40
Less annualized cost of the rectifier[c]	−2.14
Net cost of electrolyzer w/o rectifier, using a capital recovery factor of 16%, plus 2% of the installed capital cost per year for O&M costs plus another 2% for ad valorem taxes and insurance.	2.26
Net cost of electrolyzer w/o rectifier using a capital recovery factor of 7.2% (Note A.5-B), plus 2% and 0.3% of the installed capital cost per year, respectively, for O&M costs and insurance, but neglecting taxes.	1.07
Net cost converted from 1977 to 1981 dollars	1.49

(a) For industrial unipolar technology as of 1983. See Figure 5, page 375, in R.L. Leroy and A.K. Stuart, 1979.
(b) For a rectifier efficiency of 96% and an electrolysis efficiency of 80%.
(c) In R.L. Leroy and A.K. Stuart, 1979, the annualized rectifier cost is given by

$$(777 \times K \times Vi \times XR)/(365 \times CF \times 100 \times nr)$$

where
K = annual capital charge rate = 0.20 (see p. 366 in R.L. Leroy and A.K. Stuart, 1979).
Vi = cell voltage = 1.86 volts (for a current density of 450 mA/cm^2—see Figure 1 in R.L. Leroy and A.K. Stuart, 1979).
XR = capital cost of rectifier = $65/kW (see Table 1 in R.L., Leroy and A.K. Stuart, 1979).
nr = rectifier efficiency = 0.96 (see Table I in R.L. Leroy and A.K. Stuart, 1979).
Thus the annualized rectifier cost is $2.14/GJ.

Note A.6-E. Estimated electrolytic hydrogen production cost for a 50 MW PV power plant located near El Paso, Texas[a]

	1981 dollars per GJ	
	w/o O_2 credit	w/ O_2 credit
With solar cell panels @ $0.30 per peak watt[b]	8.50	6.70
With solar cell panels @ $0.15 per peak watt[c]	7.00	5.30

(a) The production cost for hydrogen (CH_2), (in dollars per GJ) is given by:
CH_2 = C_{elect} − 0.056CrO_2 + Cpve/(n × 0.0036 GJ per kWh),
where
C_{elect} = capital and O&M contributions to the cost of electrolysis,
CrO_2 = credit for byproduct oxygen, in $ per tonne (0.056 tonnes of O_2 are produced per GJ of H_2),
Cpve = busbar cost of DC electricity produced with solar cells,
n = efficiency of electrolysis.
Here it is assumed (Note A.6-D) that Celect = $1.49 per GJ and n = 0.8. It is also assumed that before the oxygen market is saturated, the byproduct oxygen can be sold for $31.3 per tonne, the wholesale price of oxygen in 1979, expressed in 1981 dollars.
(b) In this case the estimated busbar electricity cost is $0.020 per kWh (Note A.6-C).
(c) In this case the estimated busbar electricity cost is $0.016 per kWh (Note A.6-C).

APPENDIX B: Glossary of Terms, Useful Data, Energy Unit Definitions, and Energy Conversion Factors

B.1 Glossary of Terms

Breeder reactor: A nuclear reactor is which plutonium is produced through the capture of neutrons in uranium-238 in such quantities that more fissile isotopes are produced than are consumed through fissioning of uranium-235 and plutonium.

Commercial energy: Fuel or electricity that is purchased.

CMEA: The Council for Mutual Economic Assistance, which might be loosely categorized as the world's industrialized countries with centrally planned economies, but actually includes as members the Soviet Union, German Democratic Republic, Poland, Czechoslovakia, Hungary, Romania, Bulgaria, Cuba, Vietnam, and Mongolia.

Energy service: The objective of energy use, e.g. illumination, heating of a room, mechanical power.

Estimated Ultimately Recoverable Resources: Estimated remaining ultimately recoverable resources are proved reserves plus discovered resources that are not yet economical to recover but judged to be economically recoverable sometime in the future, plus estimated undiscovered resources that are judged to be recoverable at some time in the future.

Final Energy: Final energy is defined as the total fuel (including bunkers) and electricity consumed by "final consumers". Excluded are losses in the generation, transmission and distribution of electricity, the consumption of fuels by refineries and other energy processing plants, and the transport of fuels. [For the U.S. the convention is instead to define final energy use as primary energy use less losses in the generation, transmission, and distribution of electricity. Other losses are counted as final energy use in the industrial and transport sectors].

IEA: The International Energy Agency (IEA) is an autonomous body which was established in November 1974 within the framework of the OECD to implement an International Energy Program. Its members are: Australia, Austria, Belgium, Canada, Denmark, Germany, Greece, Ireland, Italy, Japan, Luxembourg, Netherlands, New Zealand, Norway, Portugal, Spain, Sweden, Switzerland, Turkey, United Kingdom, United States.

Light Water Reactor (LWR): A nuclear reactor where the neutrons released in the fission process are moderated by light (ordinary) water. There are two kinds of LWRs: boiling water reactors (BWR) and pressurized water reactors (PWR), referring to whether the cooling reactor water is permitted to produce steam directly, or if the pressure is kept high to prevent boiling in the reactor. In the latter case steam is produced in steam generators located outside the reactor vessel.

Non-commercial Energy: Fuelwood, animal wastes or agricultural residues used for fuel purposes but gathered/collected outside the market economy.

Nuclear fuel reprocessing: The technology whereby the spent fuel discharged from present day LWRs or breeder reactor is processed to extract the contained plutonium and unused uranium by chemical means for recycle in fresh nuclear fuel.

OECD: The Organization for Economic Cooperation and Development, which might be loosely categorized as the world's industrialized countries with market economies, with the following members: Australia, Austria, Belgium, Canada, Denmark, Finland, France, Federal Republic of Germany, Greece, Iceland, Ireland, Italy, Japan, Luxembourg, Netherlands, New Zealand, Norway, Portugal, Spain, Sweden, Switzerland, Turkey, United Kingdom, United States.

Plantation Expansion Rate (PER): As used in Chapter 4, the average rate of starting new biomass plantations (in million hectares per year) needed to assure that enough biomass is available for harvesting in 2020 to meet the biomass supply level needed in that year for energy purposes.

Primary Energy: Energy in its naturally occurring form (coal, oil, uranium, etc.) before conversion, transmission, and distribution to end-use forms.

Proved Reserves: The amount of a resource that has been discovered and measured and is economically recoverable under present market conditions.

B.2   Useful Data

Gross National Product deflators for the United States (1984 = 100):

1984	100.00
1983	96.37
1982	92.82
1981	87.54
1980	79.86
1979	73.14
1978	67.32
1977	62.68
1976	59.23
1975	56.20
1974	51.43
1973	47.30
1972	44.76
1971	42.97
1970	40.93

Source: The Council of Economic Advisers, December 1985.

Exchange rate per U.S. dollar, mean value for each year

	Brazil (Cr$)	India (Rs.)	Sweden (SEK)
1984	1,848.00	11.363	8.28
1983	577.04	10.099	7.68
1982	179.51	9.455	6.30
1981	93.12	8.659	5.08
1980	52.71	7.863	4.23
1979	26.95	8.126	4.29
1978	18.07	8.193	4.52
1977	14.14	8.739	4.49
1976	10.67	8.960	4.36
1975	8.13	8.376	4.17
1974	6.79	8.102	4.46
1973	6.13	7.742	4.37
1972	5.93	7.594	4.76
1971	5.29	7.501	5.11
1970	4.59	7.500	5.19

Source: International Monetary Fund, 1985.

Energy content of fuels (lower heating values)

		GJ
1 t.c.e.	tonnes of coal equivalent[1,2]	29.3
1 t.o.e.	tonnes of oil equivalent[2]	44.8
1 m³	crude oil[1]	36.2
1 barrel	crude oil[2]	6.12
1 m³	fuel oil[1] or diesel[1]	37.0
1 m³	kerosene[1]	35.0
1 m³	gasoline[1]	32.5
1 m³	methanol	15.6
1 m³	ethanol	21.2
1 tonne	LPG[1]	45.5
1,000 m³	natural gas[2]	37.3
1 m³t	roundwood[3], 30% moisture	5.2
1 m³f	firewood[4], 30% moisture	7.4
1 tonne	biomass, dry[5]	18.0
1 tonne	peat, 35% moisture	12.2
1 tonne	urban refuse	10.5
1,000 m³	biogas[6], appr.	18
1,000 m³	high Btu gas, appr.	>30
1,000 m³	medium Btu gas, appr.	10 – 20
1,000 m³	low Btu gas, appr.	<8
1,000 kWh	of hydro, nuclear or geothermal electricity	3.6

1. These are standard values from United Nations, 1983.
2. These are from IIASA, 1981.
3. Piled cubic meter; external volume of a well-piled stack of firewood.
4. Solid cubic meter; actual volume of wood when cavities between the pieces of wood are excluded.
5. The energy content may vary between 18.0 and 24.5 GJ per tonne.
6. If approximately 50% of the produced gas is methane.

B.3 Energy Unit Definitions

Prefix	Factor
E (Exa-)	10^{18}
P (Peta-)	10^{15}
T (Tera-)	10^{12}
G (Giga-)	10^9
M (Mega-)	10^6
k (kilo-)	10^3

1 TW-year per year is abbreviated 1 TW

1 kW-year per year is abbreviated 1 kW

B.4 Energy Conversion Factors

1 TW-year	=	5.16 billion barrels of oil = 0.704 billion t.o.e.
	=	0.846 trillion m³ natural gas = 1.08 billion t.c.e.
	=	29.9 quadrillion BTU = 31.56 EJ
1 kW-year/year	=	31.6 GJ/year = 1.08 t.c.e./year = 0.70 t.o.e./year
	=	20.6 Mcal/day = 86.6 MJ/day
1 BTU	=	1.055 kJ
1 quadrillion BTU	=	1.055 EJ
1 million barrels of oil per day (mbd)	=	49.9 million t.o.e. per year
	=	2.23 EJ/year = 0.0708 TW

References

The Council of Economic Advisers, *Economic Indicators,* United States Government Printing Office, December 1985.

International Institute for Applied Systems Analysis, *Energy in a Finite World,* Ballinger, Cambridge, 1981.

International Monetary Fund, *International Financial Statistics, Yearbook 1985,* Vol. 38, Washington D.C. 1985.

United Nations, *1981 Yearbook of World Energy Statistics,* United Nations, New York, 1983.

Notes for Section 1.1
Section 1.1

Note 1.1-A. Global Distribution of Energy Use (in per cent) by Fuel and Region

	1925[a]				1972[b]			
	Oil	Coal	Nat. Gas	Hydro	Oil	Coal	Nat. Gas	Hydro
Industrialized Countries								
Market Economies								
US and Canada	18.9	74.5	6.0	0.6	43.9	17.0	32.2	6.9
Western Europe	3.2	96.0	—	0.7	60.0	21.0	9.7	9.2
Japan	4.4	92.4	0.1	3.1	75.4	16.2	1.2	7.1
Oceania	6.9	92.6	—	0.5	49.6	34.6	5.0	10.8
SUBTOTAL	12.2	83.6	3.4	0.7	52.2	18.6	21.4	7.8
Centrally Planned Economies								
USSR	34.2	64.9	0.7	0.1	36.2	37.9	21.9	4.0
Eastern Europe	6.5	91.2	2.2	0.1	18.0	68.2	11.6	1.2
SUBTOTAL	15.2	82.9	1.7	0.1	31.1	46.9	18.8	3.2
Subtotal, Industrialized Countries	12.4	83.6	3.3	0.7	46.8	25.8	20.7	6.6
Developing Countries								
Market Economies	32.8	64.3	2.2	0.7	58.0	23.3	11.2	7.4
Centrally Planned Economies	6.0	94.0	—	—	13.0	83.6	1.5	2.0
Subtotal, Developing Countries	25.9	71.9	1.6	0.5	42.2	44.5	7.8	5.6
WORLD	13.3	82.9	3.2	0.7	46.1	28.9	18.6	6.4

(a) *Source:* Joel Darmstadter, 1971.
(b) *Source:* The British Petroleum Company, 1981.

Note 1.1-B. Historical Data on Oil Consumption by Region (in million barrels per day)

	1925[a]	1938[a]	1955[a]	1965[b]	1972[b]	1981[b]	1982[b]	1983[b]
Industrialized Countries								
Market Economies								
US and Canada	1.886	2.966	8.552	11.803	17.645	17.310	16.330	16.125
Western Europe	0.223	0.636	2.069	7.024	14.155	13.045	12.500	12.190
Japan	0.018	0.095	0.189	1.468	4.735	4.690	4.395	4.360
Oceania	0.013	0.050	0.160	0.374	0.660	0.760	0.730	0.705
SUBTOTAL	2.140	3.747	10.970	20.669	37.195	35.805	33.955	33.380
Centrally Planned Economies								
USSR	0.115	0.586	1.413	3.598	6.115	8.985	9.075	9.115
Eastern Europe	0.0047	0.075	0.189	0.605	1.360	2.480	2.450	2.425
SUBTOTAL	0.162	0.661	1.602	4.203	7.475	11.465	11.525	11.540
Subtotal, Industrialized Countries	2.30	4.41	12.57	24.87	44.67	47.27	45.48	44.92
(WORLD SHARE) %	(87.8)	(88.0)	(86.4)	(86.4)	(84.7)	(79.0)	(77.9)	(77.6)
Developing Countries								
Market Economies	0.299	0.581	1.938	3.714	7.240	10.895	11.265	11.275
Centrally Planned Economies	0.018	0.020	0.044	0.230	0.855	1.705	1.660	1.705
Subtotal, Developing Countries	0.32	0.60	1.98	3.94	8.10	12.600	12.925	12.980
(WORLD SHARE, %)	(12.2)	(12.0)	(13.6)	(13.7)	(15.3)	(21.0)	(22.1)	(22.4)
WORLD	2.62	5.01	14.55	28.78	52.77	59.87	58.405	57.900

(a) *Source:* Joel Darmstadter, 1971.
(b) *Source:* The British Petroleum Company, 1984.

Note 1.1-C. Historical Data on Commercial Energy Consumption by Region (in TW).

	1925 (a)	1938[a]	1955[a]	1965[b]	1972[b]	1981[b]	1982[b]	1983[b]
Industrialized Countries								
Market Economies								
US and Canada	0.701	0.662	1.369	1.928	2.481	2.680	2.578	2.545
Western Europe	0.484	0.580	0.701	1.094	1.552	1.654	1.625	1.622
Japan	0.029	0.058	0.062	0.202	0.413	0.469	0.453	0.453
Oceania	0.015	0.017	0.035	0.057	0.085	0.117	0.118	0.117
SUBTOTAL	1.229	1.317	2.167	3.281	4.531	4.920	4.774	4.737
Centrally Planned Economies								
USSR	0.024	0.165	0.432	0.786	1.110	1.594	1.651	1.707
Eastern Europe	0.052	0.063	0.216	0.384	0.523	0.669	0.687	0.688
SUBTOTAL	0.076	0.228	0.648	1.170	1.633	2.263	2.338	2.395
Subtotal, Industrialized Countries	1.305	1.545	2.815	4.451	6.164	7.183	7.112	7.132
(WORLD SHARE, %)	(93.9)	(92.2)	(89.5)	(85.0)	(82.5)	(79.0)	(78.1)	(77.6)
Developing Countries								
Market Economies	0.064	0.105	0.239	0.481	0.866	1.242	1.296	1.327
Centrally Planned Economies	0.022	0.026	0.092	0.305	0.444	0.672	0.694	0.734
Subtotal, Developing Countries	0.086	0.131	0.331	0.786	1.310	1.914	1.990	2.061
(WORLD SHARE, %)	(6.2)	(7.8)	(10.5)	(15.0)	(17.5)	(21.0)	(21.9)	(22.4)
WORLD	1.390	1.676	3.145	5.237	7.474	9.097	9.102	9.193

(a) *Source:* Joel Darmstadter, 1971.
(b) *Source:* The British Petroleum Company, 1984.

Note 1.1-D. Oil Revenues Paid to Oil Producing Countries in Excess of Revenues at the 1972 Oil Price

	Oil Imports[a] (billion barrels)		World Oil Price[c] (1984 $/barrel)	Oil Price Increment Above 1972 Level (1984 $/barrel)	Extra Oil Revenues[d] (billion 1984 $)	
	Industrialized Countries [b]	Other			Industrialized Countries	Other
1973	9.82	2.51	8.58	1.43	14.0	3.6
1974	9.55	2.53	24.19	17.04	162.8	43.1
1975	8.53	2.47	24.64	17.49	149.3	43.2
1976	9.53	2.91	22.65	15.50	147.8	45.1
1977	9.95	2.64	23.09	15.94	158.7	42.1
1978	9.59	2.70	21.56	14.41	138.2	38.9
1979	9.71	3.01	29.56	22.41	217.6	67.5
1980	8.39	3.06	42.44	35.29	296.0	107.9
1981	7.32	2.92	42.59	35.44	259.5	103.4
TOTALS	82.39	24.75			1544	495

(a) Source: The British Petroleum Company, 1981.
(b) This is the net oil imported by Western Europe, the U.S., and Japan.
(c) This is the refiner acquisition cost for oil imported into the U.S. Sources: Energy Information Administration, 1981, and Energy Information Administration, June 1982.
(d) These extra revenues to foreign oil producers are the product of imports and the oil price increment above the 1972 oil price.

Note 1.1-E. Global Patterns of Crude Oil Production (million barrels per day)

	1978	1979	1980	1981	1982	1983	1984
Persian Gulf	15.3	17.8	16.3	13.8	9.9	8.6	8.6
Total OPEC	29.8	30.9	26.9	22.6	18.9	17.6	17.6
Non-OPEC West	17.0	18.0	18.8	19.4	20.5	21.3	22.3
USSR	11.2	11.5	11.7	11.9	12.1	12.0	11.8
China	2.1	2.1	2.1	2.0	2.0	2.1	2.3
World Totals	60.1	62.5	59.5	55.9	53.5	53.0	54.0

Source: Energy Information Administration, April 1985.

Note 1.1-Fa. Estimates in a 1978 RAND Corporation Report (Richard Nehring, June 1978) of remaining ultimately recoverable crude oil resources (in TW-years)

Region	Proved Reserves as of 12/31/75	Additions in Known Fields Through Extensions, Full Development, and Enhanced Recovery	Additions Through New Discoveries	Total Ultimately Recoverable Crude Oil Resources
North America	11.2	8.3 – 18.4	11.1 – 20.4	31 – 50
South America	5.3	3.9 – 7.8	6.2 – 10.1	15 – 23
Western Europe	4.2	1.0 – 1.9	3.9 – 6.8	9 – 13
Eastern Europe/Soviet Union	10.0	3.9 – 7.8	8.3 – 16.1	22 – 34
Africa	10.6	2.9 – 5.8	5.8 – 12.4	19 – 29
Middle East	82.4	48.5 – 77.6	19.4 – 44.6	150 – 205
Asia/Oceania	7.3	2.9 – 4.9	7.6 – 15.3	18 – 28
TOTAL	131	71 – 124	62 – 126	264 – 381

Total remaining crude oil resources projected in this RAND report thus range from 70 to 100 percent of the total presented in Note 1.1-H.
1 TW-year = 5.154 billion barrels or 0.7045 billion tonnes of oil = 0.8464 trillion cubic metres of natural gas.

Note 1.1-Fb. Exxon estimates (Exxon, 1982) of remaining world oil and gas resources, as of 1981 (in TW-years):

Proved Resources	194 – 233
Probable Reserves	58 – 116
Undiscovered Potential	194 – 485
TOTAL	446 – 834

Total remaining oil and natural gas resources projected by Exxon thus range from 60 to 120 percent of the total presented in Note 1.1-H.

Note 1.1-Fc. Estimates of world ultimate crude oil recovery

Date	Estimator	Organisation	Billion barrels
1942	Pratt, Weeks & Stebinger		600
1946	Duce		400
1946	Pogue		555
1948	Weeks		610
1949	Levorsen		1,500
1949	Weeks		1,010
1953	MacNaughton		1,000
1956	Hubbert		1,250
1958	Weeks		1,500
1959	Weeks		2,000
1962	L.G. Weeks	Consultant	2,000
1965	T.A. Hendricks	USGS	2,480
1967	W.P. Ryman	Esso (Exxon)	2,090
1968	—	Shell Oil Company	1,800
1969	M. King Hubbert	National Academy of Sciences	1,350 – 2,100
1969	L.G. Weeks	Consultant	2,200
1970	J.D. Moody	Mobil	1,800
1971	H.R. Warman	BP	1,200 – 2,000
1971	Weeks		2,290
1971	U.S. National	Petroleum Council	2,670
1972	J.D. Moody and H.H. Emmerick	Mobil	1,800 – 1,900
1972	Richard L. Jodry	Sun	1,952
1972	Linden		2,950
1972	H.R. Warman	BP	1,800
1972	Weeks		3,560
1973	Wim Vermeer	Shell Oil Company	1,930
1973	H.R. Warman	BP	1,915
1974	J.D. Moody and R.W. Esser	Mobil	2,000
1974	M. King Hubbert	USGS	2,000
1975	J.D. Moody and R.W. Esser	Mobil	2,030
1975	—	Exxon	1,945
1975	B. Grossling	USGS	2,600 – 6,500
1975	P. Odell	Erasmus University, Rotterdam	3,575 – 4,233
1977	M. King Hubbert	Congressional Research Service	2,000
1977	—	World Energy Conference	1,900
1978	Richard Nehring	Rand Corporation	1,700 – 2,300
1979	A. A. Meyerhof	Consultant	2,230
1981	Colitti	AGIP	2,082
1982	—	Exxon	1,800 – 3,000

Source: Timothy Greening, September 1982.

Note 1.1.-G. Estimated remaining ultimately recoverable resources are proved reserves plus discover resources that are not yet economical to recover but judged to be economically recoverable sometime in the future, plus estimated undiscovered resources that are judged to be recoverable at some time in the future.

Note 1.1-H. Oil and Natural Gas Consumption (in TW-years per year) and Estimated Remaining Ultimately Recoverable Resources (in TW-years)

Region[a]	1978 Consumption[c]		Resources	
	Oil	Gas	Oil[d]	Natural Gas[e]
I NA	1.42	0.78	40	61
II SU/EE	0.75	0.50	66	96
III WE/J ANZ	1.48	0.28	23	23
IV LA	0.30	0.06	33	21
V Af/SEA	0.30	0.04	30	18
VI Me/NAf	0.11	0.04	155	117
VII C/CPA	0.12	0.02	18	13
TOTALS	4.47	1.72	365	349

(a)   The 7 world regions used in the IIASA study, ref. (b), are defined as follows:
   I North America
   II The Soviet Union and Eastern Europe
   III Western Europe; Japan; Australia; New Zealand; South Africa
   IV Latin America
   V Africa, except northern Africa and South Africa; South Asia and Southeast Asia
   VI Middle East; northern Africa
   VII China; other centrally planned Asian economies.
(b)   *Source:* Wolf Haefele, 1981.
(c)   *Source:* The British Petroleum Company, 1981.
(d)   From Table 2.6, p. 57 in ref. (b), which is based on data from P. Despairies, Report on Oil Resources, 1985-2020, Executive Summary, Tenth World Energy Conference, London, 1977.
(e)   From Table 2.7, p. 65 in ref. (c), which is based on data from: World Oil, Thirty-third International Outlook issue, vol. 187, no. 3, 1978; and World Energy Resources 1985-2020. Oil and Gas Resources. Worldwide Petroleum Supply Limits. The Future of World Natural Gas Supply, World Energy Conference, Guilford, United Kingdom: IPC Press, 1978.

Note 1.1-I. Per Capita Energy Use (in Gigajoules per Capita per Year) for Selected Developing Countries[a]

Country	Commercial Energy[b]	Biomass Energy[c]	Total	Percentage of Energy from Biomass
Bangladesh	1.2	3.0	4.2	71%
Niger	1.1	8.0	9.1	88%
Gambia	3.1	7.0	10.1	69%
Morocco	8.4	2.3	10.7	21%
India	5.2	6.0	11.2	54%
Ethiopia	0.6	11.7	12.3	95%
Nepal	0.3	13.5	13.8	98%
Somalia	2.9	15.0	17.9	84%
Bolivia	10.7	8.3	19.0	44%
Sudan	5.0	20.0	25.0	80%
Thailand	9.6	16.5	26.1	63%
Tanzania	1.9	25.5	27.4	93%
China	24.5	10.0	34.5	29%
Brazil	23.2	11.7	34.9	34%
Mexico	40.5	4.0	44.5	9%
Libya	55.3	3.0	58.6	5%
Developing Countries (Average)	17.3	13.1	30.1	43%

(a) *Source:* D.O. Hall, G.W. Barnard, and P.A. Moss, 1981.
(b) Commercial energy use data are for 1978.
(c) Mainly fuelwood and charcoal.

Notes for Section 1.2

Note 1.2-A The seven IIASA regions are:
   I. NA (North America)
  II. SU/EE (The Soviet Union; Eastern Europe)
 III. WE/JANZ (Western Europe; Japan, Australia; New Zealand; South Africa, Israel)
  IV. LA (Latin America)
   V. Af/SEA (Africa, except northern Africa and South Africa; South Asia; Southeast Asia)
  VI. ME/NAF (Middle East; Northern Africa)
 VII. C/CPA (China, other Centrally Planned Asian Economies)

Note 1.2-B. Per Capita and Total Energy Projections for Three Major Global Energy Studies

	Base Year Data				Projections					
	1972		1975		2000		2020		2030	
	E/P (kW)	E (TW)	E/P (kW)	E (TW)	E/P (kW)	E/(TW)	E/P (kW)	E (TW)	E/P (kW)	E (TW)
Industrialized Countries										
Market Economies										
WAES	7.0	5.0	—	—	10.3–12.5	9.0–10.9	—	—	—	—
WEC (1978)	6.9	4.8	—	—	9.8–10.5	8.2–8.7	12.5–13.3	11.7–12.5	—	—
IIASA	—	—	6.2	4.9	7.0–8.5	6.7–8.2	—	—	8.2–12.2	8.9–13.2
Centrally Planned Economies										
WEC (1978)	4.6	1.7	—	—	8.4–9.7	3.9–4.5	12.9–14.6	6.8–7.7	10.4–15.3	5.0–7.3
IIASA	—	—	5.1	1.8	7.6–8.5	3.3–3.7	—	—	—	—
SUBTOTAL										
WEC (1978)	6.1	6.5	—	—	9.3–10.2	12.1–13.2	12.7–13.8	18.5–20.2	8.9–13.1	13.9–20.5
IIASA	—	—	5.8	6.8	7.2–8.5	10.0–11.9	—	—	—	—
Developing Countries										
Market Economies										
WAES	0.4	0.7	—	—	0.7–0.9	2.5–3.4	—	—	—	—
WEC (1978)	0.7	1.3	—	—	0.8–0.9	2.9–3.5	1.0–1.3	5.6–7.6	—	—
IIASA	—	—	0.42	0.8	0.8–1.1	2.6–3.5	—	—	1.3–2.3	6.2–10.7
Centrally Planned Economies										
WEC (1978)	0.8	0.7	—	—	1.1–1.5	1.4–1.9	1.5–2.6	2.3–4.1	1.3–2.6	2.3–4.5
IIASA	—	—	0.50	0.5	0.7–1.1	1.0–1.4	—	—	—	—
SUBTOTAL										
WEC (1978)	0.8	2.0	—	—	0.9–1.1	4.3–5.4	1.1–1.6	7.9–11.7	1.3–2.4	8.5–15.2
IIASA	—	—	0.45	1.3	0.8–1.1	3.6–5.0	—	—	—	—

TOTAL Market Economies								
WAES	2.3	5.7	—	—	2.6– 3.2	11.3–14.6	—	—
WEC (1978)	2.4	6.1	—	—	2.5– 2.7	11.1–12.2	2.6– 3.0 17.3–20.1	—
IIASA	—	—	2.1	5.7	2.2– 2.7	9.3–11.7	—	2.6– 4.1 15.1–23.9
TOTAL Centrally Planned Economies								
WEC (1978)	1.9	2.4	—	—	3.1– 3.7	5.3– 6.4	4.3– 5.6 9.1–11.8	—
IIASA	—	—	1.8	2.3	2.4– 2.9	4.3– 5.1	—	3.3– 5.4 7.3–11.8
TOTAL WORLD								
WEC (1978)	2.2	8.5	—	—	2.6– 2.8	16.4–18.5	3.0– 3.6 26.4–31.9	—
IIASA	—	—	2.1	8.2	2.2– 2.8	13.6–16.8	—	2.8– 4.5 22.4–35.7

Note 1.2-C. Average Annual Growth Rates (Per Cent Per Annum) for Total energy Demand
Projections Advanced in Three Major Global Energy Studies

	1972 – 2000	1975 – 2000	2000 – 2020	2000 – 2030
Industrialized Countries				
Market Economies				
WAES	2.1 – 2.8	—	—	—
WEC (1978)	1.9 – 2.1	—	1.8	—
IIASA	—	1.3 – 2.1	—	1.0 – 1.6
Centrally Planned Economies				
WEC (1978)	3.0 – 3.5	—	2.8 – 2.7	—
IIASA	—	2.5 – 2.9	—	1.4 – 2.3
Subtotal				
WEC (1978)	2.2 – 2.6	—	2.1	—
IIASA	—	1.6 – 2.3	—	1.1 – 1.8
Developing Countries				
Market Economies				
WAES	4.7 – 5.8	—	—	—
WEC (1978)	2.9 – 3.6	—	3.3 – 4.0	—
IIASA	—	4.8 – 6.1	—	2.9 – 3.8
Centrally Planned Economies				
WEC (1978)	2.5 – 3.6	—	2.5 – 3.9	—
IIASA	—	2.8 – 4.2	—	2.8 – 4.0
Subtotal				
WEC (1978)	2.8 – 3.6	—	3.1 – 3.9	—
IIASA	—	4.2 – 5.5	—	2.9 – 3.9
TOTAL Market Economies				
WAES	2.5 – 3.4	—	—	—
WEC (1978)	2.2 – 2.5	—	2.2 – 2.5	—
IIASA	—	2.0 – 2.9	—	1.6 – 2.4
TOTAL Centrally Planned Economies				
WEC (1978)	2.9 – 3.6	—	2.7 – 3.1	—
IIASA	—	2.5 – 3.2	—	1.8 – 2.8
TOTAL WORLD				
WEC (1978)	2.4 – 2.8	—	2.4 – 2.8	—
IIASA	—	2.0 – 2.9	—	1.7 – 2.5

Note 1.2-D. Historical Data on Global Commercial Energy Use (in TW)[a]

	Industrialized Countries						Developing Countries						World					
	ME[b]		C P E[b, c]		SUBTOTAL		ME[b]		C P E[b, d]		SUBTOTAL		ME[b]		C P E[b, e]		TOTAL	
	Oil	Energy	Oil	Energy	Oil	Energy	Oil	Energy	Oil	Energy	Oil	Energy	Oil	Energy	Oil	Energy	Oil	Energy
1981	2.44	5.27	0.78	2.33	3.22	7.60	0.78	1.42	0.12	0.71	0.90	2.13	3.22	6.69	0.90	3.04	4.12	9.73
1980	2.61	5.40	0.76	2.30	3.37	7.70	0.76	1.36	0.12	0.73	0.88	2.09	3.37	6.76	0.88	3.03	4.26	9.79
1979	2.83	5.57	0.75	2.24	3.58	7.81	0.73	1.29	0.13	0.76	0.86	2.05	3.56	6.86	0.88	3.00	4.44	9.85
1978	2.83	5.43	0.74	2.19	3.57	7.62	0.69	1.22	0.12	0.69	0.81	1.91	3.52	6.65	0.86	2.88	4.38	9.51
1977	2.77	5.32	0.70	2.11	3.47	7.43	0.65	1.14	0.12	0.65	0.77	1.79	3.42	6.46	0.82	2.76	4.24	9.21
1976	2.71	5.22	0.67	2.02	3.38	7.24	0.62	1.07	0.11	0.61	0.73	1.68	3.33	6.29	0.78	2.63	4.11	8.93
1975	2.54	4.96	0.65	1.93	3.19	6.89	0.58	1.01	0.10	0.58	0.68	1.59	3.12	5.97	0.75	2.51	4.15	8.47
1974	2.64	5.10	0.62	1.83	3.26	6.93	0.57	0.99	0.09	0.55	0.66	1.54	3.21	6.09	0.71	2.38	3.92	8.46
1973	2.78	5.21	0.57	1.75	3.35	6.96	0.55	0.94	0.08	0.51	0.63	1.45	3.33	6.15	0.65	2.26	3.97	8.41
1972	2.59	4.96	0.53	1.69	3.12	6.65	0.50	0.87	0.06	0.47	0.56	1.34	3.09	5.83	0.59	2.16	3.68	7.99
1971	2.42	4.76	0.48	1.61	2.90	6.37	0.47	0.82	0.05	0.46	0.52	1.28	2.89	5.58	0.53	2.07	3.43	7.65
1970	2.31	4.62	0.45	1.53	2.76	6.15	0.44	0.78	0.04	0.41	0.48	1.19	2.75	5.40	0.49	1.94	3.24	7.34
1969	2.13	4.41	0.41	1.45	2.54	5.86	0.41	0.74	0.03	0.36	0.44	1.10	2.54	5.15	0.44	1.81	2.98	6.96
1968	1.95	4.08	0.38	1.42	2.33	5.50	0.38	0.69	0.02	0.33	0.40	1.02	2.33	4.77	0.40	1.75	2.74	6.52
1967	1.79	3.90	0.35	1.42	2.14	5.32	0.35	0.74	0.02	0.26	0.37	1.00	2.14	4.64	0.37	1.68	2.51	6.32
1966	1.68	3.75	0.32	1.37	2.00	5.12	0.32	0.69	0.02	0.36	0.34	1.05	2.00	4.44	0.34	1.73	2.34	6.17
Average Annual Growth Rates (% per year)																		
1966-73	7.3	4.8	8.2	3.6	7.4	4.5	7.5	4.2	24.0	7.8	8.6	5.3	7.3	4.7	9.3	4.4	7.6	4.6
1973-79	1.0	1.4	4.9	4.2	1.7	2.2	4.8	5.3	7.9	6.3	5.2	5.7	1.7	2.1	4.9	4.7	2.1	2.8
1979-81	-7.0	-2.8	2.0	2.0	-5.3	-1.4	3.3	4.8	4.0	-3.4	2.3	1.9	-5.0	-1.3	1.1	0.7	-3.7	-0.6

Source: The British Petroleum Company, 1984.

(a)
(b) ME = market economies; CPE = centrally planned economies.
(c) The Soviet Union and Eastern Europe.
(d) China only.
(e) The Soviet Union, Eastern Europe, and China.

Note 1.2-E: Projections of Primary Energy Demand (in TW)

	1978	2020, WEC ('78)		2020, WEC ('83)		2030, IIASA[g]	
		Low	High	Low	High	Low	High
Oil	4.24[a]	4.28	6.15	4.25	5.79	5.02	6.83
Natural Gas	1.61[a]	2.70	2.79	3.41	4.57	3.47	5.97
Nuclear	0.18[a]	6.72	7.32	2.30	3.20	5.17	8.09
Coal	2.59[a]	8.82	11.45	6.05	7.73	6.45	11.98
Hydro	0.61[a, b]	1.77	1.77	1.40	1.89	1.46	1.46
Biomass and Other Solar	1.41[c]	2.19	2.37	2.65	2.65	0.82	1.30
TOTALS	10.6	26.4	31.9	20.1	25.8	22.4	35.7

(a) *Source:* United Nations, 1979.

(b) This is the primary energy equivalent for hydroelectrity generation, taken to be the fuel input required to produce the same amount of electricity at a thermal power plant.

(c) Biomass energy use in 1978 is estimated to be as follows (in TW):

Developing Countries	1.32[d]
United States	0.05[e]
Other Industrialized Countries	0.04[f]
TOTAL	1.41

(d) It has been estimted (D.O. Hall, *et. al.,* 1982) on the basis of national energy consumption surveys that biomass energy consumption in developing countries amount to about 415 Watts/capita. This rate is applied to the 3.17 billion people living in developing countries in 1978.

(e) It has been estimated (Office of Technology Assessment, 1980) that biomass usage for energy purposes in the U.S. is presently as follows (in TW):

Fuel for the forest product industry	0.043
Household fuel	0.007
TOTAL	0.050

(f) Fuelwood consumption, *Source:* United Nations, 1981.

(g) See Table 17-6, page 531 and Table 17-7 page 532 in Wolf Haefele *et al.,* 1981.

Note 1.2-F. Projections of Installed Nuclear Generating Capacity, for Alternative Energy Scenarios [in GW(e)]

		1979	1985	2000	2020	2030
1.	Base Year	121				
2.	WEC (1978)					
	High Growth		—	1650	4300	—
	Low Growth		—	1300	3200	—
3.	WEC (1983)					
	High Growth			770	1800	
	Low Growth			640	1300	
4.	IIASA					
	High Growth					
	LWR		—	850	—	1800
	FBR		—	20	—	2600
	TOTAL			870		4400
	Low Growth					
	LWR		—	620	—	910
	FBR		—	10	—	1700
	TOTAL			630		2610
5.	WAES					
	Minimum Likely		291	913	—	—
	Maximum Likely		412	1772	—	—

Note 1.2-G. Worldwide coal resources recoverable at a cost of less than $90 per tonne (1984 $) are estimated to be **1580 TW—yrs.**
Source: Table 17.6 p. 531 in Wolf Haefele *et al.,* 1981.

Note 1.2-H. Total electrical generating capacity in 1975 was
$$P\ (1975)\ =\ 1600\ GW(e),$$
while IIASA analysts project (see p. 483 in Wolf Haefele *et al.,* 1981)
$$P(2000)\ =\ 3550\ to\ 4390\ GW(e)$$
$$P(2030)\ =\ 6320\ to\ 9845\ GW(e)$$
To estimate the average construction rate, 1975-2030 implicit in these scenarios, it is assumed that the all net new capacity constructed after 2000 is built once, while all net new capacity built before 2000 is built twice. Thus:
$$C\ =\ \{[P(2030-P(2000)]+2\times[P(2000)-P(1975)]+2\times P(1975)\}/55$$
$$=\ [P(2030)+P(2000)]/55.$$
Thus
C = 179.5 GW(e) per year for the IIASA low scenario.
and
C = 258.8 GW(e) per year for the IIASA high scenario

Note 1.2-I. Total nuclear generating capacity in 1975 was
P (1975) = 72 GW(e),
while IIASA analysts project (see Note 1.2-F):
$$P(2000)\ =\ \ \ 630\ to\ \ 870\ GW(e)$$
$$P(2030)\ =\ 2600\ to\ 4400\ GW(e)$$
Following the procedure outlined in Note 1.2-H, the average construction rate, 1975 to 2030, would be:
C = 58.7 GW(e) per year for the IIASA low scenario
and
C = 95.8 GW(e) per year for the IIASA high scenario.

Note 1.2-J. Here it is assumed that total fossil fuel production capacity P(t) satisfies.

$$dP(t)/dt = -(1/T) P(t) + C(t)$$

where

C(t) = rate of addition of new capacity
T = mean life of new capacity.

If

C(t) = C = constant during the period, 1975-2030,

then

$$C = (1/T)/[1 - exp(-55/T)] \times [P(2030) - P(1975) exp(-55/T)].$$

It is assumed that T = 20 years for fossil fuel production facilities, so that:

$$C = 0.0534 \times [P(2030) - 0.064 \times P(1975)]$$

Lumping petroleum, natural gas, and coal together one obtains 1975 fossil fuel production as:

P(1975) = 107.4 million B/D oil equivalent,

while IIASA analysts project (see Note 1.2-E):

P(2030) = 211 to 350 million B/D oil equivalent.

Thus

C = 10.9 million B/D oil equivalent new capacity per year the IIASA low scenario.
C = 18.3 million B/D oil equivalent new capacity per year for the IIASA high scenario.

Note 1.3-A. World Trade, including some energy-intensive materials, 1978 (billion US $ f.o.b.)

	SITC (revised)	Export from IME to:		Export from DME to:		
		IMC	DME	IMC	DME	
		(1)	(2)	(3)	(4)	(2)/(3)
Total trade	0 – 9	618	207	212	74	0.98
Crude materials excl. fuels, oils, fats	2.4	45	8.5	18	6.3	0.47
Metalliferous ores and metal scrap	28	8.3	0.65	4.9	0.48	0.13
Mineral fuel and related materials	3	34	3.4	120	35	0.28
Chemicals	5	59	20	2.0	2.6	10
Machinery and transport equipment	7	213	97	8.8	6.9	24
Passenger road vehicles and their parts	subgroups of 7	44	6.4	0.15	0.60	43
Other manufactured goods	6.8	191	54	34	14	1.6
Iron and steel	67	28	13	1.3	1.2	10
Non-ferrous metals	68	16	2.6	4.7	1.4	0.55

Here SITC = standard international trade classification; IME = industrialised market economies;
DME = developing market economies.
Source: United Nations, 1980.

Note 1.3-B. Global distribution of energy use, population and food supply (1978)

	Industrialised Countries	Developing Countries	World
Population (billion)	1.1 (26%)	3.1 (74%)	4.2 (100%)
Energy Use			
Total	208 EJ (69%)	92 EJ (31%)	300 EJ (100%)
Per Capita	189 GJ (6.0 kW)	30 GJ (0.95 kw)	71 GJ (2.3 kW)
Commercial Energy Use			
Total	206 EJ (80%)	52 EJ (20%)	258 EJ (100%)
Per Capita	187 GJ (5.9 kW)	17 GJ (5.4 kW)	61 GJ (1.9 kW)
Percent of Total Energy	(98%)	(57%)	(86%)
Biomass Energy Use			
Total	2 EJ (5%)	40 EJ (95%)	42 EJ (100%)
Per Capita	1 GJ (0.03 kW)	13 GJ (0.41 kW)	10 GJ (0.32 kW)
Percent of Total Energy	(2%)	(43%)	(14%)
Per Capita Food Supply	3353 Kcal	2203 Kcal	2571 Kcal
Percent of Daily Requirement*	(129%)	(96%)	(106%)

*The average daily is estimated to be 2,600 Kcal in industrialized countries and 2,300 Kcal in developing countries, with 2370 Kcal as a world average.

Note 1.3-C. Persons Below the Poverty Level in the United States

Year	Persons Below the Poverty Level (millions)	(Percent)	Year	Persons Below the Poverty Level (millions)	(Percent)
1983	35.3	15.2	1974	23.4	11.2
1982	34.4	15.0	1973	23.0	11.1
1981	31.8	14.0	1972	24.5	11.9
1980	29.3	13.0	1971	25.6	12.5
1979	26.1	11.7	1970	25.4	12.6
1978	24.5	11.4	1969	24.1	12.1
1977	24.7	11.6	1968	25.4	12.8
1976	25.0	11.8	1967	27.8	14.2
1975	25.9	12.3	1966	28.5	14.7
1978	24.5	11.4	1965	33.2	17.3
1977	24.7	11.6			
1976	25.0	11.8	1960	39.9	22.2
1975	25.9	12.3			

Source: U.S. Census Bureau, December 1984.

Note 1.3-D. The following are historical civilian unemployment rates in industrialized countries (in per cent), measured approximately according to U.S. conventions

	U.S.	Canada	Japan	France	F.R.G.	Italy	U.K.
1971	5.9	6.2	1.3	2.7	0.6	2.9	3.9
1972	5.6	6.2	1.4	2.8	0.7	3.4	4.2
1973	4.9	5.5	1.3	2.7	0.7	3.2	3.2
1974	5.6	5.3	1.4	2.9	1.6	2.8	3.1
1975	8.5	6.9	1.9	4.2	3.4	3.2	4.6
1976	7.7	7.1	2.0	4.6	3.4	3.6	6.0
1977	7.1	8.1	2.0	5.0	3.5	3.6	6.3
1978	6.1	8.4	2.3	5.4	3.4	3.7	6.2
1979	5.8	7.5	2.1	6.1	3.0	3.9	5.6
1980	7.1	7.5	2.0	6.5	2.9	3.9	7.0
1981	7.6	7.6	2.2	7.7	4.1	4.3	10.5
1982	9.7	11.0	2.4	8.7	5.9	4.8	12.0
1983	9.6	11.9	2.7	8.8	7.3	5.1	13.1
1984	7.5	11.3			7.4	5.6	13.5

Source: Council of Economic Advisers, 1985.

The following are OECD unemployment data, as measured by national conventions

	Average for the Above Countries	Other OECD Countries	North America	OECD Europe	Total OECD
Unemployment Rate (%).					
1981	6.5	8.8		8.2	6.9
1982	7.9	10.2		9.5	8.4
1983	8.2	11.9		10.5	9.0
1984	7½	12½		11	8½
1985	7½	12¾		11½	8½
Unemployment Level (millions)					
1981			9.2	13.7	24.6
1982			12.0	16.0	30.0
1983			12.1	17.9	32.4
1984			10	19	31 ¼
1985			9¾	19¾	31 ¾

Source: Organization for Economic Co-Operation and Development, December 1984.

Note 1.3-E. Extrapolating from data for the first 9 months of 1981 one finds that in 1981 the 5.5 million households in the U.S. having cars and incomes less than $5000 spent $940 per household on 730 gallons of gasoline for the automobile (Energy Information Administration, February 1983). Averaged over the total population of 10.4 million households (1/8 of all U.S. households) in this income class (including those without cars) the cost is reduced to an average of $500 per household or $246 per capita per year (189 gallons per year or 0.790 kW per capita), since there are 2.04 persons per household in this income class. This amounts to 15 per cent of income for this income group.

In the period April 1980 through March 1981, the 10.4 million households in the U.S. having incomes less than $5000 spent on fuel and electricity for their homes $750 per household or, since there were on average 2.04 persons per household in this income class, $370 per capita (for 1.28 kW of fuel and 0.33 kW of electricity or 2.38 kW of primary energy per capita). Source: Energy Information Administration, September 1982.

The 0.79 kW per capita for the automobile and the 2.38 kW per capita for residential energy add up to 3.17 kW per capita for households with incomes less than $5000.

In contrast, the average U.S. household with cars (88% of all households) spent $1400 on gasoline in 1981. Averaged over all households (including those without cars) expenditures amounted to $1235 per household or $450 per capita per year (339 gallons per year or 1.420 kW per capita), since on average there are 2.76 persons per household. This amounts to 5 percent of income for the average household.

In the same period the average US household spent $920 on household fuel and electricity, or, since there were on average 2.76 persons per household, $330 per capita (for 1.02 kW of fuel plus 0.37 kW of electricity or 2.24 kW of primary energy per capita). This amounts to 4 per cent of income for the average household.

The 1.42 kW per capita for the automobile and the 2.24 kW per capita for residential energy add up to 3.66 kW per capita for the average U.S. household.

Note 1.3-F. Appliance Ownership by Income Class in the US, 1980 (Percent)

	Average Household	Poor Household
Range		
Electric	54	40
Gas	46	58
Water Heater	99.7	98.2
Refrigerator		
none	0.2	0.7
Only 1	85.7	92.6
2 or more	14.0	6.7
Freezers		
None	61.9	70.8
Only 1	34.9	27.1
Dishwasher	37	11
Clothes Washer		
Automatic	72	54
Wringer	4	8
Clothes Dryer		
Electric	47	28
Gas	14	6
2 or more	3.2	2.1
Air Conditioning		
Room	30.0	26.4
Central	27.2	14.8
None	42.8	58.8
TV		
Color	82	61
Black and White	51	57

Source: Energy Information Administration, June 1982.

Note 1.3-G. Current estimated depositions of SO_2 in ECE countries (thousands tonnes per annum)[a]

Receivers: \ Emitters:	Austria	Belgium	Bulgaria	Czechoslovakia	Denmark	FRG	Finland	France	GDR	Greece	Hungary	Ireland	Italy	Luxembourg
Austria	125	7	—	84	—	77	—	48	58	—	34	—	134	—
Belgium	—	161	—	—	—	58	—	67	5	—	—	—	—	—
Bulgaria	—	—	367	19	—	10	—	5	15	17	30	—	20	—
Czechoslovakia	53	24	5	1159	5	259	—	108	468	—	163	—	94	—
Denmark	—	—	—	7	94	27	—	7	30	—	—	—	—	—
FRG	19	89	—	115	10	1346	—	259	283	—	15	—	58	13
Finland	—	5	—	20	7	36	185	12	46	—	7	—	5	—
France	—	79	—	19	—	235	—	1509	48	—	—	5	98	10
GDR	—	20	—	134	7	204	—	50	1193	—	10	—	8	—
Greece	—	—	—	82	10	7	—	7	7	223	12	—	34	—
Hungary	22	—	—	110	—	31	—	17	43	—	466	—	65	—
Ireland	—	—	—	—	—	—	—	5	—	—	—	43	—	—
Italy	14	5	—	29	—	53	—	127	26	—	26	—	1903	—
Luxembourg	—	—	—	—	—	5	—	7	—	—	—	—	—	7
Netherlands	—	38	—	5	—	108	—	36	14	—	—	—	—	—
Norway	—	10	—	19	14	48	5	26	53	—	5	—	—	—
Poland	17	17	7	326	19	192	—	62	511	—	103	—	43	—
Portugal	—	—	—	—	—	—	—	—	—	—	—	—	—	—
Romania	8	—	72	89	—	38	—	17	62	10	170	—	65	—
Spain	—	5	—	—	—	48	—	91	10	—	—	—	5	—
Sweden	—	12	—	43	38	84	24	36	101	—	12	—	10	—
Switzerland	—	—	—	5	—	31	—	55	7	—	—	—	113	—
Turkey	—	—	67	12	—	—	12	7	14	48	17	—	31	—
USSR[b]	34	43	166	454	60	456	120	161	679	53	382	5	226	7
UK	—	16	—	10	—	50	—	65	26	—	—	16	5	—
Yugoslavia	34	5	77	91	—	53	—	48	60	14	173	—	314	—
Other areas	31	122	204	254	125	605	72	708	514	190	163	48	1075	15

Receivers: \ Emitters:	Netherlands	Norway	Poland	Portugal	Romania	Spain	Sweden	Switzerland	Turkey	USSR	UK	Yugoslavia	Unidentified	Total
Austria	5	—	31	—	—	5	—	7	—	—	19	84	72	790
Belgium	10	—	—	—	—	—	—	—	—	—	43	—	21	365
Bulgaria	—	—	17	—	82	—	—	—	7	30	5	108	70	802
Czechoslovakia	17	—	228	—	34	10	—	5	—	34	77	127	218	3088
Denmark	—	—	10	—	—	—	—	—	—	—	26	—	26	223
FRG	55	—	43	—	—	15	—	15	—	7	173	36	192	2743
Finland	5	5	36	—	5	—	34	—	—	103	34	7	132	684
France	29	—	10	—	—	158	—	19	—	—	237	14	398	2868
GDR	15	—	55	—	—	—	—	—	—	7	53	14	62	1832
Greece	—	—	5	—	19	5	—	—	10	9	—	65	84	579
Hungary	—	—	50	—	38	—	—	—	—	9	7	180	48	1056
Ireland	—	—	—	—	—	—	—	—	—	—	29	—	62	139
Italy	—	—	14	—	7	41	—	17	—	5	21	156	223	2657
Luxembourg	—	—	—	—	—	—	—	—	—	—	—	—	—	19
Netherlands	96	—	—	—	—	—	—	—	—	—	65	—	26	388
Norway	7	48	24	—	—	—	24	—	—	19	96	5	178	581
Poland	16	—	1356	—	31	5	9	—	—	82	77	96	185	3154
Portugal	—	—	—	48	—	41	—	—	—	—	—	—	70	159
Romania	—	—	91	—	689	—	—	—	10	115	14	276	144	1870
Spain	7	—	—	23	—	881	—	—	—	—	36	—	266	1372
Sweden	12	19	74	—	7	—	199	—	—	58	84	17	271	1101
Switzerland	—	—	—	—	—	5	—	34	—	—	12	7	41	310
Turkey	—	—	12	—	29	7	—	—	420	41	—	53	190	960
USSR[b]	55	12	926	—	492	24	91	7	108	8664	247	499	264	13971
UK	12	—	5	—	—	9	—	—	—	—	1620	—	173	2007
Yugoslavia	5	—	50	—	62	16	—	—	—	19	16	1337	209	2583
Other areas	127	27	312	22	182	420	108	24	204	542	1466	518	281	—

(a) *Source:* N.H. Highton and M.J. Chadwick, 1983.
(b) Total depositions shown for the USSR are those resulting from emissions in ECE countries alone.

Note 1.3.-H. There are many options available for controlling SO_2 and NOx emissions from both stationary and mobile
sources.

Consider first stationary sources. Commercially available flue gas desulfurization systems ("scrubbers") are capable
of removing 90% of SO_2 from stack gases at a cost which translates into an increase on the order of 10% in the cost
of producing electricity from a coal-fired power plant. A variety of technologies are available or may soon be
commercialized for reducing NOx emissions far below current NOx emissions standards in the U.S. (Ralph Whitaker,
January/February 1982). Commercially available fluidized bed combustion technology is capable of keeping both SO_2
and NOx emissions to low levels in small industrial installations and is expected to be available in a few years for large
utility-scale operations. A new coal-burner capable of keeping both SO_2 and NOx emissions to low levels and suitable
for retrofit applications may be commercially available by the late 1980s (O.W. Dykema, September 1983).

For gasoline-powered automobiles NOx emissions can be reduced by about a factor of 10 relative to uncontrolled
emissions by using the 3-way catalytic converters, as are required in California. An alternative way to reduce NOx emissions
by a comparable amount without a catalytic converter would be to shift to methanol as an automotive fuel (U.S.
Environmental Protection Agency, September 1982).

Complementing such control strategies is the potential for emissions reduction through energy efficiency improvement.
The projection cited above that uncontrolled emissions of SO_2 would increase 30% by the turn of the century, is based
on a projected 60% increase in energy use in this period. In Chapter 2, we shall argue that if cost-effective opportunities
for energy efficiency improvement were fully exploited total energy use would actually decline in this period.

Note 1.3-I. A switch to methanol fuel for automobiles would serve many purposes, in addition to the benefit of reduced
NOx emissions:

- A shift to methanol as motor vehicle of fuel would lead to reduced hydrocarbon and carbon monoxide emissions.
- Where methanol would substitute for diesel fuel, the problems of particulate pollution would also be avoided.
- Because of the possibility of operating at a higher compression ratio, fuel economy would be improved relative
  to operation on gasoline.
- A shift to methanol vehicles would lead to a reduction on oil imports.
- Methanol production based on high sulfur coal would permit shifting power plants now burning high sulfur
  coal to low sulfur coal, while still providing high sulfur coal producers a market for their product—thereby
  enabling simultaneous reductions in both NOx and SO_2.
- Methanol is an attractive energy carrier for the Post Petroleum Era, which can be readily produced from a
  wide range of feedstocks—coal, urban refuse, wood, etc.

Since with current technology methanol can probably be derived from coal at a cost which is competitive with the
wholesales cost of gasoline derived from petroleum (U.S. Environmental Protection Agency, September 1982).

Note 1.3-J. In 1978 George Woodwell and his collaborators suggested that man may be adding an amount of carbon
dioxide to the atmosphere through disruption of the biosphere (mainly deforestation) that is comparable to or even
greater than the amount arising from the combustion of fossil fuels. They estimated that the total carbon dioxide release
rate from terrestrial biota was probably in the range of 4 to 8 billion tonnes of carbon per year (Woodwell et al., 1978).
It is certainly not inconceivable that biospheric sources could be this large: the terrestrial biospheric carbon reservoir
is 2½ times as large as the atmospheric reservoir and the carbon dioxide exchange rate between the biosphere and the
atmosphere is 10 times the rate of CO_2 release in fossil fuel combustion. However, oceanographers have maintained
that the net biospheric release rate could not possibly be anywhere near as large as suggested by these ecologists. They
argue that because the ocean as a geo-chemical sink for CO_2 can account for no more than about 40 percent of the
CO_2 released in fossil fuel combustion, the biosphere cannot possibly be a major source of CO_2 and is probably instead
of minor CO_2 sink (Broecker, et al., 1979).

More recent analyses indicate that Woodwell's initial estimte of the biospheric CO_2 contribution may be much too
high. One reason is that Woodwell's estimate of the size of the tropical forest biomss may be too high, perhaps by
a factor of 2 or more J.S. Olson et al., 1978; S. Brown and A.E. Lugo, 1980). Another is that the production of decay-
resistant charcoal in forest fires (e.g., fires to clear for urbanization, fires for slash-and-burn agriculture, and wild fires)
greatly reduces the CO_2 release rate (Seiler and Crutzen, 1980). Still another factor is that deforestation in the tropics
is offset in part by regrowth of temperate forests, for which it has becn estimated that over the last several decades
carbon as CO_2 has been withdrawn from the atmosphere at a rate of some 1 ± 0.5 billion tonnes per year (T.V. Amentano
and J. Hett, 1980). Also, Lugo argues in a recent review that even with deforestation tropical forests may be CO_2
sinks rather than sources, because some of the removed biomass is not released immediately as CO_2 but is stored as
wood products or charcoal and because succession following removal can be characterized by a much higher rate of
net carbon uptake than is the case for a climax forest (Lugo, 1980). Moreover, a recent analysis of carbon-13 and carbon-14

concentrations in tree rings indicates that while the net flux of carbon to the atmosphere from the biosphere averaged 1.2 billion tonnes per year, 1850-1950, twice the average fossil fuel release rate in that period, the net biospheric flux has been small in recent decades (Stuiver, 1978). Even Woodwell's own more recent estimate involves a lesser biospheric role—a net flux of about 1.8 to 4.7 billion tonnes per year (Woodwell, 1983). At the time of this writing, the most recent measurements of the biomass contained in tropical forests indicate that previous estimtes were much too high, so that when tropical forests are cleared or burned, much less CO_2 is released than previous estimates would suggest (Sandra Brown and Ariel Lugo, March 23, 1984).

Thus, the most recent work suggest that Woodwell's initial estimates of the biospheric role in the global carbon budget are too high and that the biosphere is probably either a small source or a small sink for CO_2 (Kerr, 1980). However, the issue is by no means resolved.

To determine the net biospheric carbon flux within ± 1 billion tonnes per year or less requires much better measurements than are now available. Remote sensing, using a combination of satellite and aircraft imagery, holds forth the potential for making measurements to this degree of precision (Woodwell, 1979).

Note 1.3-K. Carbon Release Rates (as CO_2) for Fossil Fuel Processing and Combustion

	kg per GJ	billion tonnes/TWY
Natural Gas[a]	13.5	0.43
Oil[a]	19.9	0.63
Coal[a]	24.6	0.78
Synthetic Fuels from Coal[a]		
Gas	40.1	1.26
Oil	37.2	1.17
Oil from Shale[b]		
Low Temperature (500 Degree C) Retorting with Indirect Heating	28	0.88
High Temperature (700-1100 Degree C) Retorting with Direct Heating[c]	66 – 104 (d)	2.08 – 3.28

(a)   *Source:* Gordon J. MacDonald, 1982.
(b)   *Source:* E.T. Sundquist and G.A. Miller, 1981.
(c)   With high temperature retorting extra CO_2 (in excess of what would be released from the combustion of organic matter in the shale) would be released from the decomposition of carbonate minerals in the shale.
(d)   The lower (upper) value is for shale that assays at 100 (40) liters/tonne.

Note 1.3-L. The first suggestion that comes to mind for dealing with the rapid build-up of atmospheric CO_2 is to scrub the carbon dioxide out of combustion gases and dispose of it somewhere, much the way technology has been developed to scrub SO_2 out of stack gases. But this would probably be impractical. In the first place, only about one-third of all fossil carbon released world-wide (and only one-sixth of that released in the U.S.) comes from large concentrated sources such as power plants (C.F. Baes et al., 1980). And secondly, the amount of carbon dioxide that would have to be disposed of this way would be enormous—about 4 tonnes for each tonne of coal that is consumed. Nevertheless this approach to the CO_2 problem has been seriously advanced—it has been proposed that CO_2 be scrubbed from stack gases and transported to the deep ocean, which has a large capacity to sequester CO_2 (C. Marchetti, 1977). A careful technical assessment of this and related technical proposals shows that very high energy and dollar costs would be involved. To scrub 90 percent of CO_2 from the stack gas of a coal-fired plant by the most efficient means presently available would involve an energy penalty equal to 43% of the combustion energy of the coal; and the cost of capturing the CO_2, compressing it, and transporting it 40 km by pipeline (e.g., to the deep ocean) would cost $24 per tonne of CO_2 or about $80 per tonne of the consumed coal (C.F. Baes et al., 1980). This is twice the average price paid by U.S. electric utilities for coal in 1980.

A more imaginative proposal by Freeman Dyson is to raise fast-growing trees and create a "carbon bank" (Freeman Dyson, 1977). The challenge here is to remove enough carbon from the atmosphere through photosynthesis to offset the 2-2.5 billion tonnes per year net increase in carbon as CO_2 in the atmosphere. The idea is to use fast-growing species such as American Sycamore, with a productivity of 15 tonnes of dry matter per hectare per year, or about 7.5 tonnes of carbon per hectare per year. [This is about twice the average net primary productivity of the world's forests and woodlands (Earl, 1975)]. With such productivity about 300 million hectares of new forestland would be required—about 8 percent of the world's forest area (3800 million hectares) as of 1972 (SCOPE, 1979). Dyson suggests this not as a permanent solution but not only a strategy to pursue over a period of several decades during a transition to a Post-Fossil Fuel Era. While in principle this carbon bank proposal is feasible, it would require an unprecedented level of social organization for the sole objective of dealing with an "externality." Some 2 trillion surviving seedlings would have to be planted, and some two years would be required for planting, "if the planting were done Chinese style, with every man, woman, and child on earth planting one seedling each day..."

Note. 1.3-M. The Atmospheric CO_2 Buildup for Alternative Energy Scenarios[a]

	Atmospheric CO_2 Level, Relative to Preindustrial Level (%)			Annual Atmospheric CO_2 Buildup (Percent of Preindustrial Level Per Year)			Date at which Preindustrial CO_2 Is Doubled, With Zero Growth in Fossil Fuel Use After[b]:	
	1978	2020	2030	1978	2020	2030	2020	2030
Reference	116	—	—	0.4	—	—	—	—
WEC (1978)								
Low Growth	—	141	—	—	0.9	—	2066	—
High Growth	—	145	—	—	1.1	—	2084	—
IIASA								
Low Scenario	—	—	147	—	—	0.6	—	2111
High Scenario	—	—	159	—	—	1.3	—	2054
Average Growth Rates for Fossil Fuel Use (Percent per year)								
WEC (1978-2020)								
Low Growth	1.5							
High Growth	2.1							
IIASA (1978-2030)								
Low Scenario	1.0							
High Scenario	2.1							

(a) Assuming that 50% of the released CO_2 stays in the atmosphere.
(b) Assuming that ¾ of fossil fuel use after this date is coal and that the remaining fossil fuel use is divided equally between petroleum and natural gas.

Note 1.3-N. IIASA Liquid Fuel Production, and Consumption Projections (Million B/D)

Region	1983(a)			2030, IIASA High Scenario				2030, IIASA Low Scenario			
	Oil Production(b)	Imports	Consumption	Oil Production(b)	Coal Liquids Production	Imports	Consumption	Oil Production(b)	Coal Liquids Production	Imports	Consumption
I N/A	11.7	4.0	16.1	12.0	14.2	0	26.2	7.1	10.6	0	17.7
II SU/EE	12.9	− 1.8	11.5	10.2	9.9	0	20.1	9.4	4.9	0	14.3
III WE/JANZ	4.0	12.1	17.3	4.3	17.6	9.4	31.3	1.4	3.8	17.1	22.3
IV LA	6.4	− 2.1	4.6	24.9	0	0	24.9	16.0	0	0	16.0
V Af/SEA	4.4	0.1	4.3	9.3	2.5	11.2	23.0	3.9	1.4	9.9	15.2
VI ME/NAf	14.9	−12.2	2.4	33.6	0	−20.6	13.0	33.6	0	−27.0	6.6
VII C/CPA	2.1	− 0.4	1.7	1.7	16.7	0	18.4	0	10.5	0	10.5
WORLD	56.4	0	57.9	96	61	0	157	71	31	0	102

(a) *Source:* British Petroleum Company, 1984.
(b) "Oil production" in the IIASA scenarios consists of conventional and tertiary recovery of petroleum, and heavy crude oil, shale oil, and tar sands production.

Note 1.3-O. Installed World Nuclear Generating Capacity (1979)

	GW(e)
U.S.A.	54.59
Western Europe	31.32
Japan	13.80
U.S.S.R.	10.00
Canada	6.53
Eastern Europe	2.71
India	0.64
Brazil	0.63
Korea	0.59
Argentina	0.37
Pakistan	0.14
	121.3

Source: United Nations, 1981.

Note 1.3-P. The Plutonium Management Problem with Nuclear Power

	Fissile Plutonium
	(kg)
Plutonium Required to Make 1 Bomb[a]	5 – 10
Annual Fissile Plutonium Discharge in Spent Fuel from World's Power Reactors in 1979[b]	15,000
Annual Fissile Plutonium Discharge for Alternative Projections of Nuclear Power Growth[c]	
WEC Projections for 2020[d]	
High Growth	4,100,000
Low Growth	3,200,000
IIASA Projections for 2030[d]	
High Growth	4,200,000
Low Growth	2,600,000

(a) See, for example, M. Willrich and T.B. Taylor, 1974.

(b) Nuclear electricity generation in 1979 amounted to 603 billion kWh (United Nations, 1981). A present-day 1 Gw(e) light water reactor operating on a once through fuel cycle at 65 percent capacity factor discharges in spent fuel about 141 kg of fissile plutonium each year or 24.8 kg per billion kWh (H.A. Feiveson, Frank von Hippel, and R.H. Williams, 1979).

(c) From mass flow data sheets for alternative nuclear fuel cycles prepared by C.E. Till et al at Argonne National Laboratory, one obtains a fissile plutonium discharge rate of 142 kg per billion kWh for a light water reactor operated on natural uranium plus recycled plutonium. The fissile plutonium discharge rate for a liquid metal fast breeder reactor is 181 kg per billion kWh (H.A. Feiveson, Frank von Hippel, and R.H. Williams, 1979).

(d) For the projections of generating capacity shown in Note 1.2-F, and the plutonium discharge rates given in (c).

Section 2.1

Note 2.1-A. Structure of Energy Demand and Supply in OECD Countries in 1980[a]

	OECD North America		OECD Europe		OECD Pacific		TOTAL OECD	
	E	E/C	E	E/C	E	E/C	E	E/C
	(EJ)	(kW)	(EJ)	(kW)	(EJ)	(kW)	(EJ)	(kW)
Final Energy Demand								
Residential/Commercial	19.3		14.2		3.2		36.6	
Transportation	19.9		8.5		3.2		31.6	
Industry	21.8		15.8		6.6		44.5	
TOTAL	60.9	7.65	38.5	3.10	12.9	3.05	112.7	4.55
Electricity Generation	9.5	1.19	6.0	0.48	2.5	0.59	18.3	0.74
Primary Energy Use[b]								
Oil	36.6 (35%)		26.8 (81%)		11.7 (92%)		75.1	
Natural Gas	22.1 (1%)		7.6 (11%)		1.3 (67%)		30.9	
Coal	18.0 (−15%)		11.7 (19%)		4.1 (27%)		33.8	
Nuclear	3.2		2.2		0.9		6.3	
Hydro/Other	5.4		3.8		1.3		10.4	
TOTAL	85.2	10.70	52.1	4.19	19.3	4.54	156.5	6.36

(a) *Source:* International Energy Agency, 1983.
(b) The numbers in parentheses are the percentage of primary fuel use provided by imports.

Note 2.1-B Structure of Energy Demand and Supply in CMEA Countries

	Soviet Union (1975)		Eastern Europe (1975)		Total CMEA (1975)		Total CMEA (1982)	
	E	E/C	E	E/C	E	E/C	E	E/C
	(EJ)	(kW)	(EJ)	(kW)	(EJ)	(kW)	(EJ)	(kW)
Final Energy Demand[a]								
Residential/Commercial[b]					10.7			
Transportation					6.9			
Industry[b]					24.0			
TOTAL					41.7	3.7		
Electricity Generation					5.0	0.44		
Primary Commercial Use[c]								
Oil	16.1		3.5		19.3		23.0	
Natural Gas	9.8		2.2		11.7		18.9	
Coal	13.6		10.4		24.0		25.9	
Nuclear	0.3		—		0.3		0.9	
Hydro/Other	1.3		0.3		1.6		2.2	
TOTAL	40.4	5.1	16.1	4.8	56.8	5.0	71.0	5.9

(a) *Source:* A.M. Khan and A. Holzl, April, 1982.
(b) Includes 1.4 EJ of non-commercial energy use in households.
(c) *Source:* The British Petroleum Company, 1983.

Note 2.1-C. The International Energy Agency has estimated that the weighted average real energy prices in 1980 (relative to 1973 prices) for 7 OECD countries (Canada, France, Germany, Italy, Japan, United Kingdom, United States) are the following (IEA, 1982)

- For industry: a 2.7-fold increase in the average oil price and a 2.0-fold increase in the price of non-oil energy forms.
- For the residential/commercial sector: a 2.6-fold increase in the average oil price and a 1.5-fold increase in the price of non-oil energy forms.
- For transporation: a 1.5-fold increase in the gasoline price.

Note 2.1-D. A History of Recent Energy Prices in the United States

Sector and Carrier	Average Price in 1982 (1982 $ per GJ)	P(1982)/P(1972)
RESIDENTIAL		
Natural Gas	4.78	2.04
Heating Oil	8.08	2.91
Electricity	18.89	1.35
AVERAGE	9.36	2.00
COMMERCIAL		
Natural Gas	4.48	2.65
Electricity	19.19	1.44
AVERAGE	9.97	2.16
TRANSPORTATION		
Gasoline	9.83	1.74
Diesel Fuel	7.71	2.86
AVERAGE	8.90	1.90
INDUSTRY		
Natural Gas	3.44	3.98
Residual Fuel Oil	4.13	3.18
Coal	1.93	1.81
Electricity	13.66	2.04
AVERAGE	5.62	2.95
US AVERAGE ENERGY PRICE	8.02	2.22

Source: U.S. Energy Information Administration, April 1985.

Note: 2.1-Ea. Annual Average Growth Rates for Primary Energy Use and Real GDP in Industrialized Countries, the OECD (percent per year)[a]

	OECD-N. America		OECD-Europe		OECD-Pacific		Total OECD	
	1965-73	73-84	1965-73	73-84	1965-73	73-84	1965-73	73-84
Oil	5.01	−1.19	8.11	−2.18	12.85	−1.79	7.13	−1.65
Natural Gas	3.77	−1.63	23.54	3.52	20.39	16.09	5.72	0.00
Coal	0.96	1.84	−4.48	0.11	2.13	2.32	−1.19	1.72
Nuclear	—	13.36	—	18.22	—	26.53	—	16.42
Hydro	4.71	2.18	1.00	1.45	−0.55	2.45	2.63	1.92
Total	4.0	−0.1	4.7	0.1	9.2	0.9	4.7	0.2
GDP	3.7	2.3	4.6	1.8	8.5	3.7	4.7	2.3
Energy/GDP	0.3	−2.4	0.1	−1.7	0.7	2.8	0.0	−2.1

(a)   *Sources:* Energy data are from the British Petroleum Company, 1985, and earlier issues, GDP data are from OECD, 1984b and OECD, 1985.

Note 2.1-Eb. Annual Average Growth Rates for Primary Energy Demand in Industrialized Countries, the CMEA (percent per year)[a]

	Eastern Europe		USSR		CMEA-total	
	1965-73	73-84	1965-73	73-84	1965-73	73-84
Oil	11.02	2.57	7.39	2.94	7.99	2.86
Natural Gas	10.35	5.64	6.45	7.48	7.05	7.18
Coal	1.14	1.67	- 0.50	1.14	0.18	1.39
Nuclear	—	20.10	—	21.26	—	20.98
Hydro	5.88	7.17	4.64	4.81	4.77	4.92
Total	3.7	2.6	3.7	3.8	3.7	3.5

(a)   *Sources:* Energy data are from the British Petroleum Company, 1985, and earlier issues.

Note 2.1-F. The Value of Oil Imports of Selected OECD Countries[a, b]

	1972	1973	1974	1975	1976	1977	1978	1979	1980	1981	1982	1983
Japan												
10^9 Current $	4.41	6.64	20.93	20.85	23.21	25.68	25.54	37.85	57.65	58.42	51.26	45.60
Percentage of Merchandise Exports	15.6	18.2	37.8	37.3	34.8	32.2	27.2	36.8	45.0	35.1	36.7	31.6
Sweden												
10^9 Current $	0.68	1.01	2.43	2.63	2.89	3.05	2.83	5.21	6.49	5.77	5.03	4.02
Percentage of Merchandise Exports	7.8	8.3	15.3	15.2	15.8	16.0	13.1	19.0	21.1	20.3	18.9	14.7
United States												
10^9 Current $	3.85	7.10	23.42	23.86	30.76	40.21	40.04	57.56	74.11	74.83	55.57	50.76
Percentage of Merchandise Exports	7.8	10.0	23.9	22.4	27.0	33.7	28.5	32.2	33.7	32.2	26.6	25.6

(a)   The value of net petroleum imports in current U.S. dollars, at current exchange rates. *Source:* OECD, 1984a.
(b)   Current dollar values of total exported goods and services are from OECD, 1984b. The fraction of exported goods and services accounted for by merchandise exports is from OECD, 1984c.

Note 2.1-G. Percentage Increases in Real Consumer Energy Prices, 1980-84

	OECD Europe	United States	Japan	OECD
Electricity	8.0	11.4	5.4	9.8
Natural Gas	27.5	33.5	- 2.4	30.2
Oil	11.9	- 10.3	- 21.6	- 3.9
Total Energy	14.1	1.1	- 8.6	5.1

Source: International Energy Agency, 1985.

Section 2.2
Note 2.2-A. The Historical Trend in Energy Demand Forecasts to the Year 2000 of the U.S. Government

Year of Forecast	Forecasting Agency	Documentation	Forecast (EJ per Year)
1975	U.S. Bureau of Mines	See note[a]	172
1977	Energy Information Administration, USDOE	Annual Report to Congress	146
1978	Energy Information Administration, USDOE	Annual Report to Congress	132
1979	Energy Information Administration, USDOE	Annual Report to Congress	118
1980	Energy Information Administration, USDOE	Annual Report to Congress	108
1981	U.S. Department of Energy	Biennial Report to Congress[b]	105
1983	U.S. Department of Energy	Biennial Report to Congress[c]	98

(a) W.G. Dupree and J.S. Corsentino, December 1975.
(b) Office of Policy, Planning, and Analysis, U.S. Department of Energy, July 1981.
(c) U.S. Department of Energy, October 1983.

Note 2.2-B. Using official U.S. Department of Energy cost estimates for new central station coal and nuclear power plants leads to an average busbar cost of electricity in the U.S. from an appropriate mix of new baseload, cycling, and peaking power plants of 5.7 cents per kWh, in 1982$ (a,b,c,d). With 8% transmission and distribution losses this corresponds to an average delivered cost of 6.1 cents per kWh. To this must be added the average cost of transmission and distribution, which averaged 1.5 cents per kWh in 1982 (e), bringing the total average cost of electricity from new plants to 7.6 cents per kWh, which is about ⅓ higher than the average US electricity price in 1982.
(a) Here we consider a model summer peaking utility (Electric Power Research Institute, July 1979), where 55% of the electricity is produced in baseload coal or nuclear plants, 42.7% in coal cycling plants, and 2.3% in peaking power plants fueled with distillate oil. This distribution of output would arise for a utility characterized by a 59% load factor, a 25% reserve margin, baseload plants that operate at 65% capacity factor, and the following distribution of installed capacity: 40% baseload; 45% cycling; 15% peaking. Under these conditions the cycling and peaking plants would operate at 45% and 7% average capacity factors, respectively.
(b) Government and industry estimates of the busbar costs (in 1982 cents per kWh) of new central station plants are the following:

	Baseload Nuclear[c]	Baseload Coal [c]	Cycling Coal[d]	Gas Turbine Peaking[d]
Capital	3.39	2.01	3.17	5.67
Fuel	0.99	2.30	2.40	11.06
Operation and Maintenance	0.60	0.48	0.57	0.50
TOTALS	4.98	4.79	6.14	17.23

(c) U.S. average values, for 65% average capacity factor, as estimated in (Energy Information Administration, August 1982). The capital costs assumed in this USDOE analysis were $1800 and $1100 per kW for nuclear and coal plants, respectively.
(d) These are based on capital and operations and maintenance cost estimates and performance data given in (Electric Power Research Institute, July 1979), future coal prices as estimated by the USDOE (Energy Information Administration, August 1982), and a distillate price of $7.50 per GJ (the average price in 1982).
(e) This is the average value for 1980, expressed in 1982 dollars (U.S. Department of Energy, 1981).

Note 2.2-C. The actual cost of electricity in the U.S. from new plants is probably greater than the estimate given in Note 2.2-B, because the costs for new nuclear power plants are often much higher than the $1800 per kW value assumed by the USDOE in its calculations. While $1800 per kW is much higher than the actual cost of $360 per kW (in 1982$) of the Turkey Point 3 nuclear power plant which went into commercial operation in 1972 during the heyday of the nuclear power industry and even much higher than the $1000 to $1500 per kW value (in 1982 $) estimated for new plants by the USDOE in 1978 (Energy Information Administration, 1979), it is far less than costs that are being realized in many new plants now under construction in the U.S., as indicated by the tabulation in Note 2.2-D.

Note 2.2-D. Estimated Unit Costs for Various Unfinished Nuclear Power Plants in the U.S.[a]

Plant Name	Original Cost Estimate (current $ per kW)	Most Recent Cost Estimate (current $ per kW)	Expected completion Date
Bellefonte 1 & 2 (Ala.)	300	2230	
Diablo Canyon 1 & 2 (Cal.)	280	2050	March '84 and April '85
Millstone 3 (Conn.)	345	3060	April '86
Alvin W. Vogtle 1 & 2 (Ga.)	1126	2970	March '87 and Fall '88
Clintol 1 (Ill.)	460	3050	November '86
Wolf Creek (Kansas)	680	2320	March '85
Riverbend 1 (La.)	380	2680	December '85
Waterford 3 (La.)	210	2340	December '84
Midland 2 (Mich.)	430	5460	Mid '86
Enrico Fermi 2 (Mich.)	210	2810	December '84
Grand Gulf 1 (Miss.)	360	2400	Late '84
Callaway 1 (Mo.)	490	2540	December '84
Seabrook 1 & 2 (N.H.)	380	2420	July '85 for 1; future of 2 uncertain
Hope Creek 1 (N.J.)	560	3550	December '86
Shoreham (N.Y.)	320	5010	July '85
Nine Mile Point 2 (N.Y.)	350	3820	October '86
Shearon Harris 1 (N.C.)	280	3110	October '86
Perry 1 & 2 (Ohio)	260	2160	April '85 for 1; April '88 for 2
Beaver Valley 2 (Pa.)	390	3700	April '86
Susquehanna 2 (Pa.)	1050	3900	December '86
Limerick 1 & 2 (Pa.)	800	3030	April '85 for 1; future of 2 uncertain
South Texas Project 1 & 2 (Tx.)	400	2200	June '87 for 1; June '89 for 2

(a) Based on a compilation of data from the Atomic Industrial Forum, individual utilities, and state regulatory agencies (Matthew L. Wald, February 26, 1984).

Note 2.2-E. The data shown in Note 2.2-D. are not strictly comparable to the cost estimates given in Note 2.2-C, since the costs given in Note 2.2-D are in mixed dollars—some pre-1982 and some post-1982 dollars are mixed in with 1982 dollars. However, Charles Komanoff has calculated the average cost in constant 1982 dollars, exclusive of interest during construction to be $2000 per kW for the 35 nuclear projects in the U.S. which in late 1983 were scheduled for completion after 1981 (personal communication from Charles Komanoff to Robert Williams, March 8, 1983).

To this "overnight" construction cost one must add the cost of interest during construction. If the interest rate is "i" and there are 10 payments, with fraction "fn" of the capital cost paid off with the "nth" payment, then the total installed unit capital cost (including interest during construction) is given by:

$$\$2000 \times \sum_{n=1}^{10} fn \times (1+i)^{T(1-n/10)}.$$

Using this formula one obtains a total average cost of $2350 per kW in 1982$, assuming a 5% real interest rate, a construction period of 7.8 years (Energy Information Administration, August 1982) and the following payment schedule (USAEC, October 1974):

n	fn	n	fn
1	0.01	6	0.33
2	0.01	7	0.09
3	0.03	8	0.10
4	0.13	9	0.04
5	0.24	10	0.02

Note 2.2-F. The Historical Trend in Electrical Demand Forecasts by *Electrical World,* the Trade Journal of Electrical Utility Industry in the United States[a]

Year of Forecast	US Electricity Demand (trillion kWh per Year)		Annual Average Growth Rate for Forecast (from Year of Forecast to 1995)	Extra Electrical Generating Capacity That Can Be Deferred, Relative to Previous Year's Forecast [GW(e)][b]
	In Year of Forecast	Forecast for 1995		
1974	1.701	5.695	5.9	—
1975	1.733	5.266	5.7	81
1976	1.850	4.938	5.3	61
1977	1.951	4.348	4.6	112
1978	2.018	3.869	3.9	91
1979	2.084	3.772	3.8	18
1980	2.126	3.546	3.5	43
1981	2.151	3.454	3.4	21
1982	2.094	3.156	3.2	57
1983	2.131	3.041	3.0	22

(a) Forecasts of future electricity demand are made each year in the September issue of *Electrical World,* which is published by McGraw-Hill.

(b) This is the central station equivalent capacity deferral, assuming a 65% capacity factor and 7% transmission and distribution losses.

Section 2.4

Note 2.4-A. Domestic energy use (Watts per capita) in industrialized economies, 1978/1979[a]

	Domestic Energy Use Rate[b]					Rate Total Energy Use	Domestic Energy Fraction
	Space Heat (c)	Hot Water	Cooking	Appliances	Total		
Canada	1276(0.73)	270	33	159	1738	9610	0.18
U.S.	1030(0.64)	256	80	236	1602	9250 (e)	0.17
Sweden	932(0.66)	290	37	149	1408	5300	0.27
Denmark	876(0.71)	211	32	107	1226	5090	0.24
U.K.	722(0.60)	277	114	98	1211	4700	0.26
Germany	888(0.81)	127	29	56	1100	5540	0.20
France	642(0.73)	113	61	58	874	4040	0.22
Italy	441(0.75)	58	43	49	588	3020	0.19
Japan	105(0.35)	126	51	73	355	3220	0.11

(a) The data presented here are for secondary energy. Losses in electrical generation, transmission, and distribution are not taken into account.
(b) *Source:* Lee Schipper and Andrea Ketoff, April 1982.
(c) The number in parenthesis is the space heating fraction of total residential energy use.
(d) *Source:* United Nations, 1981, except for the U.S.
(e) *Source:* U.S. Energy Information Administration, September 1982.

Note 2.4-B. The saturation of energy-intensive household activities

	Sweden[a]	France[b]	W. Germany[c]	Italy[d]	United Kingdom[e]	United States[f]
Central Heat	99 (1978)	67 (1980)	66 (1978)	57 (1978)	54 (1978)	76 (1980)
Hot Water	99 (1978)	86 (1978)	94 (1978)	69 (1978)	119 (1978)	100 (1980)
Refrigerator	99 (1979)	95 (1980)	87 (1973)	94 (1980)	84 (1975)	114 (1980)
Freezer	77 (1979)	20 (1980)	32 (1973)	10 (1980)	14 (1975)	41 (1980)

(a) *Source:* Lee Schipper, August 1982.
(b) 1980 data are from Lee Schipper and Andrea Ketoff, April 1982. The 1978 datum is from Lee Schipper, Andrea Ketoff, and Stephen Myers, May 1981.
(c) The 1978 data are from Lee Schipper, Andrea Ketoff, and Stephen Myers, May 1981. The 1973 data are from Florentin Krause, April, 1982.
(d) The 1978 data are from Lee Schipper, Andrea Ketoff, and Stephen Myers, May 1981. The 1980 data are from Lee Schipper and Andrea Ketoff, April 1982.
(e) The 1978 data are from Lee Schipper, Andrea Ketoff, and Stephen Myers, May 1981. The 1975 data are from David Olivier *et al.*, 1983.
(f) From U.S. Energy Information Administration, June 1982.

Note 2.4-C. The Kitchenaid Model KDC-58 front loading portable dishwasher uses 45 liters of hot (60 degrees C) water plus 0.23 kWh of machine energy per cycle (one cycle holds approximately 10 place-settings or roughly the number of dishes used by a family of 4 in one day. To heat the water with an 80% efficient electric water heater requires 3.1 kWh, bringing the total operational energy required per cycle to 3.3 kWh. The hot water required to clean this many dishes by hand is probably comparable in typical applications. [A typical kitchen sink holds about 20 liters of water. To provide warm (46 degree C) dishwater requires 15 liters of hot (60 degree C) water per sink.]

Considering the energy required to manufacture the dishwasher doesn't change the comparison very much. The primary energy required to manufacture a dishwasher amounts to only about 2 MJ per cycle (R. Herendeen and A. Sebald, 1975). By comparison, the primary energy required to produce the electricity for dishwasher operation amounts to about 40 MJ per cycle for a thermal power plant or 13 MJ per cycle for a hydroelectric power plant.

Note 2.4-D. Using regression analysis it can be shown that during the period 1970-1980 industrial Gross Product Originating (GPOi), the value added measure for U.S. industry, grew in relation to Gross National Product (GNP) as follows;

$$GPOi = 1.1075 \times GNP^{0.832}, r = 0.9639$$

Here the industrial sector consists of agriculture, mining construction, construction, and manufacturing. Economic product values are given here in billion 1972 dollars.

Note 2.4-E. Energy/Output relationships for U.S. industry, 1978

Industrial Subsector	Gross Product Originating[a] (billion 1972 dollars)	Energy Consumption[b] (EJ per year)				Energy Intensity (MJ per 1972 $)			
		Fuel	Elect.	Final	Primary	Fuel	Elect.	Final	Primary
Mining, Agriculture, and Construction	117	3.50	0.35	3.85	4.70	29.9	3.0	32.9	40.2
Basic Materials Processing Industries[c]	114	16.56	1.72	18.28	22.46	145.3	15.1	160.4	197.0
Other Industries	243	1.94	0.80	2.74	4.69	8.0	3.3	11.3	19.3
All Industry[d]	473	22.00	2.87	24.87	31.85	46.5	6.1	52.6	67.3

(a) Gross Product Originating is the value added measure tabulated by the Bureau of Economic Analysis of the U.S. Department of Commerce.

(b) The energy consumption data presented here were compiled in Solar Energy Research Institute, 1981.

(c) The basic materials processing industriers consist of: food; paper, chemicals; petroleum refining; stone, clay, and glass; primary metals.

(d) The industrial sector consists of all economic activities except for transportation, public utilities, wholesale and retail trade, services, and government. In 1978, industry accounted for ⅓ of total economic output in the U.S. economy.

Section 2.5.1

Note 2.5.1-A. From the second law of thermodynamics the maximum amount of work W that can be obtained from a heat source Q at absolute temperature T when the ambient temperature is To is;

$$W = Q \times [1 - (To/T)].$$

Note 2.5.1-B. From the second law of thermodynamics the minimum amount of available work W required to provide an amount of heat Q at temperature T when the ambient temperature is To is that required to run an ideal heat pump, i.e.:

$$W = Q \times [1 - (To/T)].$$

Section 2.5.3.1

Note 2.5.3.1-A. The number of heating degree days (HDD) is a measure of the severity of the winter heating season, defined here as:

$$HDD = (To - Tj).$$
$$Tj < To$$

Tj is the average of the minimum and maximum outdoor temperatures for day j, and To = 18°C, the reference temperature.

Note 2.5.3.1-B. For the 97 houses built in Minnesota's Energy Efficient Housing Demonstration Program described in Table 2-3, the additional first cost (I) was rather high—it averaged $6100. But even for these houses the cost of saved energy was less than the alternative fuel cost. The measured average load for these houses was 32 GJ per year, compared to 76.4 GJ for a house of standard construction (R.H. Williams, G.S. Dutt,. and H.S. Geller, 1983). Heated with natural gas in 69% efficient furnaces the annual fuel savings (FS) for these houses was 64.3 GJ.

The cost of saved energy (CSE) is given by:

$$CSE = i/[1 - (1+i)^{-T} \times (I/FS),$$

where

i = the interest or discount rate,
T = the expected lifetime for the investment

For 5% real interest and a 25 year lifecycle for the improvements, the cost of saved energy would be $6.70 per GJ. For comparison, a reasonable estimate of the average gas price over the life of this investment would be $8 per GJ (the average 1982 price for heating oil).

For the 9 electrically heated houses in Eugene, Oregon, shown in Table 2-3, the extra first cost was much lower—averaging $2050. For these houses (heated with electric resistance heat) the average annual electricity use for space heating is 3920 kWh per dwelling per year. For a comparable conventional house conforming to present construction practice the space heating requirements would be 7900 kWh per year. The cost of saved energy is thus 3.7 cents per kWh, with 5% real interest and a 25 year lifecycle for the improvements. This is far less than both the average U.S. electricity price in 1982 (6.5 cents per kWh) and the cost of electricity from new power plants (8.5 cents per kWh for residential customers—see Note 2.2-B).

The extra cost of conforming new single family Swedish dwellings in 1977 to the 1975 Swedish building standard has been estimated to be approximately 7500 SEK, or $1700 (1977 U.S.$). If escalated at the U.S. inflation rate this would become $U.S. 2500 in 1982$. See Lee Schipper, August 1982.

Note 2.5.3.1-C. Northern Energy Homes (NEH) sell in Vermont for about the same prices as houses of conventional construction—typically about $410 to $490 per square meter of finished area when construction is done by a hired builder [personal communication, December 1982, to Howard Geller from Candice Cruz, Northern Energy Homes, Norwich, Vermont]. This is the final price to the purchaser for the finished house, which includes the excavation and foundation but not the lot. For comparable conventionally built house in the area—wood frame construction, 15(24) cm of fiberglass insulation in the walls (ceiling), double glazing—the cost on the same basis is $380 to $480 per sq. meter (personal communication, December 1982, to Howard Geller from Roma Jean Douglass, Vermont Homebuilders Association).

The competitive first cost of Northern Energy Homes may reflect in part the modular construction techniques involved (see Table 2-3). Major cost savings are achieved with the NEH designs by using prefabricated wall sections, containing thick (20 cm) rigid polystyrene insulation, which are mounted on the outside of post-and-beam framing. The prefabricated pieces are made to fit together easily for rapid construction on site.

Note 2.5.3.1-D. The Wolgast house in Sweden (see Table 2-3) is so well designed that it can be heated largely by a few small (100 W) electric resistive heaters, providing about 1500 kWh of heat per year. Supplemental heat (including heat during the peak demand period) is provided via a wood stove that burns about 1 cubic meter of wood per year. The total initial investment was actually slightly less than that for an equivalent oil heated house of conventional construction, because the more costly heating system was not needed. It is noteworthy that the house was not built with the primary objective of saving energy. The good energy performance followed from the goals of removing particulate matter from the ventilation air for medical reasons and of building a house free of draughts for comfort considerations.

Note 2.5.3.1-E. The two different versions of the prefabricated, electric resistively heated houses offerd by Faluhus (see Table 2-3) cost 3970 and 3750 SEK per square meter respectively in 1984. Thus the more energy efficient house costs 24,640 SEK (U.S. $3200) more. The annual electricity savings for the more efficient house would be 8960 kWh per year. The cost of saved energy (assuming a 6% discount rate—the value used by the Swedish Energy Commission for assessing alternative energy technologies—and a 30 year life for the extra investment) would be 0.20 SEK per kWh (U.S. $0.026 per kWh). For comparison, electricity rates for residential customers in Sweden consist of a large fixed cost independent of consumption level (about 1200 SEK per year) plus a variable cost of 0.25 SEK per kWh ($0.032 per kWh).

Note 2.5.3.1-F. These furnaces achieve high efficiency (about 95%, compared to the average of about 69% for new furnaces in the U.S. in 1980) largely as a result of cooling the exhaust gases down to a low temperature, considerably reducing sensible heat losses through the flue, and recovering some latent heat by condensing water vapor in the exhaust. Because of their ability to recover this latent heat of condensation these are called condensing furnaces.

In 1982 the Lennox pulse combustion condensing furnace (95% efficient) sold in the U.S. for an installed cost of about $2100, compared to an installed cost of about $1350 for a conventional efficient furnace (those with electronic ignition and vent dampers, @ 78% efficiency). With the condensing furnace, however, a chimney is unnecessary, because cool exhaust gases can be vented to the side of the house through a simple PVC (plastic) pipe, leading to an installation savings of at least $100. With a 5% real interest rate and a 23 year furnace life, the cost of saved energy for the more efficient furnace would be less than a gas price of $8 per GJ as long as the heating demand were greater than about 25 GJ per year. For comparison, in 1980 the use of natural gas for space heating averaged somewhat more than 100 GJ in the state of New Jersey in the U.S. (where climatic conditions are about average) in 1980 (Margaret Fels and Miriam Goldberg, July 1982), corresponding to a heating requirement of about 60 GJ per year, for an average furnace efficiency of 60%.

Note 2.5.3.1-G. Pulse combustion space heaters that use outdoor air for combustion and which ventilate their low temperature exhaust gases directly to the outside and have shown fuel use efficiencies of 93.5% in laboratory tests are being developed by the American Gas Association Laboratories with the support of the Gas Research Institute. A prototype unit having a fuel input rate of 19 MJ per hour was developed in 1982, and 8 gas utilities in the U.S. participated in field tests with prototype units in the 1983-84 heating season (Robert J. Hemphill, May/June 1983).

Note 2.5.3.1-H. Heat pumps in Sweden that extract heat from warm exhaust air are used primarily to preheat domestic hot water, with the excess heat used for space heating. However, it is customary to assign all the energy saving to space heating. For prefabricated houses offered by Gullringshus (with a floor are of 127 square meters) calculations indicate that the use of a heat pump would reduce the electricity requirements for space heating in Stockholm from 7620 to 3175 kWh per year. Since the heat pump costs about 10,000 SEK and is expected to last about 15 years, the cost of saved energy would be about 0.23 SEK per kWh, which is less than the present variable part of the Swedish residential electrical rate (Note 2.5.3.1-E).

Note 2.5.3.1-I. The 12 MW (th) heat pump in the city of Lund, Sweden, extracts heat from sewage water @ 13 degrees C and produces hot water @ 80 degrees C, with a seasonal average COP = 3.3. The Lund system had the following installed capital cost (in million 1982 SEK):

Heat Pump	9.4
Building	1.9
Installation	1.9
TOTAL	13.2

For a 6% discount rate, a 20-year life, and a 50% annual capacity factor, this becomes 0.022 SEK per kWh of produced heat. To this capital cost one must add 0.059 SEK per kWh of produced heat, to cover the cost of electricity (the 1982 industrial electricity price was 0.195 SEK per kWh). The total cost of heat produced this way (0.08 SEK per kWh) can be compared to a cost of 0.17 SEK per kWh for heat derived from an oil boiler [assuming the 1982 price of industrial fuel oil (36 SEK per GJ) and a boiler efficiency of 85%].

Note 2.5.3.1-J. Princeton University's Buildings Energy Research Group measured heat losses in insulated attics of many Northeast U.S. houses. The measured losses averaged three times the values predicted by traditional calculations which do not account for thermal short circuits or "bypasses" in the building envelope nor adequately characterize the location or magnitude of air leakage (G.S. Dutt, J. Beyea, and F. Sinden, 1978). Typically, there are many such defects in each house contributing to these anomalous heat losses—defects which are not easy to find from visual inspection.

Note 2.5.3.1-K. Retrofits of gas-heated houses in the state of New Jersey

Module Name[a]	Gas Use for Space Heating (GJ/year)[b]	Post Retrofit Gas Use, As a Fraction of the Pre-retrofit Level		Initial Investment for Shell Modification (dollars)[b]	Real Rate of Return on the Initial Investment (per cent per year)	
		Actual, After Shell Modification[b]	Hypothetical, also with Condensing Furnace[c]		Shell Modification[d]	Condensing Furnace[e]
Edison (EG)	113.4	0.68	0.43	1370	20	14
Freehold (NJNG)	117.7	0.68	0.43	2562	8	15
Wood Ridge (PSEG)	130.7	0.72	0.46	961	29	18
Oak Valley (SJG)	72.0	0.69	0.44	911	18	8
Whitman Square (SJG)	139.3	0.73	0.46	664	44	19
AVERAGES	113.8	0.70	0.44	1315	19	15

(a) These are the modules of houses involved in the Modular Retrofit Experiments, a project designed by the Building Energy Research Group at Princeton University and carried out by the four gas utilities in the state of New Jersey, to provide a commercial demonstration of the "house doctor technique." The initials in parentheses after each module name represent the gas utility responsible for that module. There were 6 houses in each module except the Whitman Square module, which contained 5 houses.

(b) See Gautam Dutt *et al.*, June 1982.

(c) Assuming an efficiency of 60 (95) percent for the existing (condensing) furnace.

(d) This is the real rate of return in fuel savings, assuming a levelized gas price of $8 per GJ and a lifetime of 15 years for the shell improvements.

(e) This is the real rate of return that would be realized in fuel savings if a worn out furnace is replaced by a 95% efficient condensing furnace for an installed cost of $1100, as an alternative to replacement with a conventional 69% efficient furnace for an installed cost of $2100.

Note 2.5.3.1-L. An experiment carried out by the Buildings Energy Research Group (BERG) at Princeton University in a townhouse in the neighboring community of Twin Rivers gives an indication of the technical potential for improving the building shell of existing buildings. Typical townhouses in this community are already fairly "tight" by U.S. standards—they already have celing and wall insulation and double glazing in the windows. The average fuel consumption for space heating, corrected for floor area and climate, is about half the U.S. average. The BERG retrofits included: adding additional attic insulation; providing movable insulated indoor shutters for use at night on south-facing windows; adding translucent insulating covers to north-facing windows for the heating season; adding perimeter insulation to the basement walls; making special efforts to eliminate thermal "bypasses." The net effect of pursuing these measures was to reduce the fuel consumption for space heating by 2/3 to 1/3 of the level for typical Twin Rivers townhouses (F.W. Sinden, April 1978). What is significant about this effort is that most of the measures pursued are "unconventional" and have not yet been taken into account in present day efforts to improve the thermal shell of existing houses.

Note 2.5.3.1-M. Becuase of large flue losses, conventional natural gas fired storage water heaters used widely in the U.S. are very inefficient for the average unit in use only about 40-45 percent of the fuel energy is converted to usable hot water. The development of condensing furnaces and boilers permits large improvements in water heating efficiency as well. Some condensing furnace manufactures offer optional hot water tanks which are heated by the furnace or boiler without the need for a separate flue. These new water heaters have efficiencies of 80 to 90 percent, twice as high as the efficiencies of existing units (R.H. Williams, G.S. Dutt, and H.S. Geller, 1983).

The possibility of achieving high efficiency in water heating with these combined furnace/water heater units extends the range of economic viability for condensing furnaces in new super-insulated houses down to much lower space heating demand levels than would be the case if space heating were considered in isolation from water heating.

Note 2.5.3.1-N. The heat pump water heater (HPWH) is a new technology which has radically improved the potential for energy savings with electric water heating. The most efficient heat pump water heater available in the U.S. in 1982 had a measured annual average COP (in an unheated Wisconsin basement) of 2.2—which means that it uses only about 1/3 as much electricity per unit of hot water produced as the average electric water heater in use, which is about 73% efficient.
While this HPWH is very expensive ($1550 compared to $350 for a new resistive unit) it would still be cost effective. Assuming an annual heat load of 12 GJ for water heating for a family of 4 [59 litres of hot (49 degrees) water per capita per day], the electricity savings that would result from using this unit (COP = 2.2) compared to a new resistive unit (efficiency = 0.8) is 2650 kWh per year. The annualized cost (assuming a 5% discount rate, a 13 year life, and an incremental first cost of $1200 is $130. Thus the cost of saved energy would be about 5 cents per kWh. For comparison the U.S. average electricity price was 6.5 cents per kWh in 1982.

Note 2.5.3.1-O. Recent technological developments have led to the introduction of more efficient refrigerators, refrigerator/freezers, and freezers, through a combination of design changes and the use of better components, including: replacement of fiberglass insulation with polyurethane foam, which has twice the insulating value per cm; use of thicker insulation; use of improved door seals; use of more efficient compressors and motors; evaporator coils for air isolation and reduced frost formation in 2-door refrigerator/freezers; removal of evaporator fan motors from the cooled space.

Note 2.5.3.1-P. The most efficient, large (510 liter), frost-free refrigerator available in the U.S. (see Table 2-4) costs $100 more than an otherwise identical unit which consumes 460 kWh per year more. The cost of saved energy associated with the purchase of the more efficient unit (for a 19 year lifetime and a 5% real discount rate) is less than 2 cents per kWh or about 1/4 of the U.S. residential electricity price in 1982.
Similarly, the most efficient freezer available in the U.S. is a 453 liter unit that consumes 765 kWh per year, compared to an otherwise identical unit that consumes 1080 kWh per year but costs $40 less. In this case the cost of saved energy is only 1 cent per kWh.

Note 2.5.3.1-Q. Performance characteristics of alternative lightbulbs[a]

Bulb Type	Status[b]	Efficacy	Output	Lifetime	First Cost
		(Lumens/Watt)	(Lumens)	(Hours)	(Dollars)
40 W Incandescent	A	11	470	1500	$ 0.50
60 W Incandescent	A	15	900	1000	$ 0.50
75 W Incandescent	A	16	1200	850	$ 0.50
13 W Matsushita Compact Fluorescent[c]	A	36	470	6000	$ 9.20
22 W GE Circlite w/Ordinary Ballast	A	40	870	12000	$16.00
55 W GE Halarc High Intensity Discharge, Resistive Ballast	A	41	2250	5000	$12.00
28 W Westinghouse Compact Fluorescent w/Ordinary Ballast	A	42	1170	7500	$22.00
30 W Litec Electrodeless Fluorescent	E	50	1500	10000	$15.00 (est.)
16 W Osram Compact Fluorescent w/Electronic Ballast[c]	A (1984)	56	900	?	?
High Intensity Discharge w/Magnetic or Solid State Ballast[d]	T	60	>1700	?	?
18 W Philips/Norelco Compact Flourescent w/Electronic Ballast	A	61	1100	7500	$25.00

(a) Except where indicated otherwise, the performance data presented here were complied in H.S. Geller, 1983.
(b) A = commercially available; E = experimental; T = development target.
(c) Company brochure.
(d) *Source:* R.R. Verderber and F.R. Rubinstein, June 1983.

Note 2.5.3.1-R. The most efficient bulb available in the U.S. in 1982 was the Philips Norelco compact fluorescent bulb (Note 2.5.3.1-Q). This 18 w bulb has lighting efficacy 4 times that of the 75 W incandescent it could replace and it lasts 9 times as long. Even though this bulb costs $25 (50 times more costly than the incandescent bulb it would replace), it saves so much electricity that the cost of saved energy for this bulb (assuming a 5% real discount rate) would be less than the averge U.S. residential electricity price in 1982 (6.5 cents per kWh) as long as the bulb is used more than 2 hours a day.

Section 2.5.3.2

Not 2.5.3.2-A. Site energy intensity by end-use and energy carrier in buildings (GJ per square meter per year)

End use	Average, US Res. Bldgs.[a]			Average, US Comm. Bldgs.[b]			Folksam Building, Sweden[c]			Enerplex South, NJ[d]		
	Fuel	Elect.	Total	Fuel	Elect.	Total	Fuel	Elect.	Total	Fuel	Elect.	Total
Space Heat	0.529	0.026	0.555	0.825	0.025	0.850	0.054	0.079	0.133	—	0.089	0.089
Air Cond.	—	0.029	0.029	0.012	0.175	0.187	—	—	—	—	0.077	0.077
Lighting	—	0.026	0.026	—	0.177	0.177	—	0.238	0.238	—	0.105	0.105
Hot Water	0.144	0.029	0.173	0.042	0.003	0.045	0.018	—	0.018	—	0.040	0.040
Other	0.034	0.118	0.152	0.034	0.066	0.100	—	0.076	0.076	—	—	—
TOTALS	0.707	0.228	0.935	0.913	0.446	1.359	0.072	0.393	0.465	—	0.311	0.311

(a) The totals of residential energy use are for 1980 (see Section 2.6.2). The total floor area in residences in 1980 was 11.3 billion square meters, the product of the average heated space for residence (139 square meters) and the number of dwellings (81.6 million). See U.S. Energy Information Administration, June 1982.

(b) For 1980. The estimate of energy use is from R. W. Barnes *et al.*, May 1980. The total commercial area in 1980 is estimated to be the 1979 value 4.13 billion square meters (U.S. Energy Information Administration, March 1983) plus the estimated 1980 increment of 0.10 billion square meters (R. W. Barnes *et al.*, May 1980).

(c) Measured values for 1979 (K. Welmer, 1981).

(d) Results of a computer simulation (Leslie K. Norford, June 1984).

Note 2.5.3.2-B. The hypothetical buildings shown in Figure 2.23 is assumed to be located in Kansas City, Missouri, which has a typical U.S. climate [2620 heating Degree-days and 790 cooling degree days; for comparison the population weighted U.S. average values are 2700 and 610 degree days respectively]. The indoor temperature of the building is maintained at 22 to 24 degrees, and the annual energy budget is 1.9 GJ per square meter of fossil fuel energy (for space heating by a 75% efficient boiler) plus 0.9 GJ per square meter of electricity [for cooling (by an air conditioner with a COP of 3.25), lighting, fans, and office equipment]. For comparison the corresponding numbers for the average commercial building in the U.S. in 1980 were about half as large [0.91 GJ and 0.45 GJ per square meter for fuel and electricity (Table 2.6)]. But there are many buildings with comparable or worse energy performance than the Kansas City prototype (Figure 2.24).

In contrast to the situation for residential buildings, where the heat required of the heating system is in general less than the conductive heat losses through the building shell plus the heat required warm fresh air (because of the "free heat" from people, appliances, and sunshine streaming in the windows) the requirements of the heating system for this commercial building are nearly 3 times what is required to make up for these heat losses; similarly the cooling load of the building is 7 times the heat flowing into the building in summer. A clue to this mystery is provided in Figure 2.25, which shows that in the middle of summer, when there should be zero need for heating, the average monthly heating requirement is ⅓ as large as in the middle of winter; and in the middle of winter, when there should be zero need for cooling, the average monthly cooling requirement is ½ as large as in the middle of summer. Clearly, substantial demand for both heating and cooling is being created within the building.

For cooling, part of the explanation for this anomaly is rather easy to understood. In large buildings, with their low ratio of building skin to volume, the large internal loads from lights, people, and equipment, can increase the temperature beyond acceptable levels, to the point where cooling is required even in winter.

Another part of the explanation is that, unlike the situation in single family residences, where the indoor air mixes well and evens out the temperature differences in different parts of the building, the different zones of commercial buildings are subject to different heat gains and losses and thus have quite different heating and cooling requirements.

The core zone (Figure 2.23), for example, is insulated from heat losses and gains through its walls by the fact that it is surrounded by perimeter zones filled with air held at the same temperature. Its only connection with the outdoors is ventilation air. Even when the outdoor air is at 0 degrees, the cooling effect (@ the prescribed ventilation rate of one air exchange per hour) is adequate to cover only ½ of the cooling load in the building core during working hours, thus requiring air conditioning even in the middle of winter.

The perimeter zones have more contact with the out-of-doors. During working hours, heating would be required on the north side of the building when the outdoor temperature falls below about 0 degrees, while on the south side the solar gains through the windows would be so large that cooling would still be required on an averge day in January.

Most of the heating is required at night, when free heat is minimal in all zones. That this heating requirement persists even in the summer arises from the nature of the space conditioning system of the building, which produces one air stream that is heated by the furnace to 30 to 50 degrees (depending on the outside temperature) and another which is maintained at 13 degrees by a chiller. Separate mixtures of these two streams are then produced for each zone, depending on its needs for heating and cooling. Consequently, whenever there is not enough free heat (e.g., at night) in a zone to raise the temperature of the chilled air to 22°, heated air is mixed in with the chilled stream even if it is warm outside!

Note 2.5.3.2-C. Minimum lighting standards are recommended in the U.S. by the Illuminating Engineering Society (IES). They range from about 30 lumens per square meter ($1/m^2$) in hospital corridors at night to 200 $1/m^2$ in office corridors, to about 1000 $1/m^2$ for general office work, to 10,000 $1/m^2$ for "extra fine bench and machine work." For comparison, the average level of illumination averaged over all commercial space (including such categories as warehouses and garages) is 400 to 500 lumens per square meter (Note 2.5.3.2-D).

The IES minimum standard have sometimes been criticized as being unnecessarily high. They are at a level where quite large changes in the illumination level result in rather small changes in performance. For a task such as reading, for example, a decrease in the level of lighting from 1000 to 500 $1/m^2$ would have the equivalent effect of increasing the size of the smallest visible detail in the type by 3 percent and would reduce the maximum feasible reading speed by about 10 percent. Recent evidence indicates, however, that in addition to visibility, the intensity and spectral properties of lighting have both physiological and psychological effects (Jane E. Brody, June 23, 1981). The whole IES approach to lighting standards may therefore have to be reconsidered.

Note 2.5.3.2-D. From Note 2.5.3.2-A, the average level of power consumption for lighting in commercial buildings in 1980 was 0.177 GJ per square meter per year or 5.6 Watts per square meter. Assuming that lamps in commercial buildings are in use ⅓ of the time on average, that the average efficacy of commercial lighting systems in 65 lumens/Watt, and that the average lighting fixture makes 40% of its light available at the working surfaces, the average level of commercial lighting at working surfaces in 1980 would have been 440 lumens per square meter.

Section 2.5.3.3

Note 2.5.3.3-A. OECD Consumption of Refined Petroleum Products for Transportation, 1979.

	Millions of Tonnes[a]		Percentage of Total
Road			
Motor Gasoline	468.3		61
Diesel + Other	115.8		14
Rail			
Diesel	20.7		2.6
Air			
Jet Fuel + Other	79.7[b]		10.0
Water	98.8		12.3
Diesel		25.3[b, c]	
Residual Fuel Oil		73.5[b, c]	
Total	783.3		100.0

(a)   Page 11 in International Energy Agency, 1981.

(b)   Includes fuel which left the OECD countries as "bunkers" in the amounts (in millions of tonnes): jet fuel (1.0), diesel fuel (9.5), and residual fuel oil (60.0).

(c)   U.S. domestic consumption obtained from Table 3.8 in G. Kulp *et al,* 1981.

Note 2.5.3.3-B. The OECD fuel consumption numbers (International Energy Agency, 1980) are organized in a matrix by transport mode (air, road, railway, internal navigatoin, and bunkers) and by fuel (crude petroleum, liquified gas, aviation gasoline, jet fuel, kerosenè, diesel/fuel oil, residual fuel oil, naptha and others). We have made the following approximations: all motor gasoline is used for passenger transport in passenger cars and light trucks; all aviation fuel and jet fuel is used for passenger transportation; and all other transport fuels are used for freight transport.

Note 2.5.3.3-C. Motor Vehicles in the OECD Countries in 1979 (millions)

	Passenger Cars[a]	Trucks and Buses[a]	Total Vehicles Per Capita
OECD Europe	101.7	11.3	0.30
France	18.5	2.5	0.39
W. Germany	22.6	1.5	0.39
Italy	17.1	1.2	0.32
Sweden	2.8	0.2	0.37
United Kingdom	14.9	1.9	0.30
All Others	25.6	4.0	0.27
Canada	10.0	2.9	0.54
U.S.A.[b]	104.5	32.4	0.62
Japan	22.7	13.6	0.31
Australia and New Zealand	6.8	1.7	0.49
Total OECD	245.8	61.9	0.40

(a)   All but U.S. data are from p. 35ff of Motor Vehicles Manufacturers Association of the United States, 1981.

(b)   The numbers from p. B-1 in Energy and Environmental Analysis, 1982, are 13 percent lower than those cited in Motor Vehicles Manufacturers Association of the United States, 1981, probably because the latter are not corrected for scrappage.

Note 2.5.3.3-D. The U.S. Environmental Protection Agency (EPA) combined driving cycle is a weighted averge of the EPA highway (45%) and urban (55%) driving cycles.

Note 2.5.3.3-E. The relevant physical parameters of the 1981 VW Rabbit were taken as: inertia mass (vehicle mass plus 136 kg to represent two passengers), 1080 kg.; frontal area, 1.88 m^2; coefficient of aerodynamic drag, 0.42; and coefficient of rolling resistance, 0.012. These values were then substituted in formulae for the tractive energy requirements of the EPA driving cycles given in G. Sovran and M.S. Bohn, 1981, to obtain combined cycle energy of 0.372 million joules/km. Some of this energy goes to overcome air resistance (42%) and rolling resistance (30%) during the portion of the driving cycle when power is required at the wheels. The remainder (29%) is stored as kinetic energy during these periods and subsequently dissipated primarily through braking as well as through air and rolling resistance during periods of unpowered deceleration.

Note 2.5.3.3-F. The following table (Frank von Hippel and Barbara Levi, 1983) gives the effects on fuel economy of various modifications of the Volkswagen Rabbit diesel, based on a computer simulation.

Vehicle	Inertia Weight[a] (kg)	Drag Coefficient	Rolling Resistance Coefficient (Fraction of gravity)	Engine (kw)	Transmission	Composite Fuel Economy[b] [l/100 km (mpg)]
World Average	—	—	—	—	—	13.0 (18)
1981 VW Rabbit (gasoline, sp. ig)	—	—	—	55	5-speed man	7.9 (30)[c]
1981 VW Rabbit (pre-ch. diesel)	—	—	—	39	5-speed man.	5.3 (45)[c]
Computer Estimates[d] (modifications of VW Rabbit Diesel)						
Base Case	1080	0.42	0.012	(")[e]	(")[f]	5.3 (44)
Reduce Aero. Drag	1080	0.30[g]	"	"	"	5.0 (47)
Reduce Rolling Res.	1080	"	0.0085[h]	"	"	4.8 (49)
Shift to Open Chamber Diesel	1080	"	"	39[i]	"	4.3 (55)
Shift to CVT	1080	"	"	"	CVT (5:1 range)[j]	3.7 (64)
Reduce Pk. Power	1080	"	"	29[k]	"	3.3 (71)
Reduce Weight	910[l]	"	"	25[k]	"	3.0 (79)
Expand CVT Range	910	"	"	"	CVT (10:1 range)[j]	2.8 (83)
Add Engine-off During Coast an Idle[m]	910	"	"	"	"	2.6 (89)

(a)   Curb weight plus 135 kg (300 1b).
(b)   U.S. Environmental Protection Agency (EPA) Composite Driving Cycle (55% urban, 45% highway).
(c)   U.S. EPA Test List for 1981 cars.
(d)   The following assumptions are common to all cases where numbers are estimated using a computer simulation: a projected frontal area of 1.88 square meters, an effective tire radius of 0.28 m, an axle radio of 3.89, consumption by the accessories of 0.37 kw of the engine's output power, and a 10% loss of the remaining engine output in the drive-line between the engine and the tires.
(e)   The pre-chamber diesel of the VW Rabbit has been presented by the thermal efficiency map for a 4 cylinder naturally aspirated diese engine shown in Fig. 36 of B. Wiedemann and H. Hofbauer (VW), 1978. It has been assumed that the zero power fuel demand is 0.13 milligrams per revolution per peak engine horsepower at 1000 rpm, rising linearly with rpm to twice that level at 5000 rpm.

(f) Using the gear ratios: 3.45, 1.94, 1.29, 0.97, and 0.76 (VW Rabbit, 1980) and the standard EPA shift schedule (18/34/51/67 km/hour or 11/21/32/42 mpg).

(g) A number of the prototypes shown at the 1981 Frankfurt auto show had aerodynamic drags in the range 0.24–0.30. See p. 1 in Richard Feast, September 28, 1981.

(h) Recent EPA tests found that a number of commercially available radial tires have rolling resistances of approximately 0.12 at 0.24 million pascal (35 psi) inflation pressure, 80% of rated load, and tested on a 1.7 m radius dynamometer drum (Gail Klemer, August 1981). Dividing by the standard correction factor, $[1 + r/R]^{1/2}$, (where r, the radius of the tire, is assumed to be 0.33 m (13 inches) and R is the radius of the drum), the flat surface rolling of these tires is calculated to be about 0.0085.

(i) The engine map has been scaled from the 52 kw (peak output) engine map shown in Figure 10 of U.G. Carstens *et al.,* 1981. The same idle flow discussed in (e) has been used.

(j) An international consortium made up of the Dutch company, Van Doorne Transmission BV, Borg-Warner, Fiat, and the Dutch government has produced prototype CVTs with a ratio range of 5 for automobiles, 5. 9 for light trucks, and "with the addition of a liquid cooled Borg-Warner torque converter on the input side," 10 for medium trucks (Jan Norbye, Sept. 4, 1981). In another analysis (P. Baudoin, 1979), it was found that with the substitution of a Van Doorne 4.7 ratio coverage Transmatic for a 4 speed manual transmission, the fuel economy of a Renault 14 was increased by 28% on the EPA urban cycle and by 21% at a constant 88 km/hour (55 mph.). There was an associated increase of the lowest (1000 rpm) in-gear speed from 6.9 to 9.5 km/hour. (In the calculations presented here, the corresponding speed has been fixed at 7.9 km/hour.)

(k) Since the CVT makes full power available at all road speeds above 40 km/hour this horsepower would allow the vehicle specified to accelerate from 0-80 km/hour (0-50 mph) and 64-96 km/hour (40-60 mph) in less than 13 and 11 seconds respectively [assuming that the accessories draw 0.37 kw (0.5 hp) throughout the acceleration period].

(l) The VW Research Vehicle 2000 has been described as being "between the compact and sub-compact class" (the Rabbit is in the sub-compact class) and having a weight of 786 kg (1730 lb), corresponding to an inertia weight of 920 kg (2030 lb) (Ulrich Seiffert, Peter Walzer, and Herman Oetting (VW), 1980).

(m) It is assumed that 0.37 kw (0.5 hp) is drawn from storage by the accessories when the engine is not delivering power to the wheels during the driving cycle. The storage is replenished at a constant rate during the engine-on period. It is assumed that the "round trip" efficiency of the energy storage system is 70%—i.e., that the engine has to produce 1/1.4 kwh of energy for every kwh drawn from storage.

Note 2.5.3.3-G. The "adiabatic diesel" now under development is an engine even more efficient than the open chamber diesel. Made from ceramic-lined combustion chambers, the adiabatic diesel can operate at a much higher temperature and hence higher efficiency than the ordinary diesel. Because the engine requires no bulky cooling system and auxiliary pumping requirements, the engine is inherently more reliable and may turn out to cost little more than (or perhaps even less than) conventional diesels. Both Opel AG and Isuzu Motors, Ltd., have developed prototypes of adiabatic diesels (*Ward's Engine Update,* October 24 and November 1, 1983), and it is expected that Toyota will go into commercial production with a ceramic engine in the 1990s (Philip Burgert, April 25, 1983).

Note 2.5.3.3-H. See P. 2-17 in Energy and Environmental Analysis, November 7, 1981, and Edward A. Barth and James M. Kranig, October 1979.

Note 2.5.3.3-I. One safety improvement which has been opposed with particular vehemence by the U.S. automotive industry is the installation of "passive restraints"—either air bags which would quickly inflate to cushion front seat passengers in a crash or "automatic" seat belts. The estimated extra installed price of air bags in mass production would be less than $200 per car and that of automatic seat belts would be $75-100. For an annual number of 10 million cars sold per year, the total cost would be $750 million to $2 billion per year. The associated benefits estimated by the National Highway Traffic Safety Administration would be about 10,000 fewer car-caused deaths per year and an even larger reduction in serious injuries. This corresponds to a cost of about $100,000 per life saved. According to automobile insurance industry, this savings in lives and injuries would translate into "several hundreds of dollars [in] reduced insurance premiums." The automobile industry contests the estimates in the case of automatic seat belts because it claims that most would be disconnected (, *Automotive News,* November 2, 1981;   , *Automotive News,* November 30, 1981; and National Highway Traffic Safety Administration, July 1980).

Note 2.5.3.3-J. The cost of saved energy for automotive fuel economy improvements will be evaluated here from the perspective of the car buyer, under U.S. conditions. The cost of saved energy (CSE) can be written as:

$$CSE = PVEC/PVFS,$$

where

PVEC	=	present value of the extra initial cost (over the average 6 year period of ownership by the first owner), and
PVFS	=	present value of the future fuel savings over this 6 year period.
PVEC	=	$[DP - (1-DP) \times CRF(r, L)/CRF(i, L) - RV \times (1+i)^{-6}] \times EC$,

and

PVFS	=	$D(t) \times (1+i)^{-t} \times [FIo - FI]/100$.

Here

DP	=	average down payment fraction in 1981 = 0.15 (U.S. Department of Commerce, December 1, 1981).
CRF (r, L)	=	$r/[1-(1+r)^{-L}]$ = capital recovery factor.
L	=	average auto loan duration in 1981 = 3.76 years (U.S. Department of Commerce, December 1, 1981).
r	=	real average interest rate for auto loans in 1981 = 0.07 [0.16 in current dollars (U.S. Department of Commerce, December 1, 1981)].
i	=	real discount rate = 0.10 (assumed).
RV	=	resale value fraction = 0.23 (Note 2.5.3.3.-K).
EC	=	increased first cost.
D(t)	=	distance (in km) travelled in year t = 20,000 − 640t (assumed).
FI	=	fuel intensity of the improved car (in 1/100 km).
FIo	=	fuel intensity of the base car.

Thus

PVEC	=	$0.818 \times EC$,

and

PVFS	=	$83,400 \times (FI - FIo)/100$, or
CSE	=	$0.00098 \times EC/(FI - FIo)$.

In the case where the Rabbit Diesel is substituted for the gasoline-powered Rabbit:

EC	=	$525 (in 1981 $),
FI	=	5.3 1/100 km,
FIo	=	3.9 1/100 km,

so that:

CSE	=	$ 0.20 per liter ($0.75 per gallon) of gasoline equivalent.

Note 2.5.3.3.K. From Energy and Environmental Analysis, December 23, 1980, the averge resale value (in constant dollars) of a 1975 model used car in 1981 was 23%, 17% and 13% for small, mid-size, and large cars, respectively.

Note 2.5.3.3-L. The following table (Frank von Hippel and Barbara Levi, 1983) gives the estimated increases in the purchase price of the automobile (in 1981 $ per automobile) as a consequence of adopting the fuel economy improvement measures considered in Note 2.5.3.3-F.

Technology Change	
Gasoline to Pre-chamber Diesel Engine	$525[a]
Tire Rolling Resistance Reduction	0[b]
Reduction in Coefficient of Aerodynamic Drag to 0.3	100[c]
Pre-chamber to Open Chamber Diesel	0[d]
Five Speed Manual to Continuously Variable Transmission (CVT): 5:1	400[e]
Weight Reduction (upper bound estimate)	400[f]
Extended Range of CVT: 10:1	100[c]
Engine-Off During Idle and Coast	200[c]

(a) Based on the differential between the U.S. list prices of the VW gasoline and diesel powered Rabbits (*Automotive News*, June 8, 1981, p. 2).

(b) Based on the absence of correlation between radial tire rolling resistance and price. See Table 2, in Gail Klemer, August 1981.

(c)   Rough estimate by von Hippel and Levi.

(d)   *Source:* TRW Energy Systems Planning Division, 1979.

(e)   Borg-Warner expects the initial cost of the new CVT to be comparable to present automatics and that the price will come down with production (,*Automotive News,* March 1980, p. 37). The VW Rabbit equipped in the U.S. with automatic transmission costs $400 more than the same model with a 5-speed manual transmission (_____, *Automotive News,* June 8, p. 20).

(f)   This estimate, which is likely to be high, is obtained from the estimate of $2.20/kg weight reduction in Richard H. Shackson and James H. Leach, 1980.

Note 2.5.3.3-M. OECD Highway Diesel Fuel Consumption in 1979

	Million Tonnes		Per Capita (kg)	
OECD Europe	50.9		146	
France		8.8		165
W. Germany		9.7		159
Italy		8.6		151
Sweden		1.4		174
United Kingdom		6.1		109
All Others		16.3		145
Canada	3.8		155	
U.S.A.	46.4		208	
Japan	7.6		65	
Australia and New Zealand	1.6		90	
Total OECD	110.2		151	

Source: International Energy Agency, 1981.

Note 2.5.3.3-N. In 1978 an estimated 99 percent of all U.S. highway diesel fuel consumption was by heavy trucks which consumed 150 barrels of diesel fuel per vehicle on average—more than 10 times as much as the average diesel-powered light vehicle (p. B-1 in Energy and Environmental Analysis, Inc., January 30, 1982). In West Germany, 26 percent of all diesel vehicles registered in 1978 fell into the "heavy" weight range. In France, 28 percent of all diesel commercial vehicles (i.e. not including passenger cars) made in 1978 were in this weight range. In Japan, however, only 11 percent of all diesel commercial vehicles made in 1978 were "heavy" (Motor Vehicles Manufacturer's Association of the U.S. 1981).

Note 2.5.3.3-O. In 1978 U.S. intercity trucks carried 880 billion tonne-km of freight (p. 80 in U.S. Department of Transportation, 1980). The distances involved were presumably great circle kilometers, which correspond approximately to 1.15 times as many route-km (p. 6-4 in A.B. Rose, 1979). In 1978 "combination cargo vehicles" consumed an estimated 47.3 billion liters of fuel in the U.S. (p. 47, in Federal Highway Administration, 1978). If this fuel consumption were associated with the freight movements cited above, the resulting fuel intensity would be 4.67 1/100 t-km. This estimate is in agreement with one obtained by quite a different method by Rose (*op cit,* p. 6-11).

Note 2.5.3.3-P. Diesel engine fuel consumption can be reduced by an estimated 9 percent by lining the combustion chamber with heat resistant ceramic insulation. So doing could also make water cooling of the engine unnecessary. Elimination of the "parasitic" loads associated with a cooling system (radiator fan and water pump) would reduce fuel consumption by another 6 percent. Recovery and use of some of the energy in the exhaust gases with a gas turbine (turbo-compounding) would reduce fuel consumption by another 10 percent, for a total estimated fuel consumption reduction in an "adiabatic turbo-compound" diesel engine of 23.5 percent. Such an engine would have a peak fuel-to-mechanical energy conversion efficiency of 47 percent (R. Komo and W. Bryzik, 1978). Further fuel consumption reductions of 15 percent are projected as possible with the addition of a heat driven freon vapor turbine engine driven by the "waste heat" in the exhaust—raising the peak fuel-to-mechanical energy conversion efficiency to 55 percent (Thermoelectron, November 1980).

Note 2.5.3.3-Q. For the base case, it is assumed that only about 25 percent of tires on heavy trucks are radials (p. II-83, in Transportation Task Force, Solar Energy Research Institute Solar/Conservation Study, 1980). This would correspond to an average equilibrium rolling resistance of about 0.0075 (W.A. Leasure, 1979). A reduction in the average value to 0.004 or below might be possible as a result of the introduction of "advanced technology" radial tires, and the replacement of some pairs of tires by "wide singles", thus halving the losses in the tire sidewalls [see (R.E. Knight, 1979) and (W.L. Giles, 1979). Given an average truck empty weight of 13 tonnes and an average load (including 30 percent empty kilometers) of 10 tonnes (p. 6-9 in A.B. Rose, 1979), the average savings in rolling energy losses which would result from a reduction in the rolling resistance by 0.0035 would be 7.9 million joules/100 t-km. A savings of at least 0.052 liters of diesel fuel consumed per million joules of mechanical energy not dissipated in the tires is assumed (using the 55 percent peak fuel to mechanical energy converstion efficiency discussed in Note 2.5.3.3-P and assuming 10 percent engine-to-wheel transmission losses).

Note 2.5.3.3-R. An average load of 10 tonnes (Note 2.5.3.3-Q), a frontal area of 9 m2 (A.B. Rose, 1979), and a reduction in aerodynamic drag coefficient from 0.75 to 0.45 [recently Renault announced that "We expect to get down to drag coefficient of 0.45 by 1986" (Jan P. Norbye, November 30, 1981)] are assumed. See also Hans Drewitz, 1979. At a speed of 100 kmph (62 mph), the corresponding reduction in energy dissipated would be 12.5 million joules/100 t-km per km. Then, as in Note 2.5.3.3-Q, it is assumed that 0.52 liters of diesel fuel are saved per million joules of mechanical energy not dissipated.

Note 2.5.3.3-S. In 1983 a 14,600 km test run was made with a White truck powered with a 250 HP, direct injection, turbo-charged, after-cooled Caterpillar 3306B engine (_____, *Automotive News,* November 7, 1983). The truck was designed with low aerodynamic drag and used low profile Michelin radial tires. The truck ran at an average speed of 76 km/hour (47.1 mpg). To simulate typical conditions 2/3 of the run was on interstate highways, and 1/3 on two-lane roads. Fully loaded with a gross weight (for truck and cargo) of 33.7 tonnes, the vehicle fuel intensity averaged 32.4 1/100 km &7.25 mpg); without a load the gross weight was 14.1 tonnes and the vehicle fuel intensity averaged 24.8 1/100 km (9.5 mpg). Assuming that in typical operations the truck is empty 30% of the time (A.B. Rose, 1979), the freight fuel intensity averaged.

$$[0.7 \times 32.4 + 0.3 \times 24.8]/[0.7 \times (33.7 - 14.1)] = 2.2 \ 1/100 \text{ t-km.}$$

The load in these tests (19.6 tonnes when full) was higher than the average load for U.S. trucks (some 10/0.7 = 14.3 tonnes—see Note 2.5.3.3-Q). When the performance is adjusted (by interpolation) to this lower load, the result is a vehicle fuel intensity of 30.3 1/100 km when loaded, so that the average freight fuel intensity becomes instead.

$$[0.7 \times 30.3 + 0.3 \times 24.8]/[0.7 \times 14.3] = 2.9 \ 1/100 \text{ t-km.}$$

Note 2.5.3.3-T. Air Passenger Travel in the OECD in 1978

	Billion p-km	% International	Per Capita (km)
OECD Europe	171.7	88	493
France	30.2	78	564
W. Germany	17.6	87	290
Italy	13.3	83	234
Sweden	4.7	78	566
United Kingdom	45.1	95	807
All Others	60.8	89	541
North America			
Canada	29.3	47	1196
U.S.	363.9	18	1630
Japan	43.8	43	376
Australia and New Zealand	26.3	64	1486
Total OECD	635	42	870
Total World	796	47	180
World Minus U.S.	432	72	103

Source: Table 151 in United Nations, 1981.

Note 2.5.3.3-V. Energy Consumed by the OECD for Shipping in 1979[a]

| | Millions of Tonnes of Oil | | |
	International	Internal and Coastal	Per Capita (kg)
OECD Europe	33.7	7.84	119
France	4.9	0.59	103
W. Germany	2.9	0.56	57
Italy	5.1	0.35	96
Sweden	0.9	0.11	121
United Kingdom	2.7	1.26	71
All Others	17.2	5.97	198
Canada	n.a.	2.5	n.a.
U.S.	25.3	8.5[c]	
Japan	9.0	10.2	151
Australia and New Zealand	1.7	n.a.	n.a.
Total OECD	69.7	29	135
Total World	123.6[b]		

(a) Except where indicated otherwise all data are from IEA, 1981.
(b) Tables 38 and 39 in United Nations, 1980.
(c) Table 3.8 in G. Kulp *et al.,* 1981.

Note 2.5.3.3-V. The U.S. paid $4.6 billion in import freight payments to foreign carriers for shipments totalling 532 million tonnes in 1977. See Tables 1169 and 1177 in U.S. Department of commerce, 1980.

Note 2.5.3.3-W. International Shipping in 1977 (millions of tonnes)

	Imports (% Petroleum)		Exports (% Petroleum)	
OECD Europe	1333.(58)		470.2 (29)	
France		232(62)		58.72(22)
W. Germany		104(48)		32.4(6)
Italy		218(70)		36.3(44)
Sweden		55(58)		32.4(9)
United Kingdom		158(51)		77.6(38)
All Others		566(56)		232.8(32)
Canada	58.9(31)		119.8(3)	
U.S.	568.1(76)		250.2(0)	
Japan	582 (45)		78.6(0)	
Australia and New Zealand	35.8(44)		175.5(2)	
Total OECD	2578 (58)		1094(13)	
Total World	3442(55)			

Source: Table 149 in United Nations, 1981.

Note 2.5.3.3-X. In the U.S. 47 percent of the tonnage moved by domestic shipping in 1978 was petroleum and petroleum products (p. 121 in U.S. Department of Transportation, 1980).

Note 2.5.3.3-Y. On December 31, 1976, the average deadweights (displacement when fully loaded) of the world's freighters, bulk carriers and tankers were 8100, 35,700 and 62,100 tons respectively (p. 2 in U.S. Department of Commerce, 1977).

Note 2.5.3.4-A. Primary Energy Use by Industry in OECD Countries in 1979[a]

[Million Tonnes of Oil Equivalent (MTOE)]

	Prim. Met.	Chem.[b]	Paper	Food Proc.	Energy Str.[c]	Other	Total	Industry Share of Total Primary Energy Use(%)	Industrial Energy Use Per Capita (TOE per Year)
OECD Europe									
France	18.03	29.36	2.87	4.94	12.35	26.56	94.11	47.5	1.76
W. Germany	30.69	43.14	5.42	5.66	21.25	26.53	132.69	46.2	2.16
Italy	13.86	22.68	2.68	3.35	10.79	22.90	76.26	52.9	1.34
Sweden	3.90	3.97	8.15	0.86	0.82	5.92	23.62	45.8	2.85
UK	18.54	24.03	4.37	6.18	13.74	29.19	96.05*	43.5	1.72
All Others	40.80	55.10	12.25	9.87	14.15	51.39	184.02		1.18
SUBTOTALS	125.8	178.3	35.7	30.9	73.1	162.5	606.8	47.0	1.55
OECD North America									
Canada	16.02	18.73	13.07	2.73	20.28	27.15	97.98	45.1	4.13
USA	121.44	189.38	36.60	17.45	121.98	270.79	757.64	40.3	3.37
SUBTOTALS	137.5	208.1	49.7	20.2	142.3	297.9	855.6	40.8	3.44
OECD Pacific									
Japan	63.19	79.54	11.83	4.79	33.32	68.53	261.20	69.3	2.25
Australia	13.53	3.84	1.63	1.60	7.54	10.52	38.66	50.3	2.68
New Zealand	0.94	0.27	0.49	0.61	0.22	2.01	4.54	40.7	1.46
SUBTOTALS	77.7	83.7	14.0	7.0	41.0	81.1	304.4	65.4	2.28
TOTAL OECD	340.9	470.0	99.4	58.0	256.3	541.5	1766.8	45.0	2.29

(a) *Source:* International Energy Agency, 1983. Here losses in the generation, transmission and distribution of electricity are allocated to industries in proportion to their total electricity consumption.

(b) Includes feedstocks.

(c) Consumption by the energy sector refers to consumption and losses in the energy conversion industries —refineries, coal mines, natural gas production loss and use of natural gas for pipeline compressors, conversion losses in coking and the production of blast furnace and coke oven gas.

Note 2.5.3.4-B. An example of a particularly effective newspaper serving the energy management information needs of both industrial and large commercial energy users is *Energy User News*, a weekly, New York-based paper put out by Fairchild Publications, which reports improved energy management case studies, lists and periodically reviews energy management technologies and services that are offered commercially, and reports technological, economic, and institutional developments of general interest to major energy users.

Note 2.5.3.4-C. During the period 1973-1984, the ratio of industrial energy use to GNP in the U.S. declined 3.6% per year, of which 1.6% per year can be attributed to a shift in the mix of industrial output and 2.0% per year can be attributed to a reduction in energy intensity (M. Ross, Eric D. Larson, and Robert H. Williams, July 1985).

Note 2.5.3.4-D. Global Distribution of Production of Selected Basic Materials (million tonnes per year)

	Raw Steel (1978)[a]	Paper and Paper-board (1978)[b]	Cement (1978)[c]
OECD Europe			
France	22.8	5.0	28.1
West Germany	41.3	6.9	34.0
Italy	24.3	4.6	37.8
Sweden	4.4	5.7	2.4
United Kingdom	20.3	4.2	15.9
All Others	46.7	14.6	84.4
SUBTOTALS	159.8	41.0	203.0
OECD North America			
Canada	14.9	13.3	10.5
United States	124.3	53.4	77.6
SUBTOTALS	139.2	66.7	88.1
OECD PACIFIC			
Japan	102.1	16.5	85.0
Australia	7.5	1.3	5.0
New Zealand	—	0.6	0.8
SUBTOTALS	109.6	18.4	90.8
TOTAL OECD	408.6	126.1	381.9
Centrally Planned Europe			
Soviet Union	151.5	9.2	127.0
Eastern Europe	59.0	5.5	67.0
SUBTOTALS	210.5	14.7	194.7
CENTRALLY PLANNED ASIA			
China	31.8	7.7	65.3
Other	3.2	0.1	7.9
SUBTOTALS	35.0	7.8	73.2
OTHER			
South Africa	7.9	1.0	6.8
Brazil	12.1	2.5	22.2
India	10.0	1.0	19.6
Mexico	6.7	1.5	14.0
Other	13.4	5.4	139.4
SUBTOTALS	50.1	11.4	202.0
WORLD TOTALS	704.2	160.7	851.8

(a) *Source:* United Nations, 1981.
(b) *Source:* FAO, 1980.
(c) *Source:* U.S. Bureau of Mines, 1980.

Note 2.5.3.4-E. In the Elred process, the iron ore concentrate is first partially reduced with coal powder in a fluidized bed, and the final reduction is performed in a direct current-electric arc furnace, the off-gases of which are used for electricity production in a gas turbine/steam turbine combined cycle power plant.

Note 2.5.3.4-F. The Plasmasmelt process is similar to the Elred process, except that the final reduction is performed instead in a very high temperature plasma, where very rapid heat transfer takes place.

The plasmasmelt process, while electricity intensive, has very low fuel requirements and is only ⅓ as energy intensive, on a *final* energy basis, as the average for the U.S. Where electricity production is based on thermal power plants, it may be more appropriate to compare energy intensities on a *primary* energy basis, where about 3 units of fuel input would be required to produce each unit of electricity. On this basis plasmasmelt would require only 55% as much energy per tonne of steel as the U.S. average in 1980.

Note 2.5.3.4-G. Most existing facilities that produce solid direct-reduced iron (the world capacity in 1980 was about 20 million tonnes per year) are based on the use of natural gas as reductant, for which the energy use associated with the iron-making operation is only about ½ of that required by a conventional blast furnace (Energy and Environmental Analysis, February 1983a). Coal-based direct reduction systems are also being introduced.

Note 2.5.3.4-H. Direct casting represents an advance beyond the process of continuous casting, which bypasses the ingot casting and soaking pit stages of steel-making and which has already been adopted in modern steel mills. (In the U.S., about 20% of the steel was produced in 1980 with continuous casting; in Sweden 80%).

In the future it may be feasible to eliminate the secondary rolling mills as well, with direct casting to the final steel forms. While the surface quality of steel produced this way has never been satisfactory in the past, it should be feasible to achieve the desired quality with good sensors, predictive and feedback models, and the variety of microprocessor controls that are now becoming available (Lawrence C. Long, December 1981).

Note 2.5.3.4-I. Direct steel-making (Figure 2-30a) is a complementary way to achieve process integration. In direct steel-making, the integration which would be accomplished with the advanced Swedish technologies would be extended beyond iron-making to include steel-making as well. While there are major technical hurdles to be overcome (e.g., involving the durability of the refractory linings in the high-temperature oxidizing environment), the resolution of these problems would lead to major capital, labor, and energy cost savings (G.J. Hane *et al.,* September 1983).

Note 2.5.3.4-J. Dry steel-making (Figure 2-30b) is a direct reduction process that eliminate the need for the energy intensive coke ovens, blast furnaces, and steel-making furnaces associated with conventional processing and ends up with the final product in powder form—i.e., there would be no melting.

Note 2.5.3.4-K. In the U.S., where expenditures on energy account for such a large fraction of total production cost in the U.S. paper industry (Table 2.11), the industry has in recent years made major strides to improve its energy productivity. Between 1972 and 1979 (Energy and Environmental Analysis, February 1983b):

- Value-added increased 20% in the paper and paper-board industry. (The value-added measure referred to here is Gross Product Originating, as tabulated by the Bureau of Economic Analysis of the U.S. Department of Commerce.)
- The physical production of paper and paper-board (in tonnes) increased 12%.
- Total energy use increased 3%.
- Purchased energy declined 7%, as the fraction of energy requirements derived from wood wastes increased from 40 to 46%.

Thus in the period 1972 to 1979, the energy required to produce a dollar of output (tonne of output) in the U.S. paper industry decreased at an average rate of 2.1% per year (1.2% per year).

But much more can be done to improve energy productivity further. In 1979 some 39 GJ (of which 6% was electricity) was required to produce a tonne of paper in the U.S. (Energy and Environmental Analysis, February 1983b), far more than what was required for typical processes in Sweden in the early 1970s (Table 2.12).

A major study of energy productivity trends in the U.S. paper industry envisages that specific energy requirements in the U.S. industry could be reduced to 29 GJ per tonne by the turn of the century (Energy and Environmental Analysis, February 1983b), which is still far even from what has already been achieved in Sweden.

While feedstock supply constraints provide a powerful incentive for the Swedish paper to pursue energy-saving innovations (Section 2.4.1), no such constraints motivate the paper industry in the U.S., where the forest resource is considerably under-utilized, little attention is given in the R&D process to revolutionary process change, and the industry is generally slow in changing its production practices and processes.

This situation may change somewhat if biomass energy sources come to play an important role in the U.S., competing with the paper industry for the wood feedstock (Appendix A: Renewable Energy Resources), but without more emphasis on fundamental process improvement in R&D and institutional change with regard to technologies such as cogeneration, technological change in the U.S. paper industry may be slow.

Note 2.5.3.4.-L. The following are now available technological options that would result in major energy savings but which were not taken into account in the 1977 Swedish design study:

- *Counter-pressure drying:* Here drying is carried out with steam instead of air as the drying medium. 1.3 tonnes of steam @ 12 atmospheres will dry 1 tonne of pulp from 55% to 10% moisutre by condensing on the outside of heat exchange tubes carrying the wet pulp, along with some transport steam. The tubes yield 1.1 tonnes of steam @ 3 atmospheres and 1.3 tonnes of condensate @ 188 degrees C. If the low quality by-product steam can be utilized [e.g., for bleaching or (with chemical pulping) for evaporating water from the spent chemical slurry used to extract the pulp, so that the chemicals can be recovered], the result is that net energy requirements are reduced more than 5-fold, from 2.6 GJ per tonne of pulp in an ordinary oil-fired pulp dryer to 0.47 GJ per tonne with this method. Although the capital cost is 30% higher with the counter-pressure drying technique, the total cost is about half of what it is with conventional drying. The first counter-pressure drying installation went into operation in 1978. The technique has been developed into a reliable method of drying.
- *Pressurized ground-wood pulp:* Here pulp is produced by grinding the wood at high pressure, producing a pulp of higher strength than ordinary mechanical pulp. With this process electricity requirements are reduced to 1100 kWh per tonne of pulp, compared to about 2000 kWh per tonne for thermo-mechanical pulping.(*) The steam produced from the heat of grinding can be used for drying pulp or for electricity generation (Arne Lindahl and Pekka Hailala, July 1978; Arne Lindahl and Harald Wikstrom, December 1979; Hannu Paulapuro, May 1981; Mattio Aario and Hannu Salakari, 1983).

- *Oxygen bleaching:* Bleaching is often required to remove residual lignin in the pulp. Today bleaching in the U.S. is predominantly based on the use of chlorine processes, the chemicals for which require some 350-480 kWh to produce, per tonne of pulp; by switching to oxygen bleaching these requirements are reduced to 230 kWh per tonne. Most bleaching today in Sweden is based on oxygen, supplemented by the use of chlorine bleaching. A switch to oxygen bleaching also would have major environmental benefits; oxygen bleaching reduces significantly the effluent load on pollution control facilities in general and allows meeting biological oxygen demand (BOD) and effluent color requirements in the U.S. even in locations where it is otherwise difficult to meet regulatory requirements (Energy and Environmental Analysis, February 1983b).

Note 2.5.3.4-M. Advanced paper process still under development that are promising include bio-degradation of lignin (i.e., use of microorganisms to extract cellulose from wood), recovery of aromatic chemicals from the pulping waste streams, and the dry forming of paper (G.J. Hane *et al.,* September 1983). Dry forming, which involves eliminating the energy-intensive drying processes now needed to evaporate the water from the wet-formed sheets of paper (Figure 2.31), would be an especially important advance.

Note 2.5.3.4-N. The growth rate for chemicals given here is a weighted average of the physical production levels (in tonnes per year) of 20 selected chemicals that collectively account for about 50% of energy use by the chemical industry in the U.S. The different chemicals are weighted in the aggregate index according to the energy intensity of their production (in GJ per tonne). See Marc Ross, 1982.

---

* In the U.S., 70% of all pulp is produced by means of the Kraft sulphate process. The classical mechanical pulping process was the atmospheric pressure ground-wood pulp process, where the fibers are torn apart by abrading the roundwood against a water-cooled grindstone. While mechanical pulping yields more fiber (90-95%, compared to about 50% for chemical pulp—largely because the pulp still contains lignin), the pulp produced with the atmospheric ground-wood process has low strength, short fiber length, and low capacity. The recently commercialized thermo-chemical pulp process, which involves first softening chips by pre-heating, produces a stronger pulp with longer fibers, making this an attractive alternative to chemical pulps in many instances.

Note 2.5.3.4-0. The following are annual energy consumption levels (in EJ per year) for the U.S. chemical industry (SIC 28), as provided to Robert Williams by B.R. Brown, of the Dupont Chemical Company, December 1983

	1972		1979	
FEEDSTOCKS				
Natural Gas	0.71		0.64	
NGL	0.85		0.88	
Oil	0.58		1.17	
SUBTOTALS	2.14		2.69	
FUEL CONSUMPTION				
Natural Gas	1.71		1.45	
Other Fuel Gas	0.31		0.46	
Oil	0.23		0.27	
Coal	0.35		0.32	
Electricity[a]	0.28	(0.81)	0.36	(1.05)
Steam	0.12		0.12	
Other	0.07		0.08	
SUBTOTALS[a]	3.06	(3.59)	3.05	(3.75)
TOTALS[a]	5.20	(5.73)	5.74	(6.44)

(a) Here the numbers in parentheses are the fuel requirements for power generation, assuming production in thermal power plants @ 10.55 MJ/kWh.

Note 2.5.3.4-P. In 1979 the energy value of feedstocks consumed by the chemical industry amounted to 2.7 EJ (Note 2.5.3.4-O). For comparison the harvests in U.S. commercial forests in the late 1970's amounted to about 5.6 EJ per year, of which 1.8 EJ was recovered as products of the forest products industry, 1.6 EJ was recovered as fuel, and 2.3 EJ (mainly logging residues and forest stand improvement cuttings) was unused and available for other purposes. Moreover, present harvests are much less than annual production on commercial forestlands—some 9.5 to 19 EJ per year (Office of Technology Assessment, 1980).

Note 2.5.3.4-Q. Three strategies for improving the energy efficiency of cement making involve using suspension pre-heaters, shifting to cements that are less energy intensive than portland cement, and pursuing the possibilioty of "cold processing."

The suspension pre-heater was invented in Czechoslovakia in 1933 and commercialized in 1950. With the suspension pre-heater, the dry kiln feed is heated and partially calcined by being entrained in the hot flue gases leaving the kiln. A recent modification of the suspension pre-heating system is the German and Japanese designed flash calcining system, which involves two stages of firing, with a flash calcining vessel installed between the rotary kiln and the suspension pre-heater. The intimate contact of the hot gases with the suspended particles from the pre-heater results in rapid heat exchange and a further improvement in fuel efficiency (M.H. Chiogioji, 1979). As shown in Table 2-14, such commercially available innovations would give rsie to roughly a 40% savings relative to current U.S. practice.

Another way to significantly reduce energy requirements is to use cement that are less energy intensive than portland cement. Alternative cements are widely used in South America and Europe and have properties that make them competitive with portland cement. Such cements include slag cement (using blast furnace slag from the iron and steel industry) or pozzolanic cement (e.g., flyash) as additives to portland cement. When these materials enter the cement manufacturing process at the final grinding step, the large energy consuming processing steps of clinkering in the rotary kiln and raw material grinding are avoided (M.H. Chiogioji, 1979). Although less than 1% of the cement produced in the U.S. involves such blends, the percentage is much higher in other industrialized countries. For example, in France, blended cements account for about 60% of total production (J.T. Dikeou, 1980).

Beyond these commercially available alternative technologies there are still further opportunities for energy efficiency improvement; indeed, the theoretical minimum fuel requirement is only 0.93 GJ/tonne (E.P. Gyftopoulos, L.J. Lararidis, and T.F. Widmer, 1974).

One interesting possibility, which represents an entirely new direction in cement-making technology, is cold processing, in which no high temperature processing is involved. Invented by C.J. Schifferele and J.J. Coney, this process involves chemically combining quicklime and a clay component by grinding them in a conventional ball mill. Cold processing has been demonstrated in a pilot plant and is reported to have produced an hydraulic cement with properties which compare favorably with portland cement (M.H. Chiogioji, 1979).

Note 2.5.3.4-R. Electricity Use by U.S. Industry, 1972

	10 kWh	%
Industrial Motor Drive (except HVAC)		
Pumps	143	24
Compressors	83	14
Blowers and fans	73	12
Machine tools	40	7
DC drives	47	8
Other	72	12
Subtotal	458	76
Electrolytic (mostly aluminium, chlorine)	91	15
Direct Heat (Steel Industry)	14	2
Other	37	6
Subtotal	142	24
Total	600	

Source: A.D. Little, Inc., August 1976.

Note 2.5.3.4-S. How much savings S (a fraction) would have to be achieved to realize a real rate of return "i" on an investment in VSD technology, for an industrial motor of typical size and operated at a typical capacity factor CF?

The savings fraction S is the solution of the following equation:

$$(0.75 \text{ kW per HP}) \times CF \times 8760 \times S \times P/(C \times E) = i/[1 - (1+i)^{-N}],$$
where

P = the electricity price in \$ per kWh)
C = the initial capital cost (in \$ per horsepower)
E = the motor efficiency
N = the expected life of the equipment

In the size range 50 to 125 HP, the costs of VSD systems in the US range from \$100 to \$440 per HP, and average about \$210 per HP, according to a survey of a 15 manufacturers (*Energy User News,* February 7, 1983). Thus we assume here that C = \$210 per HP, and that P = \$0.05 per kWh (the average US industrial price in 1982), that CF = 0.40 [the average for motors in the size range 50 – 125 in 1977 (A.D. Little, February 1980)], that E = 0.91 (the average for motors in this size range), and that N = 20 years. Thus for i = 0.10, S = 0.17—i.e. in order to realize a 10% real rate of return, the VSD device much provide a 17% savings.

Note 2.5.3.4-T. The large variation in the cogeneration potential with the technology can be understood by examining a critical parameter characterizing cogeneration systems, viz., the electricity to heat output ratio (E/H). This is an important parameter because the industrial heat or steam load is the given "resource base" for cogeneration, so that the electricity production potential for a given cogeneration technology is proportional its E/H.

The most familiar cogeneration technology involves the use of a stream turbine. With this technology high pressure steam raised in a boiler is used to drive a steam turbine for the generation of electricity, and low pressure steam is exhausted from the turbine at the temperature and pressure required for heating applications (Figure 2.5.3.4-T1). In a steam turbine cogeneration system, typically about 85% of the fuel energy is converted to steam and electricity, but the fraction of the fuel converted to electricity is small (5 to 20 percent for most industrial process steam applications), giving rise to a very low value for E/H—typically less than 0.2 (Figure 2.5.3.4-T2).

An alternative cogeneration technology involves use of a gas turbine/waste heat boiler combination (see Figure 2.5.3.4-T1). With this system hot combustion product gases are used to drive a gas turbine; the gases exhausted from the gas turbine are then used to raise steam in a waste heat recovery boiler. In a gas turbine cogeneration system, less fuel is converted to heat and electricity (70 to 80 percent), but a much higher fraction of the fuel energy (25 to 30%) is converted to electricity, thus giving rise to a much higher E/H (typically 0.5 to 0.7) than is possible with steam turbine systems (Figure 2.5.3.4-T3).

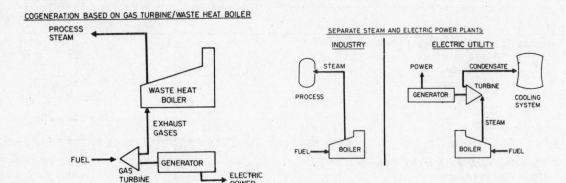

COGENERATION BASED ON GAS TURBINE/WASTE HEAT BOILER

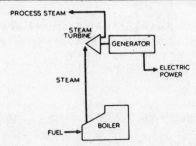

COGENERATION PLANT BASED ON BACK-PRESSURE STEAM TURBINE

Figure 2.5.3.4-T1. The top figure indicates the conventional arrangement, where electricity is produced at a centralized thermal power plant, with the waste heat from power generation dissipated in a cooling tower, and steam for industrial process use is raised separately in industrial boilers. In the figure below this, these operations are combined at an industrial site, where the steam is discharged from the steam turbine at a temperature and pressure appropriate for process use. The bottom figure shows an alternative cogeneration arrangement, involving a gas turbine and waste heat boiler. The hot exhaust gases from the gas turbine are used to raise steam in a heat recovery boiler for industrial process use.

In the gas turbine/steam turbine combined cycle cogeneration system (Figure 2.5.3.4-T4), the steam raised in the waste heat recovery boiler from the heat in the gas turbine exhaust is used to drive a steam turbine, the steam exhaust of which is then delivered to the heating application. The fraction of the fuel converted to electricity is quite high (30 to 38%) for combined cycle systems, with an E/H ratio of 0.9 to 1.3 (Figure 2.5.3.4-T5).

Diesel cogeneration sysetms are also characterized by a high electrical conversion efficiency (30 to 40%), but less high grade heat can be recovered in a waste heat boiler from the diesel engine exhaust gases (typically 20% of the heating value of the fuel). However, additional low grade heat (in the form of hot water or very low quality steam) is available from cooling the cylinder walls of the engine (Figure 2.5.3.4-T6).

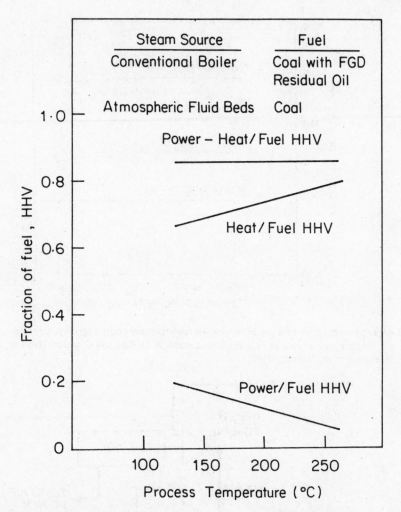

Figure 2.5.3.4-T2. Characteristics of a steam turbine cogeneration system. For this system the steam is delivered to the turbine at 100 bar and 540 degrees C. Size: 7.5 to 100 Mw.

Source: H.E. Gerlaugh *et al.,* January 1980.

Thus if cogeneration efforts are directed to the high E/H technologies, the electric power production potential of cogeneration would be 4 to 8 times what it would be with the more familiar steam turbine technology. In general, the high E/H technologies are also more thermodynamically efficient and their use leads to greater overall fuel savings than is possible with low E/H steam turbine-based cogeneration (R.H. Williams, 1978).

Note 2.5.3.4-U. High E/H cogeneration technologies based on gas turbines and reciprocating (diesel or spark-ignited) internal combustion engines usually require the use of high quality liquid or gaseous fuels because the use of low quality fuels such as coal or biomass would tend to degrade the equipment. However, it is possible to either modify the equipment to accomodate a low quality fuel or to modify the fuel to accomodate the equipment. Three possibilities are: indirect firing of gas turbines with low quality fuels; direct firing of gas turbines with wood; and gasification of low quality fuels, for use either in gas turbines or reciprocating internal combustion engines.

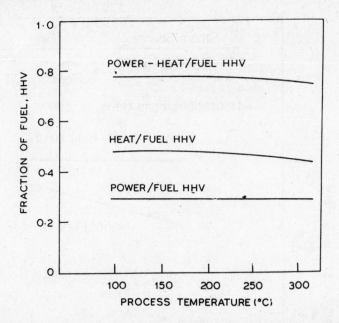

Figure 2.5.3.4-T3. Characteristics of a gas turbine/waste heat recovery boiler cogeneration system. For a gas turbine with a pressure ratio of 10, a firing temperature of 1100 degrees C, and distillate fuel. Size: 13 to 72 Mw.
Source: H.E. Gerlaugh *et al.,* January 1980.

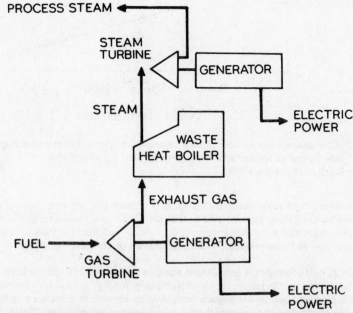

Figure 2.5.3.4-T4. A gas turbine/steam turbine combined cycle cogeneration system. With this system the hot combustion product gases are first used to produce electricity in a gas turbine, the hot exhaust gases from the gas turbine are used to raise steam in a heat recovery boiler, this steam is used to produce additional electricity in a steam turbine, and after passing through the turbine this steam is exhausted to the industrial process application at the appropriate temperature and pressure.

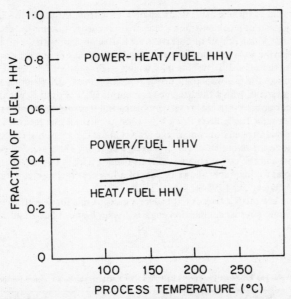

Figure 2.5.3.4-T5. Characteristics of a gas turbine/steam turbine combined cycle cogeneration system. For the gas turbine shown here, the pressure ratio is 12, the firing temperature is 1200 degrees C, and the fuel is residual fuel oil. Steam is delivered to the steam turbine at 100 bar and 540 degrees C. Size: 14 to 143 Mw.

Source: H.E. Gerlaugh *et al.*, January 1980.

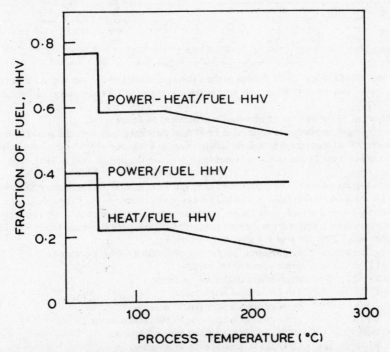

Figure 2.5.3.4-T6. Characteristics of an oil-fired diesel cogeneration system. Heat available from jacket cooling water at 80 degrees C. Gases are exhausted from the heat recovery boiler at 150 degrees C. Sizes: 0.3 to 10 Mw.

Source: H.E. Gerlaugh *et al.*, January 1980.

One solution to the problem of turbine blade degradation arising from the use of low quality fuels in a gas turbine is to use a hot air turbine instead of a combustion turbine. With the latter system the turbine is driven directly by the gaseous combustion products; with a hot air turbine clean air is used instead, with the air heated through use of a heat exchanger interposed between the combustor and the turbine inlet. Systems of this nature are not as efficient as state-of-the-art directly-fired gas turbines (because turbine inlet temperatures must be restricted), but this is a cost-effective, commercially ready technology for use with coal, urban refuse, biomass, petroleum coke, or any low quality fuel. Westinghouse offers such a system, with a fluidized bed combustor (R.W. Foster-Pegg, and J.S. Davis, March 1983).

While direct firing of a gas turbine with coal is very problematical (because of the high sulfur content, abrasive ash, and various other impurities in the fuel), direct firing with wood, a much cleaner feedstock appears to be feasible. This concept is currently being demonstrated in a commercial project in the U.S by Aerospace Research Corporation, which is building a 3 Mw unit, the electrical output of which is being sold to the Tennessee Valley Authority. The technology involves firing with sawdust and the use of commercially available cyclones to clean particles out of the combustion product gases. This technology remarkably allows the use of a low cost fuel at high-efficiency, in a low capital cost conversion system (J.T. Hamrick, June 1984).

A third strategy for using low quality fuels is to produce a gaseous fuel that is sufficiently clean that it can be used in gas turbines or reciprocating internal combustion engines. Again biomass has an advantage over coal here, being inherently easier to gasify.

Section 2.6.2

Note 2.6.2-A. Demographic and Economic Parameters Relating to the Energy Scenarios Developed for the U.S. Country Study

	Population[a]		Household Parameters[b]		Employment[c]			Economic Product (billion 1972$)					
	Total	Age ≥16	# of HH	HH Size	Industry	Service	Total	High Growth Scenario[c]			Low Growth Scenario		
	(million)	(million)	(million)	(persons)	(million)	(million)	(million)	GPOi	GPOs	GNP	GPOi	GPOs	GNP
1980	227.7	169.4	81.6	2.76	26.4	59.6	85.9	464.7	1010.3	1475.0	464.7	1010.3	1475.0
1990	249.7	191.9	93.1	2.61	28.4	73.3	101.7	635	1431	2066	581	1306	1887
2000	268.0	208.2	101.5	2.57	28.3	82.0	110.3	776	1826	2592	674	1536	2210
2010	283.1	221.1	109.9	2.50	27.0	90.2	117.2	908	2270	3178	747	1776	2523
2020	296.3	234.9	119.2	2.42	26.0	98.5	124.5	1074	2813	3887	837	2042	2879

(a) This is the "middle series" of the Bureau of the Census, in which fertility averages 1.9 births per woman (up from 1.83 births per woman in 1980) and net immigration averages 450,000 per year. See Bureau of the Census, October, 1982.

(b) The projection of the number of households is obtained as follows:
  - The household population is assumed to be 97% of the total population, the average percentage in the 1970s.
  - The size "S" of the average household is determined as a function of "A" the adult fraction of the population (the fraction aged 18 and over) by regressing "A" over the period 1970 to 1981. The result,

$$S = 7.205 - 6.187 \, A, \qquad r = 0.9969,$$

is used here to estimate future household size, and from this the future number of households is estimated.

(c) A high growth scenario is defined in which (1) the economy returns to full employment (4% unemployment) and (2) the average annual growth rates for the productivities of the goods producing and services sectors have the same average values in the future as the long term (1953-1978) historical average values—2.07% per year for goods production and 1.29% per year for services production.

The numerical values for employment, output and productivity—the quantities:

L_i	=	employment in industry,
L_s	=	employment in the service sector,
L	=	$L_i + L_s$ = total employment,
GPO_i	=	gross product originating in industry,
GPO_s	=	gross product originating in the service sector,
GNP	=	$GPO_i + GPO_s$ = gross national product,
P_i	=	GPO_i/L_i = productivity in industry,
P_s	=	GPO_s/L_s = productivity in the service sector,

can be determined for each year, assuming the initial values for 1980, the above assumptions about productivity growth and the return to full employment, and the following empirical relationship between industrial output and GNP reflecting the ongoing shift to services, obtained via regression analysis for the period 1970-1980:

$$GPOi = 1.1075 \times GNP^{.832}, \qquad r = 0.9639$$

Here both GNP and GPOi are measured in billion 1972$

Future total employment "L" is estimated to be:

$$L = a1 \times a2 \times a3 \times a4 \times P \geqslant 16 = 0.53 \times P \geqslant 16,$$

where

$P \geqslant 16$	=	population aged 16 and over,
$a1$	=	non-institutionalized fraction of $P \geqslant 16$ = 0.985 (historical averge value assumed)
$a2$	=	labor force participation rate = 0.65 (assumed to be the 1979 value)
$a3$	=	ratio of the number working to the size of the labor force = 0.96 (4% unemployment assumed)
$a4$	=	ratio of full time equivalent employees to total employment = 0.86 (historical value assumed)

The resulting high growth scenario involves a doubling of per capita GNP, 1980-2020.

(d) An alternative low growth scenario is defined having (1) the same employment levels and distribution as in the high growth scenario, but (2) per capita GNP increasing only 50%, 1980-2020. For this scenario the productivity growth rates would average 1.52% per year and 0.50% per year for the goods producing and services producing sectors, respectively.

Note 2.6.2-B. The following is the average floor area (in square meters) of new housing units in the U.S. in recent years[a, b]

Year	Floor Area	Year	Floor Area
1971	122.0	1978	141.9
1972	124.1	1979	140.2
1973	129.1	1980	136.0
1974	128.9	1981	135.5
1975	132.7	1982	134.0
1976	140.4	1983	135.9
1977	142.3	1984	134.8

(a) The decline from the peak in 1977 is due largely to a shift to multi-family housing. The multi-family housing share of new construction increased from 24% in 1977 to 38% in 1984.

(b) *Source:* Personal communication to Eric Larson from Steve Berman (Construction Division, US Bureau of the Census), June 3, 1985.

Note 2.6.2-C. Distribution of Housing by Heating Carrier and Vintage (millions)[a]

Vintage	1980			2020		
	Fuel	Elect.	Total	Fuel	Elect.	Total
Pre-1981	67.3	14.3	81.6	41.1	8.7	49.8
1981-1990	—	—	—	7.2	7.2	14.4
1991-2020	—	—	—	27.5	27.5	55.0
Totals	67.3	14.3	81.6	75.8	43.4	119.2

(a) The projection to 2020 is based on the demographic projection given in Note 2.6.2-A and the assumptions that: (1) half of all new households will be electrically heated and (2) the housing stock turns over with a characteristic half-life of 55 years, corresponding to a mean dwelling lifetime of 80 years.

Note 2.6.2-D. Heating and Cooling Load Norms for 2020, by Vintage (GJ per year)[a]

	Pre-1981	1981-1990	1991 and Beyond	All Vintages
Space Heating				
Fuel	35.5[b]	23.6[c]	7.5[d]	
Electricity	17.5[e]	20.3[c]	7.5[d]	
Air Conditioning				
Central and HP				20.9[f]
Room				7.0[f]
Water Heating				7.3[g]

(a) The heating or cooling load is defined as the output of the mechanical equipment providing the heating or cooling.

(b) Assuming average furnace efficiencies of 66% for oil and 60% for gas and LPG and U.S. Energy Information Administration, October, 1983, estimates of average levels of fuel use for space heating heating in 1980, the average heat loads for space heating in the U.S. in 1980 were 67 GJ, 47 GJ, and 38 GJ for oil, gas, and LPG heated houses, respectively. Heat loads in the year 2020 are assumed to be 30% lower, on average, due to energy saving retrofits—

(c) reflecting the experience in the Modular Retrofit Experiment (Note 2.5.3.1-K).

new houses being built now [39.6 GJ per year for fuel-heated units and 33.0 GJ per year for electrically heated units—this is 139/149 times the average for new single family units (Williams *et al.*, 1983), to account for the smaller floor area (139 sq. m.) of the average housing unit, compared to that of single family units (149 sq. m.)].

(d) This is the heat load that would result if an EEHDP house with a heated floor space of 139 square meters (the average for the U.S. housing stock in 1980) were constructed in a U.S. climate typical for new construction in the 1980s. An EEHDP house is one with a median level thermal performance (the heat output required of the heating system, per degree day per square meter of floor area) for the 97 houses that were built and monitored under Minnesota's Energy Efficient Housing Demonstration Program. See Table 2-3 and Williams *et al.*, 1983.

(e) Electrically heated houses have traditionally been thermally tighter than fuel heated houses in the U.S.; accordingly no improvements in the building shell are assumed for electrically heated houses built before 1981. The value given here is the average heat load for electrically heated houses in 1980 (U.S. Energy Information Administration, October 1983).

(f) This is the average heat load in 1980, assuming that the average COP = 1.9 and that the room air conditioner cooling load is ⅓ of the central air conditioning load (U.S. Energy Information Administration, October, 1983).

(g) This is the estimated average demand level for hot water for an average household size of 2.42, assuming 100% saturation of low-flow showerheads, and energy efficient dishwashers and clothes washers. The assumed per capita hot water load is ⅔ of the average value in 1980.

Note 2.6.2-E. Energy Performance Characteristics for U.S. Heating and Cooling Devices

System	Performance Index	Average 1980 Stock	Average 1980 New	Best 1982 New
Gas Furnace	average fuel utilization efficiency	0.60	0.69	0.95
Electric Heat Pump	heating season performance factor			
Heating		1.7	1.7	2.6
Cooling		1.9	2.2	3.3
Air Conditioner	cooling season performance factor			
Central		1.9	2.1	4.1
Room		1.9	2.0	3.2
Gas Water Heater	overall efficiency			
Stand Alone		0.44	0.48	0.62
Furnace Add-On		—	—	0.83
Electric Water Heater	overall efficiency or COP	0.73	0.80	2.2

Source: Williams *et al.*, 1983.

Note 2.6.2-F. Alternative Energy Performance Levels for Miscellaneous Appliances

Appliance	Electrical Units (kWh/year)		Fuel-Fired Unit (GJ/year)	
	1980[a]	2020	1980	2020
Refrigerator/Freezer	1670	580[b]	—	—
Freezer	1340	560[b]	—	—
Range	750	750[e]	7.4 (f)	5.3[c]
Dryer	1200	920[e]	5.4 (f)	4.3[g]
Lights	1000	300[d]	—	—

(a)   See R.H. Williams *et al.,* 1983.
(b)   In 1982 the most energy efficient refrigerator/freezer (R/F) commercially available was a 510 liter (18 cf) frost-free Amana that uses 870 kWh per year. Also a prototype 510 liter Kelvinator with a compressor 30-40% more efficient than conventional compressor being field tested in 1982 consumes only 710 kWh per year. If this compressor were added to an R/F with the features already used in the Amana, the result would be an R/F with a unit consumption of 580 kWh per year. The most energy efficient freezer commercially available in 1982 was a 425 litre (15 cf) chest freezer with a unit consumption of 710 kWh per year. Adding an efficient compressor to this unit would lead to the consumption level shown (Howard S. Geller, 1983).
(c)   The most energy efficient model commercially available in 1982.
(d)   Compact fluorescent bulbs with electronic ballasts and an efficacy of 40 to 60 lumens per watt and commercially available in 1983 are assumed to replace those ordinary incandescents which are used the most. Incandescents have an efficacy of only 11 to 18 lumens per watt.
(e)   The average for new appliances in 1980 (R.H. Williams *et al.,* 1983).
(f)   The U.S. average for gas appliances in use in 1982 (American Gas Association, March 2, 1984).
(g)   The U.S. average for units with spark ignition in 1982 (American Gas Association, March 2, 1984).

Note 2.6.2-G. Projection and Distribution (by Vintage) of Commercial Floor Space
(billion square meters)

	1979	1990	2000	2010	2020[a]
Vintage					
Pre-1980	4.13[b]				2.59
1980-1990					0.90
1991-2000					0.78
2001-2010					0.93
2011-2020					1.17
	4.13	4.96[c]	5.45[c]	5.90[c]	6.37 [c]

(a)   The distribution of commercial floor by vintage in 2020 is determined 45 years.
(b)   Energy Information Administration, March 1983.
(c)   The projection of commercial floor space (A, in billion square meters) is based on a regression against service sector employment (Ls, in million full time equivalent employees in industries other than mining, and manufacturing) for the period 1970 to 1979:

$$A = 0.876 + 0.0558 \text{ Ls}, \text{ r } = 0.991.$$

Employment data are from the *Survey of Current Business,* various issues. The historical data on construction data are those used in the ORNL commercial sector model (see d), with 1.21 billion square meters of floor area added to bring the ORNL series in line with the recent survey (b).
(d)   R.W. Barnes *et al.,* May 1980.

Note 2.6.2-H. Energy Intensity Norms of Commercial Buildings for the Year 2020, by Vintage (GJ per square meter per year)

Vintage	Fuel	Electricity
Pre-1980	0.46[a]	0.23[a]
1980-1990	0.10[b]	0.43[b]
1991-2020	0.04[c]	0.28[c]

(a)   Following the discussion in the text, a 50% retrofit savings is targeted relative to the energy intensity of the existing builidng stock.

(b)   For the period 1980 to 1990, the average performance is assumed to be a simple average of the post-1990 target performance and current practice (0.16 GJ and 0.57 GJ per sq. m. per year for fuel and electricity, respectively).

(c)   The average energy performance of the American Institute of Architects Life-cycle Cost Minimum Designs (see Table 2-6) is assumed to be the norm, on the average, for the period 1991 to 2020.

Notes for Chapter 3

Section 3.4

Note 3.4-A. The Energy/GDP Correlation

Consider any energy-utilizing activity j in an economy. Let its contribution to the Gross Domestic Product (GDP) of that economy be $C_j = (a_j * GDP)$ where a_j is the fraction of the GDP stemming from that activity, and where C_j and GDP are measured in $. Also, let the Energy Intensity of the activity be I_j measured in Joules per $.

Obviously, the summation over all the energy-utilizing activities in that economy yields the Total Energy Demand (E). Thus,

$$E = \text{Sum } (C_j * I_j)$$
$$= \text{Sum } (a_j * GDP * I_j)$$
$$= \text{Sum } (a_j * I_j) * GDP$$

This expression shows that the Energy Demand will be proportional to the GDP, that is the Energy-GDP 'correlation' will be valid only if

$$\text{Sum } (a_j * I_j) \text{ remains constant}$$

If, however, there are changes either in Energy Intensity (I_j) or in contribution to the total GDP (a_j), then the changes in ($a_j * I_j$) can negate the changes in GDP and decouple Energy Demand from the GDP. For instance, even when GDP is increasing, the product ($a_j * I_j$) can compensate for the *increase* in GDP, or even over-compensate for the increase, leading to a corresponding *constancy* in Energy Demand, or its *decrease*.

· The changes in Energy Intensity arise from *technical* changes due to efficiency improvements, or changes in process or product. The changes in contribution to the GDP stem from- so-called *structural* changes in the economy due to changes in its 'product-mix', that is, the composition of goods and services. The recent history of the industrialized countries shows abundant evidence for both technical and structural changes.

R.H. Williams, G.S. Dutt, and H.S. Geller, "Future Energy Savings in U.S. Housing," *Annual Review of Energy,* pp. 269-332, Vol. 8, 1983.

R.H. Williams, "A Low Energy Future for the United States," PU/CEES Report No. 186, Center for Energy and Environmental Studies, Princeton University, Princeton, N.J., 1985a.

R.H. Williams, "Potential Roles for Bio-Energy in an Energy-Efficient World," *Ambio,* Vol. XIV, No. 4-5, pp. 201-209, 1985b.

World Bank, "Prospects for Traditional and Non-Conventional Energy Sources in Developing Countries," 1979.

Notes for Chapter 4

Section 4.2.1

Note 4.2.1-A. In 1975 per capita GNP in the WE/JANZ region was $6250 (1982$), while that of Sweden was $11,050 (Bureau of the Census, 1984). Also, the per capita GDP level for all industrialized countries in 1975 was about the same as that of the WE/JANZ region. Thus, the per capita GNP level of Sweden was about 75% greater than the average for all industrialized countries in 1975.

Note 4.2.1-B. In a study exploring the economics of electricity conservation for hydro-rich Brazil (J. Goldemberg and R.H. Williams, 1985), investments in efficient lightbulbs and refrigerators are shown to require substantially less net capital than would be required for the purchase of conventional lightbulbs and refrigerators plus the extra new hydro-electric supply that would be required with the conventional end-use technology. Moreover, the net capital savings associated with the investment in energy efficiency tends to increase with the discount rate. The increasing benefit with the discount rate is a result of two effects: (1) the lifecycle cost for the hydro-electric facility increases with the discount rate, owing to interest charges accumulated during construction, and (2) the lifecycle costs of investments in end-use efficiency decrease with the discount rate, because these investments are short-lived and spreadout over the lifecycle of the hydro-electric supply alternative.

Furthermore, a detailed study of a wide range of electricity-saving technologies for the Brazilian electrical sector (H.S. Geller, 1985) has shown that discounted investments in efficient end-use technology totalling $4 billion in the period 1986 to 2000 would result in discounted savings of $19 billion in new hydro-electric supplies in this same period.

Note 4.2.1-C. A summary of the main features of the Base Case Energy demand scenario

	Industrialized Countries		Developing Countries		World	
	1980	2020	1980	2020	1980	2020
Population (billion)[a]	1.11	1.24	3.32	5.71	4.43	6.95
Final Energy Use[b] in TW Fuels	4.77		2.77		7.54	7.23
Electricity	0.70		0.13		0.83	1.58[c]
TOTALS	5.47	3.10	2.90	5.71	8.37	8.81
Per Capita Final Energy Use (kW)	4.92	2.5 [d]	0.87	1.0[d]	1.89	1.27

(a) Assuming the 1980 U.N. low variant population projection.
(b) In these energy balances final energy use is defined as the total fuel (including bunkers) and electricity consumed by "final consumers". Excluded are losses in the generation, transmission and distribution of electricity, and the consumption of petroleum fuels by refineries.
(c) In 1980 10% of final global energy use was accounted for by electricity, which was about 0.8 times as large as the percentage for the U.S. In this global scenario it is assumed that in 2020 the global electrical fraction of final demand is again 0.8 times the value projected for the U.S. in the U.S. country study (Section 2.6.2), or 18%.
(d) Following the discussion in the text a Base Case scenario for 2020 is described wherein per capita energy use in industrialized countries is reduced in half and that in developing countries is assumed to be about the same as in 1980, or 1.0 kW.

Note 4.2.1-D. To express the Base Case global energy demand scenario in terms more familiar to the majority of analysts in the energy modeling community, a simple model has been constructed relating *commercial* final energy demand per capita (FE/P) to gross domestic product per capita (GDP/P), the average price of final energy (P_e), and a rate of energy efficiency improvement (c) which is not price-induced:

$$FE/P \ (t) = A \times [GDP/P \ (t)]^a \times [P_e(t)]^{-b}/(1+c)^t,$$

where "A" is a constant, "a" is an income elasticity, and "−b" is a long-run final energy price elasticity. This is essentially the energy demand equation underlying the Institute for Energy Analysis/Oak Ridge Associated Universities (IEA/ORAU) global energy-economy model (J. Edmonds and J. Reilly, 1983a). Here this equation is applied separately to industrialized and developing countries, relating PE/F, GDP/P, and P_e values in 2020 to those in 1972, as indicated in the following figures, for illustrative values of the parameters (a, b, and c).

These figures show per capita GDP and energy price parameters consistent with the Base Case energy demand scenarios for industrialized countries (Figure 4.2.1-D1 and for developing countries (Figure 4.2.1-D2), for alternative assumptions about income and price elasticities and the non-price induced energy efficiency improvement rate.

The year 1972 is chosen as the base year for this modeling exercise, because this is the last year before the first oil price shock, so that it is reasonable to assume that the economic system was then in equilibrium with the existing energy prices (unlike the situation in 1980, say). For this base year the values of PE/P were 4.7 kW and 0.38 kW, compared

INDUSTRIALIZED COUNTRIES

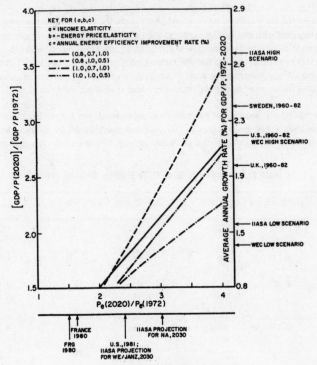

Figure 4.2.1.-D1

DEVELOPING COUNTRIES

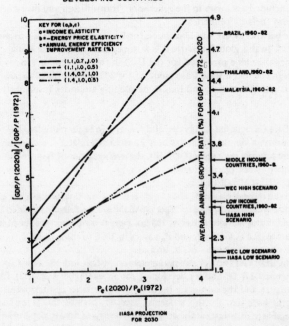

Figure 4.2.1-D2

to the 2020 Base Case scenario values of 2.5 kW and 1.0 kw, for industrialized and developing countries, respectively.

For income elasticities, alternative values of 0.8 and 1.0 are assumed for industrialized countries, while values of 1.1 and 1.4 are assumed for developing countries. The value of 0.8 would tend to capture the ongoing shift to less energy-intensive economic activity in industrialized countries (and may well understate the extent of the shift); the value of unity assumed in many modeling efforts is included for comparison. A value of 1.4 was used for developing countries in the 1983 IEA/ORAU study (J. Edmonds and J. Reilly, 1983b) and may be roughly characteristic of the historical situation in developing countries. However, as developing countries modernize in the decades ahead, the income elasticity can be expected to decline. The two assumed values may span the range of uncertainty for the income elasticity in developing countries for the time frame of interest here.

For the long run price elasticity, Nordhaus has reviewed various studies and has concluded that the range of plausible values is from —0.66 to − 1.15 (W.D. Nordhaus, 1977). The illustrative values chosen here (− 0.7 and − 1.0) span most of this range. [These elasticities appear to be high but are not. Long run price elasticities are much larger than short-run elasticities. Likewise, final demand elasticities are greater than secondary demand elasticities, which in turn are greater than primary demand elasticities (Energy Modeling Forum, 1980)].

The assumed non-price induced energy efficiency improvement rates are 1.0 and 0.5 percent per year. The higher value is the one assumed for the "CO_2-benign" scenarios developed in a MIT Energy Laboratory study (D.J. Rose, M.M. Miller, and C. Agnew, 1983), while the lower value approximately reflects the contribution from such energy efficiency improvements in a 1984 IEA/ORAU analysis (J.A. Edmonds, J. Reilly, J.R. Trabalka, and D.E. Reichle, 1984). The higher energy efficiency improvement rate is coupled with the lower price elasticity and the lower rate with the higher price elasticity in this modeling exercise, to reflect the tendency of non-price induced energy efficiency improvement policies to diminish the efficacy of prices in curbing energy demand (Energy Modeling Forum, 1980).

While assumptions about energy prices were not made explicitly in the construction of the Base Case scenario (since most of the end-use technologies underlying our analysis would be economic at or near present prices on a lifecycle cost basis, with future costs discounted at market interest rates), there may well be continuing final energy price increases to reflect rising marginal production costs, the expected continuing shift to electricity, and the levy of some energy taxes to take into account externalities (Chapter 5). Prices have already risen substantially above 1972 values: in West Germany and France average final energy prices in 1980 were 1.5 and 1.6 times the 1972 values in real terms, respectively (C.P. Doblin, 1982), and in the U.S. the average price in 1981 was 2.3 times the 1972 price (U.S. Energy Information Administration, 1984). Looking to the future, the IIASA study projects that by 2030 final energy prices will be 3 times the 1972 value in all regions but the WE/JANZ region (Western Europe, Japan, Australia, and New Zealand), for which a 2.4-fold increase is projected instead (W. Haefele *et al.,* 1981). The 1983 projection of the U.S. Department of Energy is for much larger 3.6-fold to 5.7-fold average final energy price increases for the U.S., 1972-2010, associated with an 11 to 17 percent reduction in final energy use per capita in this period (Office of Policy, Planning, and Analysis, U.S. Department of Energy, 1983). It is reasonable to associate an average in the final energy price by 2020 in the range 2 to 3 times the 1972 value with the Base Case scenario.

For industrialized countries (Figure 4.2-1-D1) and the cases (a, b, c) = (0.8, 0.7, 1.0), (0.8, 1.0, 0.5), and 1.0, 1.0, 0.5), the Base Case energy demand projection is consistent with a 50 to 100 per cent increase in per capita GDP (comparable to the values assumed in the IIASA and WEC low scenarios) and 2020 energy prices in the range 2 to 3 times the 1972 values. For the case where there would be no structural shift to less energy-intensive activities and a low price elasticity (1.0, 0.7, 1.0), the energy price in 2020 would have to be about 3½ times the 1972 value to be consistent with a doubling of per capita GDP.

For developing countries (Figure 4.2.1-D2) the Base Case scenario with 2020 prices in the range 2 to 3 times 1972 prices would be compatible with the per capita GDP growth rates assumed in the IIASA and WEC high scenarios, for the high income elasticity cases (a = 1.4). For the lower income elasticity cases (a = 1.1), the Base Case projection would be consistent with much more rapid GDP growth.

This modeling exercise shows that while the Base Case global energy demand projection for 2020 is far outside the range of most other projections, it would appear to be consistent with plausible values of income and price elasticities, and with plausible expections about energy price and GDP growth, if the non-price induced energy efficiency improvement rate can be in the range 0.5 to 1.0 percent per year. While this energy efficiency improvement rate is a measure of the public policy effort that would be required to bring about this energy future, not all of this efficiency improvement would have to be public policy-induced. Energy efficiency improvements associated with the general process of technological innovation have often been made even in periods of declining energy prices (Section 2.5.3.4), a phenomenon which led to IEA/ORAU analysts to include the non-price induced technological improvement factor in their model for the industrial sector in the first place (J. Edmonds and J. Reilly, 1983a).

At the same time, however, the energy efficiency improvement factor may not represent the full extent of needed public policy effort, if low energy demand levels were to lead to stable or even falling energy prices. If that were the case, then energy taxes may also be needed to keep gradual upward pressure on final (consumer) prices.

Note 4.2.1-E. Using the simple energy demand model described in Note 4.2.1-D we make estimates of the future trend in per capita energy use in industrialized countries under "business as usual" conditions, where little is done to promote energy efficiency with new public policy initiatives.

We consider 50 and 100 percent increases in per capita gross domestic product "GDP/P" relative to 1980 (corresponding to 75 and 135 percent increases relative to the 1972 base year).

We assume that the income elasticity is 0.8, reflecting the ongoing structural shift away from basic materials processing industries to fabrication and finishing activities.

We consider 2-fold and 3-fold increases in the average final energy price "P_e" relative to the average 1972 final energy prices. (See Note 4.2.1-D for a discussion of price projections made by various forecasting groups.)

For energy price elasticities "$-b$" we consider two alternative values at the low end of what economists expect long run final energy price elasticities to be (-0.7 and 0.8—see Note 4.2.1-D).

For the non-price induced energy efficiency improvement rate "c" we assume alternative values of 0.00 and 0.25 percent per year. The cases with non-zero values are included to take account of the phenomenon that some energy efficiency improvements will be made as a natural consequence to the process of technological innovation and can be expected even in periods of constant or declining energy prices (Note 4.2.1-D).

Under these conditions, the following alternative values are obtained for the ratio of average final energy use per capita in 2000 relative to that in 1980:

GDP/P (2020)/GDP/P (1980)	P_e(2020)/P_e(1972)	b	c	FE/P(2020)/FE/P(1980)
1.5	2	0.7	0.0000	0.92
1.5	2	0.8	0.0000	0.86
1.5	2	0.7	0.0025	0.82
1.5	2	0.8	0.0025	0.76
1.5	3	0.7	0.0000	0.69
1.5	3	0.8	0.0000	0.62
1.5	3	0.7	0.0025	0.61
1.5	3	0.8	0.0025	0.55
2.0	2	0.7	0.0000	1.16
2.0	2	0.8	0.0000	1.09
2.0	2	0.7	0.0025	1.03
2.0	2	0.8	0.0025	0.96
2.0	3	0.7	0.0000	0.88
2.0	3	0.8	0.0000	0.79
2.0	3	0.7	0.0025	0.78
2.0	3	0.8	0.0025	0.70

Section 4.2.2

Note 4.2.2-A. For the calculations of the allowable fossil fuel usage consistent with alternative ceilings on the atmospheric carbon dioxide level it is assumed: (1) that 0.63, 0.43, and 0.78 billion tonnes of carbon are released per TW-yr of oil, natural gas, and coal consumption, respectively; (2) that 50% of the released carbon dioxide remains in the atmosphere; (3) that the preindustrial atmospheric carbon level (as carbon dioxide) was 615 billion tonnes (290 ppm); and (4) that the level in 1979 was 708.5 billion tonnes (334 ppm) or 1.15 times the pre-industrial level.

It is also assumed that ultimately the amount of oil and natural gas produced is equal to the present estimates of ultimately recoverable resources [365 TW-years for oil and 349 TW-years for natural gas (Note 4.2.2-E)]. Thus ultimately the amount of carbon as carbon dioxide remaining in the atmosphere from oil and gas use would be

$$0.5 \times [0.63 \times 365 + 0.43 \times 349] = 190 \text{ billion tonnes}$$

or 0.31 times the preindustrial atmospheric level.

If half of the world's geological coal resources (8535 TW-Years—see the following table) were eventually used up, the amount of net carbon build-up in the atmosphere as carbon dioxide would be

$$0.5 \times 0.5 \times 0.78 \times 8535 = 1660 \text{ billion tonnes,}$$

or 2.7 times the preindustrial level. Thus if this much coal but no more oil and gas were used up, the atmospheric carbon dioxide level would reach 3.85 times the pre-industrial level.

The following is an estimate of the total world geological resources of coal[a]

Country	Trillion Tonnes of Coal Equivalent	TW-Years[b]	Per Cent
USSR	4.41	4096	48
USA	2.33	2164	25
China	1.31	1217	14
Other	1.14	1059	12
TOTALS	9.19	8535	100

(a) *Source:* World Energy Conference, 1978.
(b) For a heating value of 29.3 GJ per tonne of coal-equivalent.

Note 4.2.2-B. It is assumed that allowable ultimate carbon dioxide levels are reflected in alternative coal consumption levels, and that coal production in a given year is proportional to the remaining coal left in the ground that can be eventually used without exceeding the specified ceiling on the atmospheric carbon dioxide level. Thus, for the above assumptions about oil and gas use, the allowable coal consumption schedule as a function of the atmospheric carbon dioxide ceiling is the following:

Maximum CO_2 level as a fraction of pre-industrial level	Permissible total prod. (TW-yr)	1978	2000	2020	2040	2060	2080	2100
1.50	65.50	2.59	1.09	0.49	0.22	0.10	0.05	0.02
1.60	223.20	2.59	2.01	1.59	1.26	1.00	0.79	0.63
1.70	380.91	2.59	2.23	1.95	1.70	1.48	1.29	1.13
1.80	538.62	2.59	2.33	2.12	1.92	1.75	1.59	1.44
1.90	696.33	2.59	2.39	2.22	2.06	1.91	1.77	1.65
2.00	854.04	2.59	2.42	2.28	2.15	2.02	1.90	1.79

4.2.2-C. In Tables 17.6 and 17.7, pages 531-532, in Wolf Haefele *et al.*, 1981 it is estimated that the global coal resources recoverable @ a cost of up to $40 per tonne (in 1982 $) ($25 per tonne in 1975 $) amounts to 560 TW-years (606 billion tonnes), while that available in the cost range $40 to $80 per tonne is 1019 TW-years (1105 billion tonnes). Coal @ $80 per tonne is equivalent to oil @ $17 per barrel, or ½ the world oil price in 1982.

Note 4.2.2-D. The distribution of world oil production (in million barrels per day) is as follows for the Base Case (BC) scenario and the WEC and IIASA scenarios, in relation to recent world production trends:

	1979	1980	1981	1982	1983	BC	WEC (1982) Low	WEC (1982) High	IIASA (1981) Low	IIASA (1981) High
ME/NAf	25.9	22.4	19.0	16.3	14.9	15	19	26	31	34
Elsewhere	39.9	40.4	40.2	40.7	41.5	30	41	56	47	64
TOTALS	65.8	62.8	59.2	57.0	56.4	45	60	82	78	98

Note 4.2.2-E. The following are estimates of ultimately recoverable oil and natural gas resources (as of 1977) and consumption (in 1978)

Region[a]	Resources (in TW-years)		Consumption (in TW)	
	Oil[b]	Gas[c]	Oil[d]	Gas[d]
I. NA	39.7	60.5	1.28	0.74
II. SU/EE	66.3	96.4	0.56	0.47
III. WE/JANZ	22.7	23.1	1.22	0.27
IV. LA	32.6	20.9	0.23	0.07
V. Af/SEA	30.0	18.4	0.21	0.02
VI. ME/Naf	154.8	117.1	0.11	0.04
VII. C/CPA	18.1	12.5	0.12	0.01
TOTALS	364.2	348.9	3.73	1.62

(a)   The seven regions indicted here are defined as follows:
      I   NA (North America)
      II SU/EE (The Soviet Union; Eastern Europe)
      III WE/JANZ (Western Europe; Japan, Australia; New Zealand; South Africa; Israel)
      IV LA (Latin America)
      V   Af/SEA (Africa, except northern Africa and South Africa; South Asia; Southeast Asia)
      VI ME/NAf (Middle East; Northern Africa)
      VII C/CPA (China, other Centrally Planned Asian Economies)
(b)   See Table 2.6, p. 57 in Wolf Haefele *et al.*, 1981.
(c)   See Table 2.7, p. 65 in Wolf Haefele, 1981.
(d)   See J.R. Frisch, 1982.

Note 4.2.2-F. One substitution possibility is to use compressed natural gas directly as motor vehicle fuel (J.P. West and L.G. Brown, April 1979). In Italy some 270,000 vehicles were operted on compressed natural gas (CNG) in 1980; New Zealand has plans for converting 150,000 vehicles to CNG by 1985; and both Canada and Australia are gearing up for major conversions. Alternatively, natural gas might be converted to methanol.

Note 4.2.2-G. If world oil demand fell from 4.18 TW in 1980 to the base case scenario level of 3.21 TW in 2020, the required cumulative world oil production, 1981-2020, would be some 148 TW-years in this period. If in the period 1983-2020 production in the ME/NAf region were maintained at the 1983 "world oil glut level" of 1.06 TW (15 million barrels per day), cumulative oil requirements from regions other than the ME/NAf region in this period would amount to some

$$148 - (1.36 \times 3) - (1.06 \times 37) = 104.7 \text{ TW-years},$$

For comparison, world oil resources remaining outside the ME/NAf region and estimated to be ultimately recoverable at a price less than $26 per barrel (1982 $) is some 132 TW-years [see Table 17.6, p. 531 in Wolf Haefele *et al.*, 1981].

Note 4.2.2-H. The following are recent official projections of global nuclear electricity projections to the year 2000 (in TW of continuous electricity produced)

	1980[a]	2000
U.S.	0.0315	0.080 to 0.082[b]
Other OECD	0.0346	0.097 to 0.104[b]
Subtotal, OECD	0.0661	0.177 to 0.186[b]
Other Market Economies	0.0016	0.018 to 0.027[b]
TOTAL, Market Economies	0.0677	0.195 to 0.213
Soviet Union	0.0071	0.065 to 0.124[c]
Eastern Europe & Cuba	0.0027	0.026[c]
TOTAL, CPE	0.0097	0.091 to 0.124
TOTAL, WORLD	0.0774	0.286 to 0.337

(a) The British Petroleum Company, 1982.
(b) Office of Policy, Planning, and Analysis, U.S. Department of Energy, October 1983.
(c) International Energy Agency, 1982a.

Note 4.2.2-I. On the basis of economic calculations presented in R.O. Sandberg, and C. Braun, April 8-11, 1984 the uranium price would have to rise to $100 per lb. of U308 (triple the present price) before plutonium recycle would be able to compete with current once-through fuel cycles. And even if the price of uranium should increase to $150 per lb., the cost advantage of recycle would amount to less than 2% of the busbar cost of power generation.

Note 4.2.2-J. The following is the assumed global electricity supply mix for the Base Case scenario (in average TW produced at the power plants)

	1980	2020
Hydro	0.19	0.46[a]
Wind and Photovoltaics	—	0.09[b]
Cogeneration		
Biomass	—	0.14[c]
Fossil Fuel	—	0.13[c]
Central Station		
Nuclear	0.08[d]	0.30[e]
Fossil Fuel	0.66[f]	0.66[g]
TOTALS	0.93[h]	1.78[h]

(a) It is assumed that 25% of the total electricity supply (4030 TWh/year) is hydro. This is 2/5 of the global economic hydro potential (97000 TWh/year) and 1/5 of the global technically usable potential (19,400 TWh per year) estimated in 1976 for the WEC (Ellis L. Armstrong, 1978). The 1976 WEC estimates of the economic potential are probably low, in light of subsequent electricity price increases and better resource estimates—e.g., more recent estimates for Brazil and India indicate economic potentials there some 75% higher than the 1976 WEC estimates.
(b) It is assumed that 5% of the total electricity supply is wind and photovoltaics. Owing to the large uncertainties in the future of photovoltaics technology, it is not specified how the wind/photovoltaics mix might be disaggregated.
   In the event that photovoltaics technology is not commercialized, all of this electricity would be provided by wind. To put the resulting wind value (0.1 TW) into perspective, it can be compared to alternative assessments of the wind potential. The IIASA study (Wolf Haefele *et al.,* 1981) estimated that globally the technical potential for wind power is 3 TW and the "realizable" potential is 1 TW. In another analysis for North America, it is estimated that on 16% of the North American land area (mainly in the great plains) the wind energy density is in excess of 400 Watts per square meters, and that the mean electric power recoverable from this wind via Boeing Mod-2 windmills (2.5 MW each) spaced 1.5 km apart is 0.9 TW (Gordon Thompson, July 1981).
   However, if the promise of photovoltaic technology is realized (E.A. Demeo and R.W. Taylor, 1984) the photovoltaic contribution could be considerable.
(c) Here it is assumed that the cogeneration fraction of total electricity production is 15%, the same as the percentage which we have estimated could be provided in the U.S. in 2020 by the major steam-using industries. Here a 50/50 mix of biomass and fossil fuels is assumed for fuel inputs.
(d) Includes geothermal (which is small).
(e) It is assumed that in 2020 nuclear electricity is the same level officially forecast for 2000 in the early 1980s.
(f) Includes fossil fuel-based production through cogeneration (which is small).
(g) Central station power generation based on fossil fuels is assumed to be the residual.
(h) Electricity demand (Note 4.2.1-C) divided by 0.89 to account for T&D losses.

Note 4.2.2-K. The following is the assumed global primary energy supply mix for the Base Case scenario (in TW)

	1980	2020
Nuclear Power	0.22[a]	0.75[b, c]
Hydro[d]	0.19	0.46[c]
Wind and Photovoltaic Electricity[d]	—	0.09[c]
Fossil Fuels		
Coal	2.44	1.94
Oil	4.18	3.21
Natural Gas	1.74	3.21
SUBTOTAL	8.36	8.36[f]
Biomass		
Organic Wastes		0.79[g]
Plantations		0.79[h]
SUBTOTAL	1.49[i]	1.58[j]
TOTALS	10.3	11.2

(a)   It is assumed that 2.8 units of fuel are required to produce 1 unit of electricity in thermal power plants in 1980.

(b)   It is assumed that 2.5 units of fuel are required to produce 1 unit of electricity in nuclear and coal fired thermal power plants in 2020.

(c)   See Note 4.2.2-J.

(d)   The primary energy consumption associated with hydro, wind and photovoltaic electricity production is assumed to be the energy value of the output of these systems.

(e)   It is assumed that on averge 1.5 units of fuel, is required to produce 1 unit of cogenerated electricity.

(f)   Total fossil fuel consumption is assumed to be 8.36 TW, as in 1980.

(g)   In the Base Case scenario, it is assumed that half of the biomass is provided by organic wastes and half by managed biomass production. The level of organic waste use for energy purposes corresponds to about 1/5 of the estimated organic waste production level in 2020 (Note 4.2.2-L).

(h)   Assuming 18 GJ per tonne of dry wood, this implies that 1.4 billion tonnes of wood are required annually. At an average yield of 10 tonnes per hectare per year, some 140 million hectares of plantation area would be required, which is less than 4% of the land area under forest today.

(i)   Bioenergy data for less developed countries are from D.O. Hall *et al.,* 1982b. For the United States, bioenergy consumption by the paper and pulp industry in 1980 was 1.1 EJ (American Paper Institute, 1981), while wood consumption for household fuel was 0.87 EJ (U.S. Energy Information Administration, June 1982). Non-commercial energy use in other indusrialized market economies was obtained from International Energy Agency, 1982b. Fuelwood consumption data for Eastern Europe and the Soviet Union are from United Nations, 1981b.

(j)   The total final demand for all fuels is 7.23 TW (Note 4.2.1-C). The demand for biomass fuels is the difference between the this total and the final use of fossil fuels. The latter is calculated as follows. Assuming that ¼ of central station power generation is by coal converted @ 40% conversion efficiency coal used for direct purposes is:

$$41.94 - 0.75 \times 2.5 \times 0.66 = 0.70 \text{ TW.}$$

Assuming that ¼ of central station power generation is by natural gas in steam-injected gas turbines @ 50% conversion efficiency and that ½ of the fossil fuel based cogeneration is via natural gas, for which 1.5 units of extra fuel is needed to produce each unit of electricity, natural gas use for direct purposes is:

$$3.21 - 0.25 \times 2.0 \times 0.66 - 0.5 \times 1.5 \times 0.13 = 2.78 \text{ TW.}$$

Assuming that ½ of fossil fuel based cogeneration is via oil, for which 1.5 units of extra fuel is needed to produce each unit of electricity, and that 10% of gross oil use is consumed in refining operations, oil use for direct purposes is:
$$3.21 - 0.5 \times 1.5 \times 0.13 - 0.1 \times 3.21 = 2.79 \text{ TW.}$$
Thus, the direct fuel requirements from biomass sources would be:
$$7.23 - 0.70 - 2.78 - 2.79 = 0.96 \text{ TW.}$$

Assuming average conversion losses of 30% in producing final energy carriers from biomass, the total amount of biomass required for direct fuels use would be 1.37 TW. In addition, some 0.21 TW would be required to produce 0.14 TW of cogenerated electricity see (Note 4.2.2-J), assuming the same conversion efficiency as for electricity production by fossil fuels.

Note 4.2.2-L. The following is an estimate (in TW) of the global organic waste production rate in 1980 and projected to the year 2020:

	1980	2020
Forest Product Industry Wastes		
United States[a]	0.11	0.14
Rest of World[b]	0.27	0.44
SUBTOTALS	0.38	0.58
Crop Residues		
Industrialized Countries[c]	0.51	0.57
Developing Countries[c]	0.60	1.04
SUBTOTALS	1.11	..61
Manure		
Industrialized Countries[c]	0.38	0.42
Developing Countries[c]	0.73	1.26
SUBTOTALS	1.11	1.68
Urban Refuse		
United States[d]	0.046	0.060
Other Industrialized Countries[e]	0.090	0.096
Developing Countries[f]	0.034	0.058
SUBTOTALS	0.17	0.21
TOTALS	2.77	4.08

(a) In 1979 roundwood production in the U.S. was 1.53 CM per capita (United Nations, 1981c) and residue production per unit of roundwood production was (Office of Technology Assessment, 1980):

Primary and Secondary Manufacturing Residues	0.406
Logging Residues	0.377
Stand Improvement Cuttings	0.203
TOTAL	0.986

These same values are assumed to hold for 1980 (2020), when the U.S. population was (is expected to be) 228 (296) million. At 10 GJ/CM total residue production in 1980 (2020) was (would be) 3.42 EJ (4.53 EJ).

(b) In 1979 roundwood production outside the U.S. was 0.65 CM/capita (United Nations, 1981c). Here it will be assumed that residue production in the rest of the world is like that in the European Economic Community, where the ratio of residues to roundwood production is 0.32 (W. Paly and P. Chartier, 1980). For a population in 1980 (2020) of 4.14 billion (6.65 billion) residue production would be 8.61 EJ (13.83 EJ).

(c) Per capita production rates for 1975 are estimated in Theodore B. Taylor, Robert P. Taylor, and Steven Weiss, March 1, 1982 to be (in Watts):

	Industrialized Countries	Developing Countries
Crop Residues	460	182
Manure	342	221

These same rates are assumed for 1980 and 2020.

(d)   In the U.S., urban refuse with an average heating value of 10.7 MJ/kg is generated at a rate of 1.63 kg per capita per day. It is assumed that this rate persists.

(e)   In Europe, refuse with an average heating value of 8.8 MJ/kg is generated today at a rate of 1.0 kg per capita per day (U.S. Environmental Protection Agency, November 1979). It is assumed that this rate applies on average to all industrialized countries outside the U.S., in both 1980 and 2020.

(f)   It is assumed that one tenth of the population (the urban elite) generate urban refuse at the European rate.

Note 4.2.2-M. While in principle a high level of crop residue recovery might be achieved by introducing harvesting techniques which recover residues simultaneously with the primary products, only part of these wastes would be available for energy purposes. In developing countries, crop residues are often used as fodder for livestock. And some residues will have to be left behind to provide nutrients and to maintain soil quality.

To the extent that crop residues and manure are utilized for biogas production, however, it may often be feasible to return the nutrient-rich residuum from the biodigesters to the soil for such purposes.

In the case of forest residues, it may be necessary to restrict removals to the larger pieces, leaving behind the leaves or needles and twigs, in which the nutrients tend to be concentrated.

Note 4.2.2-N. Various studies indicate bioenergy productivities on managed plantations ranging from about 7 dry tonnes per hectare per year (Mesquite, Saltbush, Kochia) and arid regions, to 10 to 15 tonnes (Willow, Poplar) in Sweden, to 40 to 60 tonnes (Eucalyptus, Leucaena) in Brazil or India (see Appendix: Renewable Energy Resources).

One problem with many of these estimates is that the empirical evidence is often based on unusually good experience for limited plots where growing conditions are especially favorable. What is needed for energy planning purposes when targeting the production of billions of tonnes of biomass per year is long-term experience on large plantations.

An average productivity of 10 tonnes per hectare per year is probably not optimistic for large scale production on managed plantations or energy farms.

Section 4.3.1

Note 4.3.1-A. In 1980 the amount of reforestation in 76 developing countries (excluding China) amounted to 1.15 million hectares (FAO, 1982).

Between 1950 and 1979, 100 million hectares in China were replanted in forests, but only 28 million hectares of replanted area yielded surviving forests. Over the 10 year period 1972-1981, the rate of planting averaged 4.7 million hectares per year. In many areas the survival rate has increased recently to 50% or more. The official Chinese policy is to increase forest area from 122 million hectares in 1981 to 192 million hectares in 2000—i.e., at an annual rate of 3.5 million hectares per year (Bin Zhu, April 1984).

Note 4.3.1-B. The following are alternative U.N. global population projections (United Nations, 1981a), in billions

	1980	2020		
		Low Variant	Medium Variant	High Variant
Industrialized Countries	1.11	1.24	1.35	1.44
Developing Countries	3.32	5.71	6.47	7.14
WORLD	4.43	6.95	7.82	8.58

Note 4.3.1-C. Assuming a 65% capacity factor, the installed nuclear generating capacity for the high nuclear scenario increases from 120 GW(e) in 1980 to 300 GW(e) in 1990, to 460 GW(e) in 2000, and to GW(e) in 2020. Thus the average rate of net additions in the period 2000 to 2020 is 15.5 GW(e) per year. In addition, it is assumed that all the capacity build up through 1990 is replaced in the period 1990-2020, at an average rate of 10 GW(e) per year and for the high nuclear power scenario:

Section 4.3.2

Note 4.3.2-A. A summary of the main demand characteristics of the High Demand scenario

	Industrialized Countries		Developing Countries		World	
	1980	2020	1980	2020	1980	2020
Population (billion)[a]	1.11	1.24	3.32	5.71	4.43	6.95
Final Energy Use, in TW						
Fuels	4.77		2.77		7.54	9.67
Electricity	0.70		0.13		0.83	2.12[b]
TOTALS	5.47	6.08	2.90	5.71	8.37	11.79
Per Capita Final Energy Use (kW)	4.92	4.9[c]	0.87	1.0[c]	1.89	1.70

(a) Assuming the 1980 U.N. low variant population projection.
(b) Assuming, as in the Base Case scenario, that 18% of final energy use is accounted for by electricity in 2020.
(c) In the High Demand scenario, it is assumed that per capita energy use in industrialized (developing) countries in 2020 is about the same as in 1980 or 4.9 kW (1.0 kW).

Note 4.3.2-B. The distribution of world oil production (in million barrels per day) is as follows for the Base Case (BC) and High Demand (HD) scenarios, in relation to that for the WEC and IIASA scenarios (the averages for the high and low scenarios) and for recent years:

						2020			
	1979	1980	1981	1982	1983	BC	HD	WEC	IIASA
ME/NAf	25.9	22.4	19.0	16.3	14.9	15	15	22.5	32.5
Elsewhere	39.9	40.4	40.2	40.7	41.5	30	48	48.5	55.5
TOTALS	65.8	62.8	59.2	57.0	56.4	45	63	71	88

Note 4.3.2-C. The following is the assumed global primary energy supply mix for the High Demand scenario (in TW)

	1980	2020
Nuclear Power	0.22	0.75[a, b]
Hydro[c]	0.19	0.60[b]
Wind and Photovoltaic Electricity[c]	—	0.12[b]
Fossil Fuels		
Coal	2.44	1.95
Oil	4.18	4.46[d]
Natural Gas	1.74	4.46[d]
SUBTOTAL	8.36	10.87
Biomass		
Organic Wastes		1.01[e]
Plantations		2.00[f]
SUBTOTAL	1.49	3.01
TOTALS	10.3	15.34

(a)   It is assumed that 2.5 units of fuel are required to produce 1 unit of electricity in nuclear power plants in 2020.
(b)   See Note 4.3.2-D.
(c)   The primary energy consumption associated with hydro, wind and photovoltaic electricity production is assumed to be energy value of the output of these systems.
(d)   Oil and gas production levels (assumed to be equal) are the residual.
(e)   It is assumed that 25% of the produced organic wastes can be recovered for energy purposes.
(f)   Assuming that the plantation expansion rate, 1985-2015, is limited to 12 million hectares per year, that the average plantation yield is 10 tonnes per year, and that wood has a heating value of 18 GJ per dry tonne. Under these conditions some 3.5 billion tonnes of wood would be harvested on 350 million hectares in 2020.

Note 4.3.2-D. The following is the assumed global electricity supply mix for the High Demand scenario (in average TW produced at the power plants)

	1980	2020
Hydro	0.19	0.60[a]
Wind and Photovoltaics	—	0.12[b]
Cogeneration		
Biomass	—	0.18[c]
Fossil Fuel	—	0.18[c]
Central Station		
Nuclear	0.08	0.30[d]
Fossil Fuel	0.66	1.00[e]
TOTALS	0.93[f]	2.38[f]

(a)   As in the Base Case, it is assumed that 25% of the total electricity supply (5300 TWh/year) is hydro. By comparison, the 1976 WEC estimate of the global economic potential is 9700 TWh/year.
(b)   As in the Base Case, it is assumed that 5% of the total electricity supply is wind and photovoltaics.
(c)   As in the Base Case, it is assumed that the cogeneraion fraction of total electricity production is 15%, with a 50/50 mix of biomass and fossil fuel inputs.
(d)   As in the Base Case, it is assumed that in 2020 nuclear electricity is the same level officially forecast for 2000 in the early 1980s.
(e)   Central station power generation based on fossil fuels is assumed to be the residual.
(f)   Electricity production is equal to the electricity demand level (Note 4.3.2-A) divided by 0.89 to account for T&D losses.

Note 4.3.2-E. If world oil demand increased from 4.18 TW in 1980 to the high demand scenario level of 4.46 TW in 2020, the required cumulative world oil production, 1981-2020, would be some 173 TW-years in this period. If in the period 1983-2020 production in the ME/NAf region were maintained at the 1983 "world oil glut level" of 1.06 TW (15 million barrels per day), cumulative oil requirements from regions other than the ME/NAf region in this period would amount to some

$$173 - (1.36 \times 3) - (1.06 \times 37) = 130 \text{ TW-years}$$

By comparison, the remaining world oil resources outside the ME/NAf region and estimated in Wolf Haefele *et al.*, 1981 to be ultimately recoverable at a price less than $26 per barrel (1982 $) is some 132 TW-years see Table 17-6, p. 531 of Wolf Haefele *et al.*, 1981).

Note 4.3.2-F. Using the carbon dioxide production coefficients given in Note 4.2.2-A for alternative fossil fuels and assuming that half of the released carbon dioxide stays in the atmosphere, we obtain the following net additions of carbon (in billion tonnes) to the atmosphere in 2020:

Base Case Scenario	4.9
High Demand Scenario	6.3
WEC Scenarios (ave. of High and Low)	10.2
IIASA Scenarios (ave. of High and Low)	10.8

Note 4.3.2-G. In Wolf Haefele *et al.*, 1981, the peak oil production capacity for the ME/NAf region in the period 2020-2030 is estimated to be some 34 million barrels per day (2.4 TW).

Section 4.4

Note 4.4-A. Here we estimate the cumulative level of carbon dioxide in the atmosphere, assuming fossil fuels are phased out linearly from the 2020 levels indicated in Note 4.2.2-K., with the carbon release assumptions presented in Note 4.2.2-A. Under these assumptions, the carbon remaining in the atmosphere as a result of emissions in the period 1980 to 2020 is:

For oil:

$$0.5 \times 0.5 \times (4.18 + 3.21) \times 40 \text{ years} \times 0.63 \qquad = 46.6 \text{ billion tonnes,}$$

For natural gas:

$$0.5 \times 0.5 \times (1.74 + 3.21) \times 40 \text{ years} \times 0.43 \qquad = 21.3 \text{ billion tonnes,}$$

For coal:

$$0.5 \times 2.44 \times [1 - \exp(-40 \times 0.0068)] / 0.0068 \times 0.78 \qquad = 33.3 \text{ billion tonnes.}$$

Thus, the total atmospheric carbon increment in this period would be 101.2 billion tonnes.

If fossil fuel use were to decline linearly to zero, 2020 to 2050, the cumulative carbon dioxide build-up in this period would be:

$$0.5 \times 0.5 \times 30 \times [3.21 \times (0.63 + 0.43) + 1.94 \times 0.78] = 36.9 \text{ billion tonnes.}$$

Thus, with this fossil fuel phase out scenario the ultimate level of carbon in the atmosphere (as carbon dioxide) would be

$$708.5 + 101.2 + 36.9 = 847 \text{ billion tonnes (399 ppm),}$$

which is 1.38 times the preindustrial level.

Note 4.4-B. To estimate the solar collector area required to provide hydrogen by means of electrolysis from amorphous silicon solar cells at the level of fossil fuel consumption in 2020 for the base case scenario (8.4 TW), the following values are assumed:

- An average insolation rate of 200 Watts/square meter for sunny (desert) areas:
- An average efficiency of amorphous silicon solar cells of 15% [a practical target value for tandem, multi-layered cells, according to a 1984 EPRI review (E.A. DeMeo and R.W. Taylor, 1984)].
- An average efficiency for electrolysis of 80%.

Under these conditions the collector area required would be some 0.350 million-square kilometers—which is half the size of the state of Texas (0.691 million square kilometers) or 2% of the area of warm deserts of the world (17.8 million square kilometers).

Notes for Chapter 5

Section 5.2

Note 5.2-A. In the U.S. the most significant tax subsidies to the energy industry include the percentage depletion allowance and the expensing of intangible drilling costs for the oil industry, tax credits, and accelerated depreciation allowances. Other subsidies include loans at below market rates and loan guarantees.

Note 5.2-B. According to economic theory, there is a situation where considerations of economic efficiency could justify a subsidy to the producer. Consider the situation of a regulated monopolist whose marginal costs go down, the more he produces. If the regulator under such conditions were to set the monopolist's product price equal to the marginal cost, which could be an economically efficient policy, the result would be to put the monopolist in a chronic loss situation (since the marginal cost would be less than the average cost). A permanent government subsidy to this decreasing cost producer would lead to the least costly production for society for this output.

While the situation of declining marginal cost with output existed for many electric utilities before about 1970, the situation for utilities in most areas today (especially in industrialized countries) is one of increasing marginal costs.

Note 5.2-C. In W.J. Baumol and E.N. Wolff, 1981, the indirect energy consumption associated with the subsidized activity as well as the direct consumption is taken into account—i.e., the energy opportunity costs of non-energy inputs. Baumol and Wolff illustrate the principle behind their analysis as follows:

...Suppose that an economy has two outputs, manufactured energy and bread, and two primary resources, energy and labor; that bread can be produced by either of two processes, one labor intensive and one energy intensive; and that both processes are currently in use. Then, other things being equal, if employment is constant, every labor hour used in the manufacture of energy increases the amount of bread that must be produced by the energy-intensive rather than the labor-intensive process. The corresponding increase in energy use is the energy opportunity cost of the hour of labour...

Note 5.2-D. There are many reasons why consumer discount rates relating to investments in energy efficiency are so high.

Often consumers are inadequately informed about the opportunities for and cost-effectiveness of investments in energy efficiency. If the consumer does not expect the energy savings to be worth the bother of obtaining good information or of evaluating conflicting claims for costs and benefits, he will have no economic basis for getting better informed.

In some instances the bother of getting better informed is just one of many "hassles" that must be dealt with in implementing energy efficiency improvements, which are inherently more difficult to achieve than simply purchasing energy supplies. Advice must be obtained on what improvements are needed, the cost-effectiveness of alternative available opportunities must be evaluated, financing must be arranged, often the improvements must be carried out piecemeal (e.g., for home "retrofits"), and even when the improvements have been made it is often not easy to ascertain how much energy was actually saved.

Still another factor limiting investments in energy efficiency is the lack of adequate capital resources. In contrast to the large energy supply companies, which have ready access to enormous capital resources, energy consumers often must forgo energy efficiency investments because of capital scarcity.

For some important energy using activities, expenditure on energy are still such a small fraction of the total cost of the activity that concern about energy efficiency is minor. This tends to be true for commercial buildings (Note 5.2-E) and automobiles (Note 5.2-F).

Still another problem is that those who would invest in energy efficiency improvements are sometimes unable to capture all of the energy saving benefits of the investment. This "split responsibility problem" is a serious one for many energy efficiency investments in buildings, where responsibility is split betwteen builder and homeowner or between landlord and tenant (Note 5.2-G).

In some cases, powerful vested interests have actively discouraged investments in promising alternative energy technologies. Such efforts have certainly inhibited the development of cogeneration. In Chapter 2 we showed that the most promising cogeneration technologies are those that involve the production of much more electricity than is produced at industrial sites. In many countries, the export of electricity to the utility grid is discouraged by utilities.

Finally, the uncertainties about energy price give rise to a reluctance to make long-term investments in energy efficiency improvement. In particular, the instability of the world oil price gives consumers confusing signals about the transition to the Post-Petroleum Era.

Note 5.2-E. At the average level of energy use in existing U.S. commercial buildings (Table 2-6) and the average U.S. energy prices for the commercial sector in 1982, the average annual energy cost for commercial buildings amounts to $13 per square meter. By comparison, annual rental costs in central New Jersey amount to some $160 to $215 per square meter per year (Leslie K. Norford, January 1984), and the cost of an employee who occupies 15 square meters of office space comes to some $1400 per square meter (@ $20,000 per year).

Note 5.2-F. In Chapter 2 we showed that it is technically and economically feasible to improve automotive fuel economy significantly above the level of VW Diesel Rabbit (5.3 liters/100 km). But at this fuel economy level and the 1982 U.S. fuel price of $0.26 per liter, fuel contributes only about 8% to the total cost of owning and operating the car. Even at the much higher European prices, the cost of fuel does not make a major contribution to the total cost. The problem is compounded by the tendency for savings arising from increased fuel economy to be roughly offset by the increased first cost associated with making the improvement.

Note 5.2-G. In the case of major appliances and space conditioning equipment, an outstanding problem is that most purchases are made by builders, contractors and landlords—parties who are not responsible for operating costs. Builders, who are especially sensitive to first-cost, generally avoid purchasing equipment for which improved energy efficiency means added first cost.

Other important situations involving split economic responsibility arise in rental buildings, where renters often pay part of all their energy bills. When landlords are not responsibile for energy costs, they have no significant incentive to make investments in energy efficiency. Moreover, even when energy bills are included in the rent, the motivation for energy efficiency improvements is often limited because of allowable tax deductions for fuel costs (as an operating expense). Also, landlords are reluctant to make building improvements that might raise the assessment for property tax liability. Tenants avoid investments in energy efficiency improvements, since they typically live in the same house for three years or less.

The split responsibility problem exists even for owner-occupied housing. While the life cycles of many energy efficiency improvements are 15-20 years, owner-occupied houses are sold much more frequently—every 8 years, on average, in the U.S. Fearful that they may not be able to recover the value of their investmens at the time of resale, many homeowners are reluctant to make long-term investments in energy efficiency.

Note 5.2-H. A case study of targeted subsidies to promote energy conservation and solar energy is provided by the U.S. program of federal tax credits for household and industrial investments in certain qualifying energy conserving and solar energy technologies, established by the U.S. Congress with the passage of the Energy Tax Act of 1978.

One generic problem with these tax credits is that they must be established for well-defined sets of investment options, to limit potential abuses of the tax law. However, by creating a list of qualifying investments, policymaker order investments in ways that may not be optimal. Path-braking innovations will not be on the list because, by definition, such technologies are too new to be well-defined. Of course this problem might be remedied in part if the list of qualifying technologies were periodically updated, but the process of modifying such a list is likely to be slow and difficult.

In some instances, it may be inappropriate to allow superior technologies on a list of qualifying investments. This might be the case, for example, for investments that serve multiple purposes. Consider the tax credit for solar heating technologies. Materials and components of solar system that are structural components of a residence do not qualify for the credit. This constraint has the effect of limiting the tax credit for passive solar systems, in which the collectors also serve as south-facing windows. Yet passive systems are usually more cost-effective than active systems, which serve the single function of providing heat. This ruling does not reflect Congressional short-sightedness; rather it reflects a legitimate concern about loss of tax revenues for unitended purposes—a concern that will be raised about any technology, however promising, that serves multiple purposes.

Still another problem with the energy tax credit is that it doesn't reward energy savings—it rewards instead capital investment in nominally energy conserving devices. The problem here is two-fold. One is that actual field performance of energy-conserving systems is often quite different from theoretical predictions. A second is that the availability of the tax credit may simply drive up the prices of the devices being subsidized. One proposal which has been made to overcome these problems is to make the tax credit *performance-based*. But industrial case studies have shown that industrial managed genreally do not favor this approach: too much analysis is required to find out if the firm will qualify, and in any case the credit cannot influence the decision-making process since eligibility will not be determined until after the new technology is in operation (Alliance to Save Energy, 1983). The practical problems would of course be considerably greater for residential applications.

Also, with targeted tax credits the distribution of allowed benefits is often arbitrarily established in the political process. In the case of the residential tax credits, for example, there is no evidence that a cost-benefit calculus was used to justify the distribution of benefits between energy conservation investments (15% of the first $2000 spent, with a maximum of $300) and solar energy investments (a 40% tax credit on the first $10,000 , for a maximum of $4000).

A final generic problem with targeted tax credits is that they are not generally available. The poor, who need the financial help the most, cannot benefit from tax credits, nor can industries running in the red, which have no tax liabilities.

These problems are reflected in analysis of the accomplishments of the U.S. energy tax credit program. While the revenue losses associated with the program made this the single largest U.S. government expenditure on energy productivity, there is no evidence to suggest that the program did much good. For the residential sector, no data are available to permit a meaningful assessment of the extent to which the program led to investments that would not otherwise have been made (Eric Hirst *et al.*, 1982). For the industrial sector, case studies show that the tax credit has not been an important factor in decisions to fund individual projects (Alliance to Save Energy, 1982).

Note 5.2-I. One limitation of energy performance regulation as an instrument for effecting technological change is that regulations are not well suited for all significant energy-using activities. That the energy efficiency regulation has worked remarkably well in the case of the U.S. automotive fuel economy standard is due in large part to being able to base the regulation on an energy performance index that is easily measurable, readily understood, and widely applicable. Standards applied to major energy-using household appliances would meet this criterion as well. In the case of building energy performance, the development of such an index that is unambiguous enough that it can be used as a basis for performance regulations is a much more challenging task. In the case of buildings, this problem might be overcome by basing the standard on prescription rather than performance (e.g., indicating the allowable heat loss through walls, windows, etc.). This is the approach that has been taken in Sweden, where building codes have to a large degree been responsible for making Swedish houses the most thermally tight houses in the world.

Another problem with energy performance regulations is that they force technological change so as to meet the performance test required in the regulation, which may differ from the performance in the "real world." U.S. consumers are generally familiar, for example, with the phenomenon that the labeled fuel economy for new cars is better than the actual on-the-road fuel economy. This problem can be mitigated by periodically updating the testing procedure in light of field testing, to reduce the differences between test and field performance results. However, some discrepancies will always remain, owing to the wide range of actual use conditions that cannot be captured in testing procedures.

Perhaps the most serious limitation of performance regulations is that they tend to affect mainly the "bottom" of the market. Even in Sweden, most builders build houses that are thermally tighter than what is required by the "rigorous" Swedish building code. One reason for this phenomenon is that remarkable advances such as those we have witnessed over the last couple of years in residential energy technology cannot be readily incorporated into standards until well after they have been introduced. But perhaps more important is the fact that in the political process standards tend to be set at levels believed to be achievable with a high degree of confidence by most producers in the industry. The fact that a performance standard is a week stimulus for technological innovation is no reason to reject performance standards, however. There is good evidence for example, that an apparently "weak" California standard for refrigerators was effective in clearing the refrigerator market nationwide of energy-guzzling models (D.B. Goldstein, 1983). Policymakers should be concerned about both the top and the bottom of the market.

Note 5.2-J. In the Denver Metro Home Builders' Program, sponsored by the Solar Energy Research Institute (SERI) and carried out in cooperation with the Denver Home Builders Association, SERI solicited proposals from builder/architect teams for the design and construction of well-insulated, passive solar houses, for which the training of the architects in passive solar and improved thermal design, the designs, and the post-construction performance monitoring were all supported by SERI. Twelve winning builder/architect teams were selected from among 47 bids. The houses, completed 7 months later, required on average only ⅓ as much space heating fuel as typical new houses. During a 16-day Passive Solar Tour of Homes 100,000 people visited these houses, and in this period many sales contracts were signed by builders. The competition may have had a major long-term impact on construction industry practices in the area (K.H. Smith, 1982; D. Macmillan, 1982). An important element in the program's success was that builders were allowed considerable flexibility in design and construction; SERI's role was limited to providing information and technical support.

Section 5.3.5

Note 5.3.5-A. Audits offered to residential customers by utilities in the U.S. under the federally mandated Residential Conservation Service (RCS) involve recommendations for various energy-saving investment opportunities. Recommendations for building shell improvements to reduce heat losses (in winter) or heat gains (in summer) are based on calculations using traditional heat-loss models. Because of various anomalous heat losses, many houses do not perform as these models predict. For this reason, instrumented diagnostic audits are preferable to paper-and-pencil audits based on eyeball inspection. While such alternative audit techniques have been developed, they are not yet widely used (R.H. Williams, G.S. Dutt, and H.S. Geller, 1983).

Note 5.3.5-B. In March, 1982 H.R. 5880 was introduced in the U.S. House of Representatives and referred to the Committee on Science and Technology. This bill called for a design competition intended to get the U.S. auto industry to build cars with the following characteristics: ability to transport 4 adults with luggage; a fuel economy of 100 mpg (2.3 1/100 km); good road performance characteristics; ability to meet federal safety and environmental standards; a manufacturers' list price of between $6000 and $6500.

The winner would have to demonstrate plans to produce a minimum of 100,000 high fuel economy vehicles per year for a minimum of 3 years. The reward would be the difference between $20,000 per car and the manufacturers' list price for the first 10,000 vehicles sold (anout $140 million total). The four "runners up" in the competion would receive $20 million each.

The total cost to the government would be some $220 million. By comparison, the U.S. would be paying some $100 million less on oil imports *each day,* if the average U.S. automobile had a fuel economy of 100 mpg instead of the present 16 mpg.

Though never considered seriously by the U.S. Congress (largely because of the strong anti-market intervention frame of mind of the Reagan-dominated Congress), the bill nevertheless represents a thoughtful model of how design competitions can be shaped to promote technological leaps forward.

Note 5.3.5-C. The Public Utility Commission in California has made some rulings to motivate utilities to market energy conservation effectively. In October, 1982, the Commission awarded the Southern California Gas Co. $5 million (0.3% of its rate base) for its aggressive pursuit of energy conservation in the residential sector (California Public Utility

Commission, 1982); in December 1979 the Commission imposed $ 7 million rate penalties on the Pacific Gas and Electric Co. for each of the years 1980 and 1981 (0.2% of its rate base) for failure to make reasonable efforts to develop the cogeneration potential in its service territory (California Public Service Commission, 1979).

Note 5.3.5-D. The Public Utility Regulatory Policies Act (PURPA) passed by the U.S. Congress in 1978 requires: (1) that utilities interconnect with qualifying cogenerators and small power producers which generate electricity with renewable resources; (2) that utilities pay these power producers for the electricity they wish to sell to the utility a price equal to the cost the utility would avoid by not having to provide this power; and (3) that the rates charged by utilities for backup power not discriminate against these power producers (Alvin L. Alm and Kathryn L. Mowry, 1983).

The PURPA legislation was challenged in the courts by several utilities, but was upheld in two US Supreme Court decisions of 1982 and 1983. After the second Supreme Court decision was handed down, the cogeneration business in the US started booming (Stuart Diamond, 1984).

Note 5.3.5-E. For a discussion of various alternative institutional arrangements for cogeneration, including utility ownership, see Chapter 10, "Industrial Cogeneration: Making Electricity with Half the Fuel," in M.H. Ross and R.H. Williams, 1981.

Section 5.3.6
Note 5.3.6-A. At an average fuelwood price of $13 per cubic meter ($18 per dry tonne or $1 per GJ), the 168 million cubic meters (235 million dry tonnes of fuelwood) produced in 22 World Bank supported plantation projects, ⅔ of the output of which was fuelwood, the average rate of return was 27% (based on data presented by John Spears, Senior Forestry Advisor to the World Bank, at the Symposium on Biomass Energy Systems: Building Blocks for Sustainable Agriculture, January 29-February 1, 1985, Airlie House, Virginia). For comparison, the world oil price in 1984 was $29 per barrel or $4.80 per GJ.

Notes for Appendix A: Renewable Energy Resources

Section A.2
Note A.2-A. Solar Energy Fluxes (TW)

Solar Radiation Intercepted by the Earth[a]	175,000
Solar Radiation Absorbed by the Earth[a]	110,000
Solar Energy Involved in Evaporation[b]	40,000
Solar Energy Used in the Generation of the Atmospheric Pressure System[c]	1,800
Solar Energy Involved in Direct Sensible Heating	68,000
Solar Energy Utilized in Photosynthesis[d]	100
Man's Rate of Energy Use, 1980	10.65

(a) The solar constant (the intensity of sunlight on the top of the atmosphere) is 1370 watts per square meter and the earth's albedo (the percentage of incident sunlight reflected back into space) is 32 percent (World Meteorological Organization, 1981).
(b) M. King Hubbert, 1973.
(c) E.N. Lorenz, 1967.
(d) From Note A-4-A.

Section A.3

Note A.3-Aa. Summary of Hydro-Electric Resource Potential

	Total Electricity Production in 1980 (TWh)[a]	Hydro Electricity Production in 1980 (TWh)[a]	Theoretical Hydroelectric Potential (TWh)[b]	Technically Usable Hydroelectric Potential (TWh)[d]	Economic Potential (TWh)[d]	Ratio of Hydro Potential to Total 1980 Electricity Production	Economic Hydro Potential to Actual 1980 Hydroelectricity Production
Asia	1313.1	264.2	16,486[c]	5,340[c]	2672	2.0	10.1
South America	261.9	192.0	5,670	3.780	1892	7.2	9.9
Africa	187.0	61.1	10,118	3,140	1569	8.4	25.7
North America	2828.1	551.1	6,150	3,120	1561	0.6	2.8
U.S.S.R.	1295.0	180.0	3,940	2,190	1095	0.9	6.1
Europe	2192.3	462.9	4,360	1,430	714	0.3	1.5
Oceania	120.5	34.2	1,500	390	197	1.6	5.8
TOTAL	8197.9	1745.5	44,280	19,390	9700	1.2	5.6

(a) United Nations, 1981.

(b) World Energy Conference, September 1980.

(c) According to Zhu Xiaozhang (Associate Chief Engineer, Institute of Hydropower, Gansu Province, c/o Minister of Water Conservancy, Beijing, China, and Member of the Technical Panel on Hydropower of the Preparatory Committee for the United Nations Conference on New and Renewable Sources of Energy), this estimate may not include data from China. Mr. Zhu estimates that the theoretical hydroelectric potential for China is 600 TWh/year and that the technically usable potential is 1900 TWh.

(d) This is the economic potential estimated from the 1976 World Energy Conference Survey of Energy Resources. In light of much higher costs for alternative power generating sources, the author of the report providing these estimates (Ellis L. Armstrong, 1978) expects that the potential would be greater now.

Note A.3-Ab. Summary of Hydro-Electric Resources Potential

	Electricity Production From All Sources in 1980 (TWh)[a]	Hydro-Electricity Production in 1980 (TWh)[a]	Economic Potential (TWh)[b]	Ratio of Economic Hydro Potential to Total 1980 Electricity Production	Ratio of Economic Hydro Potential to Actual 1980 Hydroelectricity Production
Developing Market Countries	915.9	368.8	4644	5.1	12.6
OECD	5162.7	1062.8	2282	0.4	2.1
China, N. Korea, Vietnam	342.0	79.5	1546	4.6	19.8
U.S.S.R., E. Europe	1777.3	234.4	1199	0.7	5.1
TOTAL	8197.9	1745.5	9700	1.2	5.6

(a) United Nations, 1981.

(b) This is the economic potential estimated from the 1976 World Energy Conference Survey of Energy Resources. In light of much higher costs for alternative power generating sources, the author of the report providing these estimates (Ellis L Armstrong, 1978) expectes that the potential would be greater now.

Note A.3-B. Limitations of the WEC Estimates of the Hydroelectric Potential

	(TWh/year)
India	
Economic Potential	
WEC (1978)	221
CWPC (1960)[a]	221
1978 Reassessment[a, b]	396
Total with Micro/Mini[c]	420
Theoretical Potential (WEC, 1978)	750
Brazil	
Economic Potential	
WEC 1978	548
Schulman (1980)[d]	
Firm Power[e]	933
Average Power[f]	1195
Theoretical Potential[g]	1389-5256

(a) The first systematic survey of India's hydroelectric potential was made between 1953 and 1960 by the then Central Water and Power Commission. The much higher value for the economic potential given in the 1978 reassessment is due mainly to a more systematic identification of the potential in the Himalayan region.

(b) This potential is based on water inflows of 90 percent dependability, i.e., inflows likely to be available 90 out of 100 years. The averge potential is greater than this.

(c) The 1978 reassessment value does not include the potential with mini-hydro or micro-hydro schemes, which are (unofficially) estimated to be able to provide an additional 25 TWh/year.

(d) The latest official projection for Brazil is presented in M. Schulman, 1980.

(e) Firm power is the output assured at the lowest water flow levels on record, e.g., over the past 40 years for which hydrological data exist.

(f) The potential with average flow conditions.

(g) The lower estimate of the theoretical potential is that given by the WEC (Table A-1). The higher estimate is presented in Hartmut Krugman, May 1982.

Section A.4

Note A.4-A. Net Primary Production and Related Characteristics of the Biosphere[a]

	Area (million km^2)	Standing Biomass (billion tonnes)	Ratio, Standing Biomass to Net Primary Production (years)	Net Primary Productivity (tonnes/ha/ year)	Total Global Production[b] (Terawatts)
Continental					
Tropical Rain Forest	17.0	765	20.5	22.0	22.0
Tropical Seasonal Forest	7.5	260	21.7	16.0	7.1
Temperate Forest					
Evergreen	5.0	175	26.9	13.0	3.8
Deciduous	7.0	210	25.0	12.0	5.0
Boreal Forest	12.0	240	25.0	8.0	5.7
Woodland and Shrubland	8.5	50	8.3	7.0	3.5
Savanna	15.0	60	4.4	9.0	8.0
Temperate Grassland	9.0	14	2.6	6.0	3.2
Tundra and Alpine	8.0	5	4.5	1.4	0.65
Desert and Semi Desert Scrub	18.0	13	8.1	0.9	0.94
Extreme Desert—Rock, Sand,					

(Contd)

Ice	24.0	0.5	7.1	0.03	0.04
Cultivate Land	14.0	14.0	1.5	6.5	5.4
Swamp and Marsh	2.0	30	5.0	30	3.5
Lake and Stream	2.0	0.05	0.06	4.0	0.5
Subtotal	149	1837	15.6	7.8	69.3
Marine					
Open Ocean	332.0	1.0	0.024	1.25	24.5
Upwelling Zones	0.4	0.008	0.04	5.0	0.12
Continental Shelf	26.6	0.27	0.028	3.6	5.66
Algal Beds and Reefs	0.6	1.2	0.75	25.0	0.94
Estuaries (excluding Marsh)	1.4	1.4	0.67	15.0	1.24
Subtotal	361	3.9	0.07	1.55	32.4
Global Totals	510	1841	10.7	3.36	101.7

(a) *Source:* Robert H. Whitaker and Gene E. Likens, 1975.
(b) Here it is assumed that 1 tonne of dry matter has a heating value of 18.6 GJ. Thus 1 tonne/year = 590 watts.

Note A.4-B. See Note 4.2.2-L.

Note A.4-C[a]. The World's Renewable Energy From Forests

	Area (million hectares)	Annual Increment of Wood per Hectare[b] (cubic meters)	Total Increment of Wood[b]	
			(billion cubic meters)	(TW)[c]
Cool Coniferous	800	4.1	3.3	1.29
Temperate Mixed	800	5.5	4.4	1.75
Warm Temperate	200	5.5	1.1	0.46
Equatorial Rain	500	8.3	4.1	1.66
Tropical Moist Deciduous	500	6.9	3.5	1.38
Dry	1000	1.4	1.4	0.55
Totals and Averages	3800	4.7	17.8	7.1

(a) *Source:* D.E. Earl, 1975.
(b) Estimated to include all wood above ground.
(c) The energy content of the wood is taken to be 12.54 GJ per cubic meter.

Note A.4-Cb. The Utilization of the World's Incremental Forest Resources (1970)

	Annual Increment[b] (billion CM)	Annual Consumption (billion CM)			Unused Increment	
		Industrial	Fuel	Total	(billion CM)	TW[c]
Industrialized Countries	8.8	1.1	0.3	1.4	7.4	3.0
Developing Countries	9.0	0.2	0.8	1.0	8.0	3.2
TOTAL	17.8	1.3	1.1	2.4	15.4	6.2

(a) *Source:* D.E. Earl, 1975.
(b) Estimated to include all wood above ground.
(c) The energy content of the wood is taken to be 12.54 GJ per cubic meter.

Note A.4-D. Ratio of Stemwood to Total Biomass Production in Forests

Forest Type	Dry Matter Production (tonnes/ha/year)			Ratio of Stemwood to Total Production	
	Total	Stemwood Becking	Weck	Becking	Weck
Tropical High Forest					
0-500 m above sea level	63.0	13.2	8.8	0.21	0.14
500-1500 m above sea level	54.6	11.5	7.6	0.21	0.14
Temperate Rain Forest	31.0	9.3	9.3	0.30	0.30
West European Forest	18.0	5.8	5.8	0.32	0.32

(a) *Source:* D.E. Earl, 1975.

Note A.4-E. The 1982 market price in Sweden for wood feedstock for paper and pulp was 20 kr per GJ or $2.80 per GJ (1980 dollars)

Note A.4-F. Potential Yields from Biomass Plantations

Location	Species	Yield (tonnes/ha/year)
Sweden	Willow, Poplar[a]	10-15
	Reeds[b]	10
United States		
Washington	Red Alder[c]	22-45
California, Louisiana Mississippi, Florida	Eucalyptus[c]	27
Wisconsin, Illinois, Missouri, New England	Hybrid Poplar[c]	11-34
Georgia	American Sycamore[c]	18-36
North Carolina	Loblolly Pine[d]	11-18
West Texas	Mesquite[e]	7-16
West Texas	Saltbush[e]	>7
West Texas	Johnson Grass[e]	>7
West Texas	Kochia[e]	>7
Brazil		
Aracruz	Eucalyptus[d]	
First Rotation		23
Second Rotation		33
Third Rotation		44
Highest Clone		61
Oahu, Hawaii	Giant Leucaena[f]	21-33

(a) Programplan Energiproduktion, 1980.
(b) Sven Bjork and Wilhelm Graneli, 1978.
(c) James C. Bethel *et al.* April 6, 1979.
(d) J. Laurence Kulp, 1981.
(e) Richard Davids, 1982.
(f) For tree densities of 5000 to 40,000 per hectare (Molakai Study Team, 1980).

Note A.4-G. See Note 5.3.6-A.

Note A.4-H. The following are estimates of the plantation-gate costs for biomass energy produced on plantations in the U.S., in 1980 dollars per GJ[a, b, c]

	High Productivity Site		Low Productivity Site	
	Min. Cost	Max. Cost	Min. Cost	Max. Cost
Initial Capital Investment[d]				
Land Lease	0.016	0.033	0.016	0.041
Installations	0.012	0.024	0.024	0.049
Engineering & Contingency (30%)	0.017	0.017	0.012	0.027
Interest During Construction				
(4 years)	0.009	0.019	0.014	0.030
Total Capital Cost	0.044	0.093	0.066	0.147
Annual Operating Cost				
Site Preparation	0.047	0.210	0.098	0.267
Plantation	0.070	0.105	0.140	0.210
Fertilization	0.143	0.229	0.286	0.458
Irrigation	0.222	0.291	0.517	0.656
Weed Control	0.023	0.029	0.034	0.057
Protection	0.052	0.069	0.115	0.172
Harvesting	0.143	0.258	0.344	0.516
General Administration	0.006	0.011	0.011	0.023
Total Operating Cost	0.706	1.202	1.555	2.359
Total Cost	0.75	1.30	1.61	2.51

(a) These U.S. plantation cost estimates, from James S. Bethel *et al.*, April 6, 1979, are for hypothetical plantations with the following characteristics:

Annual Production	207,000 tonnes/year (dry weight)
Biomass Productivity	
Average High Productivity Site	27 tonnes/ha/year
Average Low Productivity Site	13 ½ tonnes/ha/year
Plantation Area Required	
High Productivity Site	7,700 hectares
Low Productivity Site	15,400 hectares
Expected Lifetime	32 years
Rotation Age	16 years
Coppicing	Every 4 years
Terrain	Flat to Gentle

(b) The costs in the original report are converted to 1980 U.S. dollars using the GNP deflator.
(c) Dry wood is assumed to have a lower heating value of 19 GJ per tonne.
(d) A real interest rate of 6 percent per year is assumed.

Note A.4-I. About 33% of all the world's cropland is farmed with a high degree of mechanization, and 60% with the intensive use of fertilizers (private communication from David Pimentel, February 19, 1985). However, fertilizers are used much less intensively in developing than in industrialized countries. Averaged over all cropland, the annual fertilizer input rates (for N, P_2O_5, K_2) are 115 kg per hectare in industrialized countries and 48 kg per hectare in developing countries. Moreover, the avergve is only 11 kg per hectare in Africa. The incremental crop yield per added kg of fertilizer is several times greater in developing countries where the present application rate is low than in industrialized countries (D. Pimentel *et al.*, 1985).

Section A.5

Note A.5-A. Estimates of wind electricity production costs

	1981 cents per kwh
Alternative Wind Systems[a]	
Present day small (50 kw) turbines[b]	4.6
Present day large (2.5 Mw) turbines[c]	6.5
Target for mass-produced large turbines[d]	2.8 – 3.4
Benchmark Costs	
Cost of electricity from a new, coal-fired baseload plant in the U.S.[e]	4.6
Average cost of oil used in electric power generation in the U.S. in 1981[f]	5.5

(a) Utility financing is assumed, for which the appropriate average cost of capital is assumed to be 0.038 (Note A.5-B). This corresponds to a capital recovery factor of 0.072 for a 20 year system life (assumed for small wind turbines) and 0.056 for a 30 year system life assumed for large wind turbines). In addition the annual insurance cost is assumed to be 0.003 times the initial capital cost. Following James I. Lerner, September 1982, the annual O&M cost is assumed to be 2% of the initial capital cost. The annual average capacity factor is assumed to be 35%.
(b) In James I. Lerner, September 1982, it is estimated that the installed cost for systems being built today is typically about $1500 per kw.
(c) The estimated unit costs for 40 MOD-2 wind turbines for the U.S. Bureau of Reclamation's proposed 100 Mw Medicine Bow Wind Project is $2500 per kw (L.L. Nelson, 1981).
(d) It is estimated that in mass production the unit insalled cost of the MOD-2 wind turbine will be $1100-$1300 per kw. See Note A.5-C.
(e) See Note A.5-D.
(f) For a heat rate of 10,900 KJ per kwh and an average 1981 oil price for electric utilities in the U.S. of $5 per GJ (Energy Information Administration, U.S. Department of Energy, October, 1982).

Note A.5-B. The cgCapital recovery factor for utilities is
$$CRF(i, N) = \text{capital recovery factor} = i/[1-(1+i)^{-N}],$$
where i = real cost of capital, and N = expected system life, in years.

The average cost of capital is:
$$i = (a \times cd) + (b \times cp) + (c \times ce),$$
where

      cd = debt cost
      cp = preferred stock cost
      ce = common stock cost

and

      a = debt fraction
      b = preferred stock fraction
      c = common stock fraction

In Technology Assessment Group, 1979, it is recommended that in evaluating utility investments the appropriate (inflation-corrected) costs of capital from different soruces and the mix of capital sources should be the following:

      cd = 0.019 a = 0.5
      cp = 0.024 b = 0.15
      ce = 0.071 c = 0.35

Thus for utility investments, i = 0.038, so that:
$$CRF(0.038,20) = 0.072.$$
$$CRF(0.038,30) = 0.056.$$

Note A.5-C. Estimated capital cost for mass-producted MOD-2 windmills

	1981 dollars per kw
Site preparation	90.5
Transportation	16.2
Erection and Checkout	76.2
Rotor assembly	182.9
Drive train	210.8
Nacelle subassembly	102.2
Tower subassembly	150.9
Initial spares and maintenance equipment	19.6
Non-recurring costs	19.6
SUBTOTAL	868.9
10% profit	86.9
TOTAL TURNKEY COST	957
Site-related costs	94- 255
Interest during construction	79- 91
TOTAL INSTALLED COST	1100 – 1300

(a) *Source:* James I. Lerner, September 1982.

(b) The MOD-2 is a 2.5 Mw horizontal axis wind turbine designed and fabricated for the U.S. Department of Energy by the Boeing Engineering and Construction Company.

(c) The cost estimates presented here are for the 100th unit produced in a dedicated production facility that makes 20 turbines per month.

Note A.5-D. Busbar cost for electricity from a new 500 MW coal-fired baseload power plant

	1981 cents per kwh
Capital[a, b, c]	1.22
Fuel[d, e]	2.67
O&M[f]	0.76
TOTAL	4.65

(a) The cost for 2-500 Mw units averaged over 15 different U.S. regions has been estimated to be $900 per kw in mid-1978 dollars, which becomes $1175 per kw in 1981$ (Bechtel Power Corporation, May 1981).

(b) For utility financing of a plant with a 30-year life the capital recovery factor is 0.056 (Note A.5-B). In addition, the annual insurance cost is assumed to be 0.003 times the initial investment. All taxes are neglected.

(c) Over the last decade, the capacity factor averaged about 70% for 300 Mw coal plants and 62% for 600 Mw units (Charles Komanoff, 1981). On the basis of this experience, a 65% capacity factor is assumed for 500 Mw units.

(d) The real price of coal to utilities in the U.S. increased in the period 1973 to 1982 at an average annual rate of 8.6%, from $0.70 per GJ to $1.50 per GJ, in 1981 dollars. The U.S. Department of Energy has projected that the utility|coal price will increase further, to $2.30 per GJ by 1990 and at an average growth rate of 0.9% per year thereafter (Energy Information Administration, March 1981). For a utility discount rate of 3.8% per year this implies|a price levelized over the period 1990 to 2020 of $2.60 per GJ.

(e) For a high efficiency, supercritical steam plant with wet lime/limestone flue gas desulphurization, for which the heat rate would be 10,270 KJ per kwh (Technical Assessment Group, 1979).

(f) See Technical Assessment Group, 1979.

Section A.6

Note A.6-A. Busbar cost for electricity from a new peaking power plant

	1981 cents per kwh
Capital[a, b]	0.334/CF
Fuel[c]	8.70
O&M[d]	0.30
TOTAL	9.0 + 0.334/CF

(a) In Sierra Energy and Risk Assessment, Inc., and Robert Foster, Consulting, September 1982, a survey of the capacity costs for new high efficiency peaking turbines estimated an average capital cost of $356 dollars per kw in 1980 dollars, which becomes $390 per kw in 1981 dollars.

(b) For utility financing of a plant with a 20 year life the capital recovery factor is 0.072 (Note A.5-B). In addition the annual insurance cost is assumed to be 0.003 times the initial investment. All taxes are negelected. Here CF is the annual average capacity factor. In peaking applications CF is on the order of 0.1.

(c) In Sierra Energy and Risk Assessment, Inc., and Robert Foster, Consulting, September 1982, the average heat rate for new efficient combustion turbine was estimated to be 12,900 KJ per kwh. Here the peaking fuel price is taken to be the average price of distillate fuel for utility peaking applications in the U.S. in 1981, or $6.75 per GJ (U.S. Energy Information Administration, November 1981).

(d) See Sierra Energy and Risk Assessment, Inc., and Robert Foster, Consulting, September 1982.

Note A.6-B. Cost estimates for a 50 MW amorphouse silicon PV plant, located in Princeton, New Jersey[a, b, c]

	1981 dollars per peak watt	
	Near Term	Intermediate Term
Wire	0.037	0.026
Site Preparation	0.030	0.015
Electrical Protection	0.009	0.004
Array Structure	0.566	0.354
Power Conditioning	0.141	0.094
Solar Cell Panels	0.585	0.292
Land	0.088	0.088
TOTALS	1.46	0.87

(a) These estimates were prepared at the RCA Laboratories in Princeton, New Jersey (A.J. Stranix and A.H. Firester, 1983).

(b) The sources of these cost estimates were the following:

Support Structures	Work done by Bechtel
Field wiring and electrical components	Direct quotations from suppliers
Installation of field wiring	Studies by Burt Hill Kosar Rittleman Associates and private discussions with power engineers
Power conditioning	Personnel at Sandia Laboratories, EPRI, and United Technologies
Solar Cell Panels	RCA's own R&D

(c) The cost estimates given in A.J. Stranix and A.H. Firester, 1983 were converted to 1981 dollars using the GNP deflator.

(d)   Some of the important parameters for the proposed plant are:

Solar cell efficiency	10%
Plant capacity factor	19.4%
Tilt of solar cell panels	30 degrees
Shadowing losses	>1%
Wire losses	1.4%
Annual electricity production	85.5 million kwh
Total solar cell area	0.50 million square meters
Total plant site area	94 hectares
Annual O&M costs	$0.45 per square meter
Land cost	$50,000 per hectrare

Note A.6-C. Estimated busbar electricity cost for alternative PV power plants

	1981 cents per kwh
AC electricity from a plant located in Princeton, New Jersey[a]	4.1
AC electricity from a plant of the same design located in El Paso, Texas[b]	2.5
AC electricity from the El Paso plant if the land cost were $ 2500 instead of $50,000 per hectare	2.3
DC electricity from the El Paso plant if the land cost were $2500 per hectare[c]	2.0
DC electricity from the El Paso plant if the land cost were $2500 per hectare and if the solar cell panel costs were reduced from $0.30 to $0.15 per peak watt[d]	1.6

(a)   For a capital recovery factor of 0.072 (Note A.5-B), an annual insurance cost equal to 0.003 times the installed capital cost, an installed capital cost of 0.87 dollars per peak watt (the intermediate term estimate in Note A.6-B), and an annual O&M cost of $0.45 per square meter or $0.0026 per kwh (Note A.6-B).

(b)   The electricity cost in El Paso, Eep, is estimated to be:

$$Eep = Ep \times (In/Iep),$$

where

Ep	=	the electricity cost in Princeton
In	=	the annual average insolation in Newark, N.J., on a collector tilted at the latitude angle
Iep	=	the annual average insolation in El Paso on a collector tilted at the latitude angle.

Annual averge insolation values for these cities are obtained from Jan F. Kreider and Frank Kreith, 1982:

In	=	166 watts per square meter (collector tilted at 40.7 degrees)
Iep	=	271 watts per square meter (collector tilted at 31.8 degrees)

(c)   The DC power that is produced by a PV array is usually converted to AC power, so that it can be utilized in the existing electrical system. In addition the power level is not constant throughout a day, so that the voltage is usually adjusted to a constant level to assure campatibility with the existing electrical system. For hydrogen production applications, only DC power is needed. Here it is assumed that for hydrogen production applications, the power conditioning costs (accounting for 11% of the total capital cost, as shown in Note A.6-B) can be neglected.

(d)   In 1981 researchers at the SANYO Electric Co. Research Center in Hirakata City, Osaka, Japan, projected that at 10% efficiency SANYO could produce cells for $0.15 per peak watt by the time the annual production volume reached about 350 Mw per year (Y. Kuwano and M. Ohnishi, 1981). From Note A.6-B it can be seen that a 50% reduction in the solar cell panel cost would lead to a 21% reduction in the total installed capital cost.

Note A.6-D. Capital and operating cost for an electrolyzer without a rectifier, for operation at 25% capacity factor

	Dollars per GJ
Total cost of hydrogen produced from AC electricity costing 1.5 cents/kwh @ 25% capacity factor[a]	9.80
Less cost of electricity[b]	−5.40
Less annualized cost of the rectifier[c]	−2.14
Net cost of electrolyzer w/o rectifier, using a capital recovery factor of 16%, plus 2% of the installed capital cost per year for O&M costs plus another 2% for ad valorem taxes and insurance	2.26
Net cost of electrolyzer w/o rectifier, using a capital recovery factor of 7.2% (Note A.5-B), plus 2% and 0.3% of the installed capital cost per year, respectively, for O&M costs and insurance, but neglecting taxes.	1.07
Net cost converted from 1977 to 1981 dollars	1.49

(a) For industrial unipolar technology as of 1983. See Figure 5, page 375, in R.L. Leroy and A.K. Stuart, 1979.
(b) For a rectifier efficiency of 96% and an electrolysis efficiency of 80%.
(c) In R.L. Leroy and A.K. Stuart, 1979, the annualized rectifier cost is given by

$$(777 \times K \times Vi \times XR)/(365 \times CF \times 100 \times nr)$$

where

K	=	annual capital charge rate = 0.20 (see p. 366 in R.L. Leroy and A.K. Stuart, 1979)
Vi	=	cell voltage = 1.86 volts (for a current density of 450 mA/cm^2—see Figure 1 in R.L. Leroy and A.K. Stuart, 1979)
XR	=	capital cost of rectifier = $65/kw (see Table I in R.L. Leroy and A.K. Stuart, 1979).
nr	=	rectifier efficiency = 0.96 (see Table I in R.L. Leroy and A.K. Stuart, 1979).

Thus the annualized rectifier cost is $2.14/GJ.

Note A.6-E. Estimated electrolytic hydrogen production cost for a 50 MW PV power plant located near El Paso, Texas[a]

	1981 dollars per GJ	
	w/o O_2 credit	w/ O_2 credit
With solar cell panels @ $0.30 per peak watt[b]	8.50	6.70
With solar cell panels @ $0.15 per peak watt[c]	7.00	5.30

(a) The production cost for hydrogen (CH_2), (in dollars per GJ) is given by:

CH_2	=	Celect − 0.056CrO$_2$ + Cpve/(n × 0.0036 GJ per kwh),

where

Celect	=	capital and O&M contributions to the cost of electrolysis,
CrO$_2$	=	credit for byproduct oxygen, in $ per tonne (0.056 tonnes of O_2 are produced per GJ of H_2),
Cpve	=	busbar cost of DC electricity produced with solar cells,
n	=	efficiency of electrolysis.

Here it is assumed (Note A.6-D) that Celect = $1.49 per GJ and n = 0.8. It is also assumed that before the oxygen market is saturated, the byproduct oxygen can be sold for $31.3 per tonne, the wholesale price of oxygen in 1979, expressed in 1981 dollars.
(b) In this case the estimated busbar electricity cost is $0.020 per kwh (Note A.6-C).
(c) In this case the estimated busbar electricity cost is $0.016 per kwh (Note A.6-C).

Index